Probing the
Atmospheric Boundary Layer

Donald H. Lenschow, Editor

American Meteorological Society
Boston, Massachusetts

Current Addresses of Contributors

R.B Chadwick, Meteorological Wave Propagation Laboratory, Environmental Research Laboratories, NOAA, Boulder, Colorado 80303

R.L. Coulter, Argonne National Laboratory, Argonne, Illinois 60439

W.F. Dabberdt, National Center for Atmospheric Research, Boulder, Colorado 80307

R.N. Dietz, Environmental Chemistry Division, Brookhaven National Laboratory, Upton, New York 11973

C.A. Friehe, Department of Mechanical Engineering, University of California, Irvine, California 92717

E.E. Gossard, Cooperative Institute for Research in the Environmental Sciences, University of Colorado, Boulder, Colorado 80301

J.C. Kaimal, Meteorological Wave Propagation Laboratory, Environmental Research Laboratories, NOAA, Boulder, Colorado 80303

L. Kristensen, Met. Fysik, Risoe National Laboratory, Roskilde, Denmark DK-4000

R.A. Kropfli, Meteorological Wave Propagation Laboratory, Environmental Research Laboratories, NOAA, Boulder, Colorado 80303

D.H. Lenschow, National Center for Atmospheric Research, Boulder, Colorado 80307

R. Lewis, Test and Evaluation Division, National Weather Service, NOAA, Sterling, Virginia 22170

W.D. Neff, Meteorological Wave Propagation Laboratory, Environmental Research Laboratories, NOAA, Boulder, Colorado 80303

R.F. Pueschel, Air Quality Division, Air Resources Laboratory, NOAA, Boulder, Colorado 80303

R.L. Schwiesow, National Center for Atmospheric Research, Boulder, Colorado 80307

W. Viezee, Atmospheric Science Center, SRI International, Menlo Park, California 94025

J.C. Wyngaard, National Center for Atmospheric Research, Boulder, Colorado 80307

ISBN 0-933876-63-7

American Meteorological Society
45 Beacon Street, Boston, MA 02108

Printed in the United States of America
by Edwards Brothers, Inc., Ann Arbor, Mich.

This book is sponsored by the American Meteorological Society in cooperation with the National Center for Atmospheric Research (NCAR). NCAR is operated by the University Corporation for Atmospheric Research and is sponsored by the National Science Foundation. Any opinions, findings, conclusions, or recommendations expressed in this publication are those of the author and do not necessarily reflect the views of the National Science Foundation. NCAR is an equal opportunity/ affirmative action employer.

Table of Contents

Foreword

The publication of this volume is the culmination of a process that began, as far as my involvement is concerned, several years ago in a recommendation by P. Arya, who was then chairman of the American Meterorological Society (AMS) Committee on Atmospheric Turbulence and Diffusion. He proposed that the committee sponsor a short course on boundary-layer measurements. During the ensuing discussion, the seventh issue of *Atmospheric Technology*, titled Instruments and Techniques for Probing the Atmospheric Boundary Layer, was used as a model for discussing what this course should cover. Partly because I edited that issue, I was asked, as a member of the committee, whether I would be willing to direct such a short course. I somewhat hesitatingly agreed, and we initially planned to hold the short course during the summer of 1982. Partly as a result of my own procrastination, the course was delayed to 1983.

In the meantime, the AMS Committee on Atmospheric Measurements, chaired by J.C. Kaimal, independently suggested that a short course such as this would be useful. An organizing committee, consisting of Kaimal, W. Dabberdt, A. Morris, and myself, was established to plan the details. As a result, the short course, "Instruments and Techniques for Probing the Atmospheric Boundary Layer," was held in Boulder, Colorado, on 8–12 August 1983, under the joint sponsorship of the Committees on Atmospheric Turbulence and Diffusion and on Atmospheric Measurements. A precursor to this volume was used as lecture notes for the course.

Throughout the planning and execution of the short course and the preparation of this volume I was assisted by many people. At the beginning, P. Arya, J. Wyngaard, and the members of the organizing committee helped to formulate the concept and content of the course. K. Spengler and E. Mazur of the AMS made the arrangements for holding the course and registering the participants. All the lecturers were very cooperative and understanding of the demands placed upon them, and all managed to deliver their lecture notes in time for the course. The lectures were well received, and the participants in the course were enthusiastic and receptive, and provided many useful comments on the lectures and the notes.

After the course, the lectures were revised by the authors, returned to me, and sent out for at least two scientific reviews. I am very grateful for the reviews provided by C. Bohren, J. Businger, R. Carbone, J. Dutton, C. Fairall, C. Friehe, G. Grams, P. Hildebrand, N.O. Jensen, L. Lading, D. Lilly, W. Mach, S. Nicholls, B. Sivertsen, G. Start, J. Telford, D. Thomson, J. Tillman, S. Twomey, A. Weill, and J. Wyngaard.

I do not know of any previous publication that attempts to do what this one does. On the other hand, this volume is based on precedents, in both style and content. *Atmospheric Technology*, No. 7 (1975), for example, also discusses instruments for boundary-layer research. However, the articles were unreviewed, and shorter and less tutorial, since they were not written initially as lecture notes. *Workshop on Micrometeorology* (edited by D. Haugen, AMS, Boston, Massachusetts, 1973, 392 pp.) and *Atmospheric Turbulence and Air Pollution Modelling* (edited by F.T.M. Nieuwstadt and H. van Dop, D. Reidel Publishing Company, Boston, Massachusetts, 1982, 358 pp.) are closer analogs in style, but their theme is the physics of the boundary layer, not observational technology. A source of personal guidance for me was *Research Problems in Micrometeorology* (H. Lettau, final report under Contract DA-36-039-SC-80063 from the U.S. Army Electronic Proving Ground, Fort Huachuca, Arizona, published by University of Wisconsin, Madison, 1959, 85 pp.). We have learned much and have made many technical advancements in the past 25 years, but the scientific questions that face us today were already evident then.

The cooperation, patience, innovativeness, and technical competence of M. Boyko of the NCAR Publications Office in producing the volume are gratefully acknowledged. H. Hamilton and M. Sime provided assistance with typing and transmitting several of the manuscripts, and with handling the review process.

Finally, I am particularly grateful for the assistance of L. Kristensen in critically reviewing several of the manuscripts and making many invaluable suggestions for improvement of the volume, and for his encouragement and willingness to act as a sounding board for ideas throughout the period while this volume was being produced.

Donald H. Lenschow
September 1984

Introduction

Donald H. Lenschow, National Center for Atmospheric Research

1 THE BOUNDARY LAYER

The nature of the atmospheric boundary layer is concisely summarized in the article by Wyngaard in this volume. Even more briefly, the atmospheric boundary layer can be considered as that layer of the atmosphere that, because of turbulence, interacts with the earth's surface on a time scale of a few hours or less. Typically over land, its depth is a few hundred to a few thousand meters in daytime, and 100 to 300 m at night. Over the ocean, its depth is relatively more constant at several hundred meters to a kilometer or more. Boundary-layer air mixes only intermittently with air in the overlying free atmosphere, mainly through convective cloud processes that are energetic enough to break through the temperature inversion that generally caps the boundary layer.

Almost the entire biosphere either is contained in, or depends on, the atmospheric boundary layer. The boundary layer, in many respects, can be considered as the circulatory system of the biosphere. It transports carbon dioxide and oxygen to plants and animals for photosynthesis and respiration, removes waste products, and cleanses the atmosphere through photochemical reactions, transport to the overlying free atmosphere, and deposition to the earth's surface. It transfers heat and moisture from the surface and disperses them both horizontally and vertically, effectively air-conditioning the biosphere and providing a conduit for energy to power weather systems on all scales—from fair-weather cumulus to severe convective storms to large-scale planetary waves—that provide the driving force for flow in the boundary layer, and the mechanisms for returning the evaporated water to the surface as precipitation.

The structure of the boundary layer, and its interactions with the earth's surface and the overlying free atmosphere, are known mainly as a result of a synergistic combination of observational studies, numerical and laboratory simulation, and dimensional analysis. Planning observational studies requires an understanding of instrument capabilities and limitations, strategies for designing and implementing a field study, and techniques for data handling, analysis, and synthesis. This volume concentrates on the first of these, but with some discussion of how instruments should be deployed and of techniques for data handling and analysis. Both direct and remote-sensing techniques are considered, we hope with sufficient detail that the reader can gain an understanding of how measurements are obtained and used, and what limitations exist in using them.

Some features of the boundary layer have been common knowledge longer than the existence of instruments to measure them. For example, the dispersive properties of the boundary layer are immediately obvious from observing smoke released near the ground into the atmosphere. We note quite different dispersal patterns between day and night; on a clear day, smoke rises in intermittent bursts and disperses rapidly, while at night it forms horizontal layers that dissipate much more slowly. Surface stress is made obvious by, for example, waves at the ocean surface, the rustling of leaves, and trees swaying in the wind. Cloud streets and cellular organization are visualizations of convection patterns that must be occurring in the boundary layer. Heat flux in the boundary layer causes scintillation and mirages in distant targets on the horizon due to fluctuations and vertical gradients in the index of refraction. Indeed, as Schwiesow points out in his overview of remote-sensing techniques, the human senses are, in many respects, superior to modern instruments for remote sensing of boundary-layer characteristics.

In some cases, people have observed the response of other living things to phenomena in the boundary layer to understand its structure. Woodcock (1940), for example, interpreted the soaring behavior of gulls to distinguish among different modes of convection in the boundary layer. Richardson (1922) estimated the surface stress over a wheat field by observing and evaluating the force required to bend wheat stalks. In some respects, these observations were far ahead of their time. Wyngaard's article in this volume points out that it is inherently easier to obtain statistically significant measurements by area averaging, as in the above examples, than by line averaging obtained from direct measurements from a tower or aircraft. Indeed, a major advantage of recently developed remote-sensing techniques such as radar (discussed by Kropfli and by Chadwick and Gossard in this volume) and lidar (discussed by Schwiesow in this volume) is their inherent area-averaging capability. Surprisingly, even with modern techniques we are sometimes still dependent on the response of other living things to boundary-layer phemonena. As Kropfli points out, for example, clear air radar sounding in the boundary layer still commonly uses insects as targets.

Quantitative information on the mean structure of the boundary layer can be traced back to the beginning of the

18th century when manned free balloons were first used for temperature sounding (Belinskii and Pobiyakho, 1962). By the late 18th century, both manned and unmanned free balloons, tethered balloons, and kites were used for soundings of temperature, humidity, and winds. Aircraft began to be used for meteorological measurements in the early 20th century, and for a number of years after 1930 they were used regularly for aerological observations. The early 20th century also saw the development and use of instrumentation for measuring turbulent fluctuations of temperature and wind velocity in the surface layer and the parallel development of the concept of turbulent exchange coefficients, perhaps most notably by G.I. Taylor in England and W. Schmidt in Germany.

In the last 40 years, instruments have continued to be developed for improving our understanding of boundary-layer structure. Measurements have been obtained from towers, free and tethered balloons, aircraft, and a bewildering array of remote-sensing probes. In this volume, we discuss a variety of instruments and techniques that are considered state-of-the-art or seem likely to be useful in the near future for observing the atmospheric boundary layer.

2   ORGANIZATION OF BOOK

Having the proper instruments to take measurements in the boundary layer is not the only requirement for undertaking an observational study. An investigator also must have some understanding of how to take the measurements and must know the limitations of the instruments in the environment where they are being used. This is the subject addressed by Wyngaard in the next article in this volume. Basically, all instruments have limitations. But even with perfectly accurate, instantaneously responding instruments, errors can result from, for example, inadequate sampling of a turbulent fluid or distortion of the flow field by the instrument or the measurement platform.

The five articles following Wyngaard's deal with direct measurements in the boundary layer. Kaimal discusses instruments used for flux and profile measurements from towers. The measurements include mean and turbulent fluctuations of horizontal wind, temperature, and humidity, as well as turbulent fluctuations of vertical velocity. He also discusses flow distortion effects of towers and ways to handle the large amounts of data generated by a well-instrumented tower.

The recommended sampling rate for resolving turbulence fluxes down to within a couple of meters of the surface is about ten per second. However, for some purposes, e.g., directly estimating the rates of dissipation of velocity and scalar variances, much higher sampling rates are required. Friehe, in his article, discusses instruments for fine-scale measurements of velocity, temperature, and humidity. "Fine-scale" in this context means, ideally, resolving wavelengths down to the order of 1 mm, or 1,000 samples per second from a tower and 10,000 samples per second from an aircraft.

In the next article, I discuss measurements of air velocity, temperature, and water substance from aircraft. A major advantage of aircraft is their mobility. However, correcting for aircraft motion is an added complexity in implementing aircraft measurement systems.

Besides its gaseous constituents, air contains a wide variety of aerosols of varying sizes and physical and chemical compositions. Since most aerosols emanate from the earth's surface, the boundary layer normally contains a higher concentration of aerosols than does the free atmosphere. At the same time, the earth's surface is also the major sink for aerosols, so studying their evolution in the boundary layer is an important observational problem that is discussed by Pueschel in his article.

Closely related to aerosol distribution in the boundary layer is atmospheric visibility, which is determined mainly by aerosol concentration and size. In the next article, Viezee and Lewis discuss instruments for quantitative measurements of visibility, which is basically a measure of the distance that the unaided eye can perceive objects in daytime or lights at night.

One of the most important properties of the boundary layer is its capacity for diffusing both gaseous and particulate constituents injected into it. This property of the turbulent boundary layer can be inferred by direct turbulence measurements. It can also be more directly estimated by releasing tracers into the boundary layer (thus closely simulating actual pollution sources) and then sampling their downwind concentration. Dabberdt and Dietz discuss a variety of conservative and quasi-conservative gaseous tracers, as well as techniques for releasing and sampling them and analyzing the collected samples.

The use of tracers can be looked on as a bridge between direct and remote measurements, since many remote-sensing techniques depend upon either natural or artificial tracers in the atmosphere. In recent years, remote sensing, which here we limit primarily to the detection of scattered energy from transmitted acoustic, radio, and light waves, has played an increasingly important role in boundary-layer studies, as ever-more-sophisticated technology is developed and applied to probing the boundary layer. There is now available such a wide variety of sensors and techniques that it seems increasingly difficult to evaluate the suitability of a particular technique, as compared to others, for a specific problem. Trade-offs need to be considered in areas of cost, equipment performance and availability, capacity for incorporating new technology, environmental impact, etc. In his overview, Schwiesow defines, discusses, and compares the remote-sensing techniques that are described in the remaining four articles.

The newest and probably the least developed of these techniques is lidar. In the next article, Schwiesow discusses the use of lidar for remote measurement of boundary-layer variables, including measurements of aerosol and other trace constituent concentrations, winds, and temperature. By sensing patterns in the distribution of trace constituents, researchers are also able to delineate discrete structures in the boundary layer.

Radar, a close relative of lidar, has similar applications for boundary-layer measurements, but other advantages and disadvantages. Following Schwiesow's articles, Chadwick and Gossard discuss boundary-layer measurements using refractive index variations as targets, and Kropfli discusses boundary-layer measurements using both natural and artificial particulates in the atmosphere as targets.

Finally, probably the oldest of the remote-sensing techniques is acoustic sounding, which is discussed in the final article, by Neff and Coulter. Sodars are particularly useful for short-range (less than about a kilometer) measurements of wind, turbulence, and discrete structures in both the stably and the unstably stratified boundary layer.

This, then, is a brief introduction to what is contained in this volume. There are, of course, many subjects that are not covered, since the field is large and multifaceted. Areas that are not covered include measurements of radiative heating and cooling in the boundary layer, i.e., incoming and outgoing solar and infrared radiation; balloon-based measurements, i.e., both free and tethered low-level sounding balloons and constant-volume free balloons for measuring air trajectories; and pressure fluctuation measurements. The decision to omit these subjects was based partly on the assessment that there seem to have been no recent important technological innovations in these areas. There is also little discussion of data-handling and analysis techniques. The volume does not provide, for example, a detailed description of digital filtering and spectral analysis techniques, which are of increasing importance in coping with the large data sets that can be generated by both direct turbulence measurements and remote sensing. Undoubtedly there are other subjects that, for completeness, should also have been addressed. I can only hope that enough has been included and that the subjects have been covered well enough that this becomes a useful source of information for those interested in learning how the boundary layer is observed.

Measurement Physics

J.C. Wyngaard, National Center for Atmospheric Research

1 INTRODUCTION

I have long been interested in what I call "measurement physics"—the science of the interaction of a sensor with its environment. It is a very important but, I think, neglected aspect of boundary-layer meteorology, and so I will devote a good deal of this article to it.

Since this article comes out of the opening lecture of the course, I will also give an introductory description of the planetary boundary layer (PBL). Experimental thrusts over the past two decades have sharpened tremendously our perceptions about the structure of the PBL, and have taught us a good deal about its turbulence; we still live with the turbulence "closure problem," but today we are much more adept at dealing with (e.g., parameterizing) PBL turbulence than we were a generation ago.

Finally, I will discuss another somewhat neglected aspect of boundary-layer meteorology—the matter of dealing with scattered data. We have learned in recent years that we must average our boundary-layer measurements over long time periods, or long flight paths, in order to generate statistically stable mean values. The research community understands why this is, and that we can get relief by averaging over more than one dimension—e.g., by averaging over an area. However, we are only slowly transmitting this message to all groups within our broad field. The air quality modeling community, for example, has only recently focused on the poor correlation between model predictions and short-term (1–3 h) observations of mean pollutant concentration. They have learned that even a perfect model would correlate poorly with perfect short-term observations because of "inherent uncertainty"—the inevitable and unpredictable differences between atmospheric behavior over a finite time period (i.e., measurements) and the most likely, or ensemble average, behavior (i.e., model predictions). This looms as a major new area of interest within air quality modeling, so I will try to introduce the subject here.

2 THE NATURE OF THE PBL

The definition of the PBL varies with the writer, and with the application; generally speaking, however, we mean by "PBL" the turbulent region adjacent to the earth's surface.

One obvious feature of the PBL is its pronounced diurnal cycle over land. In clear-air, daytime conditions it typically deepens during the day, reaching a depth ranging from a few hundred to a few thousand meters by late afternoon. Its vigorous turbulence, driven primarily by the surface heating,

gives strong vertical mixing; for this reason it is often called the "mixed layer." It tends to mix pollutants, for example, rapidly and uniformly in the vertical. Figure 1a shows an artist's conception of a convective PBL.

As sunset nears, the radiative energy input at the surface decreases to the point that the surface begins to cool; this shuts off the energy supply to the convective motions, and the PBL turbulence decays. A new, stably stratified PBL grows upward from the surface in response to this cooling. Its turbulence tends to be much less diffusive; pollutants from surface sources tend to remain near the surface, and those from elevated sources tend to remain aloft. Near-surface winds can be quite light, even with strong winds aloft, because the turbulent diffusivity is insufficient to transfer the mean momentum to the surface layer. Figure 1b shows an artist's conception of this stably stratified nocturnal PBL. At daybreak the new convection erodes the stable PBL from below, and the cycle continues.

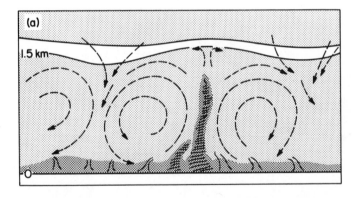

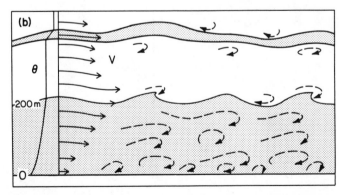

Fig. 1 An artist's conception of (a) a convective PBL capped by an inversion; (b) a stably stratified nocturnal PBL.

The top of the PBL is very often marked by a stably strati-fied layer which extinguishes the turbulence and buffers the free atmosphere above, which is only intermittently tur-bulent, from the PBL. Thus, the daytime, convective PBL depth is simply z_i, the distance to the first inversion base. This inversion is often quite conspicuous in temperature and humidity soundings. Figure 2 shows data from the Air Mass Transformation Experiment (AMTEX) which clearly reveal a sharp inversion.

At night over land, under clear skies, the entire PBL is usually stably stratified, with the intensity of the stratification (the Richardson number) increasing with height until it reaches the critical value, where turbulence vanishes, at the PBL top. The classical neutral PBL is actually quite elusive in nature, since a very slight surface-to-air temperature dif-ference makes the PBL either stably or unstably stratified. Nonetheless, when the neutral PBL does occur it is likely that its depth is also simply the height of an overlying inversion layer.

There was a good deal of excitement in the PBL commu-nity in the early 1970s when monostatic acoustic sounders clearly showed the shallow nature of the nocturnal PBL, its sharp top, and the rich variety of turbulent and wave activity within it. These devices also revealed the rapid deepening of the morning PBL. Subsequent developments with lidar and radar have expanded tremendously our ability to observe PBL structure, and in fact now threaten to swamp researchers with data.

The turbulence in the PBL is perhaps most conspicuous in the daytime. By turbulence we mean not only a random, three-dimensional velocity field, but also the random scalar fields (temperature, water vapor, pollutant) which this ve-locity field creates. Because of this turbulence, we generally study averaged measurements in the PBL; averaging mini-mizes the chaos in the instantaneous fields. As we will show, this averaging is particularly important in the daytime, when the PBL can have large, intense, long-lived convective eddies which strongly bias short-term observations.

Let us consider more precisely what we mean by tur-bulence. The basic notion is that we decompose a field of any property $\mathcal{P}$ into mean and turbulent components:

$$\mathcal{P} = P + p \qquad (1)$$

Throughout, I will use uppercase letters to denote the mean, and lowercase for the fluctuation.

There are a number of ways to generate an average. Tur-bulence theoreticians tend to use the average over an infinite ensemble of realizations; this is the ensemble average. Numerical modelers use a volume average in addition to the ensemble average. Experimentalists use both of these, plus line, area, and time averages. We will discuss each of these

Fig. 2   Profiles of potential temperature (left) and water vapor density (right) measured in the AMTEX boundary layer (Wyngaard et al., 1978). Note the strong jumps at the top.

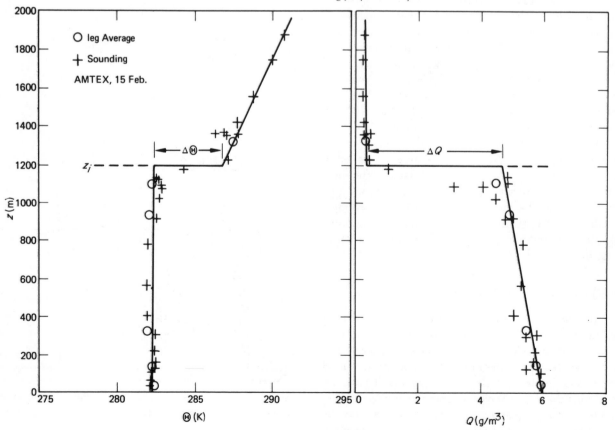

types of average in some detail later, but for now we will not be concerned with the specifics of the averaging process.

Since the mean field in Eq. (1) depends on the type of averaging used, the turbulent component does as well. For example, a synoptician deals with spatially averaged fields which have no detail on scales smaller than tens or even hundreds of kilometers; thus, he considers mesoscale circulations such as land-sea breezes and drainage flows as turbulence. The experimentalist working with time averages from a single meteorological tower would determine these same circulations to be part of the mean field. An experimentalist measuring from an aircraft, on the other hand, would see them as spatially varying fields, and if the mean were defined as the average along the flight path, then these fields would be part of both the turbulence field and the mean. It is important, therefore, that we be careful to define our mean field, so that we know what our turbulent component represents.

3 THE NATURE OF TURBULENCE

The turbulent velocity is a random, three-dimensional, time-dependent vector field. Its evolution in space and time is governed by the Navier-Stokes and continuity equations, which express momentum and mass conservation, respectively. However, for practical purposes a given realization of a turbulent velocity field is unpredictable in detail because of the complex physics embodied in these equations and their sensitivity to initial and boundary conditions.

In spite of these formidable difficulties with turbulence, we do have the benefit of the "folk wisdom" gained from several generations of experience in trying to deal with it experimentally, theoretically, and numerically. Thus, while we still have no rational scheme for solving the turbulence equations analytically, we do have an effective physical understanding of turbulence, and can deal with it quite successfully in many applications.

Let us discuss several concepts which are useful when dealing with PBL turbulence. We call turbulence stationary if its statistical properties are independent of time. In everyday terms, this means that any piece of a time record of a flow property looks like any other piece; there are detail differences, of course, but the statistics are the same. Thus stationarity implies statistical invariance with respect to translation of the time axis. In the PBL this is often the case near mid-day in fair weather.

If the field is statistically invariant to translation of the spatial axes we call it homogeneous. While PBL turbulence is never homogeneous in the vertical (it is strongly affected by the lower surface and the capping inversion) it can be horizontally homogeneous, to a good approximation.

An isotropic field is statistically independent of translation, rotation, and reflection of the spatial axes. This is clearly not the case in the PBL, since the upper and lower boundary conditions, and the effects of buoyancy, make properties different in the horizontal and vertical. However, experimental data and theoretical arguments suggest that the small-scale (meters and smaller) structure of the PBL is effectively isotropic. This is called "local isotropy," meaning isotropy confined, or localized, to the smallest-scale structure.

We just introduced the concept of length scales in turbulence. If we had a detailed air velocity record from a fast-moving aircraft in the PBL we could interpret it as a spatial record; by forming the correlation function as a function of separation along the flight path we could then determine the integral scale. This is a measure of the distance over which the signal is correlated; we interpret this length as the scale (in the flight direction) of the dominant eddies.

The eddies carrying the bulk of the turbulent kinetic energy and doing most of the turbulent transport are (in order of magnitude) integral-scale sized. In principle there is an integral scale for every flow property, and in general it depends on distance from the surface, stability, and the direction of the measurement path (lateral or streamwise). For example, some integral scales (e.g., that for vertical velocity) scale with height, z, in the surface layer and with PBL depth in the middle and upper regions. Others (e.g., that for horizontal velocity under convective conditions) scale with PBL depth, even in the surface layer. Stability can strongly restrict integral scales, particularly in the nocturnal PBL. We cannot dwell on any of these complications here, however, and will simply denote the integral scale by l.

Another important scale is λ, the Taylor microscale. It is defined by the turbulence dissipation rate per unit mass (ϵ), the turbulence kinetic energy per unit mass (q^2), and the molecular kinematic viscosity ν: $\lambda = (\nu q^2/\epsilon)^{1/2}$.

This scale does not mark the size of the dissipative eddies; rather, we can think of it as the eddy size where viscous effects begin to become significant. As such, it roughly marks the small-scale end of the inertial subrange, that broad range of scales between the anisotropic, energy-containing eddies and the isotropic, dissipative eddies which convert kinetic energy into internal energy through their viscous friction.

A third length scale, the Kolmogoroff scale η_K, does indicate the size of the viscous eddies. A turbulent fluid, having a kinematic viscosity ν and needing to dissipate kinetic energy at a rate ϵ, establishes a dissipative range of eddies scaling in size with $\eta_K = (\nu^3/\epsilon)^{1/4}$.

We can now quickly sketch the distribution of turbulent kinetic energy over horizontal wave number, $\varkappa$. The spectral density E (kinetic energy/wave number) has a peak at wave numbers on the order of $1/l$. At larger $\varkappa$ (smaller scales), Kolmogoroff argued that E depends only on ϵ and $\varkappa$; on dimensional grounds this gives the now well-known inertial range form $E \sim \epsilon^{2/3} \varkappa^{-5/3}$. This extends to wave numbers roughly on the order of $1/\lambda$, where viscous effects begin to be felt. There E begins to fall more sharply, and it cuts off very steeply near the viscous wave number $1/\eta_K$. Figure 3 shows a schematic of the turbulence spectrum.

In a convective PBL of height z_i, z_i and hence l can range from several hundred meters to a few kilometers. If the surface temperature flux H_0 is 0.2 m K/s, typical over land in fair weather, then the mid-PBL buoyant production rate of turbulent kinetic energy, which is on the order of $gH_0/(2T)$, is about 3×10^{-3} m²/s³. If $q^2 = 2$ m²/s², and using $\nu = 1.5 \times 10^{-5}$ m²/s for air, we find that $\lambda = (\nu q^2/\epsilon)^{1/2} \sim 0.1$ m. Then very roughly, the spectral region between $1/l$ and $1/\lambda$—three decades, if $l = 100$ m and $\lambda = 0.1$ m—contains the inertial range. The dissipative eddy size is $\eta_K = (\nu^3/\epsilon)^{1/4} = 0.001$ m, so the eddy size range covers roughly five decades!

4 MEASUREMENT PHYSICS

With this introduction to the planetary boundary layer, and to its turbulence, I can get to the heart of my subject. My intention in this section is to show that by dealing with the measurement process on a mathematical level we can make better instruments, better measurements, and better interpretations of the data we obtain.

4.1 Spatial and Temporal Resolution

We just saw that PBL turbulence has a tremendous range of spatial scales—from millimeters to kilometers. Fortunately, we seldom need information on this entire range; instead, we can usually use sensors and data acquisition techniques which attenuate, or filter, some spatial scales and temporal frequencies.

To illustrate the effects of this filtering, let us consider first a one-dimensional problem. Take the simplest case, where the measurement process can be characterized as a first-order, linear system obeying

$$\tau(de_0/dt) + e_0 = e_i \qquad (2)$$

where $e_0(t)$ and $e_i(t)$ are output and input signals (that is, the

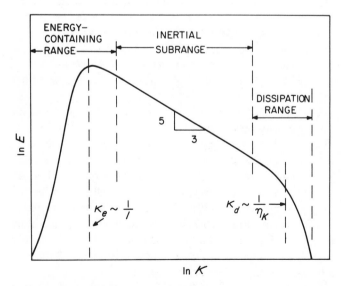

Fig. 3 A schematic of the spectral energy density of turbulence. A wave-number characteristic of the energy-containing range is $\varkappa_e \sim 1/l$, where l is the integral scale; a dissipative range wave number is $\varkappa_d \sim 1/\eta_K$, where η_K is the Kolmogoroff microscale.

measured and true variables), and τ is the time constant. This is the familiar first-order, low-pass RC filter, and is a good approximation for the effects of thermal lag in many temperature sensors, and for the effects of inertia in many wind-speed sensors. As in linear circuit theory, we can represent the signals by a combination of harmonics:

$$\begin{pmatrix} e_i(t) \\ e_0(t) \end{pmatrix} = \int_{-\infty}^{\infty} e^{i\omega t} \begin{pmatrix} E_i(\omega) \\ E_0(\omega) \end{pmatrix} d\omega \qquad (3)$$

Our representation here is not rigorously correct, for we should use stochastic integrals (Lumley and Panofsky, 1964). However, it will serve our illustrative purposes. Substituting Eq. (3) into Eq. (2) gives the relation between input and output amplitudes:

$$E_0 = \{(1 - i\omega\tau)/[1 + (\omega\tau)^2]\}E_i \qquad (4)$$

At very low frequencies ($\omega\tau \ll 1$) the input and output signals are the same, while at high frequencies ($\omega\tau \gg 1$) the output is strongly attenuated. The "break frequency" scales with $1/\tau$. Multiplying Eq. (4) by its complex conjugate gives the relation between power spectral densities:

$$\phi_0 = \{1/[1 + (\omega\tau)^2]\}\phi_i \qquad (5)$$

Now consider a different type of low-pass filtering—one in which the output is the centered average of the input over an interval $2T$:

$$e_0(t, T) = \frac{1}{2T} \int_{-T}^{T} e_i(t' + t)\, dt' \qquad (6)$$

Equation (3) then shows that

$$e_0(t, T) = \int_{-\infty}^{\infty} e^{i\omega t} E_i(\omega) \frac{\sin(\omega T)}{(\omega T)}\, d\omega \qquad (7)$$

so that the input and output amplitudes are related by

$$E_0(\omega, T) = \{[\sin(\omega T)]/\omega T\}E_i(\omega) \qquad (8)$$

Note from Eq. (8) that the ratio of amplitudes is purely real; this is because we used a centered average in Eq. (6). Had we averaged entirely over past time, the ratio would have been complex.

We can extend this to the spatial averaging of the three-dimensional turbulence fields we meet in the PBL. Here the Fourier representation is

$$\begin{pmatrix} e_i(\mathbf{x}, t) \\ e_0(\mathbf{x}, t) \end{pmatrix} = \int_{-\infty}^{\infty} e^{i\mathbf{x} \cdot \mathbf{x}} \begin{pmatrix} E_i(\mathbf{x}, t) \\ E_0(\mathbf{x}, t) \end{pmatrix} d\mathbf{x} \qquad (9)$$

Now assume we do a centered average over a path $2L$—for example, over the filament of a temperature sensor or a hot-wire anemometer, or over an acoustic anemometer path.

Then the input-output relation is

$$E_o(\boldsymbol{x},t; \mathbf{L}) = \{[\sin(\boldsymbol{x} \cdot \mathbf{L})]/(\boldsymbol{x} \cdot \mathbf{L})\}E_i(\boldsymbol{x},t) \qquad (10)$$

Note that the ratio is again purely real.

We can also average over a plane. For example, we might average over a rectangle of sides $2L_x$ and $2L_y$. Then the input-output relation is (for centered averaging)

$$E_o(\boldsymbol{x},t; \mathbf{L}) = \frac{\sin(x_xL_x) \, \sin(x_yL_y)}{(x_xL_x) \, (x_yL_y)} E_i(\boldsymbol{x},t) \qquad (11)$$

The generalization to volume averaging is straightforward.

The relations between input and output spectra then follow directly. For example, with line averaging we have

$$\phi_o(\boldsymbol{x},t; \mathbf{L}) = \{[\sin^2(\boldsymbol{x} \cdot \mathbf{L})]/(\boldsymbol{x} \cdot \mathbf{L})^2\}\phi_i(\boldsymbol{x},t) \qquad (12)$$

and the area- and volume-averaging relations follow just as easily. These relations are difficult to use, however, due to our inability to measure these three-dimensional spectra—that is, spectra which depend on the three-dimensional wave number $\boldsymbol{x}$. We measure one-dimensional spectra; for example, measurements from aircraft yield the one-dimensional spectrum along the flight path. These spectra are related by

$$\phi^{(1)}(x_1,t) = \int_{-\infty}^{\infty} \int_{-\infty}^{\infty} \phi^{(3)}(\boldsymbol{x},t) \, dx_2 \, dx_3 \qquad (13)$$

Because of the integration we generally do not have the simple, multiplicative "transfer functions" for one-dimensional spectra that we find in time series analysis, for example. This generally complicates the interpretation of spatial-averaging effects.

Covariance measurements involve an additional spatial filtering effect. If the two variables are not measured at the same spatial point, but are instead separated by $\mathbf{d}$, then one can see from Eq. (9) that the spectral transfer function has the additional factor $\exp(i\boldsymbol{x} \cdot \mathbf{d})$, which is complex. The cross spectrum of two variables is in general also complex: the real part (cospectrum) is even in $\boldsymbol{x}$ and integrates to the covariance; the imaginary part (quadrature spectrum) is odd and integrates to zero. Thus, spatial separation makes the measured cospectrum a combination of the true cospectrum and the quadrature spectrum; in principle, then, it can induce a covariance when none actually exists.

We can calculate spatial-averaging effects reliably when they involve only the smaller scales, which can be considered to be isotropic. Wyngaard (1968, 1969, 1971) has calculated the response of single and multiple arrays of hot and cold wires. Kaimal et al. (1968) and Horst (1973) have calculated the spatial-averaging properties of sonic anemometers. Gal-Chen and Wyngaard (1982) treated some of the effects of volume averaging on the vertical velocity fields measured with Doppler radar. Figures 4 and 5 illustrate the calculated one-dimensional spectral response of a sonic anemometer and

a Doppler radar, respectively.

In general we must rely heavily on our experience and intuition when assessing the effects of spatial averaging on the energy-containing range of turbulence. Our spectral models tend not to be applicable there, so we cannot reliably calculate the effects numerically. If we want to know how near the surface we can mount our 20-cm-path sonic anemometer and fine-wire thermometer in order to measure turbulent

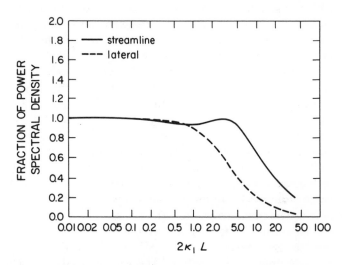

Fig. 4 The one-dimensional spectral response of a typical sonic anemometer. Solid curve, streamwise velocity; dashed curve, lateral velocity. L is the sensor averaging length. (From Kaimal et al., 1968.)

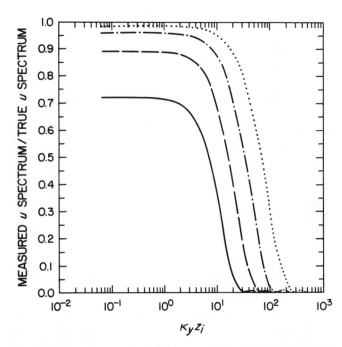

Fig. 5 The one-dimensional spectral response of lateral velocity fluctuations measured with a Doppler radar which averages over a cube of side Δ in a boundary layer of depth z_i. Dotted curve, $\Delta/z_i = 1/40$; dot-dash, $1/20$; dashed, $1/10$; solid, $1/5$. Note that the response ratio does not approach 1.0 as $x_yz_i \to 0$. (From Gal-Chen and Wyngaard, 1982.)

temperature flux, we can best answer the question experimentally (by measuring at several heights in the constant-flux layer and determining the flux loss) or use conservative rules of thumb (see discussion by Kaimal in this volume).

4.2 Taylor's Hypothesis

Up to this point we have focused on spatial structure (e.g., wave-number spectra). However, most often we measure temporal structure; for example, we use fixed sensors (e.g., mounted on towers) and observe the temporal variations as the turbulent flow passes by. Even when we use a moving probe, as with aircraft, we measure time series.

Nonetheless, in turbulence research we traditionally interpret these temporal measurements in terms of streamwise structure—that is, as if its structure were "frozen" in space and convected past the probe by the mean flow. We can trace the origin of this interpretation to Taylor (1938), who wrote, "If the velocity of the air stream which carries the eddies is very much greater than the turbulent velocity, one may assume that the sequence of changes in u at the fixed point is simply due to the passage of an unchanging pattern of turbulent motion over the point; i.e., one may assume that $u = f(t) = f(x/U)$, where x is measured upstream at time $t = 0$ from the fixed point where u is measured. In the limit when $u/U \to 0$ [this] is certainly true." This is now called "Taylor's hypothesis," or sometimes the "frozen-field hypothesis."

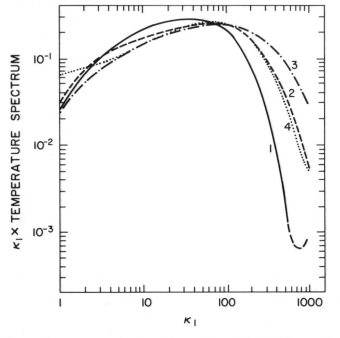

Fig. 6   Temperature spectra, measured by a probe moving horizontally at speed U_0 in laboratory free convection of characteristic velocity w_*, plotted as a function of horizontal wave number $\varkappa_1$. These were obtained from frequency spectra by using Taylor's hypothesis. Curve 1, $U_0/w_* = 4.27$; curve 2, 0.51; curve 3, 0.27; curve 4, 0. Note the aliasing of spectral density to higher wave numbers due to the fluctuating convection velocity. (From Deardorff and Willis, 1982.)

We can interpret Taylor's limit $u/U \to 0$ as follows. An eddy of length scale l and velocity scale u has a lifetime on the order of l/u, and is convected past the probe in a time l/U. Taylor's hypothesis should be valid if the eddy lifetime is much greater than its convection time, or $l/u \gg l/U$; thus $u/U \ll 1$.

Unfortunately, $u/U \ll 1$ is usually satisfied only for aircraft measurements. For tower measurements, Taylor's hypothesis can fail in different wave-number regions for a variety of reasons. I know of no way of correcting for this failure within the energy-containing spectral range.

At smaller scales there is hope, however. Lumley (1965) considered the effects of mean wind shear and turbulence level, and found that if certain criteria were satisfied, "the spectrum beyond roughly $30/l$ may be described adequately by isotropic frozen regions convected by a spatially uniform fluctuating velocity having the characteristics of the energy-containing eddies."

Lumley presented an analytical model in which the structure at inertial range and smaller scales is effectively frozen but is convected with a velocity which fluctuates about a mean U. He found that this fluctuating convection velocity causes the high–wave-number spectrum to be overestimated. The complete explanation for this overestimation is complicated, particularly for the velocity field, because one must consider the three-dimensional, tensor geometry of the problem. However, we can explain the essence of the mechanism in very simple terms. Turbulent spectral energy appears at any $\varkappa_1 = \omega/U$ from three sources: from eddies of smaller $\varkappa_1$ convected at a speed less than U; from eddies of wave number $\varkappa_1$ convected at U; and from eddies of larger $\varkappa_1$ convected faster than U. A crude model of this process is

$$
\begin{aligned}
\phi^m(\varkappa_1) &\cong \epsilon\phi(\varkappa_1 - \Delta\varkappa_1) \\
&+ (1 - 2\epsilon)\phi(\varkappa_1) \\
&+ \epsilon\phi(\varkappa_1 + \Delta\varkappa_1) \\
&\cong \phi(\varkappa_1) + \epsilon(\Delta\varkappa_1/\varkappa_1)^2\,\varkappa_1^2(d^2\phi/d\varkappa_1^2)
\end{aligned}
\tag{14}
$$

Here a superscript m denotes measured and ϵ is the fractional "leakage" from adjacent wave numbers. We expect $\epsilon(\Delta\varkappa_1/\varkappa_1)^2$ to scale with u^2/U^2, where u^2 is the variance of the convection velocity. For a power-law spectrum the second derivative term in Eq. (14) is positive, causing the measured spectrum to be overestimated.

Lumley's full calculations are more complicated because the full problem is three-dimensional. Wyngaard and Clifford (1977) have extended them and shown how convection velocity fluctuations influence both small-scale velocity and scalar statistics.

Deardorff and Willis (1982) have confirmed, through measurements of temperature spectra in their convection tank, that a fluctuating convection velocity aliases energy to higher wave numbers. This energy is borrowed from the energy-containing range, and their spectra (Fig. 6) show the distortion in that region as well.

4.3 Flow Distortion

Traditional sensors (i.e., "in situ" or "immersion" sensors) inevitably distort the flow in their immediate vicinity, and thereby also disturb the measurements. This is most obvious with velocity sensors, but it affects all types of measurements.

Micrometeorologists have for years been sensitive to the influence of masts, booms, towers, and probe bodies on measured mean velocities. The literature contains an abundance of papers concerning "tower shadow effects" on mean winds, for example. Very much less attention has been given to flow distortion in other situations (on aircraft measurements, for example) or to the effect on turbulence measurements.

Turbulent flow around obstacles is very complicated. Hunt's (1973) paper covers over 80 pages, and yet in some respects only opens the subject. Thus all we can do here is introduce the reader to the problem, and hope to make its broad outlines clear.

Let us consider a simple example. Figure 7 shows a turbulent flow approaching a two-dimensional circular cylinder. Let the approach flow have a velocity $(U_1 + u_1, u_2, u_3)$ with U_1 the mean. The Reynolds number is large enough to produce a turbulent wake behind the cylinder, while upstream of the cylinder the flow remains smooth but is distorted. We wish to know how this distortion affects the turbulence components there.

The approach flow has an integral scale l, and an important simplification occurs if $l \gg a$, the cylinder radius. In that case, judged by the scales of the flow near the cylinder, the approach flow appears spatially uniform and very slowly varying in time.

Imagine first that the upstream turbulence is purely streamwise; i.e., the turbulent velocity is $(u_1, 0, 0)$. This will induce both streamwise and vertical component fluctuations near the cylinder, because the streamlines there are distorted by the body. If the approach flow has instead a purely vertical turbulent component, the turbulence near the cylinder again has components in both directions. Clearly, then, the turbulence field, as well as the mean field, is distorted near the body.

I have developed (Wyngaard, 1981) a potential flow solution for the distorted turbulence components in this case where $l \gg a$. Denoting the distorted field with a tilde, the general solution for a three-dimensional body is

$$\widetilde{u}_i(\mathbf{x},t) = a_{i1}(\mathbf{x})u_1(t) + a_{i2}(\mathbf{x})u_2(t) + a_{i3}(\mathbf{x})u_3(t) \quad (15)$$

Here the a_{ij} are distortion coefficients which depend on the body shape and on position. They have the property

$$a_{ij}(\mathbf{x}) = \delta_{ij} + \epsilon_{ij}(\mathbf{x}) \quad (16)$$

where $\epsilon_{ij} = 0$ in the free stream. Thus in the undistorted flow far ahead of the body Eq. (15) becomes

$$\widetilde{u}_i(\mathbf{x},t) = u_i(t) \quad (17)$$

as required.

For the two-dimensional cylinder I found that $\epsilon_{31} = \epsilon_{13}$, and $\epsilon_{33} = -\epsilon_{11}$. Thus the distorted turbulence components are

$$\widetilde{u}_1 = (1 + \epsilon_{11})u_1 + \epsilon_{13}u_3 \quad (18)$$

$$\widetilde{u}_3 = \epsilon_{13}u_1 + (1 - \epsilon_{11})u_3 \quad (19)$$

We can immediately calculate from Eqs. (18) and (19) the distortion in statistics. For example, multiplying the two equations and averaging give, to first order in the ϵ_{ij},

$$\widetilde{\overline{u_1 u_3}} \cong \overline{u_1 u_3} + \epsilon_{13}(\overline{u_1^2} + \overline{u_3^2}) \quad (20)$$

Thus the shearing stress measured in the distorted flow is contaminated by the variances. Figure 8 shows an example of the serious errors this can cause.

The key to this calculation is determining the distortion coefficients a_{ij}. For simple bodies one can calculate them analytically; for more complicated bodies one can measure them (Wyngaard, 1981).

Consider next what happens when a scalar field (temperature, θ, say) in a uniform velocity field approaches our cylinder. We assume large Reynolds/Peclet number flow, so that we can neglect molecular diffusion; then θ satisfies

$$D\theta/Dt = 0 \quad (21)$$

which says simply that θ is unchanged as the flow goes around the obstacle. If the approach flow has uniform θ, then θ is uniform around the body as well.

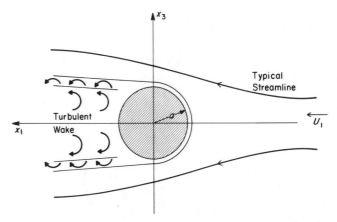

Fig. 7   A schematic flow near a circular cylinder. The flow in the trailing wake is complicated and unpredictable in detail, but the flow ahead of the cylinder can be easily calculated if the integral scale l in the free stream is much larger than the cylinder radius a.

Something interesting does happen if the approach flow has spatially varying θ. For example, let the upstream θ be periodic in x_1. Then the θ — constant curves upstream are vertical lines with constant spacing in x_1, but are strongly distorted near the cylinder because of the varying velocity along the streamlines; Fig. 9 illustrates this.

We can study this analytically by starting from the equations which govern the distortion of the scalar gradient:

$$\frac{D}{Dt}\frac{\partial \tilde{\theta}}{\partial x_1} = -\frac{\partial \tilde{u}_1}{\partial x_1}\frac{\partial \tilde{\theta}}{\partial x_1} - \frac{\partial \tilde{u}_3}{\partial x_1}\frac{\partial \tilde{\theta}}{\partial x_3} \qquad (22)$$

$$\frac{D}{Dt}\frac{\partial \tilde{\theta}}{\partial x_3} = -\frac{\partial \tilde{u}_1}{\partial x_3}\frac{\partial \tilde{\theta}}{\partial x_1} - \frac{\partial \tilde{u}_3}{\partial x_3}\frac{\partial \tilde{\theta}}{\partial x_3} \qquad (23)$$

Here the tildes remind us that the variables have been distorted from their upstream values by the presence of the

body in the flow. For weak distortion (that is, not too close to the body) we can approximate the scalar gradients on the right sides by their undisturbed values; integration along streamlines then gives

$$\frac{\partial \tilde{\theta}}{\partial x_1}(t) = \frac{\partial \theta}{\partial x_1}\left(1 - \int_0^t \frac{\partial \tilde{u}_1}{\partial x_1}dt'\right) - \frac{\partial \theta}{\partial x_3}\int_0^t \frac{\partial \tilde{u}_3}{\partial x_1}dt' \qquad (24)$$

$$\frac{\partial \tilde{\theta}}{\partial x_3}(t) = -\frac{\partial \theta}{\partial x_1}\int_0^t \frac{\partial \tilde{u}_1}{\partial x_3}dt' + \frac{\partial \theta}{\partial x_3}\left(1 - \int_0^t \frac{\partial \tilde{u}_3}{\partial x_3}dt'\right) \qquad (25)$$

We assume that the flow is incompressible, and irrotational on the scale of the obstacle; these imply

$$\int_0^t \frac{\partial \tilde{u}_1}{\partial x_1}dt' = \epsilon_1 = -\int_0^t \frac{\partial \tilde{u}_3}{\partial x_3}dt' \qquad (26)$$

$$\int_0^t \frac{\partial \tilde{u}_3}{\partial x_1}dt' = \epsilon_2 = \int_0^t \frac{\partial \tilde{u}_1}{\partial x_3}dt' \qquad (27)$$

Fig. 8 Contours of $\widetilde{u_1u_3}/\overline{u_1u_3}$, the ratio of distorted and undistorted stress, upstream of a circular cylinder in a turbulent flow typical of the atmospheric surface layer. Left panel is for flow with a narrow turbulent wake; right panel has a wider wake. (From Wyngaard, 1981.)

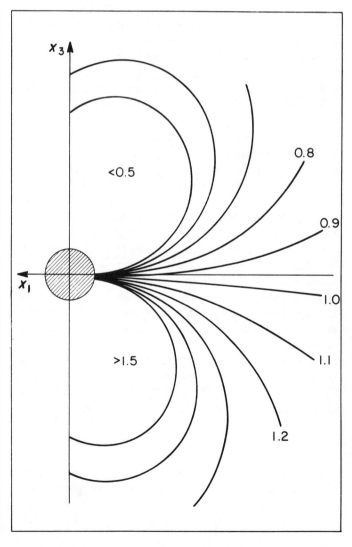

Thus, the approximate solution for weak distortion is

$$\frac{\partial \tilde{\theta}}{\partial x_1} \cong (1 - \epsilon_1)\frac{\partial \theta}{\partial x_1} - \epsilon_2 \frac{\partial \theta}{\partial x_3} \tag{28}$$

$$\frac{\partial \tilde{\theta}}{\partial x_3} \cong -\epsilon_2 \frac{\partial \theta}{\partial x_1} + (1 + \epsilon_1)\frac{\partial \theta}{\partial x_3} \tag{29}$$

These results have some interesting implications. For example, squaring and averaging Eqs. (28) and (29), and keeping first-order terms in ϵ_i, give

$$\overline{\frac{\partial \tilde{\theta}}{\partial x_1}\frac{\partial \tilde{\theta}}{\partial x_1}} = \overline{\frac{\partial \theta}{\partial x_1}\frac{\partial \theta}{\partial x_1}}(1 - 2\epsilon_1) \tag{30}$$

$$\overline{\frac{\partial \tilde{\theta}}{\partial x_3}\frac{\partial \tilde{\theta}}{\partial x_3}} = \overline{\frac{\partial \theta}{\partial x_3}\frac{\partial \theta}{\partial x_3}}(1 + 2\epsilon_1) \tag{31}$$

If the approach flow is locally isotropic, then

$$\overline{\frac{\partial \theta}{\partial x_1}\frac{\partial \theta}{\partial x_1}} = \overline{\frac{\partial \theta}{\partial x_3}\frac{\partial \theta}{\partial x_3}} \tag{32}$$

but Eqs. (30) and (31) show that the distortion introduces anisotropy. Note that it also induces a cross moment which vanishes under local isotropy:

$$\overline{\frac{\partial \tilde{\theta}}{\partial x_1}\frac{\partial \tilde{\theta}}{\partial x_3}} = -\epsilon_2\left(\overline{\frac{\partial \theta}{\partial x_1}\frac{\partial \theta}{\partial x_1}} + \overline{\frac{\partial \theta}{\partial x_3}\frac{\partial \theta}{\partial x_3}}\right) \tag{33}$$

Figures 10 and 11 show contours of ϵ_1 and ϵ_2 ahead of a circular cylinder in a uniform flow. It is surprising that the scalar gradient is affected so far upstream; the distortion is as large as 10% five diameters ahead of the cylinder.

These examples indicate that flow distortion can seriously degrade turbulence measurements; in fact, the turbulence can be affected much more seriously than the mean fields. However, the problem is a good deal more complicated than we have shown here. Hunt (1973) found that the effects on turbulence are completely different at scales which are small compared to the body size. In that case the distortion is no longer effectively instantaneous, and the nonlinear turbulent interactions play a role in the distortion physics. That problem needs a good deal more work, particularly from the experimental side.

4.4 Cross Talk

Imagine that we have a pair of propeller anemometers, each responding only to the wind component along its axis. We mount them with axes perpendicular and obtain $\mathcal{U}_1$ and $\mathcal{U}_3$ signals. If we make a small alignment error so the axes are at a small angle ϕ from the vertical and horizontal, the measured (superscript m) signals are

$$\mathcal{U}_1{}^m = \mathcal{U}_1\cos\phi - \mathcal{U}_3\sin\phi \tag{34}$$

$$\mathcal{U}_3{}^m = \mathcal{U}_1\sin\phi + \mathcal{U}_3\cos\phi \tag{35}$$

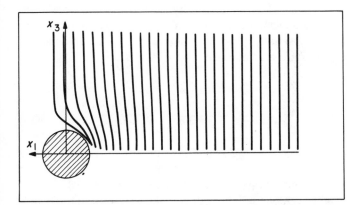

Fig. 9 *The distortion of a temperature field around a circular cylinder. A streamwise periodic temperature distribution exists in the undistorted upstream flow, so that temperature contours are vertical there. As the flow decelerates along the x_1 axis the temperature contours are compressed, while near the top of the cylinder the acceleration spreads the contours.*

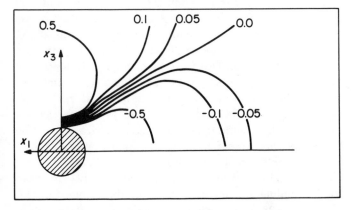

Fig. 10 *Contours of $\epsilon_1 = \int_0^t (\partial\tilde{u}_1/\partial x_1)\,dt'$ ahead of a circular cylinder. From Eqs. (30) and (31), the distortion in the temperature gradient field is proportional to ϵ_1.*

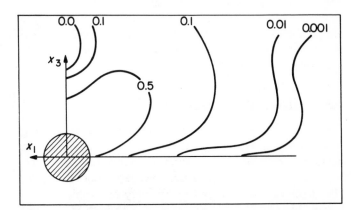

Fig. 11 *Contours of $\epsilon_2 = \int_0^t (\partial\tilde{u}_3/\partial x_1)\,dt'$ ahead of a circular cylinder. From Eqs. (28) and (29), the cross talk in the temperature gradient field is proportional to ϵ_2.*

Averaging and assuming $U_3 = 0$ give

$$U_1{}^m = U_1 \cos \phi \qquad (36)$$

$$U_3{}^m = U_1 \sin \phi \qquad (37)$$

If ϕ is small the error in U_1 can be negligible, but some "offset" will appear in U_3.

Consider now the equations for the measured turbulence:

$$u_1{}^m = u_1 \cos \phi - u_3 \sin \phi \qquad (38)$$

$$u_3{}^m = u_1 \sin \phi + u_3 \cos \phi \qquad (39)$$

Note that the instrument tilt causes "cross talk" in the turbulence components—that is, u_3 contaminates u_1 and vice versa. We met this type of cross talk earlier, in Eqs. (18) and (19), in our discussion of the distorted velocity field ahead of a body, and also met, in Eqs. (28) and (29), the analogous cross talk for components of a scalar gradient field.

As another example, consider the hot-wire sensor. In the high-overheat, constant-temperature mode it responds to heat-transfer fluctuations. These are caused principally by streamwise velocity fluctuations but also by temperature fluctuations. Thus the fluctuating output is

$$u_1{}^m = u_1 + a\theta \qquad (40)$$

where a is negative. If instead we operate the wire at a very small, constant current it responds principally to temperature but has some velocity sensitivity as well:

$$\theta^m = \theta + cu_1 \qquad (41)$$

In this case c is negative.

Cross talk is insidious because it affects all the statistics that one subsequently calculates from the signals, but it affects each statistic differently. For example, the type of velocity-field cross talk we have discussed is seldom serious for the velocity variances (or spectra) but can severely distort the stress; Kaimal and Haugen (1969) demonstrate this experimentally. As another (and rather arcane) example, I showed (Wyngaard, 1971) that velocity contamination of the temperature signal, as in Eq. (41), can lead to a very skewed temperature derivative but leave its variance virtually unaffected.

4.5 Indirect Measurements

The micrometeorological community has made good progress recently in measuring PBL structure, as you will read in detail in later articles. Fluxes of heat, momentum, moisture, and many trace constituents can now be measured directly; PBL depth can be measured by a variety of techniques; turbulent microstructure can be probed with both in situ and remote sensors. Many of these new techniques are feasible

only in a research environment, however. How then can those in the operational community have access to this new information on PBL structure?

One way is through indirect measurement techniques. By this I do not mean using remote sensors, which reveal PBL structure through its effect on propagated electromagnetic or acoustic waves. Instead, I mean inferring one property from measurements of others, through use of an empirical or theoretical relationship among the properties.

The framework of relationships which makes indirect techniques possible in the surface layer is Monin-Obukhov (M-O) similarity (Busch, 1973). This relates most mean and turbulent properties through length, temperature, constituent, and velocity scaling parameters plus a stability index. We can briefly illustrate. The length scale is z; the velocity scale is $u_* = (\tau_0/\varrho)^{1/2}$, where τ_0 is surface stress; the temperature scale is H_0/u_*, where H_0 is surface temperature flux; the constituent scale is C_0/u_*, where C_0 is the surface flux of constituent; and the stability index is z/L, where L is the M-O length, $u_*{}^3 T/(kgH_0)$. The similarity hypothesis is that nondimensionalized properties are universal (that is, the same in all surface layers) functions of z/L. For example, the mean wind and temperature gradients are

$$(kz/u_*)(\partial U_1/\partial z) = \phi_m(z/L) \qquad (42)$$

$$-(kz/T_*)(\partial \Theta/\partial z) = \phi_h(z/L) \qquad (43)$$

where $k \sim 0.4$, the von Karman constant, is traditionally used for scaling purposes. The functions ϕ_m and ϕ_h are known empirically, so measurements of dU_1/dz and $d\Theta/dz$ at height z can be used through Eqs. (42) and (43) and the definition of L to give the surface fluxes τ_0 and H_0.

As another example, Coulter and Wesely (1980) present a method for determining the surface temperature flux from acoustic backscatter and from laser scintillation, and show that it can give good agreement with direct measurements. A colleague and I (Wyngaard and Clifford, 1977) have even suggested obtaining momentum and moisture fluxes as well as the temperature flux from structure parameters of velocity, humidity, and temperature inferred from scintillations.

In this way one can use simple, statistically stable measurements to infer the surface fluxes, which are notoriously difficult to measure directly. The underlying M-O similarity relationships are now fairly well established for stable, neutral, and unstable conditions (Busch, 1973) although we now know that some parameters (notably statistics of the horizontal wind components) do not follow M-O similarity under unstable conditions (Panofsky et al., 1977).

There is also a fairly well-established similarity structure for the middle and upper portions of the convective PBL (Caughey, 1982), and therefore indirect measurement techniques could be feasible there as well. For example, the turbulent velocity scale is $w_* = (gH_0z_i/T)^{1/3}$, and in mid-PBL the vertical velocity variance is about $0.4w_*{}^2$. Thus, in principle, an acoustic sounder with Doppler could measure both z_i and

$\overline{w^2}$, and one could infer H_0. As another example, above the surface layer the temperature structure parameter behaves as

$$C_T^2 = 2.7 H_0^{4/3} (g/T)^{-2/3} z^{-4/3} \qquad (44)$$

(Wyngaard and LeMone, 1980), and acoustic sounder measurements of C_T^2 could be used to give H_0; this is one of the techniques explored by Coulter and Wesely (1980). With z_i from the acoustic sounder, this would then yield $\overline{w^2}$ from its similarity relation; one would not need Doppler acoustics with this approach.

Indirect measurements of PBL parameters can be more representative than direct measurements, because the latter can be very sensitive to local conditions, particularly in the surface layer. On the other hand, the results can only be as accurate as the underlying similarity relation, and broadly speaking these are not known to better than 20 %.

A particularly challenging task is to develop indirect techniques for obtaining PBL parameters from routine meteorological data. This is frequently required in air quality modeling, for example. Weil and Brower (1983) present methods for obtaining H_0, u_*, $U_1(z)$, and z_i under convective conditions from routine measurements, and find good agreement between their indirect values and direct measurements in field experiments.

4.6 Flux Measurements

The PBL is a region of strong turbulent fluxes, including fluxes of trace constituents. Unfortunately, at the present time we can measure only a few of the latter, even in research environments.

There is a mistaken but fairly widespread notion that in order to measure the vertical flux $\overline{u_3 s}$ of a constituent s, one needs to obtain time series of s and of u_3, each to sufficiently high frequency (i.e., covering the flux-carrying range) and then multiply the series and average. This is in fact sufficient, but not necessary. We will discuss two ways of relaxing the high-frequency time series requirement.

If one needs only the flux, and not its spectral distribution, then one needs only sufficiently fast-response (i.e., sufficiently short rise time) samples of u_3 and s; the time interval between successive samples of s can be arbitrarily large. In fact, sampling is most efficient if this time interval exceeds one integral scale, so that successive samples are statistically independent; in this way a given number of samples will produce the most stable statistics. This opens the possibility of measuring trace constituent fluxes with pulse laser detection techniques, which give very-short-rise-time concentration measurements at a slow repetition rate. If the vertical velocity is sampled during the pulse (and proper attention is paid to plumbing-induced lags) this can yield direct flux measurements.

A second way to avoid high-frequency time series is the "eddy accumulation" technique. Here we recognize that the turbulent flux can be written as a second moment of the joint probability density of u_3 and s:

$$\overline{u_3 s} = \int_0^\infty \int_{-\infty}^\infty \beta_{u_3 s} u_3{}' s' \, du_3{}' \, ds' \qquad (45)$$

This can also be written

$$\overline{u_3 s} = \int_0^\infty \int_0^\infty \beta_{u_3 s} u_3{}' s' \, du_3{}' \, ds'$$

$$- \int_0^\infty \int_{-\infty}^0 \beta_{u_3 s} |u_3{}'| s' \, du_3{}' \, ds' \qquad (46)$$

which is simply the difference of the fluxes due to upward and downward motions. In principle one can measure each of these separately by sampling air at a rate controlled by u_3 and storing it in two vessels, one for the negative u_3 and one for the positive. Subsequent analysis of the mass of s in each vessel yields the flux $\overline{u_3 s}$ by difference. Hicks and McMillen (1983) discuss this relatively unexplored technique.

5 THEORIES, MODELS, AND MEASUREMENTS

Even if we have a good working knowledge of turbulence and the PBL, and master measurement physics, we face surprises when we compare measurements with predictions. In this final section, I will discuss some of the reasons why.

5.1 Background

Turbulent flow researchers in general, and micrometeorologists in particular, have shown a subtle bias against theoretical and numerical modeling studies, and toward measurements. We need not look for any profound explanations for this; it stems simply from the closure problem in turbulence, which makes any theoretical or numerical modeling work less than completely rational and hence subject to uncertainty. By contrast, we have tended to regard measurements as the truth, the standard against which to test theories and models.

In the summer of 1968 the Boundary-Layer Branch of the Air Force Cambridge Research Laboratories, of which I was a new member, conducted a surface-layer field program at a beautifully flat, uniform site in southwest Kansas. Our objective was to study the "constant-flux" layer, the first few tens of meters above the surface, which theoreticians had predicted to have negligible vertical variation in the fluxes of heat and momentum.

The first paper (Haugen et al., 1971) from this experiment seemed to show that the fluxes were far from constant in the first 20 m for the usual averaging periods—say 15 min. Even over an hour, the fluxes were constant only within 20 %, and over the 24 h of carefully chosen, research-quality runs the fluxes were constant only to within 5–10%.

Figures 12 and 13 summarize the Haugen et al. results. The authors attributed the extreme variability of their measured fluxes to "submesoscale circulations." One can sense in the paper their disappointment in having failed to make a definitive assessment of the constant-flux-layer prediction, but at the same time their excitement in having discovered the strong time variations in surface-layer fluxes.

The root problem here, of course, is that our theories deal with ensemble-average properties, while we most often measure time averages. The prediction of a constant-stress layer, for example, comes from analysis of the ensemble-mean horizontal momentum equation; that analysis shows that a quasi-steady, horizontally homogeneous flow cannot support a significant surface-layer stress divergence in the ensemble mean. In fact, that conclusion seems so inescapable that one might wonder why it would have to be tested experimentally.

These Kansas results, and results from other experiments in the same time period, focused attention on the statistical scatter in time-averaged measurements. A convenient framework for interpreting this scatter had been provided in the Lumley-Panofsky (1964) monograph. They showed that for a property f, the variance σ^2 between the time mean $\bar{f}$ and the ensemble mean $<f>$ is

$$\sigma^2 = <(\bar{f} - <f>)^2> = (2\tau/T) <f'^2> \qquad (47)$$

where $<f'^2>$ is the variance of f; i.e., $<f'^2> = <(f - <f>)^2>$. Here τ and T are the integral scale and averaging time, respectively, and the result assumes $T \gg \tau$.

I covered in some lectures ten years ago (Wyngaard, 1973) many of the implications of this averaging-time relationship for surface-layer measurements. Rather than repeat that discussion here, I will simply emphasize that the averaging time necessary to achieve a given "accuracy" $e = \sigma/<f>$ is, from Eq. (47),

$$T = 2\tau(<f'^2>/<f>^2 e^2) \qquad (48)$$

and that it is proportional to the integral scale of f, the ratio of the variance of f to the square of its mean, and (1/required accuracy)².

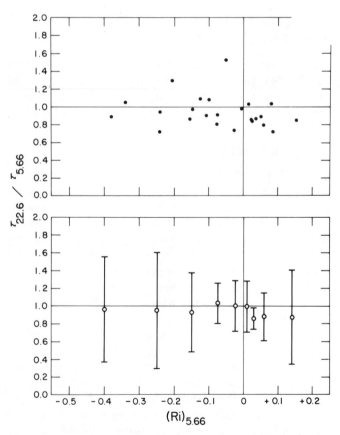

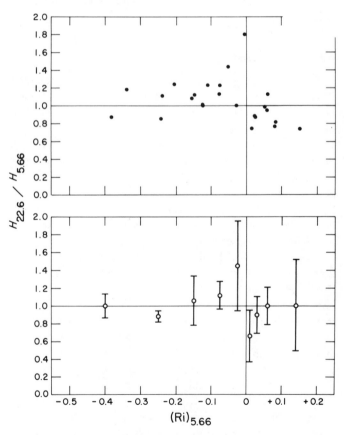

Fig. 12   Ratio of shearing stresses at 22.6 m and 5.66 m measured in the 1968 Kansas experiments. Upper plot, 15-min data in 1-h blocks. Lower plot, data grouped into blocks of Richardson number intervals, with circle indicating mean and bars plus and minus one standard deviation. (From Haugen et al., 1971.)

Fig. 13   As in Fig. 12 but for heat flux. Note that the scatter is less than for stress.

5.2 Scatter in Measurements

Of the three factors just listed which determine the required averaging time, I will focus here on the second, and call it F.

If the signal f is horizontal wind speed, for example, then F is $\overline{u^2}/U^2$, which is usually in the range 0.01 to 0.20. If f is turbulent shearing stress $\overline{uw}$, then (using a Gaussian approximation for the fourth moment $\overline{u^2w^2}$)

$$F \cong (\overline{u^2w^2} + \overline{uw}\,\overline{uw})/(\overline{uw}\,\overline{uw}) \qquad (49)$$

When the correlation coefficient between u and w becomes small (as in the convective PBL, and presumably also in the cloud layer) then F can become quite large. For example, in the convective PBL (without strong baroclinic effects) $\overline{uw}$ remains on the order of the surface stress u_*^2, while $\overline{u^2}$ and $\overline{w^2}$ scale with the convective velocity scale w_*^2. By the definitions of these scales $(w_*/u_*)^2$ is on the order of $(-z_i/L)^{2/3}$ so that $F \sim (-z_i/L)^{4/3}$. Since z_i/L can easily range over two orders of magnitude in different realizations of the PBL, F varies even more. This explains why stress is very difficult to measure reliably in the convective PBL. Figure 14, showing the large scatter in the stress data from AMTEX, demonstrates this.

F can also be very large for scalar concentration measurements. One extreme case is when the scalar signal is intermittent, alternating between one value and zero. If ϵ represents the fraction of time the signal is "on" (i.e., in the plume or cloud) then one can easily show that $F = (1 - \epsilon)/\epsilon$. This can grow large without limit as the intermittency increases. Thus concentration statistics can also require very long averaging times in order to reduce scatter to an acceptable level.

These examples illustrate that PBL measurements, even error-free ones, have their own uncertainty. No matter how carefully taken, if they are averaged over finite time or space they will inevitably scatter about the most likely, or ensemble-average, value.

In recent years the air quality modeling community has become particularly concerned about measurement scatter. This concern stems in part from attempts to verify simple diffusion models with some of the extensive data bases which have been gathered over the past ten years. Such exercises can give startling results—the model predictions often show very small correlations with observations. The first reaction is often to conclude that the models are poor, but more recently the community has come to suspect that the small correlations are due substantially to the "inherent uncertainty" in the data. The measurements give time averages, while the models predict ensemble averages. Hanna (1982) shows that there is roughly a factor-of-two scatter in the hourly averaged sulfur dioxide concentrations measured in St. Louis; this is probably due primarily to the inherent uncertainty in the 1-h average.

The laboratory simulations of Willis and Deardorff (1976), which gave the results shown in Fig. 15, give some insight into the scatter in time-averaged measurements of diffusion from a continuous point source. Note that the individual realizations (which represent averaging times of 20–30 min in the atmosphere) have large scatter. At $y = z = 0.4$, for example, measured values of C range from 0.4 to 2.2; the mean is 1.1, and the standard deviation over the seven realizations is 0.7. Thus the concentration measured at that point over 20–30 min is uncertain to about $0.7/1.1 \sim 60\%$. From Eq. (48) this inherent uncertainty is reduced to about 40% for 1-h averages, and to about 25% for 3-h averages.

5.3 Minimizing Scatter

Today's instruments and recording systems make possible measurements that we could only dream about 20 years ago. However, scatter is a fundamental obstacle to our making use of these measurements; in fact, it seems a larger obstacle than we realized earlier.

How can we minimize scatter? The most obvious way is through "brute force"—simply gathering enough data that our averages have minimal uncertainty. This is expensive.

Second, we can use (where appropriate) area averages instead of line averages. Our averaging expression, Eq. (47), rewritten for line averages, shows that the scatter variance is proportional to l/L (i.e., integral scale/averaging path length). Flying aircraft legs that are 2,000 integral scales long in order to get stable estimates of stress is not always feasible. For area averages, however, the scatter variance goes as $(l/L)^2$; thus we can estimate stress just as reliably by averaging over an area of 2,000 (integral scales)^2 (Wyngaard, 1983). This is equivalent to a square $(2,000)^{1/2} \sim 45$ integral scales on a side. In principle, area averages are possible with modern radars, and their potential is just beginning to be explored.

Finally, it may now be feasible to use laboratory and numerical (e.g., large-eddy) simulations to generate data bases, at least for first-cut studies. Willis and Deardorff (1976) have done pioneering work here in diffusion applications. This can be a valuable, low-scatter, inexpensive substitute for direct atmospheric measurements.

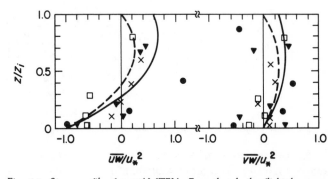

Fig. 14 Stress profiles from AMTEX. Even though the flight legs were 30–70 km long, there is so much scatter that one cannot discriminate between competing theoretical predictions (solid and dashed lines). (From Lenschow et al., 1980.)

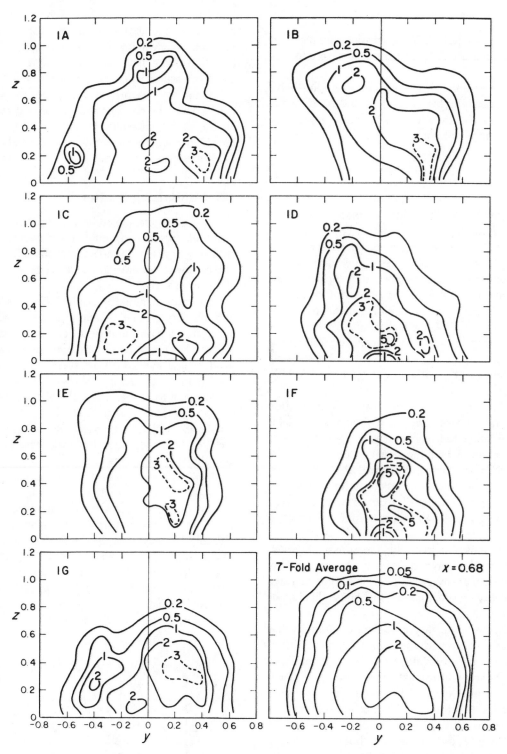

Fig. 15 Concentration contours for a laboratory simulation of diffusion from a continuous point source in a convective PBL. Individual realizations represent 20- to 30-min averages. Note the large scatter. (From Willis and Deardorff, 1976.)

Acknowledgments. I am indebted to the many colleagues with whom I have worked on measurement physics, and particularly to J.L. Lumley, who introduced the subject to me. I am also grateful to Boba Stankov, who programmed the solutions shown in Figs. 9–11, to anonymous reviewers for helpful suggestions, and to Shirley Michaels, who typed the manuscript.

Flux and Profile Measurements from Towers in the Boundary Layer

J.C. Kaimal, Wave Propagation Laboratory, NOAA

1 INTRODUCTION

Towers offer convenient platforms for observing mean and turbulent properties of flow in the atmosphere's boundary layer. Much of what we know about the structure of this layer comes from measurements made with sensors mounted on towers. All varieties of structures, from short masts to very tall television towers, have been used for this purpose. Sensors have become more sensitive and accurate over the years, methods of recording and analyzing data have advanced dramatically, but the design of towers has not changed significantly. Many of the problems associated with tower measurements, such as flow distortion induced by the structure, are therefore still with us. The Eulerian nature of the observations may be appropriate for some studies, but for others Taylor's hypothesis, with all its uncertainties, has to be invoked to convert the observed time scales into spatial scales. These limitations have seldom dissuaded researchers from using towers—indeed, the convenience of easy access to sensors for maintenance and of direct connection to recording facilities on the ground argues persuasively in favor of their use. Short towers are employed extensively for mounting sensors in air quality monitoring and mesoscale observing networks. Tall towers continue to provide the data needed for verifying theoretical models of the boundary layer. Given the wide use of towers, it is important that designers of tower-based systems be cognizant of their inherent problems.

In this article I will present a broad survey of sensors and techniques used in boundary-layer studies and some effects of tower interference on measurements. I will also discuss data acquisition and archiving strategies aimed at making the data available to users in a compressed form. Many of the ideas presented are embodied in the design of the Boulder Atmospheric Observatory (BAO) (Kaimal and Gaynor, 1983).

2 DATA REQUIREMENT FOR BOUNDARY-LAYER STUDIES

The amount of information needed to characterize flow in the boundary layer varies widely with the application. For studies of boundary-layer structure and for model verification, we need measurements of the mean and fluctuating components of the wind field, the temperature, and the humidity at several levels on a tall (100–300 m) tower. The mean components are measured with sensors that respond slowly, but accurately, to slow changes (time scales of 60 s and longer) in the variable. These sensors are well suited for vertical profile (or gradient) measurements. For the fluctuating components, we need sensors that can respond to frequencies as high as 5 Hz. The high-frequency information enables one to compute the turbulent fluxes of momentum, heat, and water vapor, as well as the variances and third-order terms that appear in the budget equations. The same information is needed to compute statistical quantities such as spectra, cospectra, correlations, and probability distributions on time scales relevant to turbulent transport in the lower boundary layer.

The required vertical spacing for the measurements depends on the height range of interest. In the first 50 m above the ground, where the boundary-layer structure is strongly height-dependent and the profiles of wind speed and temperature tend to be steep, a minimum of three observation levels is required. The spacing should be logarithmic, or approximately so (e.g., 10, 22, 50 m at the BAO). More profile levels can be added below 10 m if needed, with the vertical spacing contracting steadily, closer to the ground. Flux measurements below 5 m run the risk of underestimation from bandwidth limitations (see Section 3) and should not be attempted with the ordinary sensors used for boundary-layer measurements. Above 50 m, in the region commonly referred to as the mixed layer (under convectively unstable conditions), a linear spacing is preferred, since the turbulent lengths here scale with the boundary-layer depth and not with the height above ground. The 50-m spacing maintained between 50 and 300 m at the BAO appears to be a reasonable separation for observing changes in the spatial structure during the day as well as at night.

Surface mesoscale networks, such as the portable automated mesonet (PAM) system developed by the National Center for Atmospheric Research (Brock and Saum, 1983), meet another type of observational requirement. In such systems, measurements are made of the mean fields (1-min averages) of wind speed, wind direction, temperature, and humidity. Slow-response sensors mounted on masts are used. The standard height for wind-speed measurement is 10 m, and for temperature and humidity, 2 m. When the network is deployed over a large area, radio telemetry or even satellite linkages are needed to transmit data to the base station. The Program for Regional Observing and Forecasting Services (PROFS) mesonet system (Pratte and Clarke, 1983) operated by the National Oceanic and Atmospheric Administration uses dedicated telephone lines to link stations spread throughout northeastern Colorado with the polling computer in Boulder.

Much less demanding are the requirements for air quality monitoring, where the accuracy and sampling requirements are less rigorous than for research use. Recommended height for wind speed and direction is 10 m; for temperature and humidity, 2 m; and for vertical temperature gradient, the difference between 2 and 10 m.

3 SENSOR CHARACTERISTICS

The list of sensors used for tower-based measurements in the boundary layer would be a very long one indeed. An early classification of those sensors can be found in Middleton and Spilhaus (1953). Moses (1968) gives a comprehensive survey of meteorological instruments used in the atomic energy industry. Kaimal (1975) has listed the advantages and limitations of sensors used for profile and flux measurements. More recent surveys can be found in the review paper by Wyngaard (1981a) and in the contributions to *Air-Sea Interaction—Instruments and Methods*, edited by Dobson, Hasse, and Davis (1980). Only a hurried treatment of the subject is possible in this presentation, and only of techniques most commonly used today. The sensors discussed are either easy to construct or available commercially. For broader surveys and more details, the reader is urged to consult the references listed in this chapter. My focus here will be on the relative merits and limitations of techniques employed to measure wind, temperature, and humidity in the atmosphere's boundary layer.

Sensors in this section are separated into slow- and fast-response types, consistent with the needs for profile and flux measurements, respectively. Absolute accuracy, rather than speed of response, is the desired attribute in the former type. Response times are typically between 10 and 60 s. Among the latter type, speed of response is often achieved at the cost of long-term stability in the calibration. However, good relative calibration accuracies are maintained through such simple schemes as comparing slow changes in the variable with readings from an accurate slow-response device. There is no upper limit to the frequency response a sensor may have; the sensor with the slower response sets the limit for the response in the flux calculations.

3.1 Slow-Response Wind Sensors

Cup anemometers, propeller anemometers, and wind direction vanes head the list of sensors in this category. They have been in use for over a century and are still the sensors most widely preferred for profile measurements. Simplicity, ruggedness, and dependability are their principal virtues. Typical response times for these rotating devices are too long for most turbulence work, but for other applications, where mean wind-speed and direction readings suffice, they do a creditable job. Busch et al. (1980) and Wyngaard (1981a) pro-

vide good discussions of the response characteristics of these sensors.

Cups, propellers, and vanes appear on the market in a variety of forms. A common configuration for a cup and vane system is a three-cup wind-speed sensor and a direction vane installed side by side on a common mounting arm. The advantage of the cup anemometer is that it does not require alignment into the wind direction. Starting speeds are typically 0.5 m/s, and distance constants (63% recovery time converted to distance), between 2 and 5 m. Well-designed cup anemometers can be calibrated in a wind tunnel to an accuracy of ±1%. However, there is a tendency in cups to "overspeed," partly as a result of their nonlinear response to fluctuations in wind speed, and partly from sensitivity to the vertical wind component. Estimates of overspeeding error range from 5% to 10%.

Propeller anemometers, on the other hand, do not overspeed, but they operate most dependably when pointing directly into the wind. Two- and four-blade propellers are available, the latter for lower wind-speed ranges (0.5–35 m/s). The propellers can be flat or helicoid. Propellers typically exhibit significant deviations from axial response, so they are mounted at the end of a vane with closely matched response characteristics to keep them oriented into the wind. When they are used in a fixed configuration, corrections for deviation from the true cosine response have to be made. Otherwise, the measured wind components would be underestimated. Accuracies and distance constants for propeller-vane combinations are comparable to those of lightweight cup anemometers. Their calibration should be checked in a wind tunnel, but for most noncritical applications, periodic checks with a motor of known rotation rate are sufficient.

For discussions of the errors in cup anemometer response, see Wyngaard et al. (1974), Busch and Kristensen (1976), and Kaganov and Yaglom (1976). Corrections for response errors in three-component propeller anemometers are given by Horst (1973). Finkelstein (1981) describes a recent wind tunnel study of the dynamic performance of wind vanes. Monna (1978) discusses results of a comparative study of the dynamic properties of three propeller-vane anemometers.

3.2 Fast-Response Wind Sensors

The hot-wire anemometer is the sensor traditionally used for high-frequency velocity measurement in turbulence work. It is best suited for laboratory and wind tunnel studies where the dimensions of the sensor have to be small (on the order of millimeters) and the frequency response high (on the order of 10 kHz). No other sensor can match the high-frequency resolution of the hot-wire anemometer. Consequently, it is used also for fine-scale measurements in the atmosphere (frequencies greater than 1 Hz and spatial scales smaller than 1 m). (See the article by Friehe in this volume.)

For boundary-layer flux measurements, the excellent frequency response of the hot-wire anemometer does not make up for its two major drawbacks: susceptibility to calibration shifts from atmospheric contamination, and fragility. The scales of interest in the boundary layer range from 1 m to 10 km, and continuous measurements are sometimes needed for periods ranging from days to weeks. In such applications, the sonic anemometer is the preferred instrument; it has none of the drawbacks mentioned above and none of the response problems associated with rotating-type anemometers. However, the high cost of sonic anemometers has led many researchers to look for simpler and less expensive options such as dynamic anemometers. Dynamic anemometers measure wind velocity by sensing either the pressure or the drag force on an object placed in the flow. They include thrust anemometers, anemoclinometers, and vortex anemometers. Smith (1980) has reviewed these dynamic anemometers. They respond more slowly to fluctuations in the wind than the sonic anemometers, but more quickly than the rotating types. Most of these devices are custom-made at various laboratories and are not available commercially. The same is true of the few truly lightweight cup anemometers (Wyngaard, 1981a). The types of sensors currently used for wind-fluctuation measurement in most low-budget applications are lightweight propeller anemometers and bivanes offered by some manufacturers of meteorological instruments. My main focus in this section will be sonic anemometers and lightweight propellers and bivanes, these being the types most easily available for research and operational use.

Sonic anemometers achieve their frequency response by sensing the effect of wind on the transit times of acoustic pulses traveling in opposite directions across a known path. The only limitation to their frequency response is the spatial one, imposed by line-averaging along the path. The effects of line-averaging along different directions within the flow are discussed by Kaimal (1980). For a crosswind component the response function resembles a single-pole, low-pass filter with its half-power point at a wavelength[1] λ equal to the acoustic path length d (see Fig. 1). In the streamwise direction, it has the $(\sin^2 x)/x^2$ response of a moving average filter, which rolls off more sharply and drops to zero at $\lambda = d$. The shapes of the response functions for wind components resolved from measurements made along arbitrary coordinate directions will be more complex, but almost all of the distortion is confined to wavelengths smaller than $2\pi d$ (Kaimal et al., 1968). For an anemometer with $d = 20$ cm, this wavelength corresponds to 1.26 m.

Knowing the limiting wavelength λ_d for undistorted spectral response and the limiting wavelength λ_i for the onset of the inertial subrange ($-5/3$ power law in all velocity components), we can determine the lowest height at which a sonic anemometer can be operated without compromising the first octave of the inertial subrange. Recognizing that in the first 10 m or so of the atmosphere, $\lambda_i \cong z/2$, where z is the height above ground, we should, to satisfy the above condition, maintain $\lambda_i = 2\lambda_d$. Thus,

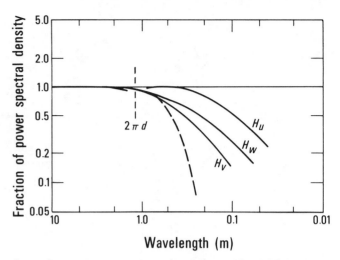

Fig. 1 *Response functions for u, v, and w for the nonorthogonal three-axis array of Fig. 2. Path length on each axis is 20 cm. Mean wind is assumed to be along the axis of symmetry. The dashed curve corresponds to the function* $(\sin \pi d/\lambda)^2/(\pi d/\lambda)^2$ *for simple line-averaging in the streamwise direction.*

$$z_{min} = 8\pi d \cong 5 \text{ m} \qquad (1)$$

for $d = 20$ cm. This criterion is important if calculations of dissipation rates and vertical velocity variances are to yield values with reasonable accuracy. There is some indication from recent experiments at the BAO and from theoretical modeling (Kristensen and Fitzjarrald, 1984) that this height limitation can be relaxed for flux calculations. This is reasonable since the cospectrum has its major contribution at $\lambda > \lambda_i$; at $\lambda \leq \lambda_i$ the turbulence is isotropic and therefore cannot produce flux. Kristensen and Fitzjarrald's heat flux averages over 20-min periods remain constant down to about 2 m, and the theoretical analysis indicates that the flux may remain constant to as low as 1 m for $d = 20$ cm in the neutral to unstable surface layer. For the stably stratified surface layer, correction for cospectral attenuation may have to be applied. We await further experimental verification of this lower limit.

Sonic anemometers can be of either the pulse type or the continuous wave type. The former measures transit time differences to compute the velocity component along the path, whereas the latter measures phase differences. Both are related directly to velocity. V_d, the velocity component along path length d, can be expressed as

$$V_d = (c^2/2d)(t_2 - t_1) \qquad (2)$$

where c is the velocity of sound in air and t_1 and t_2 are the transit times for sound pulses traveling downwind and upwind along the path. If c and d are known, V_d reduces to the measurement of $t_2 - t_1$, a relatively simple time interval measurement. In most applications, the velocity of sound is approximated by $c^2 = 403\mathscr{T}(1 + 0.32\mathscr{P}_e/\mathscr{P})$, where $\mathscr{T}$ is the absolute temperature, $\mathscr{P}_e$ is the vapor pressure of water, and $\mathscr{P}$ is the atmospheric pressure, all expressed in SI units. The

---

[1] $\lambda = U/f$ assuming Taylor's hypothesis, U being the mean wind velocity and f the cyclic frequency.

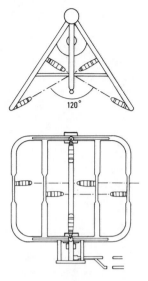

Fig. 2 *Nonorthogonal configuration in the Kaijo Denki sonic anemometer. The horizontal axes are separated by an angle of 120°.*

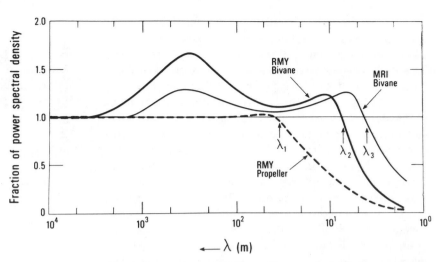

Fig. 3 *Response functions for the w component in the R.M. Young vertical propeller, the R.M. Young propeller-bivane system, and the Meteorology Research, Inc., front-damped bivane and cup anemometer system. λ₁, λ₂, and λ₃ are the cutoff wavelengths defined as the 98% power point.*

contribution from $\mathscr{P}_e$ is small and usually negligible, as are fluctuations in $\mathscr{T}$ compared with the mean absolute temperatures in the lower atmosphere. Thus, c^2 in Eq. (2) can simply be replaced by $403T$, where the capital letter denotes the average temperature for the observing period. This approach is used in the BAO sonic anemometer.[2]

The Kaijo Denki sonic anemometer (Hanafusa et al., 1980) uses an alternate approach that eliminates the dependence on $\mathscr{T}$. It invokes the relationship

$$-V_d = (d/2)[(1/t_1) - (1/t_2)] \qquad (3)$$

Microprocessor techniques are used to compute the reciprocals of t_1 and t_2 and to perform other computations as well. The BAO and the Kaijo Denki anemometers differ in two significant respects:

• The Kaijo Denki anemometer uses two transducers instead of four on each axis to reduce flow distortion along the path (see Fig. 2). The two transducers operate alternately as transmitters and receivers.

• The BAO anemometer probe has an orthogonal configuration (see Fig. 6), whereas the Kaijo Denki anemometer probe maintains a 120° separation between its horizontal axes.[3] The former is designed for fixed operation on a tower and the latter for more flexible installations where it can be rotated into the wind for virtually unobstructed flow along the horizontal paths.

The fixed orientation in a BAO-type application requires that steps be taken to compensate for the effects of wind shadowing by the transducers on the horizontal velocity measurements. These corrections, which have their basis in wind tunnel and atmospheric tests, are implemented in real time with special algorithms (Kaimal and Gaynor, 1983) incorporated in the data acquisition software. The calibration

procedure for a sonic anemometer is relatively simple. An occasional adjustment of the zero-wind setting, with the axis enclosed in a small anechoic box, is all that is usually required. If the path length is adjustable, this needs checking as well.

It is instructive to compare the dynamic responses of the bivanes and propellers used in atmospheric work with that of the sonic anemometer. Specifications list the delay distances (50% recovery) for bivanes and distance constants (for 63% of stepwise change) for propellers as being on the order of 1 m. However, these distances translate to much longer cutoff wavelengths if one were to define them in terms of deviations from true response. Figure 3 shows response functions for the vertical wind component w measured by two bivane systems and a propeller. These functions were obtained simultaneously from a sonic anemometer at the same height, 5–10 m away. Defining λ_c, the cutoff wavelength, as the point where the power spectrum drops to 98% of its true value, we find λ_c to be 32 m for the R.M. Young vertical propeller, 7 m for the R.M. Young propeller-bivane system, and 4.4 m for the Meteorology Research, Inc., front-damped bivane-cup system (for details see Lockhart et al., 1983).

The bivane's response to w appears superior to that of the propeller, but the amplification at mid-frequencies[4] raises concern as to its effect on flux calculations. To a moderate degree, the effect on the variance is beneficial, as it compensates for the high-frequency cutoff. Vertical velocity variances from the bivanes compare very closely with sonic

---

[2]This anemometer is now manufactured by Applied Technologies, Inc., in Boulder, Colorado.

[3]Kaijo Denki also manufactures an orthogonal probe very similar in concept to those on the BAO.

[4]The possibility of a gyro effect, causing "cross talk" from the lateral fluctuations, is being explored by L. Kristensen (personal communication).

anemometer values. The 10-m propeller variances, on the other hand, are severely underestimated: 25% lower under daytime conditions and up to 50% lower under nighttime conditions (Lockhart et al., 1983).

The large λ_c value for the w propeller is not surprising when one considers the off-axis degradation in response and the relationships that exist between distance constant L, half-power wavelength λ_o, and λ_c. The distance constant for the R.M. Young propeller increases from 1 m to 2 m as the flow deviates from axial to 80° off-axis (Garrat, 1975). In principle, the distance constant approaches infinity at 90° off-axis.

For a linear first-order system, the spectral transfer function can be written as

$$H(\lambda) = 1/[1 + (2\pi L/\lambda)^2] \qquad (4)$$

At the half-power point we have

$$\lambda_o = 2\pi L \qquad (5)$$

and λ_c, defined as the 98% power point, becomes

$$\lambda_c = 2\pi \lambda_o \cong 40L \qquad (6)$$

With L = 2 m (for vertical velocity fluctuation, w, measurement), λ_c would be 80 m for a linear first-order system. Fortunately, the propeller's response is better by a factor of 2.5 (see Fig. 4), which brings the actual value of λ_c down to 32 m as observed.

The need for careful alignment of the measurement coordinates with respect to a known reference direction cannot be overemphasized. The effect of leveling error in flux measurements is known. There seems to be less recognition of the fact that the y and z (lateral and vertical) axes cannot be retrieved by simply rotating the coordinates twice in order to force the lateral velocity, V, and W to be zero. A third constraint is needed to define the y-z coordinates unambiguously. This constraint could be the orientation of the sensor with respect to gravity.

3.3 Slow-Response Temperature Sensors

The sensors most often used for mean temperature and temperature-difference measurements are platinum resistance thermometers, thermocouples, thermistors, and quartz thermometers (Deacon, 1980; Cole, 1978). Profile studies require accuracies and resolutions on the order of 0.05°C and 0.01°C, respectively. These sensors are capable of offering such performance. The sensing elements are usually encapsulated in glass or metal for protection from wear and atmospheric contamination. The attendant increase in sensor size results in the time constant being long (~ 1 min), which is not a limitation for mean profile measurements. However, radiation error is a serious problem in sensors of that size, so proper shielding and aspiration are essential. The most serious limitations to overall accuracy are the uncertainties introduced by the shielding and the aspiration.

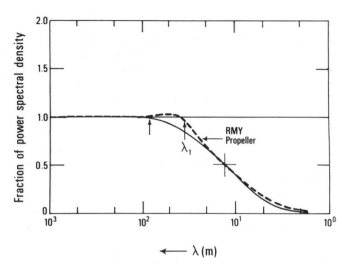

Fig. 4   Response function for a linear first-order system with distance constant L = 2 m (solid curve) compared with the actual response of the R.M. Young propeller measuring the w component (dashed curve). Their half-power points coincide, but the cutoff wavelength in the actual response is smaller by a factor of 2.5.

For applications where a high degree of stability is required, platinum resistance thermometers and quartz thermometers are the logical choices. Additionally, they have the advantages of high sensitivity, good resolution and linearity, and interchangeability. Platinum resistance thermometers in bridge circuits are an order of magnitude less accurate than quartz thermometers, but they also cost significantly less. Thermocouples require accurate reference junctions for absolute measurement but they are ideal for differential measurement. However, great care is needed to reduce parasitic junctions. Thermistors have very high (negative) temperature coefficients of resistance (about ten times those of platinum resistance thermometers), but their small thermal dissipation rates make self-heating a problem, tending to reduce drastically the available sensitivity. Nonlinearity in the output and poor long-term stability are problems commonly encountered in thermistor sensors.

The ideal choice for a temperature profile system is the quartz thermometer. Its output is a beat frequency that varies linearly with temperature and its basic accuracy (0.001°C) is degraded only by the errors (±0.05°C) introduced by the radiation shield and the aspirator. The operation of such a system on the BAO tower is described by Kaimal and Gaynor (1983).

3.4 Fast-Response Temperature Sensors

Platinum wire thermometers, thermocouples, thermistors, and sonic thermometers have all been used for temperature fluctuation measurements. Thermistor and thermocouple sensing elements cannot be made as small as platinum wires, so they tend to be less responsive and not as well suited for flux measurements near the ground. All three types are normally exposed to free air without the benefit of protection from radiation shields. They are, therefore, vulnerable to damage from winds, rain, snow, and flying debris. The chances of damage can be reduced significantly if the element is properly supported. Such supports increase the time constant of the sensor. The 12.5-μm platinum wire in the fast-response probe (Atmospheric Instrumentation Research model DTIA) used at the BAO is wound around a small helical bobbin with threads strung lengthwise. The spectral response of the wire is unaffected to a frequency of 5 Hz, but drops off more sharply above that because of conduction through the support. With the support, the sensing element often survives for weeks and even months. Freezing rain, wet snow, and very high winds are the events that usually break the wire. In this context it should be pointed out that platinum wires should not be left exposed to the elements for too long. Contamination of the wire has a direct effect on its time constant. The length of exposure the elements can stand depends on the environment. A detailed discussion of fast-response thermometry is given by Larsen et al. (1980), as well as in the article by Friehe in this volume.

The sonic thermometer (Mitsuta, 1974; Hanafusa et al., 1980) has the requisite frequency response for boundary-layer flux measurements. Temperature is sensed along the same path as w, so its spatial averaging characteristic is compatible with that of w. However, the temperature measurement is contaminated by residual sensitivities to humidity and to the wind component normal to the path. These errors are negligible under daytime unstable conditions, when the temperature fluctuations are large, but they can be significant under near-neutral and stable conditions (Kaimal, 1969). The nature of the crosswind contamination is such as to introduce a bias in the estimation of the time of transition through neutral stability; the observed transition is shifted further into the stable regime.

3.5 Slow-Response Humidity Sensors

Humidity is among the more difficult parameters to measure in the boundary layer. Reviews of methods used for measuring mean humidity are given by McKay (1978) and Coantic and Friehe (1980). Many different units are used to specify the water vapor content of air: absolute humidity (mass/unit volume), specific humidity (mass of water/mass of moist air), and relative humidity (vapor pressure of water in air/vapor pressure of water in saturated air at same temperature and pressure).

Dew point (or frost point) hygrometers provide absolute measurement of humidity, and psychrometers that measure wet-bulb depression yield specific humidity through the psychrometric equation. Hygroscopic devices that respond to changes in relative humidity are simpler to use, but usually depend on curves provided by the manufacturer for calibration.

For applications where accuracy and long-term stability are of critical importance, the dew point hygrometer would be the most logical choice. In this device, a mirrored surface is maintained piezoelectrically at the temperature where the moisture in the air begins to condense, or freeze, on it. The temperature of the mirror is sensed by a platinum resistance thermometer, thermocouple, or thermistor. The device is generally housed in an aspirated shield. The sensor requires periodic cleaning of the mirror surface and checks of the calibration adjustments, but with reasonable maintenance absolute accuracies of $\pm 0.5\,^\circ$C are possible.

The psychrometric technique, used widely in field experiments, is inexpensive (compared with the cost of a dew point device), simple in concept, and relatively easy to maintain. It consists of two identical ventilated temperature sensors, one of which is covered with a wick saturated with pure water. Great care is needed in its design to ensure proper shielding from solar radiation, adequate ventilation, and wetting of the wick. No commercial versions exist, since the device does not lend itself to self-contained packaging. The accuracy of this system is low at low humidities and temperatures, but is usually assessed in terms of the reading at 100% relative humidity. Accuracies of 0.5–1.0 °C equivalent dew point temperatures can be maintained over a 20–80% relative humidity and a 0–25 °C temperature range (McKay, 1978). The errors increase sharply below freezing, because it is difficult to form and maintain an ice bulb on the wet thermometer.

The hygroscopic sensors are also relatively inexpensive and easy to use. The Vaisala Humicap is one of the more widely used sensors of this type. It responds to humidity with a capacitance change. The sensor provides a voltage directly proportional to relative humidity. McKay (1978) describes the relative merits and drawbacks of various other devices in this category. Among their common failings are hysteresis effects, susceptibility to contamination, and loss of accuracy at high relative humidities. The hair hygrometer, one of the oldest and simplest of the hygroscopic devices, can be kept accurate to within 5% with weekly calibration checks, if the air temperature is above freezing.

3.6 Fast-Response Humidity Sensors

Three different techniques are currently used for measuring humidity fluctuations.[5] They involve the absorption of ultraviolet radiation by water vapor (Lyman-alpha hygrometer), the absorption of infrared radiation by water vapor (infrared hygrometer), and the dependence of microwave refractivity on humidity (microwave refractometer). A review of these techniques is given by Hay (1980).

The simplest of the three devices is the Lyman-alpha hygrometer. It requires a Lyman-alpha source, a nitric oxide detector, and a space between the two where the absorption takes place. Magnesium fluoride windows are needed on the source and detector tubes since most other materials are

---

[5]Shaw and Tillman (1980) have described a digital filtering technique which restores the frequency response of wet-bulb psychrometers to meet the requirements for boundary-layer turbulence work. The technique should be useful for speeding up the response of any slow-response device.

opaque to ultraviolet radiation. Buck (1983), in describing the most recent advances in this technique, points out that the cutoff frequencies of the nitric oxide detector and the magnesium fluoride windows neatly bracket the Lyman-alpha emission line of atomic hydrogen (121.56 nm). The other emission lines in the hydrogen glow discharge produced by the source are thus filtered out.

The very strong absorption of this emission line by water vapor makes measurement possible over short path lengths (~ 1 cm). (By comparison, absorption by ozone and oxygen, the only other absorbers in the atmosphere, can be neglected in the boundary layer.) The dimensions of the sensor are small enough (20 cm long × 2 cm in diameter) to permit installation close to a sonic anemometer. For best directional response, the sensor is mounted with its path oriented vertically. A configuration specifically designed for tower measurements (Buck, 1983) is now being manufactured by Atmospheric Instrumentation Research, Inc., of Boulder, Colorado.

Two sources of drift in the calibration have been noted: aging of the Lyman-alpha source and window degradation from reaction of atmospheric constituents with the window material. The effect of aging is retarded to some extent by using higher hydrogen pressures and sealing uranium hydroxide in the source body (Buck, 1983). The window degradation is reversible since it occurs on the outer surface; washing with alcohol and rubbing with a fine abrasive restores its transmission properties. Nevertheless, for maximum accuracy, this device should be operated in conjunction with a dew point hygrometer, so its calibration can be continuously updated by comparing changes in its mean readings with those derived from the slower, more accurate, dew point measurements.

The infrared hygrometer detects humidity through differential measurement of infrared transmittance at two adjacent wavelengths, one located in a region of high water vapor absorption and the other where the absorption is negligible. The transmitting path is typically 0.2–1.0 m long, and the beams are usually modulated by a mechanical chopper to permit high-gain amplification of the detected signal. Hay (1980) describes several designs used in field studies. Optical components such as narrow-band filters and beam splitters add to the complexity of this device. Calibration stability depends on the stability of the lamp output (usually a tungsten filament operated at less than its rated voltage). At humidities approaching saturation, the calibration is undependable because of scattering from water adsorbed and absorbed by particles in the air.

The microwave refractometer measures the refractive index of air in a cavity and depends on the relationship among refractivity, specific humidity, temperature, and pressure to derive humidity. Strictly speaking, simultaneous data on temperature and pressure fluctuations in the cavity are needed to remove their influence on the refractive index; in practice, pressure fluctuations are ignored since they are usually small (< 1 mb) compared with the absolute pressure.

The temperature measurement is made as close to the cavity as possible.

The sensing element in the microwave refractometer is a resonating cavity with ventilating ports. The cavity dimensions are small, but the spatial resolution of the sensor is more a function of its flushing efficiency than of its actual size. For a 5-cm cavity, in moderately light winds, spectral attenuation starts at a wavelength of 1.5 m. In a recent experiment conducted at the BAO (J.T. Priestley, personal communication), the refractometer response was found to fall off at 2 Hz, with wind speeds of 3–4 m/s. The Lyman-alpha hygrometer's response under the same conditions extends another decade, to 20 Hz.

4 TOWER INTERFERENCE AND EXPOSURE OF SENSORS

Towers, booms, and mounts used for supporting a sensor can interfere with the flow, thus introducing errors in the measured gradients and fluxes. These errors may be reduced to acceptable levels if some care is exercised in the design of the supporting structure and in the placement of sensors. There is considerable discussion on this subject in the literature; see Blanc (1983) for the most recent survey. No simple solutions or relationships for flow distortion have emerged because the geometries of the obstacles discussed vary so greatly. Wucknitz (1980), in his study of wind-field distortion around a cylindrical mast, considers two distinct regions of the disturbed flow: one close to the obstacle, at distances smaller than the diameter of the obstacle, where the flow is complicated and difficult to describe theoretically; the other at distances greater than the diameter, where potential flow can be assumed.

The first region is one the experimenter would do well to avoid. For a tower with structural members that are small and widely separated (ratio of obstructed to unobstructed area < 0.1), one might consider installing sensors at upwind distances closer than its outside dimensions. Most towers have structural member densities between 0.2 and 0.3. There the sensors should be mounted no closer than 1.5 times the largest lateral dimension of the tower. In applications where only one level of observation is needed (say at 10 m), the anemometer should be mounted on top of the tower (or mast) to avoid direct tower shadowing. To reach a flow region where potential flow can be assumed, the sensor has to be at least three lateral dimensions above the top of the tower (Wucknitz, 1980), supported by a thinner mast.

Another region of the flow to be avoided is the downwind side of the tower. At the very least, readings from this region should be treated with caution. Here, the flow is strongly influenced by the wake structure. Even on towers with low structural member density, the combined effect of wakes from the separate members can be greater than for a solid tower offering the same obstruction area to the flows. Moses and Daubeck (1961), Gill et al. (1967), Camp and Kaufman (1970), and Wucknitz (1980) have examined the

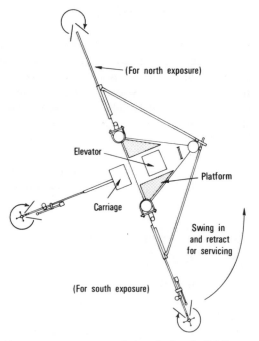

Fig. 5   *Plan view of an instrument platform level on the BAO tower showing wind direction sectors for which data are considered acceptable. A 30° clearance is allowed on either side of the tower.*

characteristics of flow in the wake region. The wake intensity and width vary with Reynolds number and are sensitive to both the roughness elements on the tower and the turbulence intensity in the undisturbed flow. For the high Reynolds numbers appropriate to masts and towers in atmospheric flows, the wake is often nonstationary. So the experimental results of the different investigators tend to be inconclusive. Wind tunnel results do not apply since Reynolds numbers achieved in such studies are smaller; more important, the turbulence intensity and roughness effects are not accounted for.

In the absence of any firm guidance on the subject, one develops useful rules of thumb. For example, the data can be restricted to a sector that excludes wind directions through the tower and 30° on either side (as a safety factor). For a sensor mounted at the end of a boom 1.5 times the tower width, this rule should leave a 270° sector from which winds can be accepted (see Fig. 5). Identical instrumentation on booms pointing in the opposite direction is essential for full 360° coverage. In many research applications, the investigator can afford to wait for favorable wind directions or confine analyses to periods when conditions are optimum.

The amount of reduction in the measured average wind speed one can expect in the upwind region has been estimated at 5% by Izumi and Barad (1970) and at 7% by Angell and Bernstein (1976). Wucknitz's (1980) plot comparing data from many published wind tunnel and field studies shows reductions of the same magnitude at 1.5 to two diameters upwind of the tower. The wind direction deflection is small directly upwind, but increases as the wind shifts to either side. The maximum deflection observed on the BAO tower is 5°, at a distance 1.5 times the tower width, for winds normal to the direction of the boom.

The influence of towers on flux measurements was first addressed by Mollo-Christensen (1979). He examined the effect of detached vortices generated upwind of a tower. The theoretical framework for calculating the effect of flow distortion on fluxes is poorly developed. Investigators have examined the effect on the measured stress of boxes stored on tower platforms (Wieringa, 1980), of a horizontal boom supporting a sonic anemometer (Dyer, 1981), and of flow distortion caused by the probe itself (Wyngaard, 1981b; Högström, 1982). Their studies suggest stress errors as high as 20%. Every effort should be made to avoid placement of obstacles above, below, and even downwind of the sensor, and to incorporate as much vertical symmetry as possible in the probe design. Figure 6 shows an example of one effort to provide vertical symmetry in the w measurements.

5 DATA ACQUISITION AND ARCHIVING

The rapid strides made in microprocessor and data acquisition technology in recent years have greatly benefited the boundary-layer researcher. It is feasible now to perform

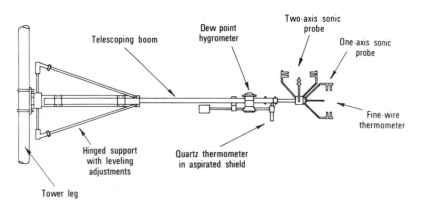

Fig. 6   *Arrangement of sensors on the BAO tower boom. The retract-able boom permits access to the sensors when the boom is swung toward the side of the tower (see Fig. 5).*

much of the data processing in real time and to avoid the months and years of delay encountered by earlier workers. Properly designed displays and printouts of relevant parameters permit evaluation of an experiment in progress. Researchers can analyze data in the field and decide on the adequacy of information collected before returning to the home laboratory.

In operations similar to those at the BAO, where data are continually processed and archived, the problems are somewhat different. Limitations on sampling rates, data transmission rates, tape storage capacity, and real-time processing ability make it imperative that strategies be developed to compress the information acquired and to store it in a form easy to retrieve and work with.

At the BAO, the problem was solved by storing the high-frequency information (0.01–5 Hz) in the form of smoothed spectral and cospectral estimates, updated every 20 min, and the low-frequency information (0–0.05 Hz) in the form of 10-s averaged time series. The choice of 20 min as the averaging period for summary listings is an arbitrary one, but it offers a convenient compromise between the need for stability in the statistical calculations and the need to follow mesoscale variations that might affect boundary-layer structure. The data-handling procedure for the BAO is outlined in Fig. 7.

The spectral data saved are the averages of ten successive 1,024-point (~2-min) fast Fourier transform (FFT) spectra. The spectral estimates are block-averaged over nonoverlapping frequency bands to produce roughly seven to ten logarithmically spaced estimates per decade (see Fig. 8). This procedure reduces the number of estimates per variable per 20-min period to slightly over 20. At present, spectra and cospectra are computed only for w and θ since they can both be treated as scalars. These estimates provide information on the heat flux as well as the dissipation rates of turbulent kinetic energy and temperature variance. See Kaimal and Gaynor (1983) for more details.

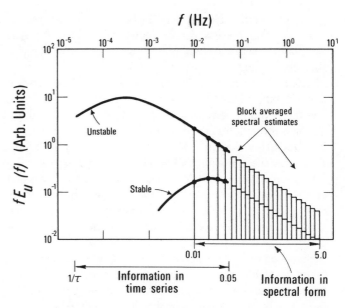

Fig. 8 *Schematic representation of the method used at the BAO for compressing the information archived. The high-frequency information is stored as spectral estimates and the low-frequency information as time series.*

The low-frequency information is stored in two parallel time series: as 10-s nonoverlapping block averages and as 10-s grab samples (last data point in each block). The two serve different purposes. The averaged time series contains the spectral information needed to fill in the low-frequency portion of the spectrum not covered by the real-time FFT computations. Block-averaging reduces spectral distortion from aliasing (present in the grab samples) and ensures a good match over the region of overlap (0.01–0.05 Hz) with the high-frequency spectrum (see Fig. 8). The grab-sample time series is useful for recomputing turbulence quantities, since the high-frequency information contained in it brings the computed values close to those obtained from the original

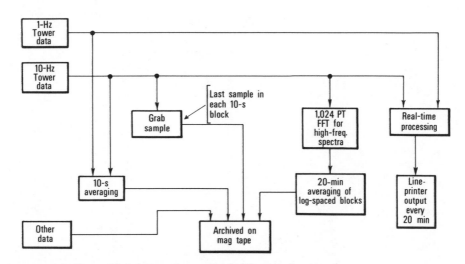

Fig. 7 *Block diagram showing data-handling procedure for real-time processing and archiving of BAO data.*

0.1-s time series. Haugen (1978) has shown that the 1-h variances and covariances computed from the 10-s grab-sample time series are within 5 % and 10 %, respectively, of their true values.

The real-time summary listings of means, variances, fluxes, and Obukhov lengths, together with the spectral estimates and the two time series, should provide the user with almost all the information needed for data analysis. These comprise the data archived at the BAO for future studies. In addition, a descriptor file containing all significant 20-min summary information acquired to date is maintained to aid in the search and retrieval of the data. Once the archival scheme and the descriptor files are established, it is possible to develop standard routines to serve a large portion of the processing and plotting needs of most users. These procedures are essential if the large amounts of data collected at an installation are to be made available to interested users. Aspects of the data acquisition and archival procedures outlined above can be used to advantage even in smaller field operations.

6 CONCLUDING REMARKS

The sensors available for measuring profiles and fluxes have improved over the years. So has the technology for acquiring and processing the data. More and more, data are being processed and displayed in real time, providing researchers the means to evaluate the quality of data collected during the course of an experiment.

Present trends indicate a mix of in situ and remote sensors in future experiments and a broader application of microprocessor techniques to coordinate the different types of information collected. These trends are nowhere more apparent than in the complex-terrain studies conducted by the U.S. Department of Energy and the Environmental Protection Agency during the past few years. Another trend seems to be toward inclusion of turbulence and flux measurements in mesoscale surface arrays and in air quality monitoring systems. Prediction models find increasing need for such data. In that context, the present interest in flow distortion on flux measurement is highly relevant.

Fine-Scale Measurements of Velocity, Temperature, and Humidity in the Atmospheric Boundary Layer

Carl A. Friehe, University of California at Irvine

1 INTRODUCTION

The atmospheric boundary layer is usually characterized by turbulent motion, and this motion extends to very small length and time scales. Important scalar quantities such as temperature, humidity, and trace gases are mixed by turbulent motion and also have fluctuations at small scales. (We do not consider pressure fluctuations.) Turbulent velocity fluctuations are detectable at scale sizes of a few millimeters or less.

Why are we interested in such small-scale measurements? Indeed, it must seem strange to the synoptic meteorologist who is concerned with scale sizes of 100 km and greater. The answer lies in the nature of turbulent motion itself: the kinetic energy of the large-scale motion is ultimately dissipated at small scales by viscous forces in the fluid (by the molecular, not "eddy," viscosity). To completely understand, model, and parameterize the rates of dissipation of kinetic energy and scalar variances in the boundary layer, we must at some point measure these fine-scale quantities. Fine-scale measurements are also necessary in the study of the scintillation of light and radio waves by boundary-layer turbulence. Finally, fine-scale measurements or estimations of the rates of dissipation of kinetic energy, temperature variance, and humidity variance are sometimes used to estimate the turbulent fluxes in the surface layer (Champagne et al., 1977; Large and Pond, 1981 and 1982). The turbulent budget equations that are used, in their simplest approximation, relate the dissipation rates to the production terms of the appropriate flux multiplied by the mean vertical gradient of velocity, temperature, or humidity.

With a flux-gradient relationship, the fluxes can then be estimated. The dissipation rates themselves can be directly measured from the averaged variances of the time derivatives of velocity, temperature, and, in principle, humidity. Or, they can be estimated from power spectral levels in the Kolmogoroff inertial subranges. This technique is particularly useful at sea, where conventional flux measurements, as described by Kaimal elsewhere in this volume, are difficult because of the wave-induced motion of a ship or buoy.

In this review, we concentrate on measurements of velocity, temperature, and humidity fluctuations for scales less than about 1 m and frequencies greater than about 1 Hz. Larger scales and lower frequencies generally correspond to direct turbulence covariance flux measurements, covered by Kaimal in his article. Most instruments discussed in this chapter also have good low-frequency response, and although they can in principle be used for flux measurements, this is not usually done because of the complexity of the instrumentation and fragility of the sensors.

The lower size limit for velocity fluctuations is set by the Kolmogoroff scale, $\eta_K = (\nu^3/\epsilon)^{1/4}$, where ν is the kinematic molecular viscosity and ϵ is the average rate of dissipation of turbulent kinetic energy, which is equal to $15\nu\overline{(\partial u/\partial x)^2}$ for locally isotropic turbulence. At η_K, the local turbulent Reynolds number is unity; i.e., inertial and viscous forces are in balance. Below η_K the large viscous forces essentially damp out all motion in the averaged sense we are considering here. In typical surface-layer turbulence, $\epsilon \sim 2 \times 10^{-2}$ m²/s³ and $\nu = 1.6 \times 10^{-5}$ m²/s, so that $\eta_K = 0.7$ mm.

For scalar fields, the fine scale length is given by the Batchelor scale, $\eta_B = (\nu D^2/\epsilon)^{1/4}$, where $D = $ molecular diffusivity of the scalar (temperature or humidity). The Corrsin scale, directly analogous to the Kolmogoroff scale, can be defined as $\eta_C = (D^3/\epsilon)^{1/4}$ (Tennekes and Lumley, 1972). The scales are related by $\eta_B = \text{Pr}^{-1/2}\eta_K$ and $\eta_C = \text{Pr}^{-3/4}\eta_K$, where Pr is the generalized Prandtl number for heat (temperature) or mass (water vapor) transfer and is equal to ν/D. (For mass transfer ν/D is often termed the Schmidt number.) For typical boundary-layer conditions, Pr = 0.71 for heat and Pr = 0.58 for water vapor, so that $\eta_B \cong \eta_C > \eta_K$; i.e., slightly lower spatial resolution is required for temperature and water vapor sensors than for velocity sensors.

The highest frequency required for resolving fluctuations is set by the speed of the large-scale motion via Taylor's hypothesis $\Delta x \cong -U\Delta t$. (Here x is positive in the downstream direction and the minus sign appears; this is in contrast to Wyngaard's statement of Taylor's hypothesis in his article in this volume.) This results in Kolmogoroff "frequencies," $f_K = U/(2\pi\eta_K)$, of 1.6 kHz for $U = 10$ m/s (as on a tower) and 16 kHz for $U = 100$ m/s (as from an airplane). As will be seen below, these high frequencies and small scales can impose problems for sensors and associated instrumentation.

2 MEASUREMENT TECHNIQUES

2.1 Velocity

2.1.1 General Principles

For fine-scale velocity measurements, the hot-wire anemometer is normally used. This device has been under development since the early 1900s for the investigation of turbulent flow, primarily in wind-tunnel research (see Freymuth [1983] for a historical review). Early methods used the constant-current technique, which we will not consider here. Hot wires were applied to the atmosphere in the early 1960s by several research groups, although a few experiments had been conducted in the 1930s. In the following, we review the essential features of hot-wire anemometers. More detailed reviews are given by Corrsin (1963), Comte-Bellot (1976), Wills (1980), Hasse and Dunckel (1980), Blackwelder (1981), and the book by Perry (1982).

The hot-wire anemometer system consists of a very small resistance wire which is electrically heated relative to the surrounding fluid. In the constant-temperature method, the wire is kept at constant temperature (resistance) in a Wheatstone bridge with a feedback circuit. As the velocity varies past the wire, the heat transfer varies and the power required to keep the wire at constant temperature varies. Usually the voltage into the bridge is used as the output signal, and turns out to be nonlinearly related to the velocity.

The wires are typically tungsten, $4\ \mu m$ ($1\ \mu m = 10^{-6}$ m) in diameter; platinum, $5\ \mu m$ or smaller in diameter; or platinum alloys. Frequently the bare tungsten is coated with a thin layer of platinum to prevent oxidation. The wires operate at about 450 K for tungsten, with higher temperatures for platinum-iridium and platinum-rhodium alloys. Tungsten requires a lower operating temperature because it becomes brittle at high temperatures.

The temperature and resistance of the sensor material are related by

$$R = R_0[1 + \alpha(T - T_0)]$$

where R_0 is the resistance at the reference temperature T_0 and α is the (temperature) coefficient of resistivity. If we take T_0 to be the ambient temperature for calibration, the overheat ratio is (Blackwelder, 1981):

$$A_T = (T - T_0)/T_0 \qquad (1)$$

or, in terms of sensor resistance,

$$A_R = (R - R_0)/R_0 = \alpha T_0 A_T \qquad (2)$$

The coefficient of resistivity is $\alpha = 0.0042 - 0.0052$ K^{-1} for annealed tungsten, 0.0039 K^{-1} for annealed pure platinum, 0.0016 K^{-1} for platinum alloyed with 10% rhodium, and 0.0007 K^{-1} for platinum alloyed with 20% iridium. Thus for a tungsten wire with a conservative operating temperature of 450 K in air at 300 K, $A_T = 0.50$, $A_R = 0.78$. (Sometimes the overheat ratios are defined as T/T_0 or R/R_0; i.e., the ratios are 1 plus the ratios defined above.)

The lengths of the wires are usually at least 200 diameters—1 mm—to approximate two-dimensional flow

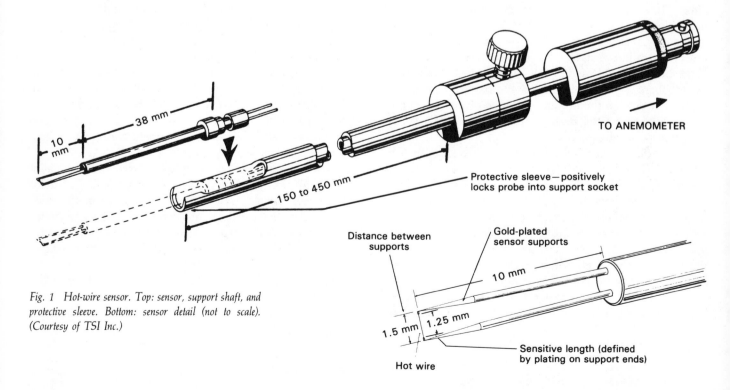

Fig. 1 Hot-wire sensor. Top: sensor, support shaft, and protective sleeve. Bottom: sensor detail (not to scale). (Courtesy of TSI Inc.)

past an infinitely long cylinder. The ends of the wires are copper- or gold-plated and then spot-welded or soldered to gold-plated sensor prongs; the prongs are embedded in a probe body usually 3–5 mm in diameter and 3–5 cm long to complete the "sensor." Cylindrical hot-film sensors have also been used since about 1963. These are quartz tubes 25 to 150 μm in diameter, with a sputtered platinum or nickel coating to form the resistive element. A thin layer of quartz is sputtered over the metal film for protection against abrasion and contamination. Hot-film sensors are considerably more rugged than hot wires, and for this reason are sometimes used in field work. Hot-film sensors are also available in other shapes—e.g., cones and wedges—primarily for use in water.

A typical commercial hot-wire sensor is shown in Fig. 1, together with the supporting probe shaft and optional shield. The shield is useful for field work, as the sensor can be withdrawn into the shield for protection against breakage while being installed on towers. The small size, and hence mass, of the wire element gives the desired fast response to velocity fluctuations past the wire. A schematic of a constant-temperature anemometer bridge is given in Fig. 2.

The relationship between heat transfer and velocity is usually governed by forced convection:

$$\text{Nu} = f(\text{Re}, \text{Pr}) \tag{3}$$

where Nu = Nusselt number = hd/k
Re = Reynolds number = Ud/ν
Pr = Prandtl number = ν/D = 0.71 for air
h = heat transfer coefficient = $Q/[A(T_w - T_a)]$
d = wire diameter
k = thermal conductivity of air
ν = kinematic viscosity of air
D = thermal diffusivity of air
Q = heat transferred per time to air from wire
A = surface area of wire
T_w = mean wire surface temperature
T_a = mean ambient air temperature

In Eq. (3), we have neglected buoyant convection effects, compressibility effects, rarefied gas effects, radiative heat transfer, temperature loading effects on physical properties, and length-to-diameter effects. The Reynolds number based on the wire diameter and flow speed U is generally in the range $0.1 < \text{Re} < 20$, so that vortex shedding of the wire itself is avoided. (Vortex shedding for an unheated cylinder occurs at Re ~ 50; heating the cylinder delays shedding to slightly higher Reynolds numbers.)

Several "laws" have been proposed to obtain a quantitative relationship among Nu, Re, and Pr. In 1914, King (see Hinze, 1975; Wills, 1980) proposed

$$W = (A' + B'\sqrt{\mathcal{U}})(T_w - T_a) \tag{4}$$

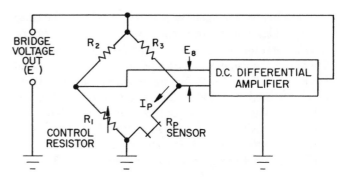

Fig. 2 Constant-temperature anemometer bridge. (Courtesy of TSI Inc.)

where W ≡ power expended in heating the wire
A', B' ≡ constants (actually weak functions of temperature difference and wire diameter)
$\mathcal{U}$ = air velocity component normal to wire axis

Since the power is proportional to the square of the voltage, $\mathcal{E}$, across the probe, for a given $(T_w - T_a)$ Eq. (4) is written for calibration purposes as

$$\mathcal{E}^2 = A + B\sqrt{\mathcal{U}} \tag{5}$$

Eq. (5) shows the basic nonlinear relationship between $\mathcal{E}$ and $\mathcal{U}$. Sometimes, specialized analog circuits ("linearizers") are employed to solve Eq. (5) for $\mathcal{U}$. If the $\mathcal{E}$ data are digitized into a computer, linearization can be performed with software. For very small fluctuations (not usually the case in the boundary layer, except perhaps from an aircraft measurement), Eq. (5) may be linearized ($e \ll \mathcal{E}$, $u \ll \mathcal{U}$):

$$\frac{e}{E} = \frac{1}{4}\frac{B(U)^{1/2}}{A + B(U)^{1/2}}\left(\frac{u}{U}\right) \tag{6}$$

where e and u are fluctuations about the means E and U, respectively.

In 1959, Collis and Williams performed an extensive set of experiments on long, heated wires in a low-turbulence wind tunnel and determined that

$$\text{Nu} = 0.24 + 0.56\text{Re}^{0.45}, \text{Re} < 44 \tag{7}$$

where we have neglected a small temperature-loading term. In principle, Eq. (7) determines the absolute calibration of a hot-wire sensor. However, the length-to-diameter ratios, l/d, of most commercial sensors are not quite equal to those used by Collis and Williams and furthermore, the diameter is difficult to measure. For these reasons, and to account for prong interference, etc., A and B are almost always determined by individual calibration.

In calibration, the quantity A is *not* the value of the square of the bridge voltage with no flow, since with no external flow around the hot wire, natural convection transfers heat. Therefore, the value of A is obtained by extrapolating data of E^2 versus $U^{0.45}$ to $U = 0$. (For commercial tungsten hot wires in air operated at ~450 K, the no-flow value of A is coincidentally only slightly larger than the extrapolated value.) Probes are usually calibrated in a low-turbulence-level facility with temperature control. Linearization, whether analog or computed, results in a voltage $\mathscr{E}_L$ proportional to $\mathscr{U}$:

$$\mathscr{E}_L = K_0 \mathscr{U} \qquad (8)$$

where K_0 is the calibration slope.

In hot-wire anemometry, the tacit assumption is that when a wire is placed in a turbulent flow, the wire responds to fluctuations exactly according to the nonturbulent flow calibration. This is difficult to check, especially up to the frequencies of tens of kilohertz to which the anemometer system is capable of responding. On most constant-temperature anemometer circuits, provision is made to inject a square wave signal into the bridge and adjust the feedback circuit for optimum electronic response. The validity of the fluctuation assumption and the interpretation of the electronic square wave test are major topics of the book by Perry (1982).

The ideal hot-wire sensor measures the magnitude of the velocity vector normal to the sensor axis, independent of azimuthal angle about the axis. Thus a vertical hot wire is subject to rectification errors in the streamwise velocity, and higher-order corrections due to lateral velocity fluctuations. Also, corrections for vertical velocity fluctuations may be important if deviations from the normal component cooling "law" occur due to finite sensor length-to-diameter ratio. Rose (1962) discussed these effects, and later Tutu and Chevray (1975) presented a detailed analysis and correction scheme for high-intensity turbulence. Recently Kawall et al. (1983) described a three-wire probe and a digital processing technique which gives corrected results for turbulent flows with intensities of up to 40%.

The above is a brief description of the essentials of hot-wire anemometers for a single wire to measure the streamwise mean and fluctuation velocity component in turbulent flow. Since the heat transfer depends mainly on the velocity component normal to the wire axis (the cosine "law"), an array of several wires inclined at different angles can resolve all three components of the turbulent velocity vector. The references cited at the beginning of this section describe the use of multiwire sensors and also the effects of deviations from the cosine law.

In summary, the hot-wire anemometer is about the only practical and accepted means of obtaining high-frequency, small-scale data in turbulent flows. (We do not consider in this review the laser-Doppler anemometer or velocimeter for application to fine-scale boundary-layer turbulence.) A major disadvantage of the hot wire is the necessity for individual calibration of each sensor for accurate work. We now turn to problems peculiar to the use of hot wires in the atmospheric boundary layer.

2.1.2 Application to the Boundary Layer

For the usual tower-based measurements in the surface or boundary layer, setting up the hot-wire sensor is not difficult; the probe and probe support combination is roughly the size of a pencil, so that it is easy to mount and does not interfere with other instrumentation. However, it is usually not convenient to operate the anemometer circuit next to the probe, so very long probe cables are required between the sensor and its associated circuitry in the instrument shelter. Commercial anemometers can be supplied with probe cables up to 100 m long and corresponding compensation in the electronics, at added expense. Some loss of frequency response occurs and noise pickup is a possibility. The resistance of a hot-wire element is about 5–10 Ω, so the resistance of 100 m of cable in series with the probe becomes significant. It is also subject to environmental temperature changes, which means that the effective calibration of the system can change if the probe cable resistance changes are large.

In contrast to most laboratory experiments, the ambient temperature in a boundary-layer experiment is subject to change; hence, when the temperature during the experiment is different from that in calibration, the difference has to be accounted for. Since tungsten hot wires are normally used for their strength and their operating temperatures are not all that high, the difference between the calibration and the ambient temperature ($T_c - T_a$) can be significant. Bearman (1971) shows that a change of +7 K in ambient temperature causes a 10% decrease in indicated mean velocity and 2% decrease in turbulence intensity (variance). Champagne (1979) shows that the fluctuations, e_L, where $e_L < E_L$, are

$$e_L = \alpha u - \beta\theta, \quad \theta = T_c - T_a \qquad (9)$$

where

$$\alpha = (\partial E_L/\partial U)_{T_c - T_a}$$

$$\beta = [\partial E_L/\partial(T_c - T_a)]_U$$

Champagne further presents formulas for α and β in case direct calibration data are not available. Ideally $\alpha = K_0$ and $\beta = 0$. Champagne showed that for $T_c - T_a$ from +2.8 to -5.7 K, $\alpha = 1.07$ to $0.89K_0$. This gives intensity corrections larger than Bearman's. Similarly, the temperature coefficient β was found to be nonnegligible, with values of 0.10 to 0.17 V/K, increasing with increasing velocity.

If we divide by K_0, Eq. (9) becomes

$$u^m = \alpha'u - \beta'\theta \qquad (10)$$

where the superscript m stands for the velocity fluctuation one would measure without correction, e_L/K_0. Champagne shows that the nonzero value of β' can seriously affect some statistics involving u^m. Fortunately, the variance of the time derivative of u^m, $\overline{(\partial u^m/\partial t)^2}$, is not severely contaminated. This variance is proportional to the rate of dissipation of turbulent kinetic energy after one applies Taylor's hypothesis.

Equation (10) also shows that the measured turbulent stress is contaminated by the heat flux:

$$\overline{u^m w^m} = \alpha' \overline{uw^m} - \beta' \overline{w^m \theta} \qquad (11)$$

where w^m is the measured vertical velocity. Champagne shows that $w^m \cong w$, since w^m is obtained from the subtraction of the instantaneous signals of two oppositely inclined hot wires, and the temperature contamination term cancels. Thus, Eq. (11) becomes

$$\overline{u^m w} = \alpha' \overline{uw} - \beta' \overline{w\theta} \qquad (11a)$$

With typical values for tungsten hot wires at low overheat [$\beta' = 0.17$ m/(s K)] and for slightly unstable conditions $(\overline{u^m w} = -0.08$ m²/s², $\overline{w\theta} = +0.03$ m K/s), $\overline{uw} = 0.93 \overline{u^m w}$. Thus the temperature correction is not quite negligible. In general, a "contamination" term like $\overline{w\theta}$ affects odd-moment statistics and cross statistics with other variables more than even-moment statistics.

To summarize the temperature contamination problem, it should be recognized in an experiment, and calibration of the hot wire should include calibration for temperature as well as velocity. If calibration for temperature is not possible, Champagne's formula for β should be used. Recently, Paranthoen et al. (1983) have shown that the dynamic sensitivity of constant-temperature hot wires to temperature fluctuations is not the same as the static effect described above. The differing responses of the wire and prongs cause a slight attenuation of the hot-wire signal above 1 Hz.

Another problem which arises with hot wires in boundary-layer experiments in the atmosphere is the possibility of fairly large changes in the mean or low-frequency flow direction. Since spatial resolution in the vertical is not a problem in the atmospheric boundary layer as it is in the wind-tunnel boundary layer, the hot wire can be positioned vertically so that the mean flow direction can swing around the axis of the hot wire to large angles before effects of the probe supports and shaft are a problem. Since the vertical velocity fluctuations are reasonably small, i.e.,

$$\overline{w^2}^{1/2}/U \cong 0.10$$

and have a mean zero, being constrained by the earth's surface, the fluctuations sensed by the wire can be taken to be those in the horizontal plane to a good approximation. This does mean that a vertical wire senses horizontal wind speed, and not the component of the horizontal velocity vector along the probe shaft axis. For inclined sensors (X-wires) used to measure both u and w, the arrangement of the wires is usually one wire inclined at $+45°$ to the horizontal, and the other at $-45°$. Thus changes in horizontal wind direction are more severe for the X-wire probe, since one wire can be in the wake of the other wire and its support for relatively small flow angle deviation from the streamline direction. For such X-wire probes, the conical angle of acceptance of the mean flow direction should be within $\sim 20°$. Direct calibration of the X-wire for velocity and angle dependence is recommended to account for deviations from the cosine law.

For the atmosphere, Larsen and Busch (1974) developed a multiple hot-wire sensor arrangement which was mounted on a carefully designed vane. The sensors were fixed in a "dog-leg" support and they rotated about the center line of the rotation axis of the vane so that vane motion did not induce translational velocities. Mercury-pool contacts in the base housing allowed for electrical connections with freedom of rotation and low friction. The device worked successfully in the 1968 Kansas micrometeorological experiment (Larsen and Busch, 1976), with good comparison at low frequencies to similar measurements made with sonic anemometers.

A final problem encountered in the atmospheric boundary layer is contamination of the wire surface by airborne particles. (Of course, the hot wire is a "fair weather" sensor: while it can survive in rain, the probability is not high that it will be useful.) Airborne particles striking the wire cause spikes in the output signal which can give problems if high-order moment statistics are sought. If the particles stick, the calibration of the wire can change and its frequency response is lowered.

Contamination can be most severe in the salt spray environment over the ocean. Schacher and Fairall (1976) show striking electron microscope photographs of salt-encrusted 4.5-μm tungsten wires, with salt particles up to ~ 10 μm thick. However, they reported that the calibration did not change significantly. The coefficient B of Eq. (5) showed no definite trend of increase or decrease for six probes exposed to salt air for two to four days. Schacher and Fairall developed a model equation for the uniformly coated wire, and predicted only a small increase in B, as the conductivity of the salt is high compared to that of air and hence does not insulate the wire. However, no mention was made of the increase in effective diameter of the wire due to the salt deposition and the consequent change in the effective Reynolds number.

Hasse and Dunckel (1980) report severe problems with salt spray on hot wires, but not with larger-diameter (150-μm) hot-film sensors. Friehe and Champagne (unpublished) observed substantial calibration drifts with 60-μm hot-film sensors when operated for a few hours over the sea.

The impact of salt spray aerosols and their subsequent evaporation is apparently not observed in the anemometer signal. The hot-wire or film signal does, however, exhibit spikes when larger particles (rain) impact the wire (Hasse and Dunckel show time series of this). Over land, Champagne et al. (1977) show a typical spike resulting from a dirt particle or spider web striking the wire.

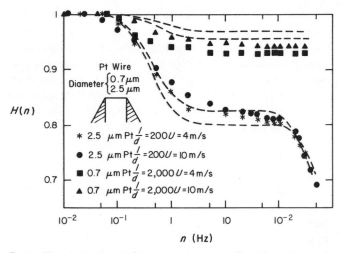

Fig. 3 Transfer function of platinum resistance wires. (From Paranthoen et al., 1982; reproduced with permission of Cambridge University Press.)

2.2 Temperature

Fine-scale temperature measurements are usually made with platinum resistance wires. Thermocouples and thermistors are not considered here, as they are considerably larger, with frequency response limited to about 5 Hz, only adequate for flux measurements. Principles of resistance thermometry are given in the text by Sandborn (1972), and atmospheric applications are reviewed by Larsen et al. (1980).

For platinum resistance wires, the goal is to make the wire diameter as small as possible to provide for high frequency response and high nominal resistance to lessen amplifier noise. Wires with diameters down to 0.25 μm have been made by the Wollaston process, in which a larger platinum wire is coated with silver and the jacketed wire drawn down until the platinum core reaches the desired small diameter. The silver jacket aids in handling and soldering the Wollaston wire to probe supports. The center silver section is then etched away with nitric acid, leaving the small-diameter platinum wire of desired length. (Wollaston wire of 0.25-μm inner diameter is apparently no longer available from the U.S. manufacturer, Sigmund Cohn, Inc., because of lack of demand.) The final probe looks much like the hot-wire probe of Fig. 1; often a hot-wire probe body is used.

Frequency response of etched Wollaston wires has been measured by LaRue et al. (1975). There was a noticeable decrease in response of the wires as they aged through use, even in the laboratory test environment. For new 0.625-μm wires, the -3-dB points are a function of the velocity, and range from 2 to 6 kHz for $0 \leq U \leq 10$ m/s. After use, the response dropped to 0.5 to 2 kHz over the same velocity range.

Recently, the response at *low* frequencies has been investigated, because of the conduction interaction between the probe support/solder/Wollaston jacket and the fine wire. Paranthoen et al. (1982) show that even for wires with l/d ratios of 2,000, the amplitude transfer function drops from

1.0 to 0.95 in the band 0.1 to 1 Hz. For shorter wires, the effect is correspondingly greater (Fig. 3). For the probe configurations commonly used in turbulent temperature measurements, the fine-scale signal is at frequencies greater than 0.1 to 1 Hz, so the change in the transfer function shape does not affect the measured spectral shapes. However, the value of the dynamic sensitivity in the measuring band has to be determined. Petit et al. (1981) give experimental results for a variety of wires and probe configurations and also formulas for the calculation of the attenuation. For fine-scale measurements, spatial resolution is not usually a problem, so that the use of wires with large l/d is recommended. The changing transfer function at low frequencies presents more of a problem for flux measurements.

The resistance wires are usually operated in a Wheatstone bridge, so that the signal due to the mean temperature can be balanced out and signal processing can be applied to the fluctuations about zero DC level. A DC bridge configuration is relatively simple to construct: some commercial thermal anemometer circuits have a temperature option or switch position. However, signal levels are usually quite low, since only a very small current is desired through the resistance wire, and the low impedance of the sensors (10 to 1,000 Ω) causes noise. DC bridges are subject to noise pickup, especially if long cables are used between the sensor and the bridge. Amplification of the bridge output signal for DC to several kilohertz is simple, but one has to contend with white noise, noise inversely proportional to frequency (1/f noise), and drift due to environmental temperature changes of the amplifier.

AC bridges have been used to lessen some of these problems. The bridge is excited at 25 to 100 kHz, and the bridge signal is then an amplitude-modulated signal at the carrier frequency. The DC-coupled temperature signal is obtained by synchronous detection, and is largely independent of 1/f noise and low-frequency noise pickup (e.g., AC power and its harmonics). System noise has approached two to three times the theoretical thermal noise (LaRue et al., 1975). However, with an AC system, bridge and amplifier design are more critical than for DC, and the capacitance of the sensor cable is also important. Deaton (unpublished) has developed battery-powered AC bridges, capable of being remotely operated and of measuring the in situ frequency response transfer function.

Up to this point, we have not specified the wire current required to detect the resistance changes. One usually sets the current "low," e.g., at 0.1 mA for a 100-Ω 0.625-μm wire, and neglects any side effects from the small heating of the wire. (Sometimes, in cases of small temperature fluctuations, one is sorely tempted to increase the current in order to lift the signal out of the noise.)

Wyngaard (1971a) showed that the small heating may not be negligible for all statistics derived from the measured signal. Wyngaard wrote

$$\theta^m = \theta - cu \qquad (12)$$

where θ is fluctuating temperature, the superscript m indicating "measured"; u is the fluctuating streamwise velocity past the wire; and c is the velocity sensitivity of the wire due to the sensing-current heating. The sensitivity c, derived from the Collis and Williams heat transfer relation for hot wires (Eq. 7), is proportional to the power dissipated in the wire (current squared), and inversely proportional to the square of the wire diameter. For a 0.625-μm platinum wire with a sensing current of 0.3 mA at 5-m/s velocity, $c = 2.4 \times 10^{-2}$ K s/m. Wyngaard showed that this is nonnegligible when calculating the skewness of the temperature derivative (which should be zero for isotropic turbulence); the finite value of c can account for the nonzero measured values that are observed through the higher-order moment correlation, $c(\partial\theta/\partial x)^2(\partial u/\partial x)$, which is nonzero in isotropic turbulence. Since Wyngaard's original paper, experimenters have varied c over a wide range by varying the wire current, and still have found nonzero values for the skewness of the time derivative of temperature as $c \rightarrow 0$ (Antonia, 1975). However, Wyngaard's analysis did point out the need for total understanding of what a sensor measures, especially for odd-moment statistics. For example, we can consider the heat flux, proportional to $\overline{w\theta}$, by multiplying Eq. (12) by a "true" w signal

$$\overline{w\theta} = \overline{w\theta^m} + c\overline{uw} \qquad (13)$$

i.e., the heat flux is contaminated by the stress, proportional to $\overline{uw}$. For typical values of c, $\overline{uw}$, and $\overline{w\theta^m}$, a 5% error in heat flux is possible. This suggests that one should use a separate large-diameter wire with a very low c for heat flux, and not the signal from a fine wire, although it may be conveniently available.

As is the case for hot wires, contamination of temperature is also a problem. The laboratory studies of Ochs (1967) and LaRue et al. (1975) found a significant decrease in the frequency response of submicrometer-size platinum wires used in a wind tunnel. The mechanism for this is not exactly known, but may be caused by aerosols contaminating the surface and adding mass, which lowers the response. Because of the small size, cleaning the sensor in water or solvent brings a high risk of breaking it. In the atmosphere, the possibilities for contamination are much greater than in the laboratory. Over land, in the study of Champagne et al. (1977), fine spider webs were observed on 0.625-μm wires when they were brought in from field exposure.

Over the ocean, very serious contamination problems are caused by salt spray aerosols in the boundary layer. Figure 4, reproduced from Schmitt et al. (1978), shows the typical spiky, high-pass filtered nature of temperature signals from a thermistor and platinum wire operated over the ocean. The "hot" and "cold" spikes are believed to be due to the condensation and evaporation, respectively, of water from salt spray aerosols deposited on the sensor surfaces. Latent heat changes associated with condensation and evaporation produce changes in the measured temperature. The condensation and evaporation are caused by local changes in the humidity

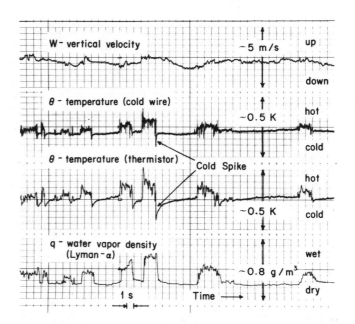

Fig. 4 *Time series of temperature and humidity in the marine boundary layer, showing hot and cold spikes that result from salt contamination. (From Schmitt et al., 1978.)*

around the sensor. As shown in Fig. 4, the humidity changes can be highly correlated with temperature changes in the turbulent atmospheric boundary layer. Schmitt et al. observed that cleaning a temperature sensor results in a "normal" signal, but the sensor becomes recontaminated in a short time, 20–30 min in their experiments. It does not appear possible to correct the measured temperature signal for the spray-evaporation/condensation contamination effect. (It may be possible to repel the salt spray from a sensor or precipitate it by electrostatic means; however, this has not been tried.) Therefore, most fine-scale (and flux-scale) temperature measurements in the boundary layer over the ocean are suspect.

2.3 Humidity

The measurement of fine-scale humidity[1] fluctuations has not advanced to the same level of spatial resolution and temporal response as for velocity and temperature. To date, the fine-scale psychrometer, microwave refractometer, and Lyman-alpha hygrometer have been developed. Several infrared devices have also been developed (Hyson and Hicks, 1975) but at their present stage of spatial resolution and response, they belong more to the flux-measuring range. The psychrometer is characterized by good spatial resolution but poor frequency response for fine-scale measurements, while the reverse is true of the Lyman-alpha and refractometer. As a result, it is possible to measure humidity sufficiently well at flux scales and in the Kolmogoroff-Obukhov-Corrsin scalar

---

[1]In the text, we use the term humidity to refer to the general measure of water vapor in the atmosphere. When a more precise measure is required, we will use more specific definitions (water vapor density, etc.).

inertial subrange, but not to the dissipative scales. Hence, the rate of dissipation of humidity variance has not been measured directly. The usual assumption is that humidity is a scalar like temperature (the molecular Prandtl numbers are 0.71 for temperature and 0.58 for humidity), so that certain fine-scale properties of temperature (e.g., its spectral constant) can be applied to humidity as well. This enables the rate of dissipation of humidity variance to be estimated from spectral data. A review of fast-response humidity instrumentation has been given by Hay (1980).

The psychrometer response has been extended to small scales by using fine thermocouples for the wet and dry elements. The basics of psychrometry have been reviewed by Wexler (1970). Shaw and Tillman (1980) describe a system using 25-μm thermocouple beads, one wrapped with cotton thread and fed by a water supply system for continuous operation. The response of the dry bead was about 4 Hz and the wet bead a factor of 10 slower. Shaw and Tillman show (by comparison with a Lyman-alpha hygrometer) that correction for the lower time response of the wet bead gives good spectral data to about 1 Hz. This psychrometer is subject to the same problems as larger devices for the measurement of mean humidity (Coantic and Friehe, 1980). Specifically, contamination of the wick can be a problem, especially by salt over the ocean. Use is limited to above-freezing conditions.

The use of microwave refractometers to measure humidity is the result of investigations of the propagation characteristics of the atmosphere. At microwave frequencies, the refractive index of air is a reasonably strong function of humidity. Various types of refractometers have been developed to measure the refractivity, $N = (R - 1)10^6$, where R is the refractive index. Bean and Dutton (1966) review those instruments developed up to the mid-1960s; Hay (1980) discusses some recent improvements. The devices are based on the fact that the resonant frequency of a cavity depends upon its dimensions and the refractive index of the gas inside. The sampling cavities are usually constructed of a material with a low dimensional change with ambient temperature (e.g., Invar). It has been found that openings to allow for passage of air can be accommodated without seriously affecting the resonance properties. The sampling cavity is usually operated together with a reference cavity, and phase shift or frequency difference is measured and converted into a signal proportional to N units. Considerable effort has gone into the development of cavities with low thermal expansion characteristics. Absorption of spray, water, or contaminations on the cavity surface also changes performance. Some experimenters coat the surface with a hydrophobic material.

The equation relating N to temperature, humidity, and pressure is reasonably well known. Hay (1980) shows that the fluctuations in specific humidity at sea level, for 75% mean relative humidity and 300-K mean temperature are

$$q \times 10^3 = 0.137n + 0.184\theta - (4.45 \times 10^{-4})p \qquad (14)$$

where n and p are refractivity and pressure fluctuations; θ is in kelvins and p is in pascals. From this equation we see that we must measure θ and p simultaneously with n to calculate q. A change of 0.02 in n is equivalent to 0.015 K in θ and 6.3 Pa in p. Simultaneous measurements of θ close to the refractometer are possible with the temperature sensors described in the preceding section. Measurements of p are far more difficult and cannot be made at extremely small scales and high frequencies (Dobson, 1980). Pressure fluctuations caused by turbulence are usually small enough that the pressure correction in Eq. (14) is not as large as the temperature correction. Equation (14) does show that incorrect compensation for the temperature and pressure terms has unknown but possibly large effects on higher-order moment statistics through correlation terms, as found for temperature contamination of hot-wire signals and velocity contamination of temperature signals.

Measurements of water vapor density of ±0.02 g/m³ are possible with refractometers. Temperature expansion of the cavity can amount to 0.03N per kelvin for compensated Invar cavities. Microwave refractometers are primarily used on aircraft (Bean and Emmanuel, 1980).

Another fast-response device is the Lyman-alpha hygrometer. Tillman (1965) outlines the basic principles of the device. It is based on Beer's law absorption of the Lyman-alpha wavelength (121.5 nm) of light by hydrogen in the sample path. Other absorbers are oxygen and ozone, but the absorption coefficient of oxygen is ~0.001 times that of water vapor, and ozone concentration is not high enough in the boundary layer to give problems. The Lyman-alpha device consists of a source tube—glow discharge of a hydrogen-neon mixture or a uranium hydride tube (Buck, 1976)—and a detector tube, usually a nitrous oxide ion chamber connected to an electrometer circuit. Because of the wavelength of the Lyman-alpha radiation, windows must be lithium or magnesium fluoride. There is less attenuation with lithium fluoride windows, but they are more soluble in water than the magnesium fluoride windows. However, sensitivity and signal strength are not usually problems with the Lyman-alpha hygrometer, so magnesium fluoride windows are usually used. Response time can be extremely fast. Priestly and Cartwright (1982) describe a simple detector amplifier and demonstrate a bandwidth of 10 kHz by modulating the source light in an operating Lyman-alpha hygrometer. Spatial resolution is the present limit of the Lyman-alpha: source tubes are ~5 cm long × 2 cm in diameter; detector tubes are somewhat smaller. When mounted on a support the total device is ~20 cm long × 8 cm high × 5 cm deep. Because of the high absorption at the Lyman-alpha wavelength, the gap between the source and detector tubes is small: 0.5 to 2 cm is common.

The absorption of Lyman-alpha radiation is described in the ideal case by Beer's law:

$$I/I_0 = \exp(-kx\varrho/\varrho_0) \qquad (15)$$

where I, I_0 = received and source intensity

x = path length

k = absorption coefficient = 387 cm^{-1} for H_2O

ϱ/ϱ_0 = concentration ratio, ϱ_0 = 0.806 kg/m^3 at STP

Equation (15) is readily differentiated to determine the sensitivity $S = d(I/I_0)/d(\varrho/\varrho_0)$ as a function of path length x. The optimum path length x^* for maximum S is (Tillman, 1965; Friehe, 1982)

$$x^* = \varrho_0/k\varrho \qquad (16)$$

and ranges from 4.3 mm at ϱ = 4.85 g/m^3 (0 °C dew point) to 0.69 mm at ϱ = 30.4 g/m^3 (30 °C dew point). Thus optimum path length coincides with good small-scale resolution for the Lyman-alpha hygrometer. In principle, path lengths on the order of the Kolmogoroff scale could be used, but the present sizes of the tube bodies introduce large flow-distortion effects. Buck (1975) considers departures from Beer's law for both conventional and uranium hydride source tube Lyman-alpha hygrometers.

In practice the Lyman-alpha hygrometers and microwave refractometers are calibrated or monitored against absolute, low-frequency humidity devices such as the cooled-mirror dew-point device (Coantic and Friehe, 1980). For the Lyman-alpha hygrometer, contamination of the windows by dirt and salt spray and etching by exposure to water tend to preclude the use of an absolute calibration. The transmission properties of the fluoride windows and source strength also change with operating time. For the microwave refractometer, the change in cavity dimensions with temperature, adsorption of spray on the cavity surfaces, and spray itself in the airstream may change the resonant frequency of the cavity. Tillman (1965) calculated that small fog aerosols should not affect the Lyman-alpha absorption. However, I have found from recent airborne observations in the marine boundary layer at high wind speeds and with a large salt spray concentration that the Lyman-alpha hygrometer gives anomalous results which may be due to salt spray (Buck, personal communication).

3 SPATIAL RESOLUTION

Up to this point, we have not considered spatial resolution requirements in detail: the Kolmogoroff and Corrsin length scales are ~ 1 mm, which is about the length of the sensing portions of hot and cold wires. This is a fortunate and fortuitous matching of length scales. It is difficult for manufacturers or experimenters to make hot wires with sensing lengths < 1 mm and maintain an l/d ratio > 200. For cold wires, it was shown above that l/d ratios > 1,000 are required to minimize the effects of support conduction, and although the wire diameters can be smaller, the total length is 0.5–1.0 mm. For engineering and research turbulent flows, the Kolmogoroff length scale is typically much smaller than

1 mm, and spatial resolution of the fine-scale eddies becomes a problem.

Uberoi and Kovasznay (1953) and later Wyngaard (1968) estimated the effect of spatial averaging of fine-scale isotropic turbulence along a hot wire using a model turbulence spectrum. Wyngaard (1971b) performed the analogous calculation for cold resistance wires and the temperature spectrum. If the ratio of wire length to the Kolmogoroff or Corrsin scale exceeds 2 to 3, then spatial averaging effects become non-negligible, especially in determining the rates of dissipation. Larsen and Højstrup (1982) have studied the combined effects of frequency response, spatial averaging, and variable turbulence intensity for cold resistance wires. Under some combinations of sensor parameters and turbulence intensities, different effects cancel and fortuitously give a correct measurement of the dissipation rate of temperature variance.

For humidity, as stated above, the present sensors are so large that they probably induce more flow distortion effects than path- or volume-averaging effects. Nevertheless, Andreas (1981) has derived the path-averaging correction for the Lyman-alpha hygrometer.

4 CONCLUSIONS

Fine-scale measurements of velocity and temperature are possible in the atmospheric boundary layer. Measurements can be obtained down to the Kolmogoroff or Corrsin scales (~ 1 mm) with correspondingly high bandwidth. For humidity, the present sensors have either a spatial limitation (Lyman-alpha hygrometers, microwave refractometers) or temporal response limitations (psychrometers) that limit measurements to flux scales and just into the inertial subrange (scales ~ 1 m, frequency ~ 10 Hz).

All of the above measurements can have problems: calibration, secondary response to another variable, contamination, reliability (breakage of wires, etc.). In the boundary layer

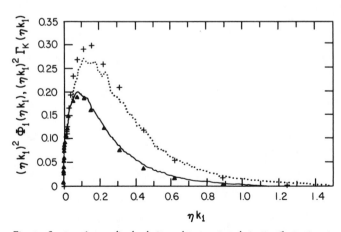

Fig. 5 Spectra of normalized velocity and temperature derivative fluctuations in the surface layer. Solid line and triangles: velocity, from Champagne et al. (1977) and Williams and Paulson (1977), respectively. Dotted line and plus symbols: temperature, from Champagne et al. (1977) and Williams and Paulson (1977), respectively.

over the ocean, the ubiquitous salt spray can cause severe measurement problems: we are gradually learning that the air in the marine layer is a two-phase fluid.

Fine-scale measurements are not quite as routine as, say, momentum flux measurements with a sonic anemometer. Yet they are important—to measure the rates of turbulence dissipation and to further our understanding of turbulence, the extremely nonlinear process so basic to the motion and mixing in the atmosphere.

To illustrate the results of some past experiments, two completely independent sets of spectral data of streamwise velocity and temperature obtained in the surface layer (Champagne et al., 1977; Williams and Paulson, 1977) are compared in Fig. 5. Here, the normalized (with Kolmogoroff scales) velocity and temperature derivative spectra are plotted on a linear-linear scale against the normalized streamwise wave number. When plotted as dimensional values, the area under the velocity derivative spectrum curve is proportional to the rate of dissipation of turbulent kinetic energy; the area under the temperature derivative spectrum is proportional to the rate of dissipation of thermal variance. In normalized form, these peak at about $0.1\eta_K$. With $\eta_K = 1$ mm, the bulk of the dissipation is then occurring at scales of about 1 cm, 10^7 times smaller than the synopticians' 100-km scale. The comparisons between the two data sets—from experiments in Minnesota and Oregon—are good. However, McBean (1982) has recently published surface-layer temperature spectra which differ. Future experiments will resolve and extend fine-scale turbulence measurements in the boundary layer.

Acknowledgments. The author would like to thank Frank Champagne, John LaRue, and Carl Gibson for the many useful discussions and ideas about fine-scale velocity and temperature measurements. The comments of Lutz Hasse, James Tillman, and John Wyngaard on the present paper are greatly appreciated.

Aircraft Measurements in the Boundary Layer

Donald H. Lenschow, National Center for Atmospheric Research

1 INTRODUCTION

Aircraft have had a long history of use for meteorological research. Indeed, the history of aircraft closely parallels the history of meteorology. This is not surprising, since flying requires accurate weather information and, conversely, aircraft provide a convenient and unique platform for collecting meteorological information. Thirty years ago, before the widespread use of pressurized aircraft, a large fraction of aircraft flight time was within the boundary layer. Even now, commercial jet aircraft must at least pass through the boundary layer on takeoff and landing. Thus, the structure of the boundary layer is still very relevant to the needs of aviation. At the same time, developments in aviation technology have led to improved tools for observing the boundary layer from aircraft.

One of the major advantages of an aircraft as a boundary-layer measurement platform is its mobility. An aircraft can probe the entire depth of the boundary layer (with the exception of the lowest 15 m or so) and can obtain statistically significant measurements about an order of magnitude faster than possible with direct fixed-point measurements. Furthermore, an aircraft can be used for either vertical profiling or horizontal traverses, or some combination of the two. Compared to fixed-point observations, which are limited to sampling air advected by the wind, aircraft can measure along a path at any arbitrary angle with respect to the wind, and can measure in a frame of reference that moves with the wind.

Unfortunately, an aircraft's mobility also creates some disadvantages. The aircraft motion needs to be accurately measured and corrections need to be applied to the data. Furthermore, most aircraft need to fly relatively fast (compared to the wind) in order to stay airborne. Therefore, corrections need to be applied for compressibility, adiabatic heating, and flow distortion induced by the aircraft. This complicates the use of aircraft and the subsequent analysis of aircraft data. Furthermore, aircraft are not suitable for long-time-series measurements. Thus, fixed-point measurements (towers, tethered balloons, etc.) are, in many cases, complementary to aircraft measurements.

A large array of sensors have been flown on aircraft for research in the boundary layer. We will review the various types of instrumentation used on aircraft and problems associated with the measurements, particularly emphasizing those instruments and problems that are unique to aircraft. Both mean and turbulence measurements of air velocity, temperature, humidity, and other trace gases will be discussed, with emphasis on air motion measurements because of their complexity and uniqueness of the techniques to aircraft.

2 MEASURING AIR VELOCITY

2.1 Correcting for Aircraft Motion

The velocity of the air is a fundamental variable that is needed for almost all meteorological studies of the boundary layer. This is not a very straightforward measurement from an aircraft because of its relative motion with respect to the earth. An aircraft has complete three-dimensional freedom to move about depending on control surface and power settings, and on velocity and density fluctuations of the air.

The velocity of the air with respect to the earth, $\mathcal{V}$, is obtained by adding the velocity of the aircraft with respect to the earth, $\mathcal{V}_p$, and the velocity of the air with respect to the aircraft, $\mathcal{V}_a$. If all the measurements are made at the same point, we have

$$\mathcal{V} = \mathcal{V}_p + \mathcal{V}_a \tag{1}$$

The components of $\mathcal{V}_a$ are measured with sensors mounted on the aircraft, usually on a boom forward of the aircraft, to reduce as much as possible the airflow distortion induced by the aircraft. An alternative technique involves using the pressure perturbations caused by the airflow distortion to obtain measurements of the airflow angles with respect to the aircraft (Brown et al., 1983). In either case, $\mathcal{V}_a$ is measured in the airplane coordinate system, not in the earth-based coordinate system in which the wind is desired.

The components of $\mathcal{V}_p$ can be obtained from integrated accelerometer outputs on an inertial navigation system (INS), from radiation transmitting and receiving devices such as Doppler radars or radar altimeters, or by radio navigation techniques. If the aircraft velocity is obtained from integrated acceleration, the velocity is measured in an inertial frame of reference. Since we want to measure winds in an earth-based coordinate system, terms involving the angular velocity of the airplane and of the earth must be added to the measured aircraft acceleration $\mathcal{A}_p$ in order to obtain the velocity of the aircraft with respect to the earth. Thus,

$$\dot{\mathcal{V}}_p = \mathcal{A}_p - (\omega_p + \omega_e) \times \mathcal{V}_p + \mathbf{g} \tag{2}$$

where ω_e and ω_p are the angular velocities of the earth and platform, respectively; $\mathbf{g}$ is the gravitational acceleration (which includes the centripetal acceleration); and the superscript dot denotes a time derivative. Since the accelerometers may not be located near the air velocity sensors, there may be an additional term $\dot{\mathbf{\Omega}}_p \times \mathbf{R}$, where $\dot{\mathbf{\Omega}}_p$ is the angular acceleration of the aircraft and $\mathbf{R}$ is the distance from the accelerometers to the air velocity sensors, which must be included. Integrating the accelerations and angular velocity terms to obtain the velocity of the air, we have

$$\mathscr{V} = \mathscr{V}_a + \int \dot{\mathscr{V}}_p \, dt + \mathbf{\Omega}_p \times \mathbf{R} \qquad (3)$$

Since the air velocity $\mathscr{V}_a$ is usually measured in the aircraft coordinate system, the measured components must be rotated to the earth-based coordinate system. Therefore, it is necessary to know the angles between the coordinates of the measuring systems and the local earth coordinates. The mea-

sured velocity components can then be rotated by means of the appropriate angular transformation to the local earth coordinate system. This would be true also for components of $\mathscr{V}_p$ if, for example, $\mathscr{V}_p$ were measured with sensors hard-mounted to the aircraft. Most presently available INSs are stabilized along the local earth axes so that the rotation angles of the platform gimbals are equivalent to the aircraft attitude angles. From the inside platform gimbal outward, i.e., from the stabilized accelerometer and gyro cluster outward to the aircraft frame, the order of the rotations usually is as follows: first, a rotation ψ (airplane heading or azimuth) about the z-axis of the earth coordinate system; second, a rotation θ (pitch) about an axis obtained by rotating the y-axis in the horizontal plane by the angle ψ; and third, a rotation ϕ (roll) about the x'-axis, where the prime denotes an axis in the aircraft coordinate system. These angles are known as Euler angles in classical physics. The aircraft axes are shown in Fig. 1; a right-handed angular rotation is positive. Details of

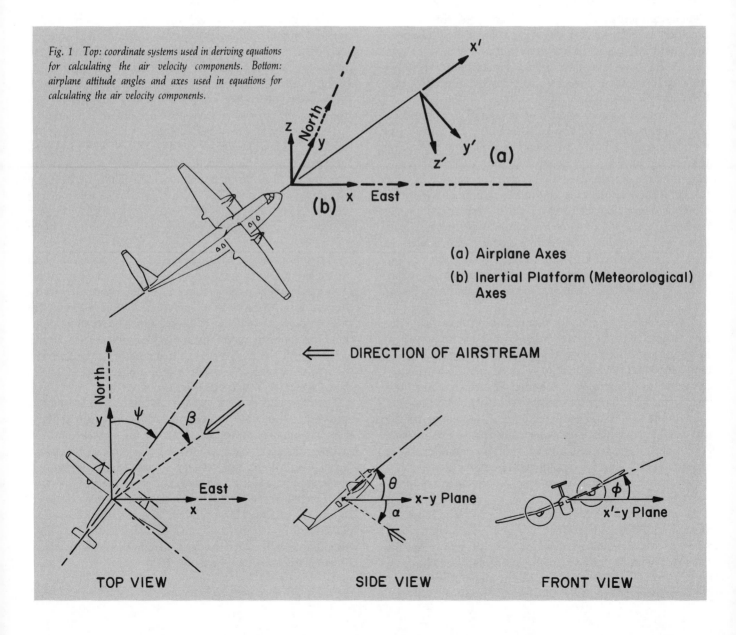

Fig. 1 Top: coordinate systems used in deriving equations for calculating the air velocity components. Bottom: airplane attitude angles and axes used in equations for calculating the air velocity components.

(a) Airplane Axes

(b) Inertial Platform (Meteorological) Axes

DIRECTION OF AIRSTREAM

TOP VIEW   SIDE VIEW   FRONT VIEW

this transformation are presented by Axford (1968), Lenschow (1972), and Nicholls (1983). Goldstein (1980) discusses angular transformations of this type from a more general viewpoint, and shows how to calculate the appropriate rotation matrix using different rotation sequences, since the rotation matrix depends on the order of rotation.

The angle of attack, α, is the angle of the airstream with respect to the aircraft in the aircraft's vertical plane, with α positive in the downward direction; the angle of sideslip, β, is the angle of the airstream with respect to the aircraft in the aircraft's horizontal plane, with clockwise (looking from above) rotation positive, as shown in Fig. 1. The air velocity components with respect to the aircraft are calculated by first correcting the measured true airspeed for angles of attack and sideslip sensitivities. Normally, the airspeed is measured with a pitot-static tube. To a first approximation, most standard aircraft pitot-static tubes measure the magnitude of the air velocity vector, independent of airflow angles. It can then be shown that the components of air velocity in the aircraft frame of reference are $-\mathcal{U}_a D^{-1}$ along the x'-axis, $-\mathcal{U}_a D^{-1}\tan \beta$ along the y'-axis, and $-\mathcal{U}_a D^{-1}\tan \alpha$ along the z'-axis, where $D = (1 + \tan^2\alpha + \tan^2\beta)^{1/2}$. (This differs from the expressions in Lenschow [1972]. For small flow angles, however, the magnitudes of the differences are small. I am grateful to A. Weinheimer and J. Leise, who each independently pointed out the error and corrected the previous derivation.) To summarize this information: in the standard meteorological frame of reference—which has the x-axis pointing east, the y-axis north, and the z-axis up; with ψ, the true heading, measured clockwise (looking down) from north—the final equations for calculating the air velocity with respect to the earth are

$$\mathcal{U} = -\mathcal{U}_a D^{-1}[\sin \psi \cos \theta + \tan \beta(\cos \psi \cos \phi \\ + \sin \psi \sin \theta \sin \phi) + \tan \alpha(\sin \psi \sin \theta \cos \phi \\ - \cos \psi \sin \phi)] + \mathcal{U}_p - L(\dot{\theta} \sin \theta \sin \psi \\ - \dot{\psi} \cos \psi \cos \theta)$$

$$\mathcal{V} = -\mathcal{U}_a D^{-1}[\cos \psi \cos \theta - \tan \beta(\sin \psi \cos \phi \\ - \cos \psi \sin \theta \sin \phi) + \tan \alpha(\cos \psi \sin \theta \cos \phi \\ + \sin \psi \sin \phi)] + \mathcal{V}_p - L(\dot{\psi} \sin \psi \cos \theta \\ + \dot{\theta} \cos \psi \sin \theta) \quad (4)$$

$$\mathcal{W} = -\mathcal{U}_a D^{-1}(\sin \theta - \tan \beta \cos \theta \sin \phi \\ - \tan \alpha \cos \theta \cos \phi) + \mathcal{W}_p + L\dot{\theta} \cos \theta$$

where L is the distance between the INS and the air-sensing platform along the x-axis of the aircraft, and the lateral and vertical separation distances are assumed to be negligible.

These are the exact equations for calculating the air velocity. A schematic diagram showing where the variables required for these computations are measured is shown in Fig. 2. For purposes of illustration, however, it can be shown that for approximate straight and level flight (i.e., where the pilot or autopilot attempts to keep the aircraft level, but air velocity fluctuations still cause perturbations in the aircraft

velocity and attitude angles) many of the terms in Eq. (4) are negligible. First we assume that terms involving the separation distance L are small. This is generally true if the separation distance is less than about 10 m and the aircraft is not undergoing a pilot-induced pitching maneuver. Next we make the following small angle approximations: for the horizontal components, $\mathcal{U}$ and $\mathcal{V}$, we assume that the cosines of α, θ, and ϕ are unity, that terms that involve the products of sines of two of the above angles are negligible, and that $\tan(\alpha,\beta) \cong \sin(\alpha,\beta)$. For the vertical component $\mathcal{W}$, we make the same assumptions for the angles β, θ, and ϕ. Equations (4) then reduce to a form commonly used for approximate calculations of the three velocity components,

$$\mathcal{U} = -\mathcal{U}_a\sin(\psi + \beta) + \mathcal{U}_p$$

$$\mathcal{V} = -\mathcal{U}_a\cos(\psi + \beta) + \mathcal{V}_p \quad (5)$$

$$\mathcal{W} = -\mathcal{U}_a\sin(\theta - \alpha) + \mathcal{W}_p$$

We can estimate the required accuracy of attitude and airstream incidence angle measurements from Eq. (5). In order to obtain the mean horizontal wind, angle $(\psi + \beta)$ must be measured accurately. For a mean wind accuracy of 0.5 m/s and an aircraft speed of 100 m/s, this angle must be measured with an absolute accuracy of 0.005 rad, or 0.3°. A heading angle of this accuracy is difficult to measure with a standard aircraft magnetic compass. An INS, however, can measure true heading to well within this accuracy.

A reasonable figure for short-term velocity accuracy necessary to estimate turbulence fluctuations is 0.1 m/s. At an airplane speed of 100 m/s, the required angular accuracy for the first terms on the right side of Eq. (5) is 0.001 rad, or

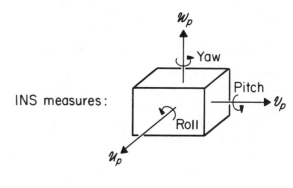

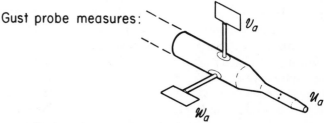

Fig. 2 Schematic diagram of an inertial navigation system (INS) and a gust probe, showing the variables measured by each.

0.06°. This can be a difficult requirement to meet. Fortunately, an INS has even more stringent requirements than this for navigation, so attitude angles of this short-term accuracy can be measured.

Alternatively, a Doppler navigation system can be used to measure the aircraft ground speed

$$\mathscr{V}_g \equiv \sqrt{\mathscr{U}_p^2 + \mathscr{V}_p^2} \qquad (6)$$

and the drift angle δ, which is the angle between the aircraft's heading and its track (positive for a clockwise rotation looking down). From Eq. (5), the horizontal wind components are approximately

$$\mathscr{U} \cong -\mathscr{U}_a \sin(\psi + \beta) + \mathscr{V}_g \sin(\psi + \delta)$$
$$\mathscr{V} \cong -\mathscr{U}_a \cos(\psi + \beta) + \mathscr{V}_g \cos(\psi + \delta) \qquad (7)$$

The angles β and δ are normally small; thus, to the first order, the effect of an error in ψ is an error in the wind direction of the same amount. Therefore, ignoring the accuracy of the aircraft velocity measurement, a Doppler navigation system can, in principle, provide more accurate horizontal winds with the same quality heading reference than, for example, a radio navigation system for measuring airplane velocity which does not measure drift angle. Although Doppler systems have previously been used for air motion measurements in the boundary layer (e.g., Lenschow, 1970), they have several limitations; among them are the long averaging time required for accurate velocity measurements (tens of seconds) and difficulty in operating over water surfaces, particularly smooth water. Doppler navigation systems are discussed extensively by Kayton and Fried (1969).

2.2 Air Velocity Corrections

2.2.1 Application of Bernoulli's Equation

Because of the speed at which an aircraft flies, the kinetic energy of the airstream relative to the aircraft is considerably enhanced. Thus, the measured pressure and temperature may be quite different from those that would be obtained from a fixed point or from a probe moving with the wind. Yet, the meteorological variables of interest are the values intrinsic to the undisturbed air. Therefore, the aircraft measurements of pressures and temperatures need to be corrected.

The three components of airstream velocity are commonly obtained from measurements of the true airspeed $\mathscr{U}_a$ and the two airflow incidence angles α and β. The true airspeed is normally calculated from measurements of a pitot-static pressure difference $\mathscr{Q} = \mathscr{P}_s - \mathscr{P}_o$, the static pressure $\mathscr{P}_o$, and the total air temperature $\mathscr{T}_t$. The dynamic pressure $\mathscr{P}_s$ is measured at the forward-facing front port of a pitot tube. The static pressure $\mathscr{P}_o$ may be measured either at the static ports of a pitot-static tube or at static ports on the fuselage of the aircraft.

We can obtain a relationship between the airspeed and temperature from Bernoulli's equation for the conservation of energy of a perfect gas undergoing an adiabatic process,

$$\mathscr{U}_1^2/2 + C_p\mathscr{T}_1 = \mathscr{U}_2^2/2 + C_p\mathscr{T}_2 \qquad (8)$$

where C_p is the specific heat capacity of air at constant pressure. Applying this equation to the aircraft, $\mathscr{U}_2 = 0$, $\mathscr{T}_2 = \mathscr{T}_t$, $\mathscr{U}_1 = \mathscr{U}_a$, and $\mathscr{T}_1 = \mathscr{T}_s$. Thus,

$$\mathscr{U}_a^2 = 2C_p(\mathscr{T}_t - \mathscr{T}_s) \qquad (9)$$

For an adiabatic process,

$$\mathscr{T}_s = \mathscr{T}_t(\mathscr{P}_s/\mathscr{P}_t)^{(\gamma-1)/\gamma} \qquad (10)$$

where $\gamma = C_p/C_v$, and C_v is the specific heat capacity of air at constant volume. Substituting Eq. (10) into Eq. (9),

$$\mathscr{U}_a^2 = 2C_p\mathscr{T}_t[1 - (\mathscr{P}_s/\mathscr{P}_t)^{(\gamma-1)/\gamma}] \qquad (11)$$

or equivalently, using $\mathscr{T}_s$

$$\mathscr{U}_a^2 = 2C_p\mathscr{T}_s[(\mathscr{P}_t/\mathscr{P}_s)^{(\gamma-1)/\gamma} - 1] = \gamma R\mathscr{T}_s M^2 \qquad (12)$$

where

$$M^2 \equiv 2[(\mathscr{P}_t/\mathscr{P}_s)^{(\gamma-1)/\gamma} - 1]/(\gamma - 1) \qquad (13)$$

and $R = C_p - C_v$ is the gas constant. The Mach number M is the ratio of the true airspeed to the speed of sound.

In practice, neither the total nor the ambient temperature is measured. Instead, the measured temperature $\mathscr{T}_r$ lies between $\mathscr{T}_t$ and $\mathscr{T}_s$. Therefore, a recovery factor,

$$r \equiv (\mathscr{T}_r - \mathscr{T}_s)/(\mathscr{T}_t - \mathscr{T}_s) \qquad (14)$$

is determined for each thermometer on the aircraft. This factor is assumed to be constant and independent of humidity.

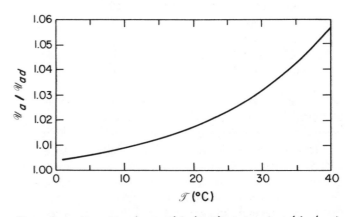

Fig. 3 Ratio of true airspeed corrected for humidity to true airspeed for dry air versus temperature for saturated air at 700 mb for $\mathscr{Q}/\mathscr{P}_o \ll 1$.

Substituting Eq. (10) into Eq. (14),

$$\mathcal{T}_s = \mathcal{T}_r / [r(\mathcal{P}_t / \mathcal{P}_s)^{(\gamma-1)/\gamma} + 1] \qquad (15)$$

and finally, substituting Eqs. (13) and (15) into Eq. (12),

$$\mathcal{U}_a{}^2 = \gamma R M^2 \mathcal{T}_r / \{[(\gamma - 1)/2]r M^2 + 1\} \qquad (16)$$

Friehe (personal communication) has pointed out that the values of γ and C_p are not completely independent of humidity; their water vapor dependence is given by List (1971). Figure 3 shows an example of the error that exists in true airspeed if dry air values are assumed; in very humid conditions, the error is significant.

2.2.2 In-Flight Calibrations

Regardless of where instruments are located and how carefully they are calibrated, errors are likely to be present in the measured variables. Ground tests are not useful for calculating velocity-related errors. Wind tunnel tests are difficult and prohibitively expensive for exact simulation of flight conditions. Therefore, in-flight calibrations play an important role in estimating errors and correcting aircraft measurements.

Because of upwash ahead of the aircraft, the airflow angles (attack and sideslip) and airspeed measured at the tip of a nose boom may differ considerably from the actual values that would be measured far away from the aircraft. Nicholls (1983), for example, determined a correction factor for attack angle from flight tests of 0.87 (i.e., the measured angle must be multiplied by this factor to obtain the correct angle) for a rotating vane at the end of a 7-m nose boom on a C-130 aircraft. The airspeed measured at the tip of a nose boom is reduced by a similar factor. The upwash affects not only the sensitivity, but also the zero offset of angle measurements, which, therefore, must also be determined from in-flight calibrations.

Maneuvers used for this purpose involve changes in aircraft speed and attitude angles. Axford (1968), Telford and Wagner (1974), Telford et al. (1977), Nicholls (1983), and Nicholls et al. (1983) discuss various flight maneuvers and techniques that can be used to estimate errors and make corrections, and give examples of variables measured during maneuvers. The following list summarizes several useful maneuvers used on NCAR aircraft equipped with INS and the information that can be obtained from them; examples of these maneuvers are shown in Fig. 4:

(1) Fly at constant altitude and heading (usually in smooth air above the boundary layer) for several minutes. Then turn 180° by first turning 90° in one direction, then 270° in the other direction at a constant rate so that the airplane flies through the same volume of air on its return track. This maneuver modulates the airspeed and sideslip angle errors, since they are measured in the airplane coordinate system.

The INS errors are not modulated, however, since they are measured in an inertial frame of reference. If the wind along the flight track is assumed to stay constant during this maneuver, differences in the two wind components between the two headings can be used to independently estimate errors in both airspeed and sideslip angle.

(2) Fly at constant altitude and heading, and smoothly vary the aircraft speed from close to stall to close to maximum cruise speed. Since the lifting force on the aircraft is directly proportional to the attack angle and the square of the airspeed, modulating airspeed also modulates attack angle. For level flight, $\mathcal{W}_p = 0$ in Eq. (5); if $\mathcal{W}_a$ is small, $\alpha = \theta$. Thus, α can be calibrated in flight by this technique, provided that θ is measured accurately. The attitude angle transducers, in contrast to airflow angle sensors, can be accurately calibrated in the laboratory.

If airspeed is measured incorrectly, temperature may also be modulated. Temperature recovery factors (Eq. 13) can also be measured or corrected with this maneuver since airspeed variations modulate the measured temperature because of dynamic heating effects. Any other measurements affected by either airspeed or attack angle variations are also modulated by this maneuver.

(3) Vary the aircraft elevator angle while holding the heading constant to obtain a sinusoidal pitching motion with a period of 10 to 20 s and a maximum rate of ascent and descent of 2.5 to 4 m/s. This maneuver modulates vertical airplane velocity $\mathcal{W}_p$, airspeed $\mathcal{U}_a$, and, to a lesser extent, attack angle α. If any of these variables have significant errors, a periodic error in vertical air velocity should be evident. Since the terms do not have the same phase angle, in practice it is often possible to determine which of the variables is in error simply by determining the phase of the error in $\mathcal{W}$ and comparing it with the phases of $\mathcal{W}_p$, θ, $\mathcal{U}_a$, and α. This maneuver can also be used to detect dynamic errors in static pressure or rate-of-climb instruments by comparing their outputs with the integrated INS vertical acceleration.

(4) Vary the aircraft rudder angle while holding the roll and altitude constant to obtain a sinusoidal skidding or sideslip motion with a period of about 10 s and a maximum amplitude of about 2° sideslip. This maneuver modulates heading ψ, horizontal airplane velocity components $\mathcal{U}_p$ and $\mathcal{V}_p$, and β. Analogous to (3), errors in any of these variables will cause a periodic variation in the horizontal wind velocity.

On the NCAR aircraft, the system performance is judged to be satisfactory if the vertical air velocity error is less than 10% of the vertical airplane velocity for maneuver (3), and the lateral air velocity component is less than 10% of $\mathcal{U}_a \sin \beta$ for maneuver (4).

Telford and Wagner (1974) and Telford et al. (1977) discuss and show examples of another maneuver—a tight circle with a large roll angle, which is useful in estimating the zero offsets in the flow angle measurements, as well as providing an overall check on the air velocity system accuracy. They estimate that short-term (i.e., not including long-term

INS drift) velocity errors can be reduced to < 0.3 m/s by in-flight calibrations, which agrees well with the analysis by Nicholls (1983).

Another technique that has been successfully used for improving the quality of measurements is intercomparison of measurements from several aircraft flying in formation. As LeMone and Pennell (1980) and Nicholls et al. (1983) have shown, intercomparisons can be used to evaluate the quality of the measurements and uncover erroneous data. Nicholls (1983) has shown that one airplane, with a superior navigation system, can be used to improve the wind measurement on another aircraft flying in formation which has a less accurate navigation capability.

2.3 Use of Aircraft Response Characteristics

An alternative to measuring all the components of $\mathscr{V}_a$ and the three attitude angles in order to obtain $\mathscr{V}$ is to measure the response of the aircraft to fluctuations in the air velocity and use the aircraft transfer function (i.e., the ratio of the airplane response to the input forcing function) to help estimate the air velocity. This process, however, has inherent

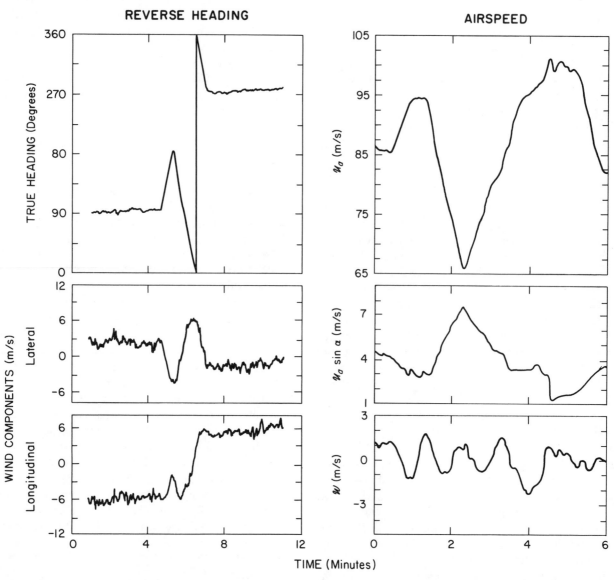

Fig. 4a Examples of reverse heading and airspeed maneuvers that are used to check the quality of air velocity measurements. The lateral and longitudinal velocity components are measured with respect to the aircraft; therefore, the measured wind should change sign but not amplitude after the 180° turn, if the wind field remains constant and is measured without error. An error in airspeed will result in a difference in the amplitude of only the longitudinal component before and after the turn, while an error in the sideslip angle will similarly affect only the lateral component, which simplifies correction procedures. The airspeed maneuver modulates the attack angle and pitch angle; if pitch angle θ is measured accurately, the error in attack angle α can be determined by comparing the vertical wind component with respect to the airplane ($\mathscr{U}_a\sin\alpha$) with the vertical wind component with respect to the earth ($\mathscr{W}$). In this example, there is little correlation between the two, so the fluctuations in $\mathscr{W}$ are presumed to result from turbulence rather than an inaccurate measurement of α. The airspeed maneuver can also be used to estimate airspeed-dependent errors in the temperature recovery factor.

limitations. First, the airplane transfer function is not known exactly; furthermore, it depends upon the airplane mass, which is not constant with time because of fuel consumption. Second, the aircraft cannot respond to high-frequency variations in the air velocity. Third, the airplane responds indistinguishably to both horizontal and vertical gusts. Nevertheless, for some applications (e.g., Lenschow, 1976) useful results can be obtained by this technique. This approach was used by Bunker (1955) for turbulence flux measurements, and more recently by Kyle et al. (1976) to estimate updraft profiles in thunderstorms.

An intermediate approach which is considerably more accurate is to measure the airspeed and pitch angle of the aircraft and use its aerodynamic characteristics to calculate angle of attack, from which the vertical air velocity can be obtained. This approach is discussed by Kelly and Lenschow

(1978). Here again, the accuracy depends upon aircraft flight characteristics and response time.

3 AIRCRAFT INSTRUMENTATION

3.1 Air Motion Measurements

Aircraft have been used to measure both mean horizontal wind and turbulent fluctuations in all three components of air velocity. It is not normally feasible, however, to directly measure the mean vertical wind, since the achievable long-term accuracy of vertical wind measurement from aircraft is probably about 0.1 m/s, which is usually greater than the mean vertical wind. It may be possible, however, to measure horizontal divergence around closed flight paths in special circumstances and thus indirectly estimate the mean vertical

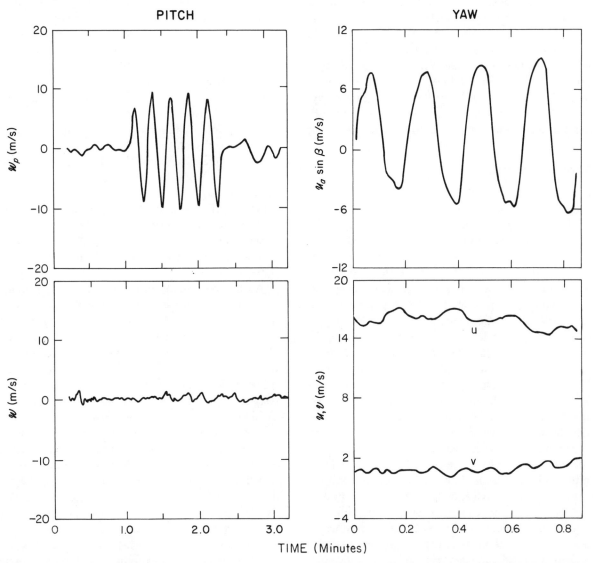

Fig. 4b Examples of pitch and yaw maneuvers. The pitch maneuver is used as an overall check on the accuracy of the w measurement; in this example, there is little modulation of w during the pitching maneuver, which implies that fluctua-

tions in w are measured accurately. Similarly, the yaw maneuver is an overall check on the lateral (with respect to the aircraft) component; there is little modulation of u and v (in geographic coordinates) by the yawing maneuver.

wind (Raymond and Wilkening, 1982, and Nicholls, 1983).

Measurements of vertical air velocity fluctuations over a bandwidth sufficient to estimate vertical turbulent fluxes in the boundary layer may have been first attempted by Bunker (1955). He estimated vertical air velocity fluctuations by measuring vertical aircraft acceleration and horizontal airspeed fluctuations, and then used the aerodynamic characteristics of the airplane to calculate fluctuations in vertical air velocity. Airspeed fluctuations were used along with the vertical velocity to estimate the turbulent momentum flux along the direction of flight. As pointed out in the previous section, this approach has serious limitations. An early approach to measuring the vertical air velocity independently of the airplane characteristics was to mount an accelerometer and a two-axis free gyro in the aircraft to measure fluctuations in the airplane vertical acceleration (in the airplane coordinate system) and pitch angle. By combining these measurements with airspeed and attack angle measurements, vertical air velocity fluctuations can be estimated (e.g., Lappe et al., 1959; Lenschow, 1965; Myrup, 1969). An improvement to this was to stabilize the accelerometer by mounting it on a gimballed platform which was slaved to a two-axis free gyro so that gravity had relatively little effect on the measured acceleration (e.g., Telford and Warner, 1962). More recently, inertial navigation systems have been used to measure both the acceleration and the attitude angles to far better accuracy.

3.1.1 Airplane Velocity

Measurement of wind entails measurement of both the velocity of the airplane with respect to the earth and the velocity of the air with respect to the aircraft. Horizontal airplane velocity has commonly been obtained either from measurement of the ground velocity—by sensing the Doppler shift of a transmitted electromagnetic signal (Doppler radar) or by using externally transmitted radio navigation signals—or from double integration of the airplane acceleration (inertial navigation). Combinations of these techniques can often provide better information than each by itself. Radio navigation techniques, for example, can provide accurate position fixes for updating positions obtained by inertial navigation, which has time-dependent errors (Kayton and Fried, 1969). In addition to airplane velocity measurements, the attitude angles of the aircraft are needed for rotating the air velocity to an earth-based coordinate system. This can best be done with an INS. A further advantage of INS is its inherent fast response; the INS accelerometers are capable of resolving the entire spectrum of aircraft motion. In contrast, Doppler radar and radio navigation techniques are limited to measurements averaged over many seconds. Thus, although they can be useful for mean wind measurements, they are of limited use for turbulence measurements. For these reasons, we discuss here only INS. Doppler radar and radio navigation techniques are discussed, for example, by Kayton and Fried (1969); a current summary of global navigation systems is presented in a special issue of the *Proceedings of the IEEE 71(10)*,

October 1983. Nicholls (1983) discusses an INS-based wind-measuring system that uses Decca and LORAN-C (Frank, 1983) radio navigation and Doppler radar information to renavigate the entire flight in postflight data processing and thereby greatly improve the accuracy of the airplane velocity measurements. Nicholls estimates the accuracies of horizontal and vertical wind measurements with this system as ±0.4 m/s and ±0.1 m/s, respectively. Without external updating, errors in an INS-based wind-measuring system increase with time and, with typical commercially available systems, errors of several meters per second can build up after several hours of flying. Commercial systems with real-time updating of INS using radio navigation information are also available.

In the future, aircraft navigation may very well be revolutionized by a satellite-based radio navigation system called the NAVSTAR Global Positioning System (Parkinson and Gilbert, 1983). This system, which is planned to be fully operational by the end of the decade, will use at least 18 orbiting satellites to obtain continuous coverage over the entire earth, with a predicted three-dimensional positional accuracy of < 20 m, and velocity accuracy of <0.1 m/s. However, even though an INS may not be needed for navigation or airplane velocity measurements, an INS is still essential for accurate air velocity measurements since the aircraft attitude angles must also be measured.

It is technically feasible to make a high-accuracy INS (e.g., Kuritsky and Goldstein, 1983; Adams and Hadfield, 1984) that accumulates position and velocity errors less than 0.5 km and 0.2 m/s, respectively, in an hour (compared to typical commercial-system errors of about 1.8 km and 0.5 m/s in the same period). In the past these systems have been too expensive for use on meteorological research aircraft. But this may change in the future as technological improvements result in smaller, more reliable, less expensive, and more accurate INSs.

An INS consists of an orthogonal triad of accelerometers whose output is integrated to obtain the aircraft velocity, and integrated again to obtain position. Basically, two approaches are used to orient the measurements to a local earth coordinate system. In one case (known as a strapdown mechanization), the accelerometers are mounted directly on the airframe, and the angular orientation of the accelerometers with respect to the earth is calculated from the outputs of gyros capable of rapid and accurate measurements of large angles. Although strapdown systems are mechanically comparatively simple, the requirements for rapid computation, wide-angle gyro accuracy, and accurate acceleration measurements at arbitrary orientations with respect to gravity has made them difficult to implement for aircraft use. The recent development of the ring laser gyro and rapid advancements in computer technology have made strapdown INSs more attractive for aircraft (Kuritsky and Goldstein, 1983). The ring laser gyro measures frequency difference between two light waves traveling in opposite directions around a closed path, which is proportional to rotation rate in an inertial frame of reference.

The other type, which is the kind most commonly used in present airborne air-motion measuring systems, contains accelerometers mounted on a gyro-stabilized gimballed platform which is kept level by continuous computation and application of the required torquing rates. These rates are determined by integrating the horizontal accelerometer outputs and applying corrections based on the earth's rate of rotation and the geographic coordinates of the platform. The geographic coordinates are obtained from a second integration of the accelerometer outputs, again with appropriate corrections. For horizontal navigation, only the two horizontal accelerometers are essential, since they are at right angles to the gravitational acceleration. Figure 5 is a block diagram of a gimballed INS.

Calculation of INS horizontal position and velocity is accomplished within a closed second-order loop, which limits the rate of growth of errors in position and velocity due to errors in measured horizontal acceleration. Basically, if one of the accelerometers senses an erroneous acceleration, due to either accelerometer error or platform tilt, this acceleration, when integrated, causes a velocity error which then is used to calculate an erroneous torquing rate. The net result is a platform rotation in a direction that tends to introduce a component of gravity that opposes the acceleration error. The period of oscillation of this undamped sinusoidal response to an error is

$$\tau = 2\pi\sqrt{R_o/g} \cong 84.4 \text{ min} \qquad (17)$$

where R_o is the radius of the earth. This periodicity, which is equal to the period of a pendulum with a length equal to the radius of the earth, is known as the Schuler oscillation. More detailed discussions of INS principles of operation, perfor-

mance characteristics, and errors are presented in Broxmeyer (1964) and Kayton and Fried (1969).

Other errors in position and velocity exist as well. Generally, the INS errors can be approximated by a monotonically increasing, time-dependent error and a periodic error at the Schuler frequency. An example of a comparison between INS positions and positions determined by a limited-area (< 100 km), ground-based, multiple-aircraft radio navigation system (Johnson and Fink, 1982) is shown in Fig. 6. Figure 7 presents a similar comparison between velocities measured by an INS and those measured by both Decca Navigator and LORAN-C radio navigation systems, which have ranges of 560 and 1,800 km, respectively (Nicholls, 1983). These errors can change drastically from one flight to another on the same system since unpredictable errors occur in INS gyros and accelerometers, and the initial alignment, which varies from flight to flight, plays an important role in subsequent INS performance. The alignment sequence, which normally takes 20 to 40 min, includes initializing the INS coordinates, leveling the platform, and gyrocompassing to determine true north. Once the INS is in its navigation mode, any interruption can cause it to lose its navigation capability for the duration of the flight.

In the vertical, no closed loop exists for inertial measurements. However, pressure altitude can be used to limit errors introduced by errors in vertical acceleration. (An offset, or bias in vertical acceleration, when integrated twice, results in an error in altitude that increases with time squared.) At NCAR, the pressure altitude is combined with the vertical acceleration in a third-order loop (see Fig. 8) to provide an absolute long-term reference for altitude and vertical velocity. The advantage of a third-order baro-inertial loop (Blanchard, 1971) over a second-order loop system is the elimination of

Fig. 5 Block diagram of an inertial navigation system. The dashed rectangle represents the gimballed stable element which is kept level with respect to local earth, and $\mathbf{R}_p$ is the position vector.

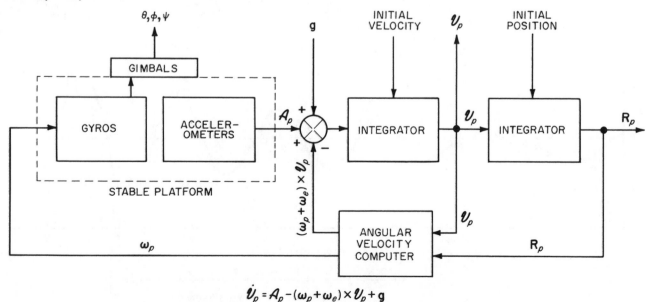

$$\dot{\boldsymbol{\mathcal{V}}}_p = A_p - (\omega_p + \omega_e) \times \boldsymbol{\mathcal{V}}_p + g$$

Fig. 6 Differences E−W (the x-direction denoted by Xs) and N−S (the y-direction denoted by Ys) in position measured by the multiple aircraft tracking system, MAPS (Johnson and Fink, 1982) and the INS aboard an NCAR Queen Air aircraft. The lines are least-squares fits of the function $a_0 + a_1t + a_2\sin(2\pi t/\tau) + a_3\cos(2\pi t/\tau)$, where τ is the Schuler period (∼ 84.4 min) and the a_1s are constants. The point-to-point scatter is due to resolution of the MAPS system, which has excellent long-term accuracy (< 1 km), but poorer resolution than the INS. (Figure courtesy of C. Biter and J. Fankhauser.)

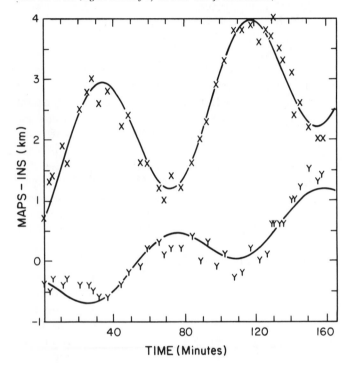

velocity and position error due to bias in the measured acceleration. For the NCAR aircraft (Lenschow et al., 1978a), a 60-s time constant is used in the mechanization of this loop. This value is based on a compromise between minimizing the effect of high-frequency noise from the pressure altitude measurement (e.g., data system resolution, transducer noise, and pressure line "ringing"); and (1) minimizing the recovery time of the loop for extraneous disturbances (i.e., glitches), and (2) improving the long-term stability of the vertical accelerometer output.

3.1.2 Air Velocity Measurements

By far the most common sensor for measuring the longitudinal component of air velocity is a pitot or pitot-static tube (Fig. 9). True airspeed is obtained from a differential pressure measurement between the dynamic (forward-facing) port on the pitot tube and either an airplane static port or the static port on a pitot-static tube, combined with measurements of ambient temperature and static pressure. The equations are presented in Section 2. The sample ports are connected to pressure transducers by pressure lines, which often limit the

Fig. 7 Differences in the N (top) and E (bottom) airplane velocity components measured by an INS and a Decca navigation system (short dashed lines), and an INS and a LORAN-C radio navigation system (long dashed lines) mounted on a C-130 operated by the Meteorological Research Flight, Royal Aircraft Establishment, England. The solid line is faired by eye to the data. The short-period oscillations are due to noise in the radio positions, which are noisy for short averaging times but have excellent long-term accuracy. The INS velocity error, on the other hand, increases with time. (Figure courtesy of S. Nicholls.)

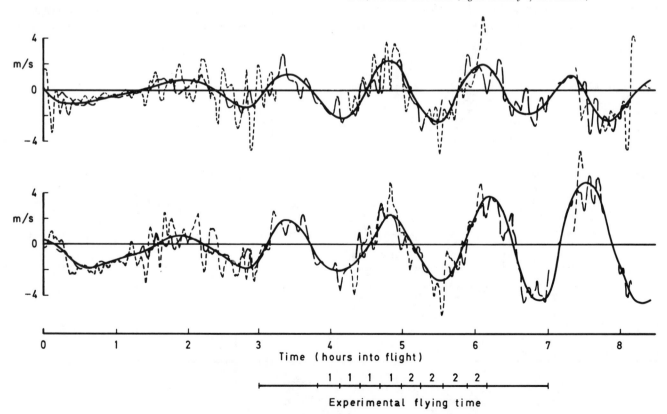

frequency response of the measurement. Generally, if the pressure lines are less than about 1 m in length, with no constrictions, negligible instrument volume and an internal diameter of 0.3 cm or greater, there will be negligible attenuation and phase shift up to at least 20 Hz (Iberall, 1950).

The frequency at which significant attenuation and phase shift begin is approximately proportional to the square of the length-to-diameter ratio. This frequency is reduced if there is a significant cavity or constriction in the system. One way of estimating the frequency response of pressure measurements from in-flight data is by measuring the phase shift between the pressure system of interest and a reference pressure measurement with short tube lengths. An example is given in Fig. 10.

Another method of measuring true airspeed is by counting vortices that are shed by a cylinder mounted transversely to the flow. The true airspeed is directly proportional to the frequency of vortex shedding; no density corrections are necessary. A commercial instrument is available that uses sonic techniques to sense the vortices, which are then counted. Hot-wire and hot-film techniques have also been used on aircraft for airspeed, as well as flow angle measurements. For the most part, however, they have been limited to high-frequency fluctuation measurements of air velocity in clear air (e.g., Lenschow et al., 1978b; Merceret, 1976) and, therefore, will not be discussed further here.

A variety of techniques have been developed to measure the transverse air velocity components. Probably the simplest in principle is to use vanes which align themselves with the airstream. The basic requirements for this technique are: (1) a rugged vane structure with a very low mass/unit area; (2) an angle transducer with minimal inertia, little or no friction, and a resolution capability of about 0.01°; and (3) accurate counterbalancing of the vane mass. The transverse air velocity components are given by $\mathcal{U}_a \sin \alpha$ and $\mathcal{U}_a \sin \beta$. Several rotating vanes have been developed and used for turbulence measurements over the years (e.g., Telford and Warner, 1962; Lenschow, 1971). A frequency response of greater

than 20 Hz with nearly critical damping is possible (Lenschow, 1971; Nicholls, 1983).

Another technique uses a vane that is constrained from rotating; the flow angle is obtained from measurements of the lifting force exerted on the vane by the airstream $\mathcal{F}$, and the dynamic pressure, according to the relation

$$\alpha = \mathcal{F}/(K_1 \mathcal{Q} A) \qquad (18)$$

where A is the vane surface area and K_1 is the sensitivity coefficient. In the sensor described by Johnson et al. (1978), the force is sensed with a strain gage bridge mounted on a stainless steel beam that is counterbalanced to remove acceleration sensitivity. The value of K_1 for this vane was found, through wind tunnel tests, to depend somewhat on the ratio

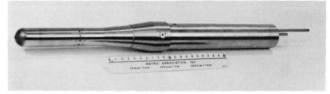

Fig. 9 Gust probes used on NCAR aircraft. Top: fixed-vane probe with two attack and sideslip angle vanes and a pitot-static probe. A fast-response thermometer (above left) and a Lyman-alpha hygrometer (below left) are mounted at the rear of the probe. Bottom: differential pressure probe, with ports for measuring attack and sideslip angles, and pitot and static pressure.

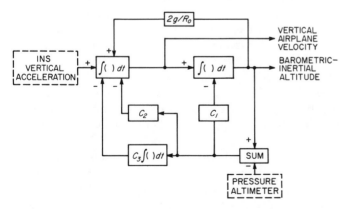

Fig. 8 Block diagram of a third-order baro-inertial loop for calculating altitude and vertical airplane velocity from a pressure altimeter and INS vertical airplane acceleration. Feedback loops involving C_1, C_2, and C_3 are set for a time constant of 60 s on NCAR aircraft. The term involving $2g/R_o$ is a correction for the change of gravity with altitude (R_o is the distance to the center of the earth).

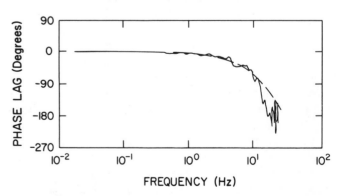

Fig. 10 Phase lag of dynamic pressure measurements using a tube 4 m long (0.43 cm inside diameter) between pitot tube and pressure transducer, with an orifice diameter of 0.16 cm. Dashed line is the phase lag $-6f$, where f is frequency in hertz. (From Brown et al., 1983.)

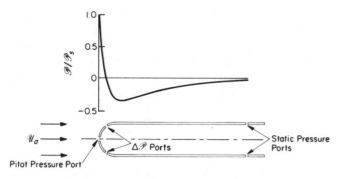

Fig. 11 Top: pressure distribution normalized by the dynamic pressure on the surface of a hemisphere-capped cylinder (bottom) in a uniform flow parallel to the cylinder. The schematic cross-section view of the cylinder shows the arrangment of pressure-sensing ports for sensing airflow angle ($\Delta \mathscr{P}$ ports), and pitot and static pressures. (From Lenschow et al., 1978a.)

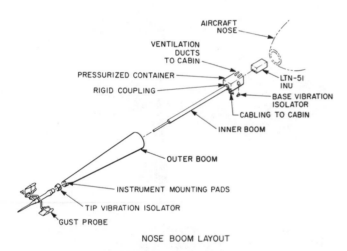

Fig. 12 An exploded view of an air motion sensing system on NCAR aircraft, showing both the inner and the outer booms, the gust probe at the tip, and the inertial navigation system (LTN-51 INU) at the base of the boom. The gust probe is rigidly coupled to the INS by the inner boom.

$\xi = \mathscr{F}/(\mathscr{Q}A)$, although assuming K_1 constant ($= 0.064 \, \text{deg}^{-1}$) introduces an error of less than $0.25°$ for $-10° < \alpha < 10°$. Figure 9 shows a gust probe using fixed-vane sensors for both attack and sideslip angles. Advantages of this technique over the rotating vane are: (1) a frequency response of several hundred hertz is possible, and (2) it has no moving parts. Reliability of the strain gage bridge, however, has been a problem.

Instead of measuring the force of the airstream on a surface, another technique is to measure the pressure difference between ports at two locations on a probe. Figure 9 shows one example of such a probe which is widely used for airstream angle measurements. The probe is constructed in the shape of a hemisphere capping a cylindrical tube, with holes drilled at angles of $\pm 45°$ with respect to the longitudinal axis in both the vertical and the lateral planes, as well as on the front and along the sides of the cylinder for dynamic and static pressure measurements, respectively. Figure 11 shows the pressure distribution along the probe surface with the

flow parallel to the probe. If the flow is at an angle with respect to the tube axis, the pressure distribution is no longer symmetric with respect to the tube axis. Thus, for a pair of ports symmetrically located above and below the tube axis as in Fig. 11, an angular rotation causes an increased pressure in one port and a reduced pressure in the other. The net effect is a differential pressure between the top and bottom ports which is a function of flow angle and indicated airspeed. By placing a second pair of ports at right angles to the first, simultaneous measurements of both flow angles with respect to the local flow are obtained. For a pressure difference $\Delta \mathscr{P}$ measured across a symmetric set of ports, both theory and observations predict a relation of the form (Brown et al., 1983)

$$\alpha = \Delta \mathscr{P}/(K_2 \mathscr{Q}) \qquad (19)$$

where K_2 is the sensitivity coefficient. For the probe in Fig. 9, $K_2 = 0.079 \, \text{deg}^{-1}$ for $M < 0.55$. As is the case for the airspeed measurement, the angle measurements are obtained from differential pressure measurements, and thus are subject to the same frequency response limitations.

Airflow measurements of this type need to be made as far away from the aircraft as possible to avoid measuring in a flow that is distorted by the aircraft. This is usually done by mounting the gust probe on a nose boom well in front of the aircraft. Boom length involves a compromise between isolating the sensors from aircraft influence and maintaining structural integrity, high resonant vibration frequency, and reasonable weight and cost. A commonly used rule of thumb is to extend the boom about 1.5 fuselage diameters in front of the airplane nose. Airflow angle measurements are particularly sensitive to a nonrigid coupling between the gust probe and the INS. Therefore, in an attempt to minimize angular differences between them, as well as reduce vibrational coupling between the airplane and the air motion sensors, the INS and gust probe are sometimes rigidly connected with an inner boom, and this whole assembly is isolated from aircraft vibration by shock and vibration isolators (Fig. 12).

An alternative to this type of air-sensing probe is to measure pressure differences on the nose of the aircraft itself. The pressure distribution around the front of a typical aircraft resembles that of the pressure probe described above. This obviates the need for a nose boom, and also allows, in many cases, the use of shorter pressure lines between the ports and the transducers; thus higher-frequency response may be possible. It also frees up the nose of the aircraft for other instrumentation such as radar. One difficulty of this approach is the need to rely on in-flight calibration, although numerical simulations may obviate this need to some extent (Chaussee et al., 1983). Brown et al. (1983) describe a system of this type.

In addition to direct sensing of air velocity, the air velocity field can be measured remotely from an aircraft. Bilbro et al. (1984) discuss use of a Doppler lidar that can be used to construct a two-dimensional horizontal wind field in clear air with 330-m resolution to a range of 30 km. Since the measurements are obtained from a moving platform, it is essential to accurately measure the airplane velocity and attitude angles, as with direct air velocity measurements. Schwiesow (elsewhere in this volume) discusses other lidar techniques that are applicable to aircraft.

3.2 Air Temperature

A variety of thermometers have been used on aircraft, including resistance wires, thermistors, and thermocouples. Basically, the techniques of measurement are similar on an aircraft and on the ground (see articles by Kaimal and Friehe in this volume). However, as discussed in Section 2, one fundamental difference is the flow speed, which has significant effects on temperature measurement. At 100 m/s, for example, the total adiabatic heating is about 5 °C and is approximately proportional to the square of the true airspeed. Therefore, care needs to be taken to measure ambient temperature. Thermometer housings have been designed to measure very nearly the total air temperature $\mathcal{T}_t$, which is then corrected to obtain ambient temperature. A widely used thermometer is shown in Fig. 13, which is designed to measure very nearly the total temperature, and also provide protection to the 0.025-mm platinum wire sensing element through inertial separation of particles by a 90° bend in the housing that might otherwise break the wire. Accuracies of 0.5 °C have been demonstrated for this probe by in-flight comparisons with tower-based thermometers. An advantage of measuring $\mathcal{T}_t$ is that it is relatively insensitive to location on the aircraft, as long as the probe is outside the aircraft boundary layer. Outside the aircraft boundary layer, even though the flow is distorted by the aircraft, viscous heating effects are negligible, so the air changes speed adiabatically

Other types of thermometer housings have been used on aircraft. One type, called the vortex thermometer (Vonnegut, 1950), measures the temperature in a vortex, which is cooled by expansion to approximately compensate for dynamic heating so that, in principle, it measures ambient temperature directly. However, in practice it is difficult to adjust properly and is location-sensitive.

The frequency response of an infinitely long 0.025-mm wire without any supporting structure aligned across the flow in the total temperature probe discussed above would be greater than 20 Hz. However, thermal conduction between the wire and its support adds thermal capacity and reduces the sensor frequency response (Wyngaard et al., 1978). This is a problem for resistance-wire thermometers in general, since lengthening the distance between supports makes the wire more susceptible to breakage or stretching. Friehe (elsewhere in this volume) discusses high-frequency temperature measurements from fixed surface sites, a technique which is applicable to aircraft as well. Various probes have been built to improve the frequency response by lengthening the distance between supports and allowing higher-speed flow by the element. An example is shown in Fig. 13. This probe

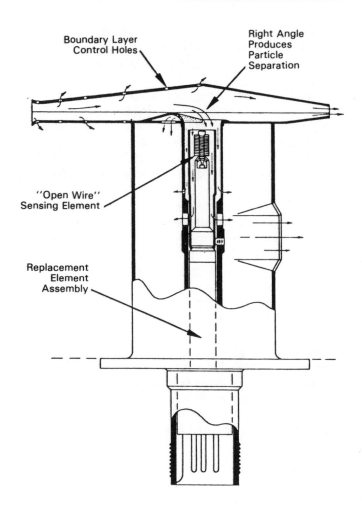

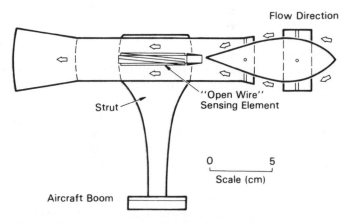

Fig. 13   Top: total temperature probe made by Rosemount Engineering Company (drawing courtesy of Rosemount). Probe is designed to measure very nearly the total temperature at typical airplane speeds ($r \cong 0.94$) and yet provide moderately fast response and protection for the 0.025-mm (diameter) platinum wire sensing element. Bottom: NCAR fast temperature probe. The 0.025-mm platinum wire is not as well protected as in the above probe and does not measure total temperature ($r \cong 0.85$), but its response is faster. (From Spyers-Duran and Baumgardner, 1983.)

has a streamlined obstruction upstream of the sensing wire. The net result is faster response, but less ruggedness; damage to the wire by particles in the airstream can still occur. Icing, in particular, is a problem.

Another problem facing direct temperature measurement is the effects of wetting by clouds and precipitation. As Lenschow and Pennell (1974) have shown, if the sensing element becomes wet, the dynamic heating is approximately as predicted by a wet adiabatic process. At 100 m/s, this can amount to as much as a 2 °C difference between wet- and dry-bulb temperatures in a saturated environment. Many times it is difficult to determine whether the sensor is wet, and often it is only partially wet, so that under these conditions temperature measurement becomes questionable at best. Related to this is the problem of salt accumulation on the sensing wire during flights in the marine boundary layer. Its effects on temperature measurement are discussed by Friehe elsewhere in this volume.

An alternative to direct temperature sensing that eliminates these problems is radiation thermometry. Although radiation thermometers are not yet operational, the concepts have been demonstrated (Albrecht et al., 1979). The technology appears to be available (Nelson, 1983) to develop accurate fast-response radiation thermometers that will eliminate the need to correct for dynamic heating and errors due to wetting and salt contamination.

3.3 Humidity

In contrast to temperature, a distinct difference in sensing technique exists between mean humidity and humidity fluctuation measurements. The most commonly used instrument for measuring mean humidity is the dew/frost point hygrometer. The technique, which is discussed in Wexler (1965), is to cool a mirror until a thin water or frost layer forms. This changes the reflectivity of the mirror surface, which is detected optically. The mirror is then heated until the water or frost starts to disappear. If the servo system controlling the mirror temperature is properly adjusted, the temperature reaches an equilibrium, which is the dew/frost point temperature. Standard aircraft dew point hygrometers can operate satisfactorily to temperature depressions of as much as 30 °C, and some units are advertised to attain even greater depressions. The time response to measure a stable mean dew/frost point temperature is several tens of seconds, and typical accuracies are ± 1 °C for air temperatures between 0 and 70 °C, and ± 2 °C between -30 and 0 °C.

Several techniques have been used to measure humidity fluctuations. The two approaches most commonly used on aircraft at present are to measure the absorption of hydrogen Lyman-alpha radiation (at 121.56 nm) and to measure the microwave refractivity of the air at about 9 GHz. Instruments for making these measurements from aircraft are not significantly different from those discussed by Friehe for measurements in the surface layer.

3.4 Water

Aircraft are used not only for clear-air boundary-layer research, but also for studying cloud-capped boundary layers. Clouds can play an important role in boundary-layer dynamics because of their effects on the solar and infrared radiation budgets, and on buoyancy due to latent heat. In addition, many pollutants cycle through cloud particles and can be physically and chemically altered, or deposited at the surface by precipitation (see the article by Pueschel in this volume). Therefore, sensors for measuring cloud droplet sizes and concentration, and total liquid water, which are essential for cloud-related studies, are also frequently used for boundary-layer studies. Even if measurement of cloud variables is not an essential part of an experiment, it is useful to monitor cloud water since flying in clouds may result in significant errors in temperature and humidity measurement, and accumulated water may clog sampling ports. A more serious operational problem is icing that occurs when flying in supercooled clouds. Ice accumulation can alter the flow characteristics of probes, seal off sampling ports, and possibly create a flight hazard through changes in the aircraft lift, drag, and mass distribution. To prevent icing, probes exposed to the airstream are commonly deiced by electrical heating, or through use of engine heat. Brown (1982) describes the principles and application of a commercially available icing rate detector.

The most commonly used type of cloud droplet probes (available from Particle Measuring Systems, Boulder, Colorado) measure droplet size by their scattering effect on a laser beam (Knollenberg, 1981; see also Pueschel's article). Baumgardner (1983) describes several cloud droplet probes and evaluates their performance based on wind tunnel tests. By far the most commmonly used cloud liquid water content meter is the Johnson-Williams heated-wire probe (available from Cloud Technology, Palo Alto, California). Strapp and Schemenauer (1982), however, have found from wind tunnel tests that calibration differences between sensing heads often exceeded 20 %. King et al. (1978) have developed a new hot-wire liquid water sensor (available from Particle Measuring Systems, Boulder, Colorado) which has a calculable system response that is fast enough for turbulence measurements and which appears to be more accurate than the Johnson-Williams probe.

3.5 Other Trace Constituents

In recent years, there has been an increasing interest in measuring a variety of trace constituents in the boundary layer. Pueschel (elsewhere in this volume) discusses methods for sizing, collecting, and analyzing aerosols. Many instruments for measuring mean concentrations of trace gases even in pristine environments have already been used on aircraft (e.g., Davis, 1980). Instruments are now available that have been developed specifically for aircraft use with sufficient frequency response that they can be used to resolve the turbulent fluctuations of concentration that contribute to the variance and vertical flux. At present, instruments have been demonstrated for measuring fluxes of carbon dioxide (Desjardins et al., 1982; Bingham et al., 1983) and ozone (Pearson and Stedman, 1980) from aircraft. Other instruments under development offer the potential of measuring fluxes of nitric oxide, carbon monoxide, and sulfur dioxide. These measurements are of interest not only for studies of the fate of trace gases in the atmosphere, but also as tracers for studying boundary-layer dynamics (Lenschow, 1982).

Profiles of aerosols and trace gases can also be measured remotely from an aircraft with lidars, as discussed by Schwiesow elsewhere in this volume.

4 APPLICATIONS

The widespread use of aircraft for boundary-layer research dates back about 20 years, and can be traced to technological advances in two areas: the development of INSs for aircraft, and the capability to rapidly record and efficiently process large data sets. For turbulence measurements, 20 or more variables may be recorded at a rate of 50/s. For a 5-h flight, this amounts to $\sim 3 \times 10^7$ data points. It was only after the development of portable magnetic tape recorders and fast digital computers that routine measurements of turbulence fluxes from aircraft became feasible.

Since then, aircraft have been used for such a wide variety of boundary-layer studies that it is impossible to discuss in detail in a single article all the applications of aircraft to boundary-layer research. I will, however, discuss in general terms, with a few specific examples, the various ways that aircraft can be used.

Aircraft have been very successful in improving our understanding of the horizontally homogeneous convective boundary layer. Using the averaging time relationship discussed by Wyngaard elsewhere in this volume (and in Wyngaard, 1973), we find that most mean and second-order moment quantities can be estimated to within about 10% accuracy by averaging over 20 min of constant level flight (i.e., 80 to 120 km). This means that a single aircraft can fly four levels in the boundary layer within about two hours. Since flight duration for many research aircraft is more than four hours, the levels can often be repeated, which permits estimation of the time changes between the two sets of traverses. Often the flight legs are split into along-wind and crosswind traverses to allow measurement of differences in turbulence structure related to wind direction. However, as Wyngaard (1973) has pointed out, shear stress (and sometimes other covariances) typically requires even more averaging time than this for 10% accuracy—sometimes more than an hour. Thus, careful consideration needs to be given to measurement requirements and operational constraints when designing a flight program. In some cases, it may not be possible to obtain the measurements with the desired accuracy even with negligible instrument error.

It is essential to estimate the contributions of instrument inaccuracies and atmospheric variability to the overall measurement error when designing an experiment. However, this is not the only way to estimate the quality of the collected data. The properties of the atmosphere itself can be used to check on the accuracy and representativeness of the measurements. This can be done by use of budget or conservation relations. If all the terms in a particular budget can be estimated, any residual is due to measurement error. For example, in certain situations, all the terms in the budgets of scalar quantities which have no internal sources or sinks in the boundary layer can be estimated in a well-designed boundary-layer experiment. In order to do this, both mean and turbulent fluctuations of the quantity must be measured with sufficient accuracy and resolution that spatial and temporal changes of its mean and its vertical turbulent flux divergence can be estimated. This has been done for temperature and humidity (e.g., Warner and Telford, 1965; Lenschow et al., 1981). Similarly, Nicholls (1983) was able to measure all the terms in the mean horizontal momentum equations and obtain a reasonable balance. Terms in the budgets of second-order moment quantities can also be estimated, although pressure correlation terms cannot, at present, be measured. Thus, if the second-order moment quantity involves a velocity component, the residual includes a pressure term and the budget cannot be measured completely. However, by normalizing the terms with mixed-layer scaling variables (e.g., Lenschow et al., 1980), one can obtain generalized curves which can then be used to check the consistency of measurements, as well as to provide guidance for estimating the magnitude of the terms in different experimental situations.

By comparison, airplane measurements in a stably stratified boundary layer are more difficult. First, the stable boundary layer is typically much shallower than the convective boundary layer—typically on the order of 100 m compared to 1,000 m for the convective boundary layer. Therefore, airplanes cannot safely resolve the vertical structure in the lowest 10 to 30% of its depth. Second, turbulent velocity fluctuations are usually considerably smaller and more intermittent than in a convective boundary layer because of damping by buoyant forces. Third, horizontal variability is relatively more important. Although aircraft are well suited for measuring horizontal variations, the additional terms are, for the most part, an unwelcome addition to the budgets. Figures 14 and 15 show examples of airplane measurements in a stably stratified marine boundary layer. The vertical sounding illustrates the detail that can be obtained from aircraft soundings—in this case, a narrow wind jet at about 200 m height. The horizontal traverse at 30 m shows an abrupt change in wind speed and in horizontal pressure gradient, which would complicate analysis of budgets.

Aircraft are useful not only in resolving mean and turbulence statistics, but also in probing discrete structures in the boundary layer, such as thermals, longitudinal rolls, mesoscale cells, and thunderstorm updrafts, downbursts, and gust fronts. They have also been used to study flow over irregular

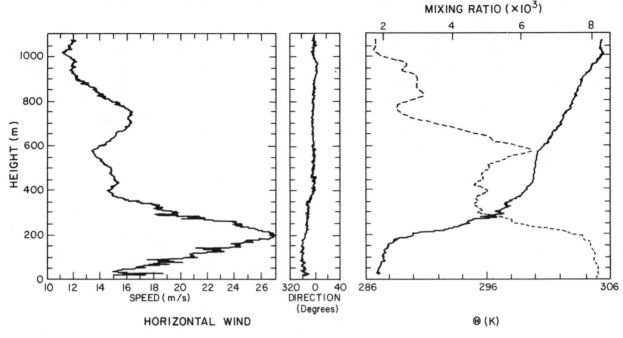

Fig. 14 Vertical soundings of wind, potential temperature (Θ), and water vapor mixing ratio obtained from the NCAR Queen Air off the northern California coast in and above a stably stratified boundary layer. The airplane measurements are able to delineate the detailed structure of the characteristic low-level wind jet and temperature inversion at the top of the boundary layer. (Courtesy of C. Friehe.)

terrain, for investigation of differences between urban and rural boundary layers, and for studies of transport and transformations of pollutants. An example of effects of chemical reactions that can be easily seen from aircraft measurements is shown in Fig. 16.

Aircraft will continue to play a pivotal role in boundary-layer research. Improvements in navigational equipment, direct sensing instrumentation for thermodynamic and dynamic measurements in clouds, and sensors for trace constituent measurements are under development. Remote-sensing capabilities such as Doppler lidar and radar for air motion measurements, and lidar for sensing trace gases and aerosols may eventually lead to revolutionary changes in the application of aircraft to boundary-layer studies.

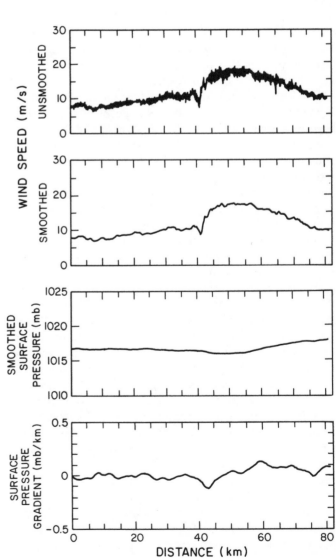

Fig. 15 Wind speed (unsmoothed and smoothed), smoothed surface pressure, and pressure gradient obtained from the same flight as in Fig. 14 at 30 m above the ocean heading upwind parallel to the coast. The pressure is obtained from the difference between the aircraft static pressure and geometric altitude (measured with a radar altimeter). The larger-scale features in the wind speed appear to be correlated to the observed pressure gradient. (Courtesy of C. Friehe; the observations are discussed in Friehe and Winant, 1982.)

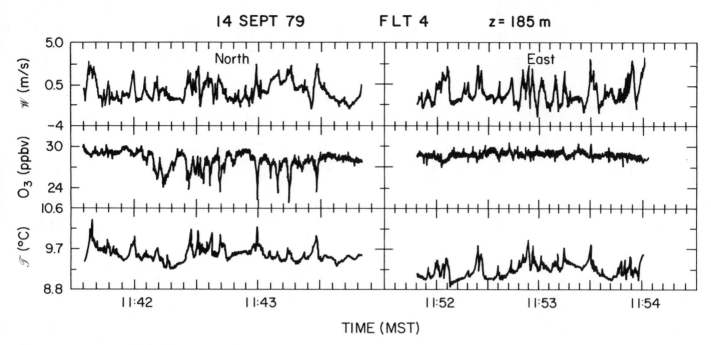

14 SEPT 79   FLT 4   z = 185 m

Fig. 16 Time series of vertical velocity, ozone, and temperature over eastern Colorado, showing a typical convective boundary layer. The convective elements or thermals are evident as temperature pulses a few tenths of a degree high and several seconds long (equivalent to a few hundred meters in length) which are well correlated with vertical velocity. The north leg is over an interstate highway. Vehicles in the highway emit nitric oxide, which is oxidized by ozone. Thus, thermals in the vicinity of the highway are deficient in ozone, relative to measurements away from the highway. (From Lenschow et al., 1981.)

Acknowledgments. I appreciate the efforts of Paul Spyers-Duran in providing material that has been used in several of the figures, and of all the personnel at the NCAR Research Aviation Facility who, over the years, have diligently worked to improve the capabilities of research aircraft. I thank Shirley Michaels for her patience and perseverance in typing the manuscript and Janell Petersen for carefully preparing the final revised version. My thanks also go to the reviewers, who have made many useful suggestions.

Aerosol Measurements in the Boundary Layer

R.F. Pueschel, Air Resources Laboratory, NOAA

1 INTRODUCTION

Aerosols, by definition, are suspensions of solid or liquid particles in air. The size range of such particles is immense: from molecules or molecular clusters on the order of 10^{-4} μm to raindrops and large dust particles of several millimeters. Aerosols also encompass a broad range of inhomogeneities in chemical composition, density, and shape. Large liquid particles are always spherical, since the surface tension is much larger than frictional forces. Small liquid and solid particles have a variety of shapes such as spheres, regular polyhedrons, straight and curved fibers, flakes, and irregular shapes. For spheres, the radius or diameter can be used to characterize the size unambiguously. For nonspherical particles, equivalent diameters must be used. If all the particles in an aerosol system have the same shape (such as cubic), the dimensions of the particles can be an accurate way to describe their size. However, when the particles vary markedly in shape, which is usually the case, a definition of their size is more difficult. The most important parameters in describing the sizes of irregularly shaped particles are the linear extension value and particle area. The linear extension value is defined as the radius of a circle equal in area to the two-dimensional projection of the particle (Giever, 1968). Figure 1 illustrates the range of equivalent sizes, typical dispersions, and methods of analysis of atmospheric aerosols. Some of the more important terms used in this discussion are defined in the Appendix to this article.

The physical and chemical complexity of aerosols constitutes the main measurement problem. Yet, the effects of

Fig. 1 Characteristics and examples of atmospheric particles.

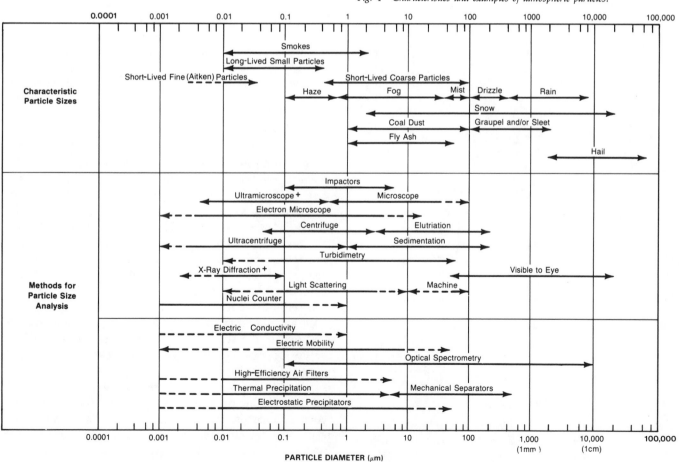

atmospheric particulates on health, environmental quality, and climate cannot be understood without adequate knowledge of the chemical and physical nature of the aerosol. Unfortunately, no single method of analysis or measurement is available today that will cover more than a small range of particles with regard to both their physical and chemical properties. For understanding aerosol properties, both physical and chemical analyses must be employed.

The amount of aerosol available for analysis usually lies in the micro- to milligram range of samples, or in the pico- to nanograms-per-cubic-meter range of suspensions. This means that only micromethods are adequately suitable. If one considers that important elements or compounds are present in the aerosol in pico-, nano-, or, at most, microgram quantities out of a few milligrams of sample, sensitivity and detectability become important requirements of the analytic technique. Because different elements to be determined can be present in a great variety of concentrations, there is a requirement for nondestructive, multielement analytic methods that can cope with a concentration range of about eight orders of magnitude, in conjunction with a particle size range of about seven orders of magnitude.

Problems associated with aerosol sampling include: (1) uncertainties in the collection efficiency of a particular device for particles of a given size and (2) changes in the physical and chemical properties of the particles (i.e., evaporation and chemical reactions) prior to analysis. An effort to overcome these problems led to the development of several physical in situ measurement methods, such as condensation nucleus counters, electric mobility analyzers, nephelometers, and optical aerosol spectrometers. However, no in situ analytic method exists that is satisfactory for most chemical analysis requirements.

In meteorology, particularly in cloud physics and atmospheric optics, the importance of particles of sizes larger than 0.1 μm radius was recognized long ago. More recently, a connection between the chemical makeup of aerosols and their effects on radiative transfer and on the formation and evolution of clouds became apparent. For example, the water solubility of an aerosol of a particular size determines its efficiency for nucleating cloud droplets; the similarity of a mineral particle's structure to that of ice determines the role it plays in the ice-forming process; the chemical makeup of an aerosol determines its refractive index, and thereby the amount of radiation it intercepts, scatters, and absorbs, which is important for visibility as well as the long-term climatic effects of aerosols.

Recent experimental investigations (Whitby, 1978; Pueschel and Mamane, 1979) and theoretical considerations (Friedlander, 1978) have shown that condensation nuclei (Aitken, 1923; Landsberg, 1938) are one of the three major categories of aerosol size distribution. Figure 2 shows an aerosol size distribution that was measured with a condensation nucleus counter in rural Utah on 18 May 1978 at 2,000 m altitude. We can see that the most abundant particle size concentration comprises condensation nuclei of about 0.01 μm in radius. Concentrations of condensation nuclei may vary considerably, however, both spatially and temporally, owing to the mechanism of formation of fine particles and their short residence time.

Condensation nuclei, usually the most numerous class of particles in the atmosphere, are usually formed by the condensation of gases. Condensation, as used here, may involve chemical reactions between the gas molecules. Examples of gases that can condense to form condensation nuclei are hydrocarbons, sulfur dioxide, and oxides of nitrogen produced during combustion by automobiles, steel plants, electric power generating plants, incinerators, forest fires, and volcanoes; and cyclic olefins emitted by vegetation. The smallest of such particles are molecular clusters 0.001–0.005 μm in diameter. Particles 0.005 μm in diameter are assumed to have reached a stable size (Schaefer, 1976).

Of great importance for the formation of clouds and precipitation in the atmosphere is a small fraction of the total aerosol population, collectively called cloud condensation nuclei (CCN) and ice nuclei (IN). Clouds are formed when air is cooled to saturation (100% relative humidity) either by radiative cooling or by adiabatic cooling through expansion when air ascends. For drops to form at relative humidities near 100% the air must contain solid or liquid particles that can serve as CCN. Without such particles, drops will not

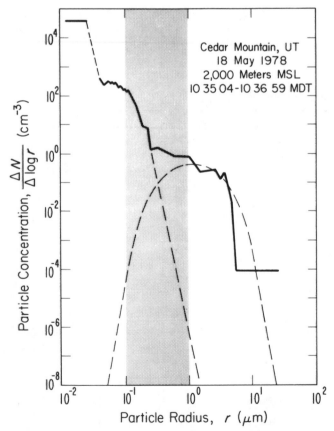

Fig. 2 Typical atmospheric aerosol size distribution showing three characteristic modes, commonly called the nucleation (smallest-particle), accumulation (small-particle), and coarse-particle modes.

begin to form until the relative humidity rises to about 400 % (300 % supersaturation).

The theory of condensation nucleation by soluble (hydrophilic) aerosols is well developed and can be found in cloud-physics textbooks (e.g., Fletcher, 1962; Byers, 1965; Mason, 1971). The humidity required for equilibrium between a solution droplet and the air can be expressed as

$$\ln S = M_w[2\sigma/(\varrho RTr)] - (n\phi f/1{,}000) \qquad (1)$$

where S is the saturation ratio (relative humidity expressed as a fraction) at equilibrium, M_w is the molecular weight of water, σ is the surface tension of the droplet relative to air (Gibbs free energy per unit area of drop surface), ϱ is the drop density, R is the universal gas constant, T is the Kelvin temperature, r is the drop radius, n is the number of ions formed when the solute is fully dissociated, ϕ is the osmotic coefficient (function of chemical species and concentration of solution), and f is the molality of the solution. It can be seen that S involves a balance between two effects: decreasing the radius increases the saturation because of the effect of radius curvature on the free energy; increasing the concentration of ions lowers the equilibrium vapor pressure in accordance with Raoult's law. The second term on the right side of Eq. (1) often exceeds the first, so that droplets of highly concentrated solution can form and equilibrate at relative humidities less than 100 %. This explains the frequent formation of clouds at subsaturation. Twomey (1954) gives the deliquescence (the equilibrium relative humidity for a saturated solution) for some common salts.

Other factors being equal, the larger the aerosol particle, the lower its critical supersaturation. However, the larger the particle, the faster it will settle out of the atmosphere. As a result, there are never more than a few aerosols per cubic centimeter with diameters larger than 1 μm. Cloud droplet concentrations, on the other hand, are typically from several hundred to several thousand per cubic centimeter in continental clouds. The CCN for the majority of the cloud drops are, therefore, of submicrometer size. The smallest-size CCN is ~ 0.01 μm; particles smaller than this require supersaturations of several percent or even tens of percents to act as CCN. This rarely happens in the atmosphere.

In middle and high latitudes, clouds often are deep enough that their tops are colder than 0 °C. These clouds consist mainly of supercooled drops. These drops can freeze only if they contain, or come in contact with, an effective ice nucleus. Depending on which freezing process the IN initiate, they are termed freezing or contact nuclei. In addition, IN can act as deposition nuclei when ice is deposited directly from the vapor phase. The effectiveness of an IN is expressed in terms of the temperature at which it causes a drop to freeze, or at which it nucleates an ice crystal directly from the vapor phase. The warmer the nucleation threshold, the greater the effectiveness. Natural IN have activation temperatures ranging from -1 °C to -30 °C or lower, but very few natural airborne IN are effective at temperatures warmer than -10 °C.

Aerosols, alone or in combination with the clouds they help to form, are responsible for a variety of optical atmospheric phenomena. Among those are rainbows and halos around the sun and the moon. In a more subtle way, aerosols interact with solar and terrestrial radiation and thus become part of the earth's climate system. The nature of a particle's optical signature depends on its diameter relative to the wavelength of the incident light, and on its refractive index at that wavelength. The general theory of the interaction of a plane wave with a homogeneous spherical particle was first elucidated (albeit in rather rough mathematical form) by Mie (1908), and in more sophisticated form by Debye (1909). We will not pursue this general theory further in this article. Modern presentations are available in Van de Hulst (1957), Deirmendjian (1969), and Kerker (1969).

The refractive index m for an arbitrary homogeneous amorphous particle is a complex number

$$m = n - ik \qquad (2)$$

The real part n is the ratio of the speed of light in air to that inside the particle; it is always greater than unity for solids or liquids. The imaginary part is a measure of absorption; it is related to the bulk absorption coefficient a (fraction absorbed per unit of distance inside the particle) by

$$k = a\lambda/4\pi \qquad (3)$$

(Van de Hulst, 1957, p. 267). For passive materials, $k \geq 0$; for an active material such as the plasma in a gas laser, $k < 0$. For most natural and anthropogenic solid aerosols, n varies from 1.5 to 1.8 in the visible range and k ranges from near zero to 0.05 or so. An exception is carbon, for which n is ~ 1.96 and k is ~ 0.66. A complication does arise for crystalline materials; in this case the refractive index is a tensor with complex components and the effective values of n and k depend on the ray path relative to the crystal axes. We will, however, not be concerned further with this complication.

We now consider the nature of the particle-light interaction as a function of the size parameter

$$x = 2\pi r/\lambda \qquad (4)$$

the ratio of particle circumference to wavelength. When x is very large, light is reflected directly from the illuminated face, or emerges after entering the particle and being reflected one or more times, so that part of the incident light is scattered in all directions. If k is not zero, part will also be absorbed and used to warm the particle.

As the size parameter decreases, the tendency for the scattered light to be concentrated in a sharp diffraction ring also decreases; the forward-scattered light occupies an increasingly wider angle with respect to the original ray path. To a backward-looking observer, the image will become increasingly "fuzzy," and the diffraction ring will broaden out into an aureole. A good example of this is the solar or lunar aureole produced by separate water clouds. If the direct beam is blocked out by interposing an absorbing disk of the right size, then the total intensity of the light in the aureole is, on the average, still proportional to the cross section of the particle. Unfortunately, the proportionality constant tends to oscillate somewhat with x; this is one source of error in optical particle spectrometers. These oscillations tend to be most severe for values of x in the range of about 2 to 20.

As x drops below unity, the aureole gradually broadens until it extends all the way to 90° from the directions of propagation. In fact, the backscattered "aureole" seen by an observer who casts no shadow and is looking along the ray path at the rear of the particle is, for small x, identical with that seen by the backward-looking observer. This is the Rayleigh-scattering domain; in the case of the clear atmosphere the "aureole" has become the familiar blue sky. In this domain, the scattered intensity becomes proportional to particle volume rather than cross section, so the scattered light becomes very weak for small particles. This sets a practical lower limit to particle sizes that can be measured with optical spectrometers.

Extinction of light in the atmosphere by aerosol particles degrades visibility and increases turbidity. Extinction is caused by both scattering σ_s and absorption σ_a, and the total extinction coefficient $\sigma_e = \sigma_s + \sigma_a$. If I/I_0 is the transmission of light in a homogeneous atmospheric path of length l, then according to the Beer-Lambert law

$$I/I_0 = \exp(-\sigma_e l) \tag{5}$$

The scattering and absorption coefficients are themselves sums:

$$\sigma_s = \sigma_{sg} + \sigma_{sp} \tag{6}$$

and

$$\sigma_a = \sigma_{ag} + \sigma_{ap} \tag{7}$$

where σ_{sg} is a term due to Rayleigh scattering by gases, σ_{sp} is due to scattering by particles, σ_{ag} is due to absorption by gases, and σ_{ap} is due to absorption by particles.

It is often assumed that $\sigma_{sp} \gg \sigma_{ap}$; however, this does not apply in air with high concentrations of carbon as soot. In the boundary layer remote from cities or in the upper troposphere, the measurement of σ_{sp} may suffice for estimating the aerosol contribution to σ_e. In this case, it is useful to have measurements of σ_s and, more specifically, of the particle-scattering extinction σ_{sp}.

Throughout this article, we will be discussing techniques for measuring aerosols which will allow us to relate the aerosols to their sources. Aerosol concentrations are a result of sources and sinks. One of the most important sinks for aerosols in the context of the boundary layer is deposition at the surface. The transfer of particles from the atmosphere to an underlying surface is usually parameterized in terms of a

deposition velocity v_d. When no horizontal gradient exists, the vertical flux F_a, taken to be positive when downward, can be approximated by

$$F_a(z) = K_a(z) \, [\partial C(z)/\partial z] + v_g(z) \, C(z) = v_d(z) \, C(z) \tag{8}$$

where K_a is the eddy diffusivity for particles (in meters squared per second), C is the particle concentration (in micrograms per cubic meter or number per cubic centimeter), and $v_g(z) \, C(z)$ is a term accounting for nondiffusional sedimentation effects. Within the surface layer at several meters above the surface, $v_d(z)$ is nearly equal to the settling velocity v_g for particles of diameters exceeding several micrometers. As the particle diameter becomes less than 5–10 μm, v_g contributes little to particle deposition in comparison with turbulent fluctuations, so that

$$v_g(z) \, C(z) \ll K_a(z) \, [\partial C(z)/\partial z] \tag{9}$$

and

$$F_a(z) = K_a(z) \, [\partial C(z)/\partial z] = v_d(z) \, C(z) \tag{10}$$

Empirical values of v_d from laboratory and field experiments are shown in Fig. 3 (Sievering, 1981).

2 AEROSOL SAMPLING

Aerosol sampling involves the deposition of a sufficient quantity of the particles on a suitable substrate. If correct

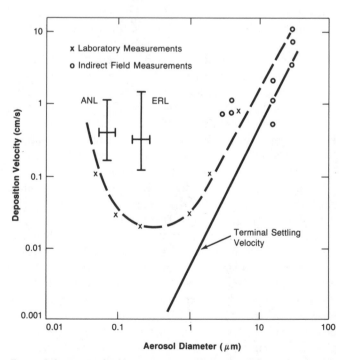

Fig. 3   Laboratory and field measurements of deposition velocities (v_d) of aerosols over land surfaces with vegetative cover and roughness length $1 < z_0 \lesssim 20$ cm. (From Sievering, 1980.)

results are to be obtained, care must be taken to minimize the differences in concentration and particle size distribution between the ambient aerosol and the sample. Such differences are partly due to phenomena taking place at the inlet to the instrument. For example, if the sampling tube is at an angle to the flow direction, larger particles, because of their inertia, will either be deposited on the inside wall of the tube or else fail to enter it. If the sampling tube is parallel to the flow but the velocity of the air inside it is smaller than that in the main flow surrounding the tube, some particles from the streamlines just outside the tube will not enter it; if the velocity in the tube is greater than that in the main flow, particles from the streamlines directed inside the tube, but passing outside it, will enter the tube. Isokinetic sampling involves a thin-walled sampling tube having its axes parallel to the airflow, and the mean gas velocity inside the tube equal to the wind velocity. Calculations of the errors, as a function of particle size, that are encountered in nonisokinetic sampling are given by Belyaev and Levin (1974), Zenker (1971), Ruping (1968), and Fuchs (1964).

Other problems in sampling occur at zero wind speeds (Zebel, 1977) and variable wind speeds (Belyaev and Levin, 1974; Zenker, 1971; Ruping, 1968). For measurements to be representative over large geographical areas, locations and times of sampling must be selected by statistical methods (Yates, 1960; Cochran, 1965).

After the aerosol has been collected inside a sampling probe, the particles are prone to wall losses by precipitation in laminar flows, and diffusional and turbulent depositions. Fuchs (1964, 1975) presents theoretical treatments of each of those cases.

For analysis purposes it is necessary to obtain a high concentration in a reasonable amount of time. The efficiency of any sampling device depends critically on the particle size. The various methods, in order of the efficiency with which they sample aerosols of a similar size range, are:

- Impaction
- Filtration
- Electrostatic precipitation and electric mobility analysis
- Centrifugation
- Thermal precipitation
- Sedimentation.

Generally, the shorter the sampling time, the more representative the sample. At long sampling times, the collected particles can undergo changes; evaporation and chemical reactions may occur. Also, the probability that the state of the atmosphere (temperature, pressure, relative humidity) changes is greater when the sampling time is longer. Descriptions of actual sampler designs can be found in the *Air Sampling Manual* (ACGITH, 1977) and in the *Handbook of Aerosol Technology* (Sanders, 1979). In the following sections, we describe these methods in more detail.

2.1 Sedimentation

The sedimentation method of sampling is to let particles settle onto a horizontal tray. This method, however, discriminates against small particles with relatively small settling velocities; the falling speed is proportional to r^2, neglecting corrections to Stoke's law. Unless precautions are taken, no controlled size classifications are possible with such samplers.

Cumulative size distributions can be achieved if a laminar flow of an aerosol through a horizontal channel is maintained (Fig. 4). Particles of a given size form deposits from the beginning of the channel to a "leading edge." Only particles smaller than that size will pass this leading edge and be deposited farther downstream. Calibrations are possible by determining the aerodynamic diameter as a function of the deposition length and of the area that is covered by particles of a given size. If this area is not covered uniformly, a density function has to be applied as correction. The product of number, or mass density, of deposited particles times the area covered by a given size results in a cumulative number, or mass distribution. The frequency distribution is obtained by differentiation.

Differential particle size separation is possible if particle-free air and a small amount of aerosol are introduced through a slit nozzle at the top of a horizontal sedimentation channel (Fig. 5). Particles of a given size have a defined deposition length. Calibration of this deposition length as a function of the aerodynamic diameter of the deposited particles results in the number frequency distribution.

2.2 Centrifugation

For particles with aerodynamic diameters smaller than approximately 1 μm, the mean displacement by Brownian motion is approximately equal to the displacement by sedimentation. Size separation of submicrometer particles requires the

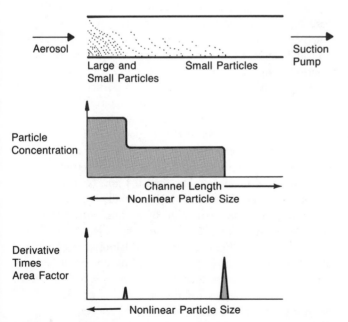

Fig. 4 *Horizontal channel forming a cumulative size spectrum. The aerosol contains particles of two sizes.*

application of forces that exceed gravity. This is the case in centrifuges that operate at speeds up to 24,000 rpm at flow rates of several liters per minute. The Goetz aerosol spectrometer (Goetz et al., 1960; Gerber, 1971) deposits particles as small as 0.1 μm in diameter. This instrument provides a cumulative size distribution; the number or mass frequency distribution is obtained by differentiation, which may result in inaccuracies if polydisperse aerosols are being sampled.

Several centrifuges providing differential particle size separations have been described in the literature (Hochrainer and Brown, 1969; Hochrainer, 1971; Abed-Navandi et al.,

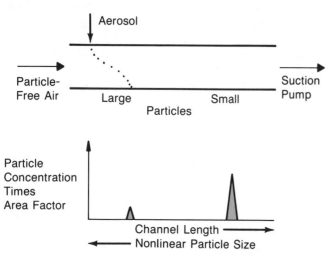

Fig. 5   *Horizontal channel forming a discrete size spectrum. The aerosol contains particles of two sizes.*

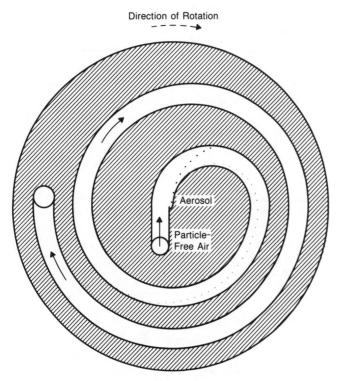

Fig. 6   *Schematic cross section of the spiral centrifuge (Stöber rotor).*

1976; Stöber and Flachsbart, 1969). The widely used Stöber rotor centrifuge consists of a channel constructed as a spiral around the axis of a rotor (Fig. 6). Particle-free air is injected into the channel at a flow rate of several liters per minute. Aerosol-laden air is added through a slit nozzle to the particle-free air at a rate of 0.1–1.0 ℓ/min. The particles that can be analyzed with this device range from 0.1 to several micrometers in radius. For particles less than 0.1 μm, the centrifugal force is too small to deposit the particles on the outer wall of a channel of practical length. For particles whose radii exceed several micrometers, inlet losses become significant.

2.3 Impaction

Impaction or impingement, also known as inertial impaction, is a widely used technique for collecting airborne particles. It is based on the principle that when a moving air parcel containing an aerosol changes direction the particles tend, because of their inertia, to continue in the same direction for some distance. At a given velocity and particle density, this distance depends on particle size.

The impaction can be produced in several ways. For example, the object on which the particles are to be collected may be moved rapidly through the aerosol by mounting it on the outside of an airplane, or whirling it with a motor. The more common arrangement, however, is to force the aerosol to flow at a very high velocity through an orifice and allow the air to impinge on a flat surface where some or all of the particles above a given mass are collected.

For size separation, cascade impactors (Fig. 7) are used, in which several impactor stages are joined in series. The size of the nozzles, determining the speed of the air at given flow rates, decreases in subsequent stages, from dimensions of several millimeters at the initial stages to tenths of millimeters at the final stages. Each collecting stage has a sigmoidal characteristic: small particles can pass with almost no deposition, a certain size range is collected with a 50% probability, and significantly larger particles are completely collected. Superimposing the deposition probability for all the collecting stages determines, for each stage, a particle size interval for which the particles are deposited. This is documented in Fig. 8 (May, 1945) for a four-stage impactor.

Impactors can be used to collect particles of sizes from a few tenths of a micrometer to several tens of micrometers. Low-pressure impactors have been developed recently to collect particles of diameters less than 0.1 μm. Neutron activation analysis (see Section 3.2) is normally carried out on substrates placed on each stage of an impactor. Most low-flow-rate impactors of 1–2 m³/h capacity collect only 2–5 mg of atmospheric particulate matter, distributed over several collection stages, during a 24-h sampling period of air with 100 μg/m³ atmospheric loading. It thus becomes apparent that analysis techniques for the study of trace elements of concentrations on the order of 10 ng/m³ are required.

2.4 Filtration

Filtration techniques are most convenient where large-particle concentrations need to be collected, and when size separation is not required. The mechanisms of impaction, diffusion, interception, and electrical forces all act in combination to determine the filter's collection efficiency. The theory of filtration by fiber and Millipore membrane filters has been reviewed by Chen (1955), Dorman (1960, 1966), Pich (1966), and Davies (1973). Spurný et al. (1969) investigated the theory of aerosol filtration by Nuclepore filters. The combination of several forces acting on the particles results in a collection efficiency that depends strongly on particle size (Fig. 9). Large particles are deposited by impaction, small ones by diffusion. Electrical forces usually increase filter collection efficiency. Between the ranges of deposition by interception, impaction, and diffusion is the so-called filter minimum. Manton (1978, 1979) contributed significantly to filtration theory by considering the effects of the inertia of the particles to impaction and of Brownian diffusion to interception.

The classic membrane filter consists of a membrane of cellulose or similar material with many irregular pores. It is difficult but not impossible to control the mean pore size during the construction process; in practice, the actual pore size may deviate from the mean or nominal pore size. A better-defined pore size is found in Nuclepore filters, consisting of a membrane of polycarbonate into which parallel holes of uniform size are etched after radioactive irradiation. Particles exceeding a given size are usually deposited on the surface of a

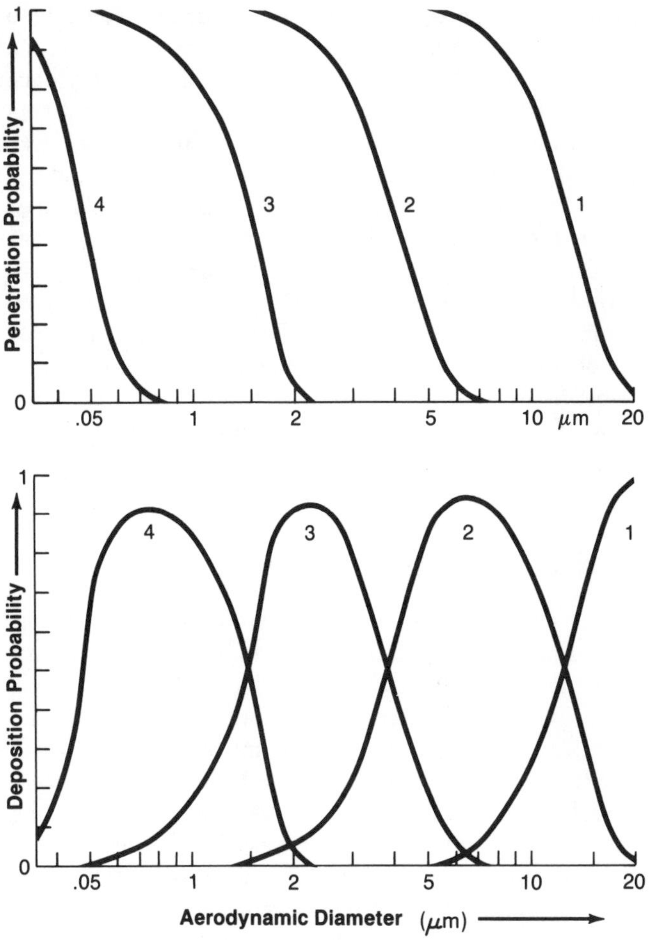

Fig. 8 Penetration and deposition probabilities of a four-stage impactor. (From May, 1945.)

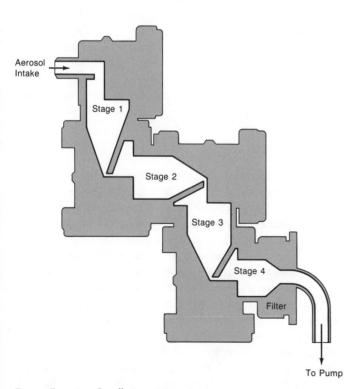

Fig. 7 Four-stage Cassella impactor.

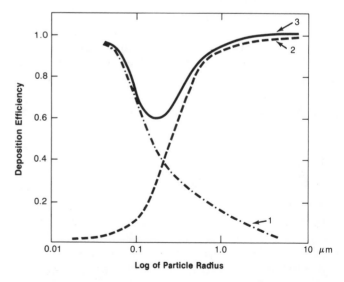

Fig. 9 Filter efficiencies: (1) diffusion, (2) impaction, and (3) total. (From Spurný and Pich, 1965.)

filter of a given pore size, rendering the particles accessible to microscopic observation.

Dams et al. (1972) concluded that cellulose fiber filters, such as Whatman 41, are the best choice for subsequent neutron activation analysis of the sample, because of their efficiency at capturing particles, their ease of handling, and the low concentration of elements in the blank filter. Another type of filter used for this purpose is the Nuclepore filter (Stafford and Ettinger, 1972); however, Liu and Lee (1976) concluded that the membrane filter Fluoropore was considerably more efficient in the collection of submicrometer aerosols than a Nuclepore filter of comparable pore size.

Fiber filters consist of a fine fiber, typically cellulose, cloth, cotton, glass, or asbestos. With the right combination of fiber diameter and air velocities, collection efficiencies of 99.99 % for any particle size are possible.

2.5 Electrostatic Precipitation

Electrostatic precipitators use a corona discharge to produce a charge on the particles to be sampled, which in turn migrate to an electrode of opposite charge. The typical electrostatic precipitator consists of a cathode producing the charge, usually a wire or metallic point; an anode on which the particles are collected, usually a metallic cylinder concentric with the wire cathode, or a metal plate adjacent to the wire cathode; and some device for passing the aerosols between the cathode and the anode. A voltage of 5–10 kV is required to produce the corona discharge. The collecting surface can have attached to it substrate specimens (e.g., foils or electron microscope grids) for subsequent particle analysis. The theory of electrostatic precipitation has been reviewed by Rose and Wood (1956) and by White (1963).

Size classification of particles, using their size-dependent mobility, is possible by separating the charging state of the electrostatic precipitator from its collecting state. This is the principle of the electric mobility analyzer, which is described in Section 5 on in situ sampling. A separation of differently shaped particles (viz., spheres versus fibers) through charging has been documented (Hochrainer, personal communication), using the combined differences in surface area, hence charging rates, and drag coefficients of particles of different shapes.

2.6 Thermal Precipitation

A temperature gradient across a particle exerts a force on the particle in the direction of decreasing temperature. Thermal precipitators operate by maintaining thermal gradients between a cold plate and a heated wire (Watson, 1936), heated ribbon (Walkenhorst, 1962), or heated plate (Kethley et al., 1952). The cold plate, on which the particles deposit, can carry suitable substrates (e.g., microscope specimen holders) for subsequent particle analysis. The temperature differences are usually on the order of 100 °C, and the surfaces are placed as closely together as is practical, to maxi-

mize the thermal gradient for a given temperature difference between the plates.

3 MEASUREMENT OF COLLECTED PARTICLES

3.1 Single-Particle Analysis

Size, shape, and chemical/elemental composition of individual airborne particles are of interest since they determine the atmospheric effects of an aerosol. The nucleation properties depend on size and on the water solubility of the particles; the ratio between scattered and absorbed light depends on particle size and chemical composition; information on particle shape and chemical composition sometimes permits the identification of sources, which is important for environmental modification studies; and the chemical composition of aerosols has to be known to determine the fate of atmospheric aerosols. Shape and morphology are closely related to material properties, which in turn are characteristic for a substance or group of substances, e.g., fibrous materials (of organic or inorganic compounds), and amorphous and crystalline substances, which usually display distinct differences in particle shape. Particle shape may also suggest the crystalline structure, which is an important clue in identification procedures.

There are, however, serious limitations associated with single-particle analysis: (1) to gain information that is representative of the atmospheric suspension it is usually necessary to investigate a rather large number of particles, which can be very time-consuming; (2) only solid substances with a sufficiently low vapor pressure can be investigated in devices with specimen chambers under vacuum and/or elevated temperature. Therefore, information on liquid or gaseous adsorption products is rarely obtainable. Table 1 lists the common methods for the characterization of individual airborne particles, the output signal, and the analytical and size resolution that is obtainable.

3.1.1 Light Microscopy

Light microscopy has been extensively described (McCrone and Delly, 1973) as a technique for determining the sizes and shapes of individual airborne atmospheric particles. Chemical identification is based on certain properties that can be observed in a light microscope which are indicative of a specific chemical composition. The most important of these properties are transparency, opacity for visible light, color, refractive index, birefringence, size, shape, and morphology. Such properties can be determined for particles as small as about 0.2 μm in radius, corresponding to a particle mass of about 10^{-9} g.

Transparency, opacity, and the colors observed with transmitted or reflected illumination yield information about the particle's identity, since many substances possess a

TABLE 1
Survey of Analytical Methods for the Characterization of Individual Airborne Particles

Analytical Method	Reagent	Signal	Analytical Information	Relative Sensitivity	Lower Limit of Particle Diameter
Light microscopy	Light	Reflected, transmitted light	Types of compounds (species, structure); size, shape, morphology	Only pure species can be identified	$\sim 0.5\ \mu m$
Electron probe microanalysis	Electrons	X-ray spectrum	Type of elements and their concentration; number of particles of a specific composition	0.X%	$\sim 0.1\ \mu m$
		Secondary electrons	Shape, size, morphology; number of particles of a specific composition		$\sim 100\ \text{Å}$
		Backscattering and Auger electrons	Shape, size, morphology; number of particles of a specific composition		$\sim 0.1\ \mu m$
Ion probe microanalysis	Ions (O^{+2}, O^-, Ar^+)	Secondary ions	Type of elements and their concentration	ppm	0.X μm
Scanning and transmission electron microscopy, scanning electron microscopy	Electrons	X-ray spectrum	Type of elements and their compounds	Major and minor compounds	$\sim 200\ \text{Å}$
		Secondary electrons	Shape, size, morphology		$\sim 50\ \text{Å}$
		Transmitted electrons	Size, shape		$\sim 10\ \text{Å}$
		Diffracted electrons	Structure and lattice parameters	Pure species	$\sim 200\ \text{Å}$
		Energy spectrum of transmitted electrons	Type of elements	Major components	$\sim 100\ \text{Å}$

distinct color. McCrone and Delly (1973) have published an encyclopedic collection of color photographs of substances that typically can be found in the atmosphere. Their book contains 609 color photomicrographs of the most important dust particle types, morphological descriptions, and ample data dealing with optical, physical, and chemical characteristics. The refractive index and the anisotropy of noncubic crystals provide further identifying information. The refractive index is determined by microscopic immersion methods in which the refractive index of the mounting liquid is varied (McCrone and Delly, 1973, Vol. 1). The anisotropy of particles is studied in polarized light. Another useful technique applied in particle identification is dispersion staining. When a transparent particle, immersed in a liquid having nearly the same refractive index, is viewed under the microscope with transmitted white light, color patterns are produced that can be used for particle identification.

Because of the large number of particles that need to be measured to obtain a statistically significant size distribution, automation of the measurements becomes a necessity. Image analysis (Zeiss, 1976; Leitz, 1976; Riediger, 1972; Levy, 1976) is an electronic area measurement technique with video scanning speed which makes particle size analysis possible in reasonable periods of time. The image scanning method not only allows rapid size classification and determination of the total amount of material (it is assumed that, on the average, the projected diameter can be used to calculate the volume of each particle), but also a differentiation of particles that have a significantly different shape (e.g.,

asbestos fibers). Modern systems allow separate counting and size analyses for particles of nonspherical shape by employing appropriate computer subroutines that identify particles on the basis of length-to-width ratios, or of circumference-to-area ratios (Malissa et al., 1974).

Automated particle imaging systems, however, do not permit a description of a particle's shape that is accurate enough to quantify morphology. Therefore, to describe the complex relationship between the physical and chemical properties of dust particles, supplementary information must be obtained. The most important physical and chemical properties that can be measured for large individual particles are density, melting point, hardness, magnetic susceptibility, infrared (IR) absorption, solubility, and chemical reactivity (McCrone and Delly, 1973, Vol. 3). The determination of these properties involves a set of analytical procedures such as hot- and cold-stage microscopic investigations, application of microdensity gradient tubes, crush tests, and observations of particle movement in a magnetic field. One of the more valuable tools is the registration of an IR absorption spectrum of very large (radius > 30–50 μm) individual particles with the aid of a microscope attachment to an IR spectrometer. Unfortunately, none of these methods is applicable to individual submicrometer particles, which are most abundant in the atmosphere, and which play a significant role in atmospheric optics, physics, and chemistry.

Microchemical reactions can be carried out under the light microscope for the chemical identification of individual particles (Brenneis, 1931; Malissa, 1951; Seeley, 1952; Malissa and Benedett-Pichler, 1958; Mamane and DePena, 1978; Mamane and Pueschel, 1979, 1980a). This method is significant for the following reasons: (1) it is applicable to submicrometer particles; (2) it is specific for anions such as NO_3^-, SO_4^-, and $S_2O_3^-$; (3) it is applicable to substances that are volatile in high vacuum and under electron or ion bombardment.

In summary, light microscopic identification of atmospheric particles is an extremely complex task, because of the mixed nature of the particles and their accumulation in the atmosphere at a size range that is at, or slightly below, the spatial resolution of the technique. A further disadvantage is encountered in morphological studies because of the low depth of field of a light microscope, which causes difficulties in the imaging of surfaces. Therefore, one needs to use analytical techniques that provide (1) a capability of direct chemical identification, (2) a possibility of characterizing particles of size less than 0.1 μm in radius, and (3) high depth-of-field images.

3.1.2 Electron Probe Microanalysis

Electron probe microanalysis characterizes individual airborne particles by analyzing their elemental composition and determining their size, shape, and morphology. Either scanning electron microscopes (SEM) or microprobes can be used for this task. The SEM, equipped with an X-ray energy dispersive spectrometer (XEDS) is advantageous for morphological studies and qualitative and semiquantitative X-ray analysis, since high lateral resolution of imaging is of primary importance. The microprobe is preferably used for the iden-

tification of particles by means of quantitative elemental analysis and X-ray valence band spectroscopy.

The general technique applied for characterizing individual particles involves imaging with the secondary electron signal and subsequent local analysis using either an energy- or a wavelength-dispersive spectrometer (EDS or WDS). The EDS has the advantages of rapid qualitative elemental analysis and multielement capacity. The WDS is preferably used for qualitative and semiquantitative analysis of second-period elements (especially O and C), for highly accurate and sensitive quantitative analysis, and for the recording of X-ray valence band spectra. The general construction of the SEM includes an electron gun, lenses to produce a fine beam of electrons at the specimen, and deflection coils to scan the beam over the specimen's surface (Fig. 10). Since the SEM operates by bombarding the specimen with electrons, X rays that are characteristic of particular elements are generated at the point of impact and can be used to determine the elemental composition of the specimen.

The particle size limit for qualitative analysis depends largely on the geometrical arrangement of the detector, the electron beam density, the substrate used, and excitation conditions. For normal SEMs (not having a variable sample-detector geometry) and metal substrates, the lower size limit for qualitative identification is on the order of 0.2 μm (Schutz, 1978) when low excitation energies (5–10 keV) are used. Maggiore and Rubin (1973), Pueschel (1976), Parungo et al. (1978), and Pueschel and Barrett (1982), using SEMs equipped with field emission guns, high beam-current densities, and variable geometry of the detector, were able to analyze particles as small as 0.05 μm in radius.

Fig. 10 Schematic of an SEM-XEDS particle analyzer.

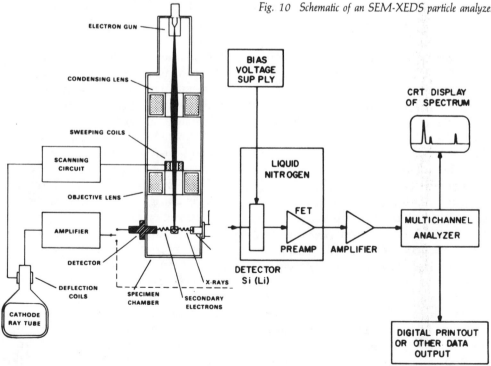

A commonly used approach to identify dust particles is based on a combination of qualitative and semiquantitative elemental analysis, and morphological features, which in combination are often characteristic of a certain chemical species. Shown in Fig. 11 (Pueschel and Barrett, 1982) are examples of the type of information gained on size, shape, and elemental composition of individual particles when they are subjected to the combined techniques of scanning electron microscopy and X-ray energy dispersive analysis (SEM-XEDA). X-ray spectra are shown in Fig. 11a for a fly ash particle collected in the plume of the Four Corners power plant (near Farmington, New Mexico) some 20 km downwind from the stacks. Its spherical shape identifies the particle as resulting from a mechanism that involves a phase change from either gas or liquid to solid. The various X-ray spectra shown in Fig. 11a correspond to different portions of the particle, as indicated. Typically, fly ash from coal combustion contains mainly light elements, e.g., Al, Si, K, Ca, and trace amounts of Fe and Ti.

X-ray spectra are shown in Fig. 11b for a soil-derived particle collected outside the plume at the same geographic location. In contrast to fly ash, this particle contains heavier elements in higher concentrations. It also follows from Fig. 11b that different portions of the same particle contain different elements in varying concentrations, indicating a large degree of chemical inhomogeneity. Figure 11c shows

an ash particle from the combustion of oil. The presence of heavy elements, typically Ni and V, differentiates this fly ash from coal combustion fly ash.

When there are distinct differences between some important anthropogenic particles, such as soot from oil firing (Fig. 11c), the mineral background (Fig. 11b), and coal-related fly ash (Fig. 11a), morphology alone can be used for the determination of the fraction of particles of each kind. This method was employed by Henry and Blosser (1970) to determine the materials and their variations with particle size in samples collected in Washington, Denver, Cincinnati, Chicago, Philadelphia, and St. Louis. Cheng et al. (1976), from morphological studies carried out with the SEM on samples from coal- and oil-fired boilers in power plants, showed that the kinds of information depicted in Fig. 11 are typical: 90% of the collected particles are spherical, but distinct differences in surface structure do exist. Particulates from oil-fired boilers contain S, Si, Ca, V, and Fe as major, and Mg, Al, P, Cr, Mn, and Ni as minor elements. Glassy spheres from coal-fired burners, on the other hand, contain primarily Si, Al, K, Ca, Ti, Fe, and S, and smaller amounts of P, Cl, Mn, and Cu. Other investigations (Pietzner and Schiffers, 1972) showed that the spherical particles from coal firings are mainly silicates. Oil soot particles are, more or less, carbon skeletons containing different elements, such as in Fig. 11c. Schutz (1978) uses the morphology of sea-salt (cubic)

Fig. 11a   Typical shape and elemental composition of a coal-derived fly ash particle. (From Pueschel and Barrett, 1982.)

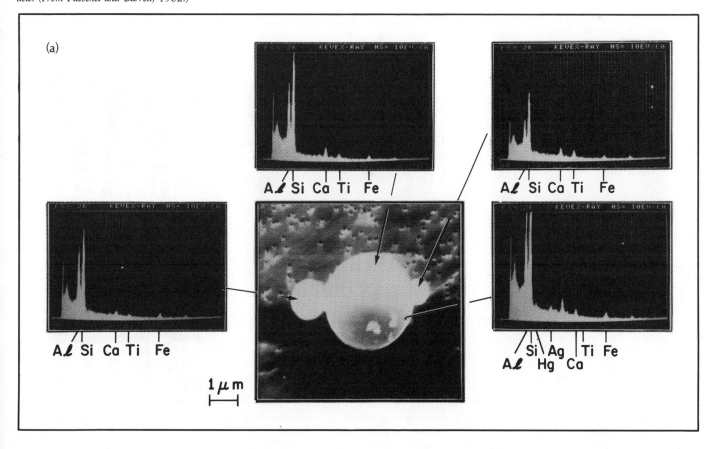

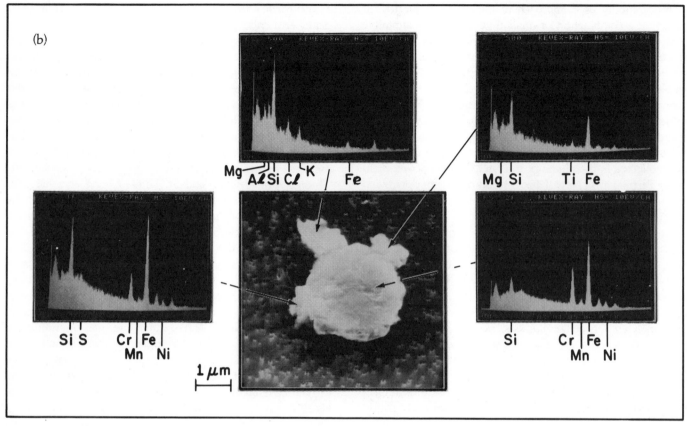

Fig. 11b Typical shape and elemental composition of a soil-derived particle.
(From Pueschel and Barrett, 1982.)

Fig. 11c Typical shape and elemental composition of an oil-derived particle.
(From Pueschel and Barrett, 1982.)

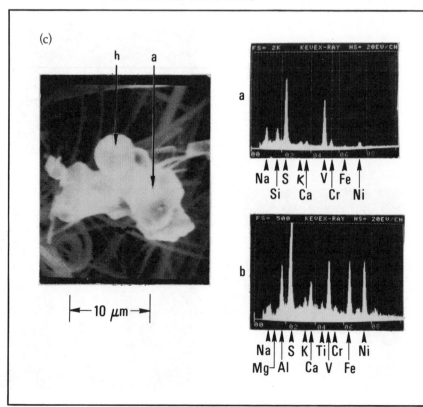

and mineral (not cubic) particles to determine the origin of particles—ocean spray versus Sahara dust—collected at the Cape Verde Islands. He emphasizes that it is not possible to identify the various mineral particles by morphology alone.

A second method of identification of dust particles is a quantitative elemental analysis that determines the compound by stoichiometric calculations from measured elemental concentrations. For particles much larger than the excited volume (volume of X-ray production) this volume can be calculated (Wittry, 1959). Quantitative analysis of their elemental composition is possible by comparing their X-ray spectra with that of a standard. For particles larger than about 10 μm in diameter, accuracy on the order of 5–10% is possible. When particles smaller than a few micrometers in diameter are excited with an electron beam, the X-ray intensity depends not only on the concentration of a specific element, but also on particle size.

The ratio method, however, is the quantitative elemental analysis procedure most widely applied to small particles. Armstrong and Buseck (1975) tested the model on a selection of silicate, oxide, and sulfide particles ranging in size from 0.5 to 20 μm. The authors established an average accuracy of $\pm 5\%$. Since the absolute X-ray intensity decreases with particle diameter, the relative standard deviation increases with decreasing particle size. The minimum particle size for which a quantitative analysis is possible is given by the statistical rule of thumb that the net signal has to be at least twice the standard deviation of the background. Landstrom and Kohler (1969) report that particles as small as 0.1 μm can be quantitatively analyzed.

Although electron probe microanalysis is an important method for the characterization of individual airborne particles of size above a few tenths of a micrometer, its limitations make it necessary to seek alternate methods. Particles, particularly of nonconductive C, N, and S compounds, can be destroyed as a result of heating in an electron beam. Heating can be minimized by using low beam currents. Lowering the beam current, however, also reduces the X-ray intensities, which is one of the limiting factors in the analysis of very small particles.

3.1.3 Ion Probe Microanalysis

Although electron probe microanalysis has a high absolute sensitivity, on the order of 10^{-15} g, its relative sensitivity in the range of 0.1% does not permit the detection of trace elements in individual airborne particles. A technique that has a high relative sensitivity for all elements is ion probe microanalysis because its signal yield is high, very little background signal is generated, and the detection sensitivity of the instruments is high. To make it work it is necessary to sample on substrates with low secondary ion yields and a low trace element level. The lateral resolution is determined by the ion beam diameter of 1–2 μm in modern instruments.

3.1.4 Electron Microscopy

Extremely high detection sensitivities are obtained in combined scanning and transmission electron microscopy (STEM). For study of size, shape, and morphology, the secondary or transmitted electron signal can be used. Imaging in the transmitted electron mode has the advantages of resolution down to approximately 5 Å, high contrast of the image, and of a possibility of observing the inner structure of the particle. The high resolution and good contrast allow correct determination of size and shape, even of particles as small as 0.01 μm in diameter. Furthermore, it is possible to determine correct particle numbers of nonvolatile substances in this size range, since the particles can easily be distinguished from the substrate. The possibility of observing the inner structure is of interest, since such information might yield clues on particle growth mechanisms and particle identity. The additional SEM mode that is available in a combined STEM system also yields information on the three-dimensional appearance of a particle, whereas the transmission electron microscope (TEM) picture is only a two-dimensional silhouette. The TEM method alone, however, has the same spatial resolution as has the STEM method. As an example, Fig. 12a is a TEM micrograph of two fly ash particles of approximately 0.25 μm diameter. The X-ray energy spectrum of one of the particles was obtained in about 100 s of counting time (Fig. 12b). The important elements—Si, Al, S, Ca, and Fe—have sufficient intensity to be evaluated.

Electron diffraction patterns are another valuable aid in identifying airborne particles. Electron diffraction of individual particles is obtainable in practically all TEM and STEM systems that are combined with wavelength-dispersion analyzers. The diffraction patterns obtained are either a regular array of points if the particle is a single crystal, or a few concentric rings if the object is polycrystalline. (For further details see McCrone and Delly, 1973, Vol. 2.)

Identification of some compounds is possible with a technique that combines electron microscopy and microchemical reaction of specific substances. Bigg et al. (1974), Mamane and DePena (1978), and Mamane and Pueschel (1979, 1980a) applied a thin film of barium chloride or of nitron to electron microscopic substrates on which aerosols were collected. Postcollection reactions of sulfates with barium chloride and nitrates with nitron resulted in characteristic patterns which, when viewed under a TEM, led to the identification of ammonium sulfate, acidic sulfate, and nitrate aerosols. Figure 12a shows fly ash particles coated with a thin layer of sulfate, and sulfate particles that formed on an aerosol other than fly ash (Mamane and Pueschel, 1980b). Figure 13 shows examples of ammonium nitrate and sodium nitrate reaction spots of particles collected in oil refinery plumes (Mamane and Pueschel, 1980a). Bigg et al. (1974) also developed chemical procedures for the identification of nitrates, persulfates, and halides. The special importance of using microchemical reactions lies in the fact that some environmentally important ions can be specifically identified.

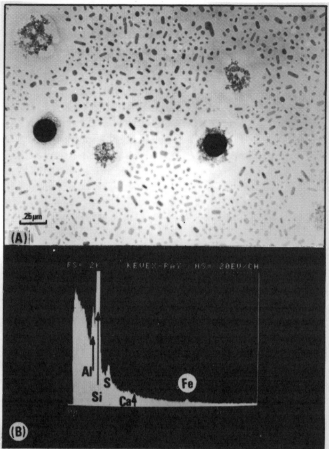

Fig 12 (A) Close-up of photomicrograph of submicrometer fly ash particles coated with a layer of sulfate. (B) X-ray spectrum of the fly ash particle on the right of part A, showing the presence of Al, Si, S, Ca, and Fe. (From Mamane and Pueschel, 1980b.)

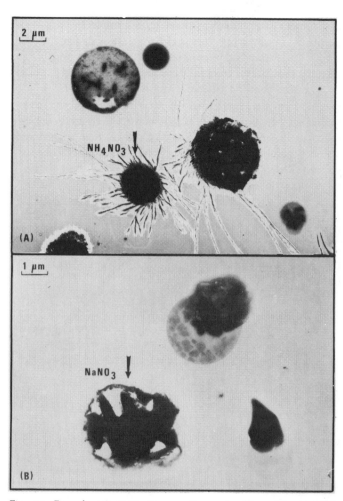

Fig. 13 Examples of nitrate-nitron reaction spots. (From Mamane and Pueschel, 1980b.)

3.2 Bulk Sample Analysis

3.2.1 Neutron Activation Analysis

An analytical technique for determining the concentrations of inorganic elements in a sample, neutron activation analysis (NAA) involves exposure of the sample to a flux of activating species, such as neutrons, charged particles, or photons. NAA, specifically, is the quantitative measurement of radiation produced by neutron-induced reactions. The technique can be nondestructive when the nuclides are detected by their γ-ray emissions. Figure 14 describes the principle of operation. Since more nuclides emit γ rays of characteristic energies than other radiation, γ-ray spectroscopy is the preferred method of detection. Quantitative analysis is also possible by radiochemical separation. The elemental concentrations are determined by comparing the activities of the sample with the activities of standards of known elemental composition.

Figure 15 shows a general approach to carrying out activation analysis. Three factors must be considered: (1) sample activation—type and flux of neutrons, irradiation, cooling,

and counting times; preparation of standards and/or flux monitors; evaluation of interferences; (2) nondestructive or destructive analysis procedures; (3) methods of activity measurements—detection of β- or γ-ray activity; types, characteristics, and calibration of detector; counting schedule.

NAA can play a key role in the study of atmospheric aerosols, because of its capability of detecting a large number of inorganic constituents. As shown in Table 2, NAA can detect windblown dust aerosols by the presence of Al and rare-earth elements, marine aerosols by Na and Cl, and anthropogenic aerosols by V, Cd, Zn, As, etc. Figure 16 shows an example of the many isotopes and elements detectable in a 24-h high-volume air sample, irradiated in a nuclear reactor and counted by a Ge(Li) detector (Ragaini et al., 1976).

Procedures for rapid radiochemical neutron activation analysis (RNAA), techniques for instrumental neutron activation analysis (INAA), and computerization of the analysis calculations have made NAA a multielement analysis technique that is useful, rapid, selective, highly accurate, and sensitive (often to better than 10^{-9} g): ideal characteristics for the nondestructive analysis of atmospheric aerosols. Up to 33 elements have been measured simultaneously. The half-lives of

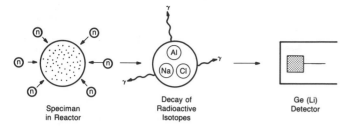

Fig. 14   Schematic diagram of instrumental neutron activation analysis (INAA) procedure.

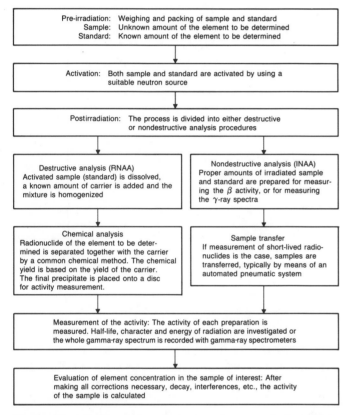

Fig. 15   Schematic diagram of a general process of neutron activation analysis. (From Tölgyessy and Varga, 1974.)

the reaction products range from 2.3 min to 5.3 years. NAA, however, is not applicable to individual particles, and does not have the ability to distinguish the chemical form of the element.

3.2.2 X-Ray Fluorescence

Another emission spectroscopic method for the qualitative identification and quantitative determination of elements in bulk samples is X-ray fluorescence (XRF). The method is generally applicable to elements of atomic number 11 to 92. With few exceptions, X-ray fluorescence provides no information concerning the chemical species to which the element belongs. For all practical purposes, it determines the bulk

elemental composition of a sample, as opposed to those techniques that provide information about the composition of the sample on the microscale, or chemical speciation of the sample.

Atoms in the samples are converted from the ground state to an excited state by an energy source. The excitation source depends on the nature of the sample, the design of the spectrometer, and the goals of the analysis. The source can be an X-ray tube, a radioactive isotope emitting X rays, an electron beam as in an SEM (Section 3.1) or a beam of accelerated charged particles, such as protons or alpha particles. The energy content of this radiation must be sufficient to promote an electron of an inner orbital, usually the K- or L-shell of an atom, to a higher level. The atom can even become ionized. The cascade transfer of outer-shell electrons to fill the inner-shell vacancy releases energy that manifests itself as a series of emission spectroscopic lines in the X-ray region of the electromagnetic spectrum. The energies or wavelengths of those lines are characteristic for the element emitting the radiation, and their intensities are related to the concentration, or amounts, of a specific element.

X-ray fluorescence is a technique that offers many advantages for the quantitative determination of the average elemental composition of dust. When the samples are collected on filters, detection limits range from several hundred nanograms per square centimeter for elements such as sodium to a few nanograms per square centimeter for the transition elements. From the volume of air drawn through the filter, the amount of an element per unit volume of air can be calculated.

The older, more conventional form of XRF spectrometers uses wavelength-dispersive X-ray fluorescence (WDXRF). These instruments disperse angularly the radiation emitted from the sample by means of an analyzing crystal. Each excited element in the sample emits a series of lines of X rays, the wavelength of which is characteristic of the element. Radiation from the sample is directed onto a crystal which diffracts the radiation and resolves the wavelengths.

The second type of XRF spectrometer uses energy-dispersive X-ray fluorescence (EDXRF). Its design eliminates the need for an analyzing crystal by use of a solid-state Si(Li) detector which has a greater resolution than the previously used gas or scintillation detectors. The pulse output from these detectors is proportional to the energy of the X-ray photon that is detected. The resolution of these pulses is sufficient to discriminate the X-ray photons of elements adjacent in the periodic table. A multichannel analyzer is used to acquire the spectra of all elements in the sample simultaneously.

Both types of spectrometers use either conventional (Coolidge-type) or transmission X-ray tubes as a source of radiation. The use of charged particles (alpha particles or protons) has found increasing interest as a mode of inducing X-ray emission. These particles are usually in the 2- to 4-MeV energy range, thus requiring an expensive source, such as a Van de Graaff accelerator. Although these accelerators are

TABLE 2
Key Chemical Constituents of Atmospheric Aerosols

Element of Molecular Constituent	Possible Origins
C (Total and organics)	Primary—natural and anthropogenic sources; secondary—photochemistry (?)
N (NO_3^-, NH_4^+, amino, and pyridino)	Sources—oxidation of NO_x, HN_3, fuel additives
Na[1]	Mainly sea salt
Al[1]	Mainly soil, some possible anthropogenic
Si	Mainly soil
S (SO_4^-, SO_3^-, S$\cdots\cdots$)	Mainly secondary production from SO_2 oxidation
Cl[1]	Mainly sea salt, but some anthropogenic
K[1]	Mainly natural (?)
Ca[1]	Cement production
Ti[1]	Anthropogenic
V[1]	Power plant, fuel oil
Cr[1]	Anthropogenic
Mn[1]	Anthropogenic
Fe[1]	Anthropogenic and natural
Ni[1]	Anthropogenic
Cu[1]	Anthropogenic
Zn[1]	Tire dust, smelting, fuel additives
As[1]	Combustion, metal production and processing
Se[1]	Combustion
Br[1]	Auto exhaust
Cd[1]	Metal production and processing
I[1]	Sea salt and (?)
Ba[1]	Diesel exhaust and lubrication oil atomization
Pb	Auto exhaust, industrial processing
H_2O	Liquid water content is a potentially important inert ingredient in visibility question.
Re[1] (Rare earths)	Mainly soil

[1]Elements detected by neutron activation analysis.

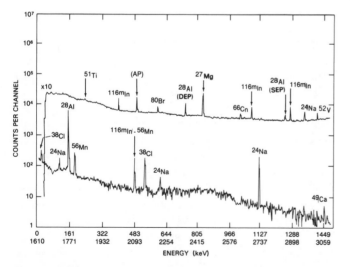

Fig. 16   Ge(Li) γ-ray spectrum of a high-volume air filter run for 24 h near a Pb smelter in Kellogg, Idaho. The spectrum was taken after two different irradiations and five different counting conditions as identified in the spectrum. The mass loading was 226 μg/m³; the mass on the filter was 216 mg. SEP, single escape peak; DEP, double escape peak; AP, annihilation peak. (From Ragaini et al., 1976.)

too expensive to be installed for the purpose of X-ray spectrometry alone, many institutions attach X-ray spectrometers to existing accelerators and rent time for analysis. In almost every case, an energy-dispersive analyzer system is used in conjunction with the accelerator source. The technique of aerosol analysis using proton-induced X-ray emission is known as PIXE (Flocchini et al., 1975). Table 3 shows the lower limits of detection of the three methods for a selected number of elements (Jaklevic and Walter, 1976).

3.2.3 Emission and Absorption Spectroscopy

These methods use the emission and absorption of photons of visible wavelengths. The process can be described by

$$M^* \rightleftharpoons M + h\nu$$

where M^* and M describe the atom or molecule in excited and ground states, respectively, and $h\nu$ is the emitted or absorbed photon of frequency ν. The direction of the upper arrow indicates emission, while the direction of the lower arrow indicates absorption.

TABLE 3
Current Lower Limits of Detection (ng/cm²) of Selected Single Elements as a Thin Film on Millipore Filters[1]

Element	MC-WDXRF[2]	TE-EDXRF[3]	PIXE[4]
Al	2.2		
Si	5.4	157	
S	6.7	38	1,100
K	3.9	13	27
V	9.5	22	8.4
Fe	17	10	3.7
Zn	6.9	6.2	2.2
As	17	3.0	1.6
Se		2.6	1.7
Sr		3.5	3.1
Cd	3.3	5.9	21
Sn	4.7	38	
Pb	5.3	8.9	5.8

[1]Millipore filters were chosen for comparison, but in many cases do not represent the best substrate.

[2]Multichannel WDXRF at EPA Laboratory, Research Triangle Park, North Carolina.

[3]Tube-excited EDXRF at Lawrence Berkeley Laboratory, California. Secondary fluorescers were used to generate exciting radiation.

[4]Proton-induced X-ray emission system at Duke University, Durham, North Carolina.

In the atomic emission (AE) methods of atomic spectroscopy the substance must be excited after evaporation. The excitation energy is applied in the form of heat, electric current, or light. Figure 17 indicates the principles involved in evaporation and excitation of a sample. The sample to be tested must be brought into the light source. In the case of emission spectrography, the powdered sample is evaporated, usually from a graphite electrode; in flame spectrometry, however, the sample solution is introduced into the flame by means of an atomizer. The energy (frequency) of the emitted light (photon) is characteristic of the atom or molecule, while the intensity of the photon current is proportional to the concentration of atoms present in the excited state. The measuring system of the emission method is demonstrated by the simple scheme shown in Fig. 18a.

Atomic absorption spectroscopy (AAS) uses light sources that emit continuous, monochromatic radiation. Usually, hollow cathode lamps are used, since these emit very intense lines of small half-widths. As a result, measurements of high sensitivity can be achieved. The measuring system of atomic absorption is shown in a simplified scheme in Fig. 18b. An area of high temperature causes the substance to evaporate and disintegrate into atoms. For flame methods, the flame serves as a thermostat as well. In comparison to the flame gases, the atomic vapors are present in very low concentrations. The medium for producing the high temperature may be strongly reducing, such as acetylene or nitrous oxide flames, which makes possible the detection of elements that form thermally stable oxides. A graphite furnace, which in-

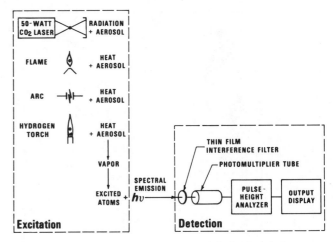

Fig. 17 Principles of spectrophotometric aerosol elemental analysis.

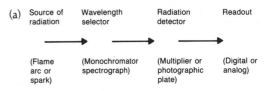

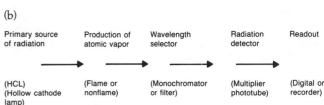

Fig. 18 The measuring system of (a) the emission method and (b) the absorption method of atomic spectroscopy.

volves a graphite tube glowing at 2,500–3,000 K, can be used as a high-temperature source area.

The inductively coupled plasma (ICP) burner is a significant recent development in emission spectrometry (Fig. 19). It is used, for example, for the detection of trace cations in precipitation as part of the National Oceanic and Atmospheric Administration (NOAA) acid rain program. The burner consists of coaxial quartz tubes placed in a high-frequency coil. Argon as a carrier gas transports the sample aerosol into the plasma in the inner tube. The aerosol is introduced into a lower-temperature zone, a so-called tunnel, around the axis of the plasma. The method is suited for simultaneous multielement trace analyses. The electrical flames have greater stability than other, more classical light sources such as the direct-current arc and the spark. They have relatively high temperatures and operate in inert atmospheres. Their accuracy is equal to that of the atomic absorption method, while their sensitivity exceeds even that of absorption flame photometry for certain elements.

Table 4 compares experimentally determined detection limits for the four different methods used in atomic emission

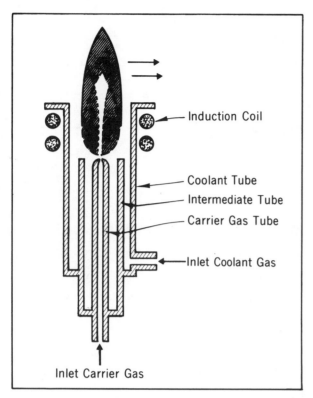

Fig. 19   An inductively coupled plasma burner. Total length of the assembly is approximately 20 cm. The shaded area illustrates the distribution of the plasma. (From Boumans and DeBoer, 1975.)

TABLE 4
Comparison of Experimentally Determined Detection Limits for Methods Used in Atomic Emission and Absorption Spectroscopy

Element	Methods of Detection			
	ICP (μg/mℓ)	AAS (μg/mℓ)	AFS (μg/mℓ)	AE (μg/mℓ)
Ag	0.004	0.005	0.0001	0.008
Al	0.992 (0.0004)[1]	0.03	0.005	0.005
As	0.04 (0.002)[1]	0.1	0.1	50
Ba	0.0001	0.5		0.002
Be	0.005	0.002	0.01	0.1
Bi	0.05	0.05	0.05	2
Ca	0.00007	0.001	0.000001	0.0001
Cd	0.002 (0.00007)[1]	0.001	0.00001	0.8
Co	0.003 (0.0001)[1]	0.005	0.005	0.03
Cr	0.001 (0.00008)[1]	0.003	0.004	0.004
Cu	0.001 (0.00004)[1]	0.002	0.001	0.01
Fe	0.005 (0.0005)[1]	0.005	0.008	0.03
Hg	0.2	0.5	0.02	40
Mg	0.0007	0.0001	0.001	0.005
Mn	0.0007 (0.00001)[1]	0.002	0.002	0.005
Na	0.0002	0.002		0.001
Ni	0.006 (0.003)[1]	0.005	0.003	0.02
Pb	0.008 (0.001)[1]	0.01	0.01	0.1
Sb	0.2	0.1	0.05	0.6
Se	0.03 (0.001)[1]	0.1	0.04	100
Si	0.01	0.1		5
Sn	0.3	0.02	0.05	0.3
Sr	0.00002	0.01	0.01	0.0002
Th	0.03			200
U	0.03			10
V	0.006 (0.00009)[1]	0.02	0.07	0.01
W	0.002	3		0.5
Zn	0.002 (0.0001)[1]	0.002	0.00002	50

[1]With ultrasonic nebulization. (From Fassel and Kniseley, 1974.)

and absorption spectroscopy (Fassel and Kniseley, 1974). By comparison, atomic absorption methods are simpler and less expensive than the emission spectroscopic types. Emission flame spectroscopy and ICP emission spectrometry, however, are more accurate. The main advantage of the emission methods lies in the fact that several elements can be measured simultaneously.

4 IN SITU MEASUREMENTS OF ATMOSPHERIC AEROSOLS

4.1 Nephelometry

The geometry of the lensless integrating nephelometer is due to Charlson (1980), following the lead of Crosby and Koerber (1963) and Beuttel and Brewer (1949). It integrates the intensity of scattered light over nearly 4π steradians, thus providing a close approximation to the scattering component of extinction. Figure 20 shows this geometry, which results in the integration of the angular scattering function $\beta(\phi)$ such that the sensor detects a flux due to a light source of intensity I_0 displaced a distance y from the sensor axis

$$B - (I_0/y) \int_{\phi_1}^{\phi_2} \beta(\phi) \sin \phi \, d\phi \qquad (11)$$

if the light source has a Lambert (i.e., $\cos \phi$) emission characteristic. If the device is long, so that $\phi_1 \cong 0$ and $\phi_2 \cong \pi$, then

$$B \cong (I_0/y) \cdot (\sigma_{sp}/2\pi) \qquad (12)$$

The scattering equations are derived by Middleton (1958) and by Butcher and Charlson (1972).

The closed chamber design of the Charlson integrating nephelometer (Charlson, 1980) allows the use of filtered air and pure gases such as carbon dioxide (CO_2) and dichlorodifluoromethane (CCl_2F_2) as standards of light scattering for quantitative comparison. Figure 21 shows the physical layout of a typical instrument, with provisions for calibration gases, a calibration object, and the possibility of a differential measurement. Several papers have appeared on the accuracy of the angular integration of these devices (Middleton, 1958; Ensor and Waggoner, 1970; Heintzenberg and Quenzel,

1973; Rabinoff and Herman, 1973), and all conclude that the accuracy is about ±10% for most atmospheric aerosols at a relative humidity of less than 90%.

The most modern photon-counting instruments use averaging times of from a few seconds to perhaps 10^3 s. With a 10-s averaging time, the uncertainty of σ_{sp} can be as low as approximately 2×10^{-6} m^{-1} (or about 10% of σ_{sg} at 525-nm wavelength), making the device attractive for use in aircraft. For routine monitoring in urban areas, an averaging time of about 100 s is often used. At Mauna Loa Observatory, Hawaii, where a differential measurement between atmospheric air and filtered air is used, it has been possible to detect changes in σ_{sp} that are only 1% of σ_{sg} (Bodhaine and Mendonca, 1974). Figure 22 shows vertical profiles of the light scatter coefficient σ_{sp}, and their relationship to other air quality parameters, in a rural area in Pennsylvania (Pueschel et al., 1981).

As a result of the application of nephelometers over the past two decades, several points of understanding about atmospheric aerosols have emerged:

• Visual range by eye at low relative humidity is usually correlated with σ_{sp}, in agreement with the theory of Koschmieder (1924). Ideal black targets of adequate size should be just detected at a distance L_v over a homogeneous path with uniform illumination, where

$$L_v \approx 3.9/\sigma_{sp} \qquad (13)$$

Experimental results (e.g., Horvath and Noll, 1969) confirm this reciprocal relationship and suggest that, at least at times, σ_s dominates σ_e and that it is not extremely variable over the sight path.

• Mass concentration (micrograms per cubic meter) of aerosol particles smaller than about 2 μm in diameter is highly correlated with light scattering if the relative humidity is lower than about 70%. Waggoner and Weiss (1980) show a correlation coefficient greater than 0.9 for a wide variety of locations. As a result of this correlation and the high sensitivity mentioned earlier, the integrating nephelometer provides a sensitive mass detector for particles with sizes in the 0.1- to 1.0-μm range.

• Because the mass concentration of particles in the size range between 0.1 and 1.0 μm is often dominated by sulfates (e.g., in the eastern United States, Los Angeles, or rural Sweden) there is also a correlation between sulfate concentration and light-scattering coefficient (Waggoner et al., 1976; Charlson et al., 1978).

• Studies of the relationship of σ_{sp} to relative humidity reveal that atmospheric aerosols in the 0.1- to 1.0-μm size range are almost always hygroscopic. Further, because of the frequent dominance of sulfates in the molecular form of sulfuric acid (H_2SO_4), and its neutralization products, with ammonia (NH_3), the shape of the curve of σ_{sp} versus relative humidity can convey chemical information (Charlson et al., 1978).

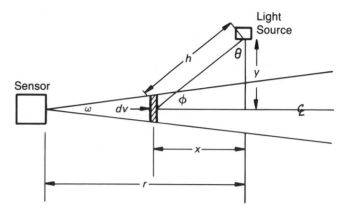

Fig. 20 Geometry of the integrating nephelometer, exaggerated in the y direction about six times. (From Butcher and Charlson, 1972.)

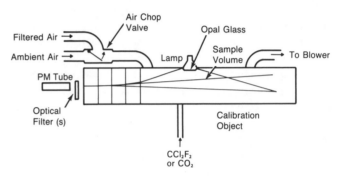

Fig. 21 A typical nephelometer with a closed chamber. (From Miller, 1975.)

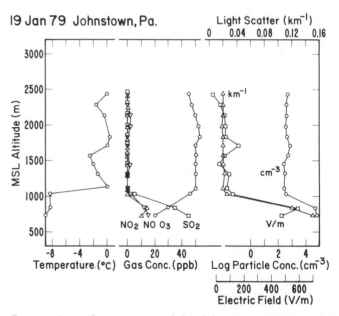

Fig. 22 Air quality parameters including light-scatter coefficient σ_{sp} (per kilometer) in industrial Johnstown, Pennsylvania.

Bodhaine (1983) summarizes long-term light-scattering data from four background sites. At Barrow, Alaska, the data exhibit a strong annual cycle with a maximum in winter and spring due to Arctic haze, and a minimum in summer; the Mauna Loa, Hawaii, results show also a strong annual cycle with a maximum in April or May caused by long-range transport of Asian desert dust; the Samoa data are representative of a clean marine atmosphere and exhibit no significant annual or diurnal cycle; and the South Pole results show a complicated annual cycle with a maximum in the austral summer and a minimum about April.

Polar nephelometers measure the intensity of light scattered from a collimated beam by the atmospheric aerosol as a function of the scattering angle (Grams et al., 1975). The resulting so-called phase function of the atmospheric aerosol is affected by the particle size distribution, by the particle's shape, and by its refractive index. Corollary measurements of several of these parameters give information on climatically important properties of the atmospheric aerosol. For example, Grams et al. (1974) measured the phase function and the size distribution of soil aerosols to estimate their light absorption index. Grams (1981) obtained data on light-scattering parameters of stratospheric aerosol particles with a laser polar nephelometer operated on board an aircraft. Patterson et al. (1982) compared volume-scattering ratios of a model aerosol determined by the polar and the integrating nephelometers and found agreement between the two methods within 10%.

4.2 Condensation Nucleus Measurements

Condensation nuclei (Aitken, 1923) are defined, for our purpose here, as particles that are detectable when vapor (usually water) condenses upon them at supersaturations of several hundred percent (Landsberg, 1938). At such high supersaturations, condensation takes place on particles having diameters as small as 0.01 μm or less. Such tiny particles are too small to either scatter or absorb light significantly, or to contribute to cloud formation in the atmosphere, where supersaturations rarely exceed a fraction of a percent. While climatologically inconsequential, these small particles nevertheless constitute a class of aerosols that play an important role in tropospheric chemistry.

Typically, condensation nuclei (CN) are relatively short-lived. They persist in the atmosphere for longer than hours only in the presence of gaseous precursors, and grow to larger particles through Brownian coagulation in highly concentrated vapors. They also interact strongly with accumulation-mode (i.e., formed by coagulation) particles of typically 0.1 μm modal radius (Whitby, 1978), altering their physical and chemical characteristics, for example, by making their surfaces hygroscopic and increasing their light-scattering cross sections (Pueschel and Mamane, 1979). This alteration of the accumulation (small-particle) mode has consequences for the environment. Accumulation-mode aerosols are large enough to cause or to affect atmospheric optical phenomena.

When hygroscopic, they begin to acquire water molecules at relative humidities between 50 and 70% and thus affect visibility. At saturation, water molecules condense on them to produce a cloud in which droplets may reach diameters of 10 μm or larger (Barrett et al., 1979). The close relationship between the formation and existence of condensation nuclei and the presence of primary gases has been used by Schaefer (1976) to detect, and in some instances identify, the gaseous precursors of condensation nuclei.

Because of their extremely small sizes (smaller than the wavelength of visible light), condensation nuclei cannot be detected by conventional optical techniques. The usual method employed in detecting and counting condensation nuclei is to condense some vapor, usually water, upon them to form droplets of visible size. The air sample to be investigated is drawn into a chamber and saturated by contact with a piece of wet felt or blotting paper. When equilibrium has been reached, the sample is suddenly expanded adiabatically to produce considerable supersaturation. This causes visible droplets to grow on all particles larger than a given size. This total number is dominated by particles of radius 0.1 μm or smaller. The expansion should be large enough to enable detection of the smallest particles of concern but should not be so large as to produce droplets upon large or ionized air molecules, which happens at about 400% supersaturation. Large expansion ratios also favor droplet growth rate and fixed droplet size.

Most counters in use today measure nuclei in discrete samples of air. Both manually operated and automated counters are available. One version of a manually operated CN counter that is frequently used today is the Gardner small-particle detector (manufactured by Gardner Associates, Schenectady, New York), "a highly reliable, rugged, self-contained, foolproof, and easily carried instrument" (Schaefer, 1976). All manually operated counters operate on similar principles and differ mostly in mechanical arrangements. The droplets grown on active nuclei after adiabatic expansion are detected by photographic recording, collected on grids for direct observation, or detected by photoelectric sensing for their light-scattering properties. Pollak (1957), in a review of methods for measuring condensation nuclei, quotes the range of an Aitken counter as 10^2 to 1.5×10^4 CN/cm^3, with an accuracy within 1–20%. The Gardner counter has a range for measuring concentrations from as low as 10^2 to an excess of 10^6 CN/cm^3.

In spite of their many advantages, manually operated CN counters have a serious drawback. It takes at least 10 s to flush the counter, draw in the sample, establish the vacuum, and let the droplets grow to terminal size. This drawback has been overcome with the availability of automated, continuously recording counters. Automatically cycling cloud chambers analyze up to 2.5 samples per second (Haberl, 1975). One unit (available from Environment One Corporation, Schenectady, New York) has the capability of analyzing one sample per second. Data are acquired by photoelectric sensing, which is presently the most practical automated means for high data-gathering rates. This method has the disadvantages, however, of not providing numbers of nuclei directly and of requiring stable sensitivity and calibration to known standards.

In airborne measurements, when ambient pressures and temperatures change with altitude, the initial conditions determining the degrees of supersaturation achievable in condensation nucleus counters are sometimes difficult to keep constant. Boundary-layer measurements, for example, may be taken up to 4,000 m above mean sea level. At such an altitude, in a standard atmosphere, the absolute pressure is about 616.6 mb, and the temperature is about $-11\,^\circ$C. To standardize to one normal sea-level atmosphere (1,013.3 mb) at 15 °C, an adiabatic compression would yield a final temperature of more than 28 °C, raising the saturation vapor pressure of water from 1.98 mm of mercury (Hg) to 26.7 mm Hg. Not only can the excess heat destroy a volatile portion of the sample, it will also change the supersaturation that is achievable and thus the smallest particle size that is detectable. Thus, although standardization is technically feasible (Haberl, 1975), care must be taken to remove the excess heat so that the temperature rise is limited to 15 °C. By the same token, to keep the supersaturation constant, independently of altitude and initial saturation vapor pressure, it is necessary to vary the expansion ratio.

Applications of condensation nucleus counts in the troposphere date back to those taken by Wigand (1919) in balloon ascents, and more recently include those by Weickmann (1957) in an airplane. Both sets of data show the same decrease in counts with increasing altitude. The decrease in counts with altitude has since been verified on many occasions (Junge, 1961; Barrett et al., 1970; Allee et al., 1970; Lopez et al., 1974), implicating the earth's surface as the source of condensation nuclei. Recent airborne measurements of condensation nuclei have given information on formation rates of fine particles in power plant plumes by the oxidation of sulfur dioxide (Pueschel and Van Valin, 1978; Mamane and Pueschel, 1980b), which can be related to the formation of fine particles by vegetation (Lopez et al., 1974). Squires (1966) and Braham (1974) measured the rate of formation of condensation nuclei in urban pollution plumes.

4.3 Conductivity Measurements

The conductivity of the atmosphere, a result of the continuous natural production of small ions, is largely controlled by the presence of fine particles. The major role of suspended particulates, with respect to the electrical state of the atmosphere, is that of providing relatively stationary surfaces that the smaller and much more mobile small ions easily strike and become attached to. Particulates that acquire a net charge due to one or more collisions with small ions are called "large ions." Because of their relatively low mobility, these large ions reduce the conductivity. The atmosphere thus contains a mixture of large and small ions and uncharged nuclei, which interact to maintain a slightly conducting environment with

the conductivity largely controlled by the nature of the ambient aerosol.

In exceptionally clean air, such as that found at the NOAA base-line stations at Mauna Loa, Hawaii, or the South Pole, there are relatively few nuclei; thus the production of small ions is largely offset by recombination between oppositely charged small ions. In air having a large number of aerosols, such as might be found in a large city, on the other hand, small ion losses due to recombination with charged and uncharged nuclei increase greatly. The extent to which suspended particulates reduce the conductivity depends on both the size and number of the particles, since a few large aerosols will present as much surface area for impaction by small ions as a greater number of small-sized aerosols.

Conductivity measurements have been used for many years in air pollution research, particularly by scientists of the Carnegie Institution of Washington. In 1967, NOAA researchers (Cobb and Wells, 1970) measured conductivity during the global voyage of the research vessel *Oceanographer* and compared them with the classical measurements of the sailing vessel *Carnegie*, made during the 1920s. Their results indicated a significant decrease in conductivity, thus an increase in suspended particulates, downwind from populated land areas, but no change in the South Pacific from measurements made 50 years earlier.

Conductivity observations at Mauna Loa Observatory, Hawaii, indicate no change since 1960 in particulate levels at the high-altitude, global base-line station. In rural Boulder County, Colorado, on the other hand, a secular decline in the conductivity of about 25 % from 1967 through 1979 has been presented as evidence of an increase in the suspended particulate burden and a resulting deterioration of the regional air quality (Cobb, 1982).

4.4 The Electrical Aerosol Analyzer

Since the mean electric mobility of an aerosol, under controlled charging conditions, is a monotonically decreasing function of particle size for particles smaller than about 1 μm in diameter, the size of a diffusion-charged aerosol can be determined from mobility measurements. Scientists at the University of Minnesota (Whitby and Clark, 1966) first described an electrical particle counter and size analyzer using this concept. Their semi-automatic system, capable of measuring particle concentration and size distribution in the 0.015- to 1.2-μm size range, is now manufactured by Thermo-Systems, Inc. (St. Paul, Minnesota) as the Model 3030 electrical aerosol analyzer. The design, operation, and performance of the electrical aerosol analyzer (EAA) have been described in detail by Liu et al. (1979) and by Liu and Pui (1975). In brief, the EAA instrument is an aerosol mobility classifier, since the electrical mobility of a charged particle is a function of the particle's size.

In operation, an aerosol is negatively or positively charged and flows at a known rate past a positively charged collecting rod (Fig. 23). Depending on the potential of the collecting

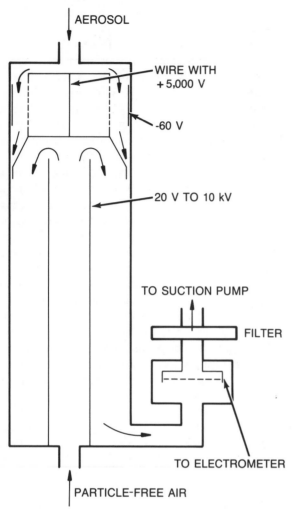

Fig. 23 Schematic of an electric mobility analyzer.

rod, particles with an electrical mobility exceeding a certain value will be attracted and collected. For particles larger than a certain size and having less than a certain mobility, the transit time past the collector will be greater than the lateral transit time toward the collector rod, and these particles will be captured downstream by a current-collecting filter. An electrometer measures the current produced as the charged particles are collected by the filter. It is readily apparent that a step increase in the collector rod voltage will result in particles of a larger size range (lesser mobility) reaching the current-collecting filter. The final output of the EAA is an "analyzer current" for each particle diameter range, which in turn is determined by 11 collector rod voltage steps.

Converting the EAA analyzer current into particle size ranges and densities involves a number of considerations such as the number of elementary charges per particle, the fraction of particles that do not become charged, and the calculation of the electrical mobility as a function of the particle diameter. These considerations and the physical principles involved are discussed in detail in the reports by Liu et al. (1979) and Liu and Pui (1975). The EAA has become a widely used sensor and is particularly useful for urban-related aerosol studies. The currently manufactured instrument (TSI Model 3030) provides particle data in ten different size ranges from 0.0032 to 1 μm in diameter. With some fairly simple data processing, the instrument provides particle concentration, surface area, and volume in each size range as well as the totals for the entire size range.

4.5 Optical Particle-Size Spectrometry

While the microscopic techniques discussed earlier can give unambiguous information about particle size and shape,

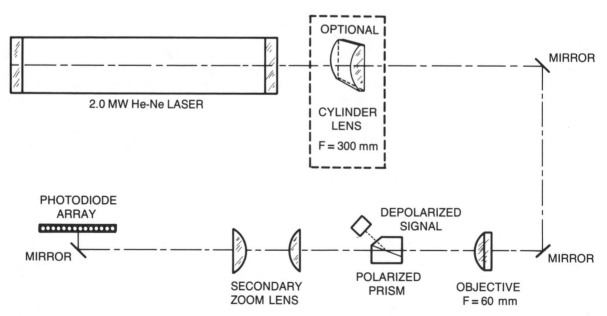

Fig. 24 Optical system diagram for two-dimensional optical array spectrometer, OAP-2D2-C. (Adapted from instrument manual, PMS, Boulder, Colo.)

they suffer from a severe limitation, viz., a very slow rate of data retrieval and reduction. It takes many hours of manual sizing, counting, and tabulating to generate a statistically significant particle size spectrum from an impactor or filter sample. For this reason, techniques that make real-time use of the interaction between light and particles in a moving stream of air have been developed in the last decade. These techniques have speeded up the rate of acquisition of particle size spectra by many orders of magnitude; the price paid is an increase in ambiguity or uncertainty in the spectra because the optical "signatures" of particles depend on other factors in addition to their physical dimensions.

This variation in particle signature with particle size gives rise to two main types of optical spectrometers (for brevity, hereafter called "probes"). These are the imaging, or optical-array, probes and the forward-scattering probes. The former are used to cover the large-size domain (radius > 25 μm), and are therefore not very useful in aerosol analysis (except possibly in dust-storm situations) other than cloud studies. They are, however, quite useful in cloud-physics research because they can distinguish between drops and ice crystals.

In an imaging probe, an expanded, collimated laser is beamed at right angles to the air stream containing particles to be sized. A magnifying optical system focuses a slice of the beam onto a linear array of photodiodes. The outputs of these are amplified and digitized to two (or in a more advanced design, four) intensity levels. In this way, the shadow area of the particle is rendered as black (or three shades of gray). A simplified schematic of the optical arrangement is shown in Fig. 24.

If the diode outputs are simply read during the transit time of a particle through the beam, then the probe measures the maximum diameter of the particle as projected on the plane normal to the beam; this constitutes a one-dimensional probe system. If, on the other hand, the diode outputs are sampled a number of times during the particle's passage and these digitized samples are stored in a digital memory, the contents of this memory can be sent to a video display or plotter in a proper sequence so as to produce a silhouette of the particle's shape. Such a probe is called a two-dimensional optical-array probe. The two-dimensional probe obviously provides more useful information about particle shape, but at the cost of much more complex electronics and greatly increased data-processing time.

The forward-scattering probes are of greater use in atmospheric aerosol measurements. The simplified optical schematic of such a probe is given in Fig. 25. In this case, the laser beam is condensed to a very small diameter (to minimize the chance of seeing more than one particle at a time). The aureole of forward-scattered light, out to 18° or so from the beam axis, is caught and refocused onto a photodiode (after passing through a narrow-band filter centered on the laser wavelength). The direct beam (and the aureole out to about 4° off-axis) are blocked by a black "dump spot" or occulting

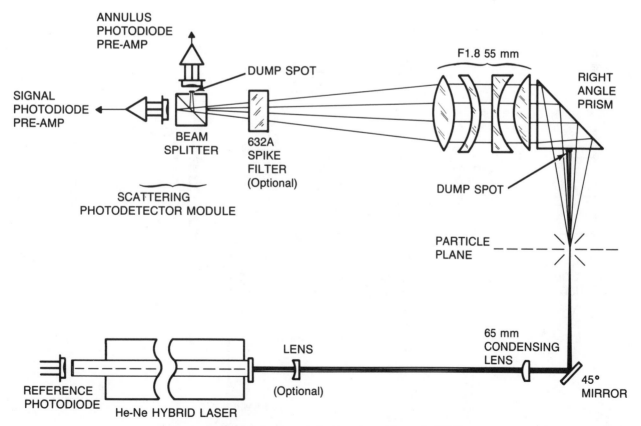

Fig. 25 Optical system diagram for a forward-scattering probe, FSSP-100. (Adapted from instrument manual, PMS, Boulder, Colo.)

disk. When a particle passes through the beam and within a prescribed distance (depth of field) from the focal plane of the condensing lens, the resulting pulse of voltage from the photodiode is sent to a pulse-height analyzer whose levels of discrimination are set so that successive "bins" contain counts of particles in equal size (radius) increments.

As the size parameter becomes smaller than unity (which occurs at a particle radius of $\sim 0.1\ \mu$m for a helium-neon laser operating at 0.6328 μm), the aureole becomes broad and the intensity weak. To obtain a good signal-to-noise ratio for these smaller sizes, a different strategy is pursued, by which the particles are introduced into the internal optical system of the laser itself.

The laser consists of a plasma tube containing the helium-neon mixture, sealed at one end by a highly reflective mirror normal to the tube axis and at the other by a clear Brewster window to eliminate unwanted reflection (Fig. 26). The air stream containing the particles enters from the side between the Brewster window and a lens system whose first (plano-convex) element has a highly reflective mirror with an opaque backing. The two mirrors comprise a cavity that resonates at one propagation mode of the characteristic wavelength. Since both mirrors are highly reflecting, very little energy escapes from this cavity; the intensity is

therefore very high (on the order of 1,000 W/cm²). The lens system is focused on a photodetector as before. If no particle is present in the cavity, then the light beam simply bounces back and forth between the mirrors and almost none reaches the detector. When a particle enters the beam, its broad aureole is caught by the lens system and is focused on the detector, and a pulse is produced as before. The strength of the scattered aureole in this case cannot be calculated from Mie theory because of the standing waves in the cavity and the detuning of the cavity by the particle; a more complicated calculation is necessary. Probes operating in this mode are called active-scattering spectrometers; they are able to size particles as small as $\sim 0.05\ \mu$m radius.

The descriptions of the operation of optical particle spectrometers given here are of necessity highly simplified. Very elaborate auxiliary optics and electronics are required in practical systems to reduce sizing errors that could arise from two causes. First, a laser beam cannot have a uniform intensity all across its diameter and zero outside. The intensity variation is actually more like a Gaussian curve, with slow change near the axis changing to a faster drop-off farther out. Therefore, only the pulses from particles close to the axis may be counted. Second, particles passing through the beam too far from the focal plane of the lens system will give pulses too

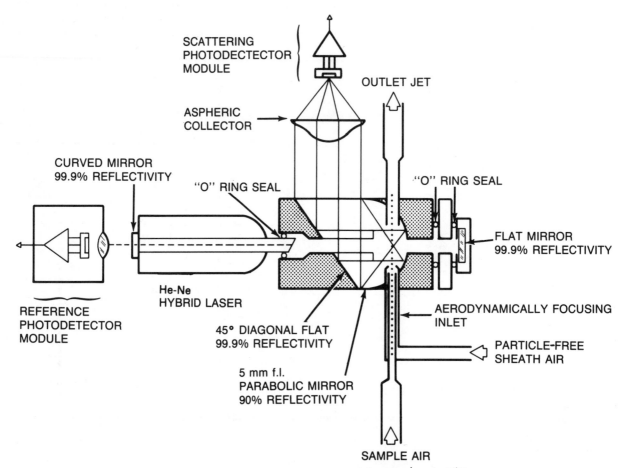

Fig. 26 *Optical system diagram for an active-scattering aerosol spectrometer, ASAS-X. (Adapted from instrument manual, PMS, Boulder, Colo.)*

weak for their size. A discussion of the details of how particles passing outside the "legal" sample volume are rejected is beyond the scope of this discussion. More complete discussions of the theory and practice of optical aerosol spectrometers may be found in Knollenberg (1970, 1972, 1973, 1976), Knollenberg and Luehr (1976), Schehl et al. (1973), and Schuster and Knollenberg (1972).

4.6 Laser Velocimetry

The ambiguities associated with optical imagery are nonexistent in a technique utilizing the interaction of particles with light that was made available commercially by Thermo-Systems, Inc., St. Paul, Minnesota. The technique does not attempt to translate optical signatures into measures of size, shape, refractive index, etc.; it uses scattered light solely to measure the time of travel of aerosol particles along a short distance.

Particles are drawn through a flow nozzle that generates an accelerating jet of air (Fig. 27). Particles accelerate within the jet at varying rates, depending on their aerodynamic size; small particles accelerate more rapidly than larger ones. Particle velocity is measured by a laser velocimeter at the exit of the accelerating flow nozzle. Since the particles accelerate within the jet at varying rates, their speed at the exit depends solely on their aerodynamic size characteristics.

The velocimeter system consists of a laser light source, beam expander and splitter, focusing and receiving optics, and a diode or photomultiplier light-detection system. An expanded laser beam is split by polarization to form two beams that are focused into two parallel planes immediately below the accelerating flow nozzle. When a particle passes through these planes, it produces two sequential pulses of scattered light. These pulses are collected and focused onto the light-detection system to produce two electrical pulses. The time interval between these two pulses is stored by a multichannel

analyzer and transmitted to a microcomputer processor for analysis.

The measured time interval between the two electrical pulses indicates the velocity of the particle under study, and thus its aerodynamic size. Aerodynamic size depends on such parameters as geometric size, shape, and density. Only for a sphere of unit density is the physical size the same as its aerodynamic size. Large, hollow, low-density spheres behave like smaller but denser spheres. Irregularly shaped particles are characterized by their aerodynamic behavior. On the other hand, optical properties of the aerosol do not affect the measurement. Hence, the data are insensitive to changes in laser power, index of refraction, and other light-scattering characteristics.

5 CLOUD CONDENSATION AND ICE NUCLEI

For cloud condensation nucleus measurements, the thermal-diffusion cloud chamber is at present the only practical technique (Fig. 28). Many varieties exist and are described in two recent workshop reports (Grant, 1971; Vali, 1976). They differ in dimensions, methods of achieving thermal equilibrium, and methods of counting the drops. Mee Industries, San Gabriel, California, and Meteorology Research, Inc., Altadena, California, have manufactured commercial versions of thermal-diffusion cloud chambers.

To reproduce conditions in natural clouds, the supersaturation with respect to water must be low, less than 0.1% if possible. The concentration of CCN is usually determined by illuminating a known small volume, photographing it, and then counting by eye the number of bright spots produced by the droplets on the film. This very time-consuming process has been overcome in an automatically operating and recording thermal-diffusion chamber developed by Radke and Hobbs (1969). They allow the nuclei, activated at a predetermined supersaturation, to grow to a predetermined, very nearly uniform size. Their droplet concentration is determined by measuring the light-scattering coefficient for about 20 cm³ of the cloud with an integrating nephelometer as described in Section 4.1. More recent designs (Hudson and Squires, 1976; Hudson et al., 1981) use optical particle counters to detect the nuclei.

The prevailing humidity in a supercooled cloud is close to water saturation at the existing temperature. Because the saturation vapor pressure over water is always greater than that over ice at temperatures below 0°C, a newly frozen droplet or embryonic ice crystal finds itself in a much more supersaturated environment than that of the unfrozen drops. The diffusional growth of ice crystals to precipitation-size hydrometeors, therefore, is very rapid and is the most important precipitation formation mechanism in middle and high latitudes (Bergeron-Findeisen process). As important as ice nuclei are for the formation of precipitation, their concentration is a very difficult quantity to measure. This becomes obvious if one considers that, in the atmosphere at −10°C,

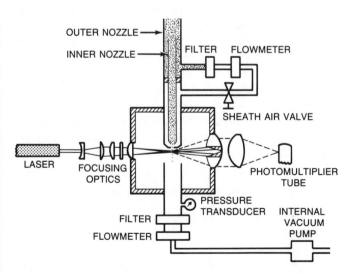

Fig. 27 Schematic of an aerodynamic particle sizer based on laser velocimetry. (After Remiarz and Johnson, 1984.)

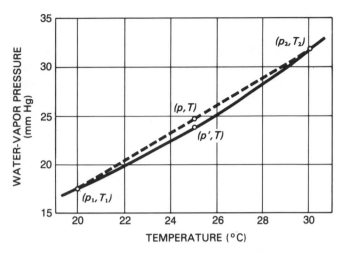

Fig. 28a Actual vapor density gradient (dashed) and saturation vapor pressure (solid) in a thermal-diffusion chamber resulting in a supersaturation $(p - p')/p'$.

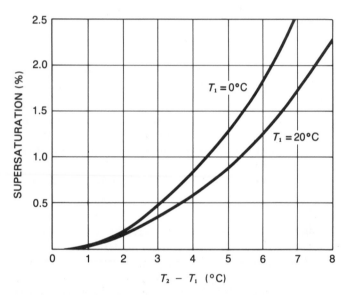

Fig. 28b Supersaturation as a function of the temperature difference between the top and the bottom of the chamber.

IN are outnumbered by factors of 10^7–10^9 by Aitken nuclei, and 10^6–10^7 by CCN.

One principal method of counting ice nuclei is to cool a known volume of air until a cloud is produced, and to count the number of ice crystals that form at a particular temperature. Cooling is produced either in expansion chambers by compressing the air and then allowing it to expand rapidly, or in mixing chambers by refrigeration. Several techniques have been used for counting the number of ice particles that appear in a cloud. It is possible to illuminate the cloud by a light beam and estimate the concentration of ice crystals visually. Alternatively, the small ice crystals forming in the chamber are allowed to settle into a disk or film of supercooled water (Cwilong, 1947), supercooled soap solution (Schaefer, 1948), or supercooled sugar solution (Bigg, 1957), where they can be

detected and counted. Automatic IN counting is possible with an acoustic technique described by Langer (1965) and applied in the so-called acoustic ice nucleus counter (Langer et al., 1967): Ice particles that form in a cold box are forced through a capillary tube where they produce audible clicks that are electronically counted and recorded. A second automatic counting method, applied in the only commercially available IN counter (Mee Industries, San Gabriel, California) has the crystals fall between two crossed polarizing filters that are placed between a light source and a photodetector. The crystals are counted by virtue of their depolarization effect on the light.

Another method which has been frequently used for counting ice nuclei in the air is to aspirate known volumes of air through membrane filters (Bigg et al., 1963; Stevenson, 1968). The number of ice crystals is determined by holding the filter at a known temperature, exposing it to a given supersaturation, and counting the number of ice crystals that grow on it. Schnell (1979) describes a modified version of this method: Aerosols are captured on hydrophobic filters, an array of nucleant-deficient water drops is placed on the filter, and the filter is cooled. The freezing of the drops at a given temperature determines the presence of IN on the filter.

Several international workshops have been held with the purpose of comparing IN measurement methods, and to find a common ground for measuring ice nucleation modes (Grant, 1971; Vali, 1976).

6 DISTRIBUTION FUNCTIONS

It is often convenient to parameterize the number, surface area, or mass versus particle size of a heterogeneous aerosol by a distribution function. If the integral of the distribution function and at least some of its moments are available in closed form, the calculation of the statistical properties of the distribution is greatly simplified. Mathematical functions are also useful in verifying models that attempt to predict aerosol size distributions. Ideally, the models themselves should define the functional form of the distributions. However, the size distribution of an aerosol is the result of a generating process, a coagulation process, and size-dependent transport and removal processes, all of which are physically and mathematically complicated and so can only be calculated by numerical methods. Therefore, the choice of a distribution function is sometimes based more on aesthetic than on mathematical-physical considerations. A few commonly used functions are described below.

6.1 The Log-Normal Distribution

This distribution has the following mathematical form (unnormalized):

$$y = A \exp[-\ln(r/r_g)^2/2(\ln \sigma_g)^2] \qquad (14)$$

This function has three parameters: A, a dimensional constant depending on the total number (surface, mass) of particles; r_g, the geometric mean radius (or diameter) of the particles; and σ_g, the geometric standard deviation. Theoretical arguments have been advanced to support the idea that this distribution arises naturally when solids are ground or crushed in a mortar or ball mill for an extended time (Kolmogoroff, 1941). However, other investigators have come up with a different distribution function for ground powders (see Section 6.3) that they claim also gives a good fit to measurements.

From the form of the function, it is clear that a plot of y against $\ln(r/r_g)$ gives a standard bell-shaped or Gaussian curve with zero skewness (on a linear scale of r it is, of course, skewed toward larger sizes). This means that measured distributions with arbitrary skewness cannot be well fitted by a log-normal function. The log-normal distribution has the property that the probability of finding a particle with a radius of, say, 0.05–0.15 r_g is the same as that of finding one of radius 5–15 r_g.

6.2 The Modified (or Generalized) Gamma Distribution

The functional form of this distribution is

$$y = Ar^a\exp(-br^c) \qquad (15)$$

again in unnormalized form. We give the distributions in unnormalized form because when they are normalized (to unit area under the curve) the coefficient A emerges as a function of one or more of the other parameters. When a function is fitted to a measured distribution by, say, least squares, the parameters are obviously not known a priori, so the unnormalized form is preferable.

This function, apparently first recommended in America by Deirmendjian (1969) for describing atmospheric aerosols, contains a total of four parameters, one more than the log-normal distribution. It therefore has greater flexibility in fitting to measured data. In particular, gamma distributions can be fitted to data with arbitrary skewness; only zero skewness is excluded. However, it can be shown by a bit of mathematical manipulation that the log-normal distribution is a limiting case of the modified gamma distribution as the skewness goes to zero when plotted versus $\ln r$.

The gamma distribution has the desirable property that it and all of its moments can be expressed in closed form. That is, the dth moment of y is

$$\int_0^\infty r^d y \, dr = A \int_0^\infty r^{a+d}\exp(-br^c) \, dr = Ac^{-1}b^{-\nu}\Gamma(\nu) \qquad (16)$$

where $\nu = (a + d + 1)/c$ and $\Gamma(\nu)$ is the gamma function. Similarly, all moments of the log-normal distribution are also obtainable in closed form. This means that statistics of distributions (mean and modal radius, spread, skewness, etc.) are easily calculated. Usually, all of the parameters A, a, b, and c of the modified gamma distribution are positive real

numbers, so that the function vanishes at zero and infinite radius (but see below for a possible exception).

6.3 The Rosin-Rammler and Nikiyama-Tanasawa Distributions

It was mentioned earlier that the log-normal distribution arises theoretically when solids are ground to powders. Two other investigators (Rosin and Rammler, 1933) studied the same process and concluded that a particular modified gamma distribution gave the best fit to measurements using sieves. Their function is derived from Eq. (15) by setting $c = a + 1$, to give

$$y = Ar^a \exp(-br^{a+1}) \qquad (17)$$

Since there are only three independent parameters in this function, the skewness is no longer arbitrary but is governed by the modal radius and the spread.

The Nikiyama-Tanasawa (1939) distribution is another special case of the modified gamma distribution, with the form

$$y = Ar^2\exp(-br^2) \qquad (18)$$

6.4 Other Distributions

Until the middle 1970s, atmospheric aerosols were collected and sized using cascade impactors or centrifugal separators, and optical and electron microscopes. The extreme tedium of such analyses discouraged workers from using fine size resolution; the particles were grouped into from four to twelve "bins." When such data were plotted logarithmically, they showed a tendency to decrease monotonically with increasing radius. The simplest function to fit to such data is obviously a straight line. This leads to the Junge (1955) power-law distribution

$$y = Ar^{-a} \qquad (19)$$

whose only virtue is simplicity. One cannot obtain moments of this distribution by integration from zero to infinity; the integrand becomes singular at $r = 0$ for all moments less than a and singular at infinity for all moments greater than a. Thus this distribution must be restricted to a finite range of particle sizes.

The problem of the singularity at $r = 0$ can be dealt with fairly simply. If, in the modified gamma distribution of Eq. (15), we allow a and c to be negative but keep b positive, we arrive at

$$y = Ar^{-a}\exp(-br^{-c}) \qquad (20)$$

in which a and c are now positive. Equation 20 may be integrated from zero to infinity. However, if $a \le 3$, the third moment of y (the total particle volume) diverges.

6.5 A Function-Fitting Procedure

As discussed in the introduction, atmospheric aerosol distributions are generally trimodal. It is therefore not good practice to try to represent the entire distribution by a single function of the Junge type; the individual modal peaks are lost. It is much better to fit each individual mode with a modified gamma or log-normal distribution, as appropriate, and then superimpose these curves to obtain the complete distribution. This can be done easily using a modern microcomputer with high-resolution graphics capability.

This is illustrated by means of aerosol particle and cloud droplet size spectra that were obtained with a forward-scattering (FSSP) and an active-scattering aerosol spectrometer probe (ASASP) at Whiteface Mountain, New York, on 4 August 1983. The FSSP and ASASP count particles with radii from 0.25 to 23.5 μm and from 0.045 to 1.5 μm, respectively, in four partially overlapping ranges of 15 bins each. The counts were recorded on magnetic tape by a data logger. After the field missions, the data tapes were read by an LSI-11 microcomputer. Particle and cloud droplet spectra were retrieved as plots and tables.

Figure 29 gives an example of a computer-produced plot of the aerosol spectrum that was taken in the absence of a cloud between 0500 and 0510 LST. The abscissa is the logarithm of particle radius in micrometers; the ordinate is the logarithm of $dN/d(\log r)$, or the number density (cm^{-3}) in a class interval $d(\log r)$. The curves to the left are the ASASP data and those to the right are from the FSSP. Two modes, a small-particle

mode around 0.1 μm and a large-particle mode above 1 μm radius, are evident from the plot.

We parameterize the spectra by a set of two or more overlapping generalized gamma distributions, using the technique of "spectrum stripping." To do this, the data around a peak (such as that near 0.1 μm) are fitted by a single gamma distribution by the use of a least-squares procedure. The function thus obtained is then subtracted from the raw data, and the remaining peak is fitted in the same way. Figures 30 and

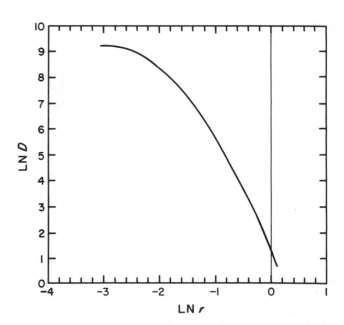

Fig. 30 *Gamma distribution fitted to the small-particle spectrum (left-hand peak) of Fig. 29. D is dN/d(log r) (per cubic centimeter), and r is particle radius (in micrometers).*

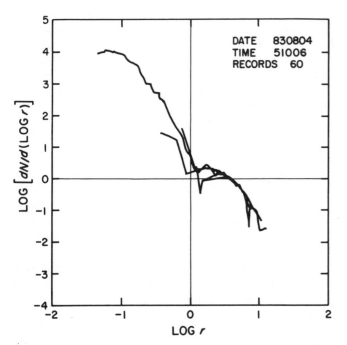

Fig. 29 *Computer plot of aerosol spectrometer data in dry air at Whiteface Mountain, New York. Ordinate is number concentration (per cubic centimeter) of particles in class interval d(log r) on a logarithmic scale from 10^{-4} to 10^{+5}. Abscissa is particle radius r (in micrometers) on a logarithmic scale from 10^{-2} to 10^{+2}.*

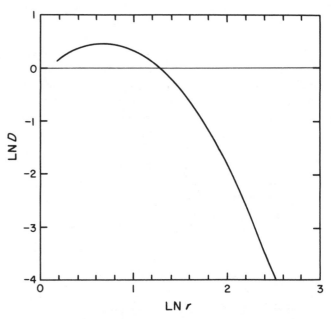

Fig. 31 *Gamma distribution fitted to the large-particle spectrum (right-hand peak) of Fig. 29. D is dN/d(log r) (per cubic centimeter), and r is particle radius (in micrometers).*

31 illustrate the application of the process to the case of Fig. 29. The small-particle spectrum, obtained by fitting a curve to all data points with $r < 1 \mu m$, is given in Fig. 30. The large-particle spectrum, fitted to the data with $r > 1 \mu m$, is shown in Fig. 31. The overall quality of the fit is quite good, as Fig. 32 attests. Figure 33 shows the overall fit of three generalized gamma distributions of the combined aerosol and cloud droplet spectra inside an orographic cloud that had formed within 1 h after the aerosol spectrum of Fig. 32 had been taken. The presence of a cloud-drop mode around 5-μm radius and a haze-particle mode at 0.7 μm is quite evident.

Some statistical properties obtained by suitable analytic integration over the gamma spectra are listed in Table 5 for the dry aerosol case of Fig. 32 and for the cloud case of Fig. 33. The dry aerosol had a number concentration of 8,042/cm³ and a mass loading of 112 $\mu g/m^3$ in the small-particle mode. In the cloud, these numbers were reduced to 2,876/μm^3 and 18.8 $\mu g/m^3$, respectively. It can be assumed that most of this aerosol has been ingested, by nucleation and in-cloud scavenging, into the cloud of 0.17 g/m³ liquid water content.

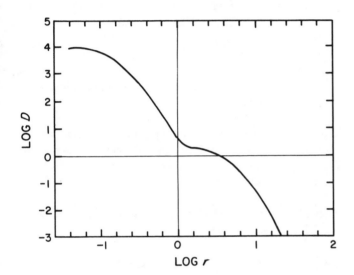

Fig. 32 *Overall fit of the two overlapping gamma distributions to the data of Fig. 29. D is dN/d(log r) (per cubic centimeter), and r is particle radius (in micrometers).*

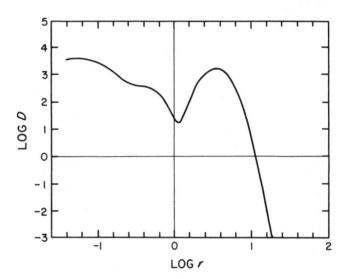

Fig. 33 *Overall fit of three overlapping gamma distributions to aerosol spectrometer data taken within a cloud at Whiteface Mountain, New York. D is dN/d(log r) (per cubic centimeter), and r is particle radius (in micrometers).*

TABLE 5
Statistics of Aerosol and Cloud-Drop Spectra,
Whiteface Mountain, New York, 4 August 1983

Mode	Parameter	Precloud Aerosol (0510 EDT)	Cloud plus Aerosol (0610 EDT)
Small particle	Mass modal radius, μm	2.74^{-1}	2.27^{-1}
	Standard deviation, μm	3.12^{-1}	2.12^{-1}
	Skewness parameter	1.48	1.38
	Number concentration, cm⁻³	8.04^{+3}	2.88^{+3}
	Mass concentration, $\mu g/m^3$	1.12^{+2}	1.88^{+1}
Haze	Mass modal radius, μm		6.57^{-1}
	Standard deviation, μm		2.88^{-1}
	Skewness parameter		1.09
	Number concentration, cm⁻³		1.42^{+2}
	Mass concentration, $\mu g/m^3$		8.44^{+1}
Large particle cloud drop	Mass modal radius, μm	6.15	4.74
	Standard deviation, μm	5.01	1.55
	Skewness parameter	1.33	9.67^{-1}
	Number concentration, cm⁻³	1.07	6.18^{+2}
	Mass concentration, $\mu g/m^3$	3.82^{+2}	1.71^{+5}

Acknowledgments. Numerous discussions with members of the Air Quality Division of NOAA's Air Resources Laboratory helped me greatly in preparing this article. Particular thanks are due to Earl Barrett and Joe Boatman, who contributed most of the material on optical aerosol spectrometry; to Bill Cobb, who supplied most of the information on electrical conductivity and mobility measurements; to Lois Stearns and Dennis Wellman, who collected the field data; and to Tony Puig, for computing the aerosol spectra and statistics. I also thank Gaye Horn and Sandy Furney for typing the manuscript.

APPENDIX: DEFINITIONS OF SELECTED TERMS

These definitions come from Huschke (1959) and Butcher and Charlson (1972).

Aerodynamic diameter: Diameter of a sphere of density 1,000 kg/m³ that has the same terminal velocity as the particle under investigation.

Aerosol: A dispersion of solid or liquid particles of microscopic size in a gaseous medium, usually air. Examples are haze, smoke, and some fogs and clouds. The particles are considered small enough to assure colloidal stability.

Albedo: The ratio of the radiant energy reflected from a surface to the total energy incident upon it.

Cloud: A distribution of hydrometeors, dust, or smoke particles dense enough to be perceptible to the human eye, and small enough to remain colloidally stable.

Colloidal instability: A property attributed to clouds by virtue of which the cloud particles aggregate into particles large enough to precipitate.

Dust: A loose term applied to solid particles predominantly irregular in shape, suspended by turbulence and winds. Dust particles are large enough to settle under the influence of gravity.

Droplet: A small liquid particle that may fall under still conditions, but remain suspended under turbulent conditions.

Fly ash: The finely divided particles of ash entrained in flue gases arising from the combustion of fuel. The particles of ash may contain incompletely burned fuel. Their trace elemental composition is characteristic of the type of fuel (e.g., coal versus oil) from which they originate.

Fog: A loose term applied to an aggregate of visible aerosols in which the dispersed phase is liquid and its base reaches the surface. Formation by condensation is usually implied. In meteorology, a dispersion of water or ice.

Fume: The solid particles generated by condensation from the gaseous state, generally after volatilization from melted substances, and often accompanied by a chemical reaction such as oxidation. Fumes flocculate and sometimes coalesce.

Gas: One of the three states of aggregation of matter, having neither independent shape nor volume and tending to expand indefinitely.

Mist: A loose term applied to dispersions of liquid particles, the dispersion being of low concentration and the particles of large size. In meteorology, a dispersion of water droplets of sufficient size to be falling.

Nuclei: Particles on which water condenses or sublimes at small supersaturations.

Nucleus counter: Any of several devices to determine the number of condensation or ice nuclei in a sample of air.

Particle: A small discrete mass of solid or liquid matter.

Smoke: Finely divided aerosol particles resulting from incomplete combustion. Consists mainly of carbon and other combustible material.

Soot: Agglomerations of particles of carbon impregnated with tar, formed in the incomplete combustion of carbonaceous material.

Terminal velocity: The constant falling speed of an object at which the drag and buoyant forces exerted by the air equal the gravitational force.

Vapor: The gaseous phase of a substance at a temperature lower than that of its critical point. It can be liquefied or solidified if sufficient pressure is applied to it.

Visibility Measurement Techniques

William Viezee, SRI International, and Richard Lewis, National Weather Service, NOAA

1 BASICS OF ATMOSPHERIC VISIBILITY

1.1 General

The term "visibility" is variously defined, but generally indicates the distance to which human visual perception is limited by atmospheric conditions. The physical and physiological mechanisms that influence visual perception during the night in distinguishing lights differ from those in the daytime in distinguishing objects illuminated by daylight. Basically, however, visibility describes the transparency of the air in the horizontal direction and represents the maximum distance that one can see in the atmosphere at any given time.

1.2 Daytime Visibility

During the daytime, the more distant a dark object is from an observer, the brighter it usually appears. At far distances, it becomes indistinguishable from the horizon sky. This limiting distance is the daytime visibility and is determined by the visual contrast threshold of the observer's eye.

The increase in brightness of an object with increasing distance from an observer is due to the scattering of sunlight by the air and to the diffuse reflection of sunlight and sky-light by underlying terrain. The scatterers of light in the atmosphere range from air molecules, condensation nuclei, dust, and fog to precipitation elements such as large water drops and snowflakes.

Air molecules and particles that are smaller than the wavelengths of light scatter light according to the Rayleigh theory with an intensity that is inversely proportional to the fourth power of the wavelength of the light. Thus, small particles scatter sunlight more in the blue than in the red wavelengths, and therefore, the presence of light haze or smoke in the atmosphere is identified by the bluish appearance of dark objects and distant terrain features or by the yellow-to-orange color of the horizon sky at sunrise and sunset.

Particles larger than the wavelengths of visible light scatter light independently of wavelength. Their presence in the atmosphere is detected by a whitish appearance of the sky, especially toward the sun. White cirrus clouds and fog and the whitish appearance of distant terrain under conditions of relatively high humidity are also visible evidence of light scattering by large particles. An atmosphere free of particles shows a decrease of spectral purity towards the horizon, and consequently the horizon sky can be white even without large particles.

Most cases of reduced and poor visibility are due to the scattering and attenuation of light by the larger particles, especially those found in fog and precipitation elements.

1.3 Nighttime Visibility

The visibility at night is usually taken as the maximum distance at which unfocused lights are visible. The illuminance at a distance R from a light source of candle power I_o in an atmosphere having a scattering coefficient k is given by

$$E = (I_o e^{-kR})/R^2$$

The light source will be visible as long as the illuminance E exceeds the threshold value at the human eye. Thus, nighttime visibility is greatly dependent upon the optical transparency of the atmosphere which, in turn, is determined by the atmospheric scattering characteristics.

1.4 Fundamentals

From the brief statements above, it appears that

• Visibility represents an observation made with the unaided eye and indicates the distance to which the visual perception of objects by day and of lights by night is limited by atmospheric conditions.

• Light scattering by air molecules and suspended particles in the intervening atmosphere is the most important factor in perceiving distant objects or lights.

• Brightness contrast between viewed targets and their background enters into a quantitative determination of visibility.

• Therefore, objective observations of atmospheric visibility by an instrumental analog to the human observer must involve accepted values of daytime and nighttime visual thresholds used in conjunction with a measure of brightness contrast and a measure of the atmospheric light scattering. These are the fundamentals on which visibility measurement techniques are based.

2 THEORY OF VISIBILITY

Since scattering of sunlight and skylight by fine particles and large particles, including hydrometeors, is the most important factor that determines atmospheric transparency, optical scattering models based on the transfer of visible radiation in the atmosphere are the obvious approach to a complete theoretical treatment of visibility. There are, however, great difficulties in formulating all human and physical factors in a real (nonstandard) atmosphere that are related to visibility. Although much progress has been made toward the formulation of workable visibility simulation models (e.g., Middleton et al., 1981; Latimer, Bergstron, et al., 1980; Ozkaynak et al., 1979), most of the complications associated with these efforts are currently avoided by the availability of a treatment devised by Koschmieder (Middleton, 1952). Koschmieder's treatment develops the concept of the visual range of objects during daylight that depends on the reduction in visual contrast between a dark object and a uniform background sky (clear or overcast). A brief presentation of the principles involved in Koschmieder's theory follows.

2.1 Principles of Visual Contrast

Referring to Fig. 1, let $(B_o - B_b)$ be defined as the natural brightness (luminance) difference between an object and its background. The natural (inherent) contrast, C, of the object with respect to its background is then defined as

$$C = |(B_o - B_b)/B_b| \qquad (1)$$

which is a fundamental measure of the discernibility of the object to the human eye (Middleton, 1952).

When the object and its background are viewed by an observer located at a distance R, attenuation by atmospheric particles and gases along the viewing path reduces B_o and B_b to the values $T(R)B_o$ and $T(R)B_b$, respectively, where $T(R)$ is the atmospheric transmittance of the path between the observer and the viewed object. At the same time, atmospheric particles and gases along the path scatter sunlight,

skylight, and diffusively reflected light from the ground into the eye of the observer, thereby producing an apparent brightness, B_a, frequently referred to as the "airlight." The combined result of attenuation and airlight yields the following apparent brightness coming from the directions of the object and its background, respectively:

$$B_o' = T(R)B_o + B_a \qquad (2a)$$

$$B_b' = T(R)B_b + B_a \qquad (2b)$$

The apparent contrast, C', of the object with respect to its background as perceived by the observer at range R is now defined as:

$$C' = |(B_o' - B_b')/B_b'| \qquad (3)$$

By use of Eqs. (1) and (2), Eq. (3) becomes

$$C' = C(B_b/B_b')T(R) \qquad (4)$$

and this is the most general expression for the law of contrast reduction by the atmosphere.

2.2 Quantitative Determination of Daytime Visibility

2.2.1 Nonreflecting (Black) Objects

For a distant ground-based target viewed in a horizontal direction against the horizon sky, Eq. (4) can be rewritten as

$$C' = C(B_h/B_h') \exp(-\bar{\sigma}R) \qquad (5)$$

where B_h = horizon-sky brightness seen at the target location
B_h' = horizon-sky brightness seen at the observer location
R = distance between target and observer
$\bar{\sigma}$ = atmospheric extinction coefficient averaged over distance R and over the visible spectrum

If constant atmospheric extinction and airlight per unit length in an infinite horizontal plane are assumed, the horizon-sky brightness is the same at the observer as it is at the target $(B_h' = B_h)$, and therefore,

$$C'/C = |(B_o' - B_h')/(B_o - B_h')| = \exp(-\bar{\sigma}R) \qquad (6a)$$

where B_o' = apparent target brightness measured at distance R
B_o = intrinsic ("close-in") target brightness
B_h' = horizon-sky brightness measured at distance R

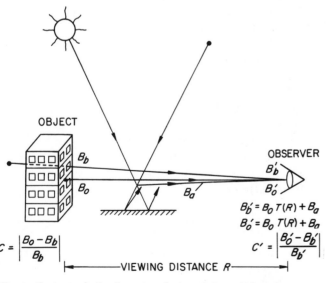

$$C = \left| \frac{B_o - B_b}{B_b} \right| \qquad C' = \left| \frac{B_o' - B_b'}{B_b'} \right|$$

$$B_b' = B_o T(R) + B_a$$
$$B_o' = B_o T(R) + B_a$$

Fig. 1 Contrast reduction from atmospheric scattering and absorption.

The natural landmark or artificial target viewed by the observer can be black [nonreflecting: $(B_o = 0)$] or nonblack $(B_o \neq 0)$. For an ideal black target, $C = 1$ and $B_o = 0$ in Eq. (6a). Then

$$C' = |(B_o' - B_h')/B_h'| = \exp(-\bar{\sigma}R) \qquad (6b)$$

If an observer recedes from a black object with the horizon sky behind it, the apparent contrast C' decreases according to Eq. (6b) until at some distance R, it becomes equal to the visual contrast threshold, ϵ. This distance R is defined as the visual range, V, and

$$\epsilon = \exp(-\bar{\sigma}V) \qquad (6c)$$

The contrast threshold, ϵ, of the observer's eye is usually assumed to be constant at 0.02 or 0.055. Actually, its precise value depends on various factors of illumination and on the distant-target angle subtended by the eye.

Using $\epsilon = 0.02$, Eqs. (6b) and (6c) can be combined to give the Koschmieder formula for daytime visual range:

$$V = 3.91/\bar{\sigma} \qquad (6d)$$

or

$$V = 3.91R/\ln C' \qquad (6e)$$

Equations (6d) and (6e) show that, in principle, it is possible to determine daytime visual range by measuring either the atmospheric extinction coefficient $(\bar{\sigma})$ or the apparent brightness (luminance) contrast (C') of one or more targets at a known distance R. If an optical instrument is used to measure apparent luminance contrast, the distance R of suitable targets is important, as shown by Fig. 2. This figure shows the apparent brightness contrast of black (nonreflecting) targets viewed against a uniform background sky as a function of visual range. It indicates that the distance of objects selected for contrast measurements should increase as the visual range increases. For example, the apparent contrast of a black target located at a distance of 5 km from the observer is very sensitive to changes in visual range between 5 and 15 km, but shows a much reduced sensitivity for higher values. When the visual range is 20 km, the contrast of a target at 15 km would provide accurate indications of changes in atmospheric transparency.

2.1.2 Reflecting (Nonblack) Objects

Most natural objects and terrain features used by an observer to determine visibility are not perfectly black. Malm et al. (1982) list values of intrinsic (close-in) target contrast C (see Fig. 1) for a large number of natural targets in the southwestern United States that range from 0.66 to 0.88. Thus, 12–34 % reflectivity is implied. For such targets, Eq. (6e) takes the form

$$V = 3.91R/\ln(C'/C) \qquad (6f)$$

where C'/C is defined by Eq. (6a).

The sensitivity of the nonblackness value (i.e., $B_o - B_h'$) to the determination of visual range according to Eq. (6f) is shown in Fig. 3 for targets at an arbitrarily selected distance of 9 km. Nonblackness (on the ordinate) is expressed as a percentage of the background-sky brightness, B_h'. The results indicate that determination of visual range is most sensitive to the nonblackness of a viewed object when the object distance is small with respect to the atmospheric visual range. For example, given the 9-km distant target with a natural brightness equal to 20 % of the background sky ($B_o = 0.2B_h'$) and a 50-km visual range, the assumption that the target is black will give a visual range of only 35 km. It is evident, however, that small deviations from blackness ($\leq 10\%$) do not introduce significant error. In general, Fig. 3 reiterates the conclusion of Fig. 2 for black targets: Eq. (6e) or (6f) is most accurate when the target or object is at a distance that is comparable to the actual visual range.

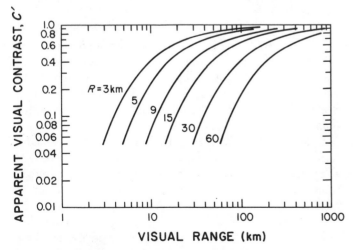

Fig. 2 Apparent brightness contrast as a function of visual range for black (nonreflecting) targets at various distances viewed against the horizon sky.

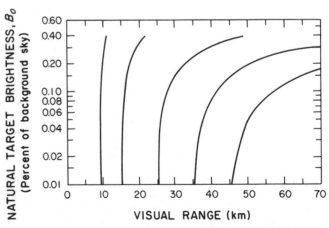

Fig. 3 Computed visual range as a function of natural (inherent) brightness of a target viewed at a distance of 9 km. Data show errors in visibility when viewed target is not perfectly black ($B_o \neq 0$).

2.3 Quantitative Determination of Nighttime Visibility

During nighttime, visibility is determined using the concept of the visual range of lights as defined by Allard's law. Allard's law gives the illumination from a point source of light at any distance in a partly transparent medium as:

$$E = (I_o/R^2)\exp(-\bar{\sigma}R) \qquad (7a)$$

where E is the illuminance at distance R from a source of luminous intensity I_o (e.g., candles) in a medium of extinction coefficient $\bar{\sigma}$.

If a threshold illuminance E_t (e.g., mile-candles) corresponding to the observer's eye is substituted for E in Eq. (7a), the corresponding value of R will be the visual range V, as follows:

$$E_t = (I_o/V^2)\exp(-\bar{\sigma}V) \qquad (7b)$$

Although the foregoing law was indeed published by Allard in 1876, Middleton (1964), in his role as a historian of science, has found that Bouguer stated this law much earlier, in 1729.

Allard expressed Eq. (7b) in terms of atmospheric transmissivity T (per unit distance) as follows:

$$E_t = I_o T^V/V^2 \qquad (7c)$$

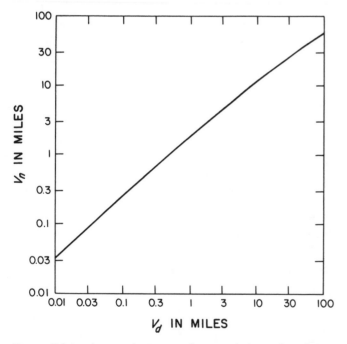

Fig. 4 Relation between daytime visual range of objects obtained from Koschmieder's law (V_d) and the nighttime visual range of a 25-candle light source obtained from Allard's law (V_n) during conditions of equal atmospheric transmissivity (Douglas and Booker, 1977).

Equation (7c) is the most common expression of Allard's law. It shows that the visual range of lights of known luminous intensity I_o can be obtained from a measurement of atmospheric transmittance in conjunction with accepted values of threshold illuminance E_t.

Threshold illuminance is not a constant. It is a function of the luminance of the background of the light, the position of the light in the field of view, and the angular size and shape of the light, its color and, if not steady burning, its flash characteristics. In addition, the observer's knowledge of the position of the light and his or her time for search have a significant influence in the threshold. For current applications of Eq. (7c), a value of $E_t = 2$ mile-candles has been chosen for the nighttime illuminance threshold, and a value of 1,000 mile-candles when light sources are viewed during daylight conditions against a relatively bright background.

Since operational application of Eq. (7c) involves measurements of atmospheric transmittance obtained with fixed-base-line transmissometers, it is useful to express Eq. (7c) somewhat differently so that data from standard transmissometers (500- or 250-ft base line) can be input directly. A convenient form is:

$$E_t = [I_o(T_b)^{V/b}]/(V/5,280)^2 \qquad (7d)$$

where b is the base line over which atmospheric transmittance is sampled, T_b is the atmospheric transmittance over the path b, and b and V are given in feet.

2.4 Comparison between Nighttime and Daytime Visual Range

Figure 4 shows the relation between the visual range obtained using dark objects during daytime (V_d) by means of the Koschmieder formula and the nighttime visual range of lights obtained by means of Eq. (7d). The intensity of the lights is taken as 25 candles in Allard's law. A value of 0.055 instead of 0.02 is used for visual contrast threshold in Koschmieder's law. When the atmospheric transmissivity, and consequently the visibility, is very low, a 25-candle light can be seen at night more than three times as far as a large black object would be seen during the day in the same atmosphere. This difference decreases as the visibility improves, becoming about two to one when objects can be seen from 0.5 mi away, and equal when the visibility is about 16 mi. When the visibility is greater than 16 mi, objects will be seen farther than 25-candle lights.

Equation (7d) can also be applied to the viewing of lights during daytime by assuming a daytime illuminance threshold E_t equal to 1,000 mile-candles. This means that when a light source of luminous intensity I_o is moved away from an observer to a distance R where its illuminance at the observer's location has decreased to 1,000 mile-candles, this distance R equals the daytime visual range for the light source. Computations by Douglas and Booker (1977) show that in daytime in low visibilities during thick fog, lights of even moderate intensity can be seen at distances farther than black objects.

The above-described relation between the visual range of lights and daytime objects is important in the design of airport runway lights and runway approach-light systems.

3 APPLICATIONS OF VISIBILITY

3.1 Weather Analysis

Visibility is part of the standard synoptic weather observation. These observations are taken daily by trained observers at designated stations at 0000, 0600, 1200, and 1800 GMT. Figure 5 shows the symbolic station model and a sample meteorological surface observation. The reported visibility is the prevailing visibility as defined below in Section 3.2. The reported prevailing visibility (VV) in Fig. 5 is 0.75 mi (statute mile) in light, continuous snow. Figure 6 shows how prevailing visibility is determined.

Visibility from a meteorological observation is used in

- Synoptic weather analysis to characterize air masses (continental vs maritime, and polar or arctic vs tropical)
- Mesoscale analysis to aid in the identification of the stability of the boundary layer, the intensity of precipitation, and the density and spatial extent of fog.

In recent years, the usefulness of visibility in weather analysis has declined because of the increasing effects from local and regional pollution sources on reported visibilities.

3.2 Aircraft Operations

3.2.1 Aviation Weather Observations

Weather observations for special application to aircraft operations are taken at all aviation reporting stations (e.g., National Weather Service [NWS], Federal Aviation Administration [FAA], U.S. Air Force, and U.S. Navy) as often as is necessary to record rapidly changing conditions, particularly those related to low cloud ceiling and low visibility. The following visibilities are recorded:

3.2.1.1 Visual Range

Visual range is determined by a local weather observer and defined as the maximum distance at which an object (preferably black or nearly black) or low-intensity light can be distinguished. Visual range and visibility are used interchangeably.

3.2.1.2 Prevailing Visibility

Prevailing visibility is determined by the local weather observer, and defined as the greatest visual range equaled or exceeded throughout at least half the horizon circle, which need not necessarily be continuous. Thus, prevailing visibility is dependent upon the variation in visual range with azimuth.

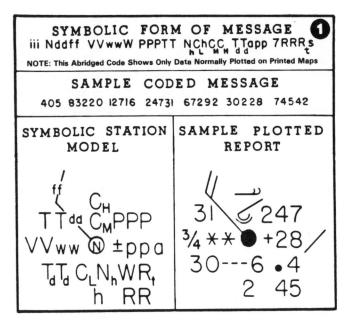

Fig. 5 Example of synoptic meteorological observation used in weather analysis. Reported prevailing visibility is 0.75 mi in continuous falling snow.

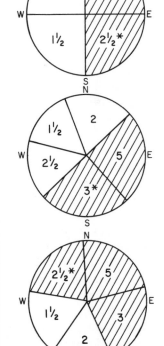

Four Sectors	
Visibility (mi)	Approximate Degrees Azimuth
5	90
2½*	90
	180
2	90
1½	90

Five Sectors	
Visibility (mi)	Approximate Degrees Azimuth
5	100
3*	90
	190
2½	60
2	50
1½	60

Five Sectors	
Visibility (mi)	Approximate Degrees Azimuth
5	72
3	72
2½*	72
	216
2	72
1½	72

Fig. 6 Determination of prevailing visibility. When the visibility is tabulated in descending order of range by sectors, and the width of the sectors is summed progressively as shown, the prevailing visibility corresponds to the value for the first sector that increases the sum to 180° or more. Prevailing visibilities are indicated by asterisks and cross hatching.

3.2.1.3 Tower Visibility

Tower visibility is defined as the prevailing visibility determined at the control-tower level whenever the prevailing visibility at the weather station level is less than 4 mi. Control-tower personnel are required to maintain a continuous meteorological watch on the visibility and report any sudden changes to the weather observer.

3.2.1.4 Runway Visibility Value (RVV)

Runway visibility value is defined as the visibility along an identified runway. Two different observing techniques can be used:

• *Visual observations.* Standing at the runway edge at the threshold of the in-use runway, reportable RVV values are obtained by counting high-intensity runway lights (HIRL) along the opposite side of the runway. If HIRL are inoperative or not installed, check points or markers are used at known distances along or near the runway.

• *Transmissometer measurements.* When a transmissometer is used along an identified runway to determine the RVV, the instrumental value is calibrated in terms of the human observer, i.e., the sighting of dark objects against the horizon sky during daylight, and the sighting of moderately intense unfocused lights on the order of 25 candles at night. Conversion tables from atmospheric transmittance to RVV are based on either Koschmieder's law for visual range in daytime ($V = \ln \epsilon / \ln T$) or Allard's law for the visual range of lights at night ($E_t = IT^V/V^2$).

3.2.1.5 Runway Visual Range (RVR)

Runway visual range is defined (in the United States) as an instrumentally derived value based on standard calibrations that represents the horizontal distance a pilot will see down the runway from the approach end; it is based either on the sighting of HIRL or on the visual contrast of other targets—whichever yields the greater visual range. RVR is normally obtained from transmissometers located alongside the runway and about 14 ft higher than the center line of the runway (cockpit height). Conversion from atmospheric transmittance along the base line of the transmissometers to RVR is made using Allard's law for various settings of the HIRL.

Figure 7 shows an example of aviation weather observations. Prevailing visibility in miles is entered in column 4, and the obscuration to visibility is noted in column 5 (F represents fog). Column 13 reports runway visibility (VV) in miles and/or fractions of miles for the designated runway (R18 is runway 180). Significant visibility differences observed in any direction (east or north) are indicated also in column 13.

3.2.2 Criteria for Instrument Landing Operations

Measurements of RVR are critical in determining whether or not a pilot can make an approach to an instrument runway. Landing criteria for instrument runways are based on the operational performance categories listed in Table 1.

TABLE 1
Performance Categories
for Instrument Landings

Category (CAT)	Decision Height	Runway Visual Range Limits[1]	
		Meters	(Feet)
I	200 ft	800+	(2,600+)
II	100 ft	400–800	(1,200–2,600)
IIIA	none	200–400	(700–1,200)
IIIB	none	50–200	(150–700)
IIIC	none	0–50	(0–150)

[1]A runway light intensity of 10,000 candles is assumed.

Under Category I, an instrument runway is served by instrument landing system (ILS) or ground control approach (GCA) aids and visual aids intended for operations down to a 60-m (200-ft) decision height, and down to an RVR on the order of 800 m (2,600 ft). Similarly, under Category II, an instrument runway is served by ILS and visual aids intended for operational use down to a decision height of 30 m (100 ft) and down to an RVR of 400 m (1,200 ft). In this case, 100 ft is the minimum height above the ground to which a pilot making an instrument aproach may descend without reference to lights or markers on the ground before executing a missed approach. Category III involves no decision height but depends only on RVR readings. A Category IIIC runway can be approached without reliance on external visual references.

3.2.3 Automated Weather Observation Systems

Extensive observations of surface weather parameters are required to support aviation operations and general weather prediction and warning services. While these observations are instrumented to some extent, a large proportion of the operation is still manual, making this function highly labor-intensive and expensive. Approximately 1,200 observational sites are operated within the United States by three government departments: NWS (Department of Commerce), the Air Weather Service (AWS—Department of Defense), and FAA (Department of Transportation). Automated observing equipment development has been under way in the three agencies for some time. Development of sensors to augment and eventually replace the traditional visual observations of sky condition, visibility, and obstructions to vision is considered to be practical (e.g., Moroz, 1977; Chisholm et al., 1980; Brown, 1980).

Test facilities have been set up in Arcata, California (FAA), Otis Air Force Base, Massachusetts, and Sterling, Virginia (NWS), for sensor evaluation. There has been continuing interagency coordination to establish test objectives and analyze and interpret test results. In general, efforts to automate visibility have fallen short of the goal of obtaining instruments that are suitable for providing representative visibility indications over a wide visibility range, and in different weather conditions. However, recent advances in elec-

Time entries on this form are __90__ th meridian time

To convert to GMT {add / sub.} __6__ hours

Height of barometer __444__ feet (MSL)

SURFACE WEATHER

TYPE (1)	TIME (LST) (2)	CEILING (hundreds of feet) AND SKY (3)	VISIBILITY (miles) (4)	WEATHER AND OBSTRUCTIONS TO VISION (5)	SEA LEVEL PRESS. (mb) (6)	TEMP. (°F) (7)	DEW PT. (°F) (8)	WIND DIRECTION (9)	WIND SPEED (knots) (10A)	MAGNETIC WIND DIR. (degrees) (10B)	CHARACTER AND SHIFTS (11)	ALTIMETER SET. (in.) (12)	REMARKS AND SUPPLEMENTAL CODED DATA (13)	(14A)	(14B)	OBSERVER'S INIT. (15)
S	0034	O	6	H				→	5	320						DR
RS	0058	O	4	GFH	129	63	61	→	3	320		990	108 84 TWR VSBY 5 GF DEP	50		DR
S	0113	O	2	F				C					R18 VV 1½			DR
S	0128	O	3	F				C					R18 VV 1½ VSBY N 2			DR
L	0143	O	3	F									R18 VV 2			DR
RS	0158	O	2	F		62	61	C				990	R18 VV2 VSBY N ¾			DR
S	0210	-X	1½	F				C					R18 VV 2 F3 VSBY N ¾			DR
L	0225	-X	1½	F									R18 VV 1½ F3			JM
L	0240	-X	2	F									R18 VV2 F4 VSBY N 1½			JM
L	0255	-X	1½	F									R18 VV2 F4			JM
R	0259	-X 80 ⊕	1½	F		62	61	↗	3	250		990	R18 VV1½ F4 VSBY ½ E			JM
L	0314	-X 80 ⊕	1	F									R18 VV 1 F4			JM
L	0329	-X 80 ⊕	1	F									R18 VV 1 F4 VSBY E ½			JM
S	0340	-X 80 ⊕	¾	F				C					R18 VV ½ F4			JM
S	0355	-X E 80 ⊕	¾	F				C					R18 VV ½ F6			JM
RS	0359	-X E 80 ⊕	⅜	F	129	61	61	C				990	R18 VV ¾ F8 302 1050			JM
S	0410	W1 X	⅛	F				C					R18 VV 3/16			JM
S	0425	W0 X	O	F				C								JM
L	0440	W0 X	⅛	F									R18 VV ⅛			JM
L	0455	W0 X	O	F												JM
RS	0458	W1 X	1/16	L--F		60	60	↗	4	250		990				JM
S	0510	W2 X	⅛	R-F				↑↗	4	230			R18 VV ⅛			JM
L	0525	W2 X	3/16	R-F									R18 VV ⅛			JM
S	0535	W4 X	½	R-F				↗	5	250			R18 VV ¾			JM
L	0550	W4 X	⅝	R-F									R18 VV ¾ VSBY N 1			JM
RS	0558	W5 X	1	R-F		60	60	↗	5	250		991	R18 VV 1½ VSBY N 2			JM
S	0612	-X 8 ⊕ M12 ⊕	1½	R-F				↑↗	3	230			R18 VV 2 F4			JM
L	0627	-X 8 ⊕ M14 ⊕	1¾	R-F									R18 VV 1½ F5			JM
S	0638	-X 14 ⊕ E 80 ⊕	2	F				↗	6	250			F2			FY
S	0653	-X 20 ⊕ E 80 ⊕	3	FH				↗	6	250			F2			FY

Fig. 7 Example of aviation weather observations recorded under low-visibility conditions that report prevailing visibility (column 4), obstruction to visibility (column 5), and the runway visibility (VV) along runway 180 (R18) in column 13.

tronic technology have given new insight into the problem of visibility measurement, and show promise of overcoming earlier sensor limitations. Some of these newer sensors are described in Sections 4 and 5 and comparisons with earlier-generation sensors are given.

3.3 Air Pollution

Amendments to the Clean Air Act issued on 7 August 1977, established as a national goal "the prevention of any future and the remedying of any existing impairment of visibility in mandatory Class I federal areas, which impairment results from manmade air pollution." Class I federal areas are international parks, national wilderness areas, national memorial parks exceeding 5,000 acres, and national parks exceeding 6,000 acres. The term "impairment of visi-

bility" included atmospheric discoloration and reduction in visual range. The Act required the EPA to recommend methods to implement this national "visibility" goal.

Subsequent to the Clean Air Act Amendments of 1977, an extensive research effort was implemented to develop techniques to monitor and evaluate the effects of local and regional air quality on atmospheric visibility. Air-pollution-related work in visibility has emphasized the human visual perception of air quality (e.g., Malm et al., 1981; Latimer, Hogo, et al., 1981; Henry et al., 1981) and the development of an instrumental analog closely paralleling the human observer (e.g., Evans and Viezee, 1982). Some measurement techniques are discussed under Section 4. The overall research effort is summarized best in the special issue of *Atmospheric Environment* titled *Plumes and Visibility: Measurements and Model Components* (Vol. 5, No. 10/11, 1981, 1785-2406).

3.4 Highway Fog

Dense fog is a threat to the safe and efficient operation of motor vehicles. The hazards of dense fog have intensified with the proliferation of freeways, expressways, and other highly improved roadway systems. Research to improve visibility guidance of motor vehicles through fog was initiated in 1967. In 1976, the Transportation Research Board of the National Research Council published a survey on available fog-measuring instrumentation for highways (Heiss, 1976). In general, the visibility measurements addressed the specific problem of determining the maximum distance at which a stopped vehicle in a traffic lane could be detected during fog conditions so that oncoming vehicles could be alerted.

Instruments that measure the atmospheric scattering characteristics appear to be the most suitable for general highway fog use. Transmissometers that measure the atmospheric extinction coefficient would be more pertinent. Transmissometers, however, are cumbersome and expensive to install and use for highway applications. The scatter types of instrument currently available are discussed in Section 5.

4 TECHNIQUES FOR MEASURING BRIGHTNESS CONTRAST

4.1 Manual Telephotometers

Quantitative measures of daytime visual range can be obtained from Eq. (6e) (Section 2) by selecting one or more dark objects or terrain features of known distance and measuring the apparent brightness contrast between the objects or terrain features and the (preferably uniform) background sky. Simple measurements of this type can be made with photometers. Specially designed instruments to measure contrast, called telephotometers, are commercially available (e.g., the VistaRanger, marketed by Meteorology Research, Inc. [MRI], Altadena, California).

A telephotometer combines the focusing optics of a telescope and the radiation detector of a photometer. The telescope objective lens focuses a distant object on the photodiode. For the MRI-type telephotometer, the detection area is 0.31 mm, an area that subtends a field of view of 0.035°. A filter turret holds four narrow-band interference filters, the effective band centers of which are set at wavelengths of 405 nm (violet), 450 nm (blue), 550 nm (green), and 630 nm (red). A beam diverter shifts (by manual operation) the position of the detection point by an elevation angle of 0.34° so that both sky and target readings can be easily made. The eyepiece allows the user to position the detection image precisely on the desired distant object, and the photodiode converts the radiant energy to a current. The instrument then transforms the current signal to a digital readout. The telephotometer can be manually positioned with respect to azimuth and, thus, can measure the radiant intensity in any particular direction from the observer. The instrument is often referred to as the contrast telephotometer, because it measures the two-point ratio of target-to-sky brightness. Thus, the instrument's response must be linear, and absolute values of radiant intensity are unimportant. The manual mode of the telephotometer instrument is being updated very rapidly to a semi-automated mode of operation.

4.2 Vertically Scanning Teleradiometer

Many of the federal Class I areas (see Section 3.3) are associated with scenic elements that require an observer to look over or through pollution "basins" or "corridors." These corridors tend to hold or funnel particulates so that layers of haze obstruct the ability to see a vista. Additionally, a suspended haze layer (plume blight) can impair the scenic qualities of the vista. Many available manually operated telephotometers are restricted to measuring radiance at one point in the sky and on the selected target, and thus, may look over or under confined layers of haze. A scanning telephotometer, measuring radiance values at many points within, for example a vertical section of a vista, would allow for determination of a more realistic impact of confined haze layers on visibility.

A continuously operating scanning telephotometer has been designed by Persha and Malm (1980). It is an integrated system consisting of a Model 3030 VistaRanger telephotometer, a computerized controller, motor-powered image scanner, hand-held terminal, digital cassette recorder, telescope, and power supply unit. This instrument allows atmospheric radiance to be measured and recorded in four different colors at 40 discrete positions along a 2.13° vertical path. The computerized controller serves the dual purpose of controlling the vertical image scanner/telephotometer functions, and working as a data logger with data recorded on a standard 300-ft digital cassette tape. Programming of the controller is done in a special high-level language designed specifically for the vertical image scanner and telephotometer. A real-time clock built into the controller makes predetermined operating sequences at selected times possible. The instrument only automates scanning along the vertical direction. Positioning in the azimuthal (horizontal) direction around the horizon circle must still be done manually.

This instrument is a great improvement over the manually operated (two-point measurement) telephotometer. It enables the measurement of a detailed profile of radiance values from terrain and background sky that can be processed in terms of contrast and contrast variations that are relevant to daytime visibility observations.

4.3 Video-Camera Application

Since 1979, work has been done to assemble, calibrate, and field-test an instrument that closely parallels the human observer's assessment of contrast and visibility. The process by which an observer determines visibility involves the spatial viewing or scanning of visibility targets with subsequent spatial integration of contrast information. The new instrument implements the concept of using a microcomputer-controlled, scanning video-camera system with digital data acquisition to quantify the visual contrast of objects by day and of lights by night.

A prototype instrument has been assembled and tested by SRI International for the Electric Power Research Institute (EPRI) in Palo Alto, California. The sensor element is a small Reticon Model LC100 solid-state, linear-array video camera placed on a precision rotary mount. The camera housing is $2.8 \times 4 \times 6$ in. in size. Under microcomputer control, the camera scans and views selected, preprogrammed visibility scenes or targets within the 360° horizon circle. Luminance (brightness) images of the scenes or targets are currently stored in digital form on flexible magnetic disks. When programmed for automated, routine monitoring of visual range and/or prevailing visibility (see Section 3.2), the sensor operates in a photopic mode to obtain scene-luminance measurements within the spectral response of the human visual system.

Data acquisition and analysis are completely automatic. Analysis programs process the stored sensor-records to provide numerical values of apparent contrast and visual range, based on a given value of contrast threshold. Stored digital luminance data can be displayed in the form of gray-scale images of the viewed visibility targets on a standard TV monitor. The image displays provide for both near-real-time monitoring of overall system performance and for visual inspection and interactive postanalysis of the recorded visibility conditions.

A more detailed description of the hardware and of the data analysis software is given by Viezee and Evans (1979, 1980, 1983).

5 MEASUREMENT TECHNIQUES FOR OPTICAL SCATTERING AND EXTINCTION

5.1 Background of Sensor Development

In the early 1940s the National Bureau of Standards (NBS) developed the transmissometer, a device that measures how much light passes through a known distance while being scattered or absorbed by particles in the intervening atmosphere. The measurement principle is termed "extinction." The 1950s saw the research and development necessary for the operational implementation of the transmissometer for the measurement of visual range along airport runways. NWS, with the support of the FAA, led this effort.

In the late 1950s Curcio and Knestrick (1958), using a pulsed-strobe light source, measured the light intensity scattered backward toward the light source by atmospheric particles and related the so-called backscatter to the visual range as determined from an NBS transmissometer. A commercially available backscatter device, the Impulsphysik Videograph, was tested by NWS in the late 1960s. George and Lefkowitz (1972) defined the concept of sensor equivalent visibility (SEV), which involved an objective technique to

report visibility with sensors. Using this concept, an effort was made to adapt the Videograph for use at automated weather stations. Comparisons of this sensor with collocated human observers led to the derivation of a best-fit regression equation by Hochreiter (1973), relating the Videograph output in fog or haze to observed visibility.

The mid-1970s saw the installation of Videographs at a few unmanned observation stations (AUTOB, for automated observation). An algorithm for processing data from a triangular Videograph array was later used to meet the requirements for reporting prevailing visibility. This method of determining visibility was included as part of a totally automated aviation observation (AV-AWOS, for aviation automated weather observation system) for an operational test in Newport News, Virginia, as reported by Bradley et al. (1978). A particular problem in these operational environments was the unrepresentativeness of the visibility reports in precipitation, and since the Videograph was calibrated for fog and haze, this disparity was not surprising. The Videograph-derived visibility was frequently too low in moderate haze.

Concurrent with NWS efforts in backscatter development, the Canadian Atmospheric Environment Service (AES) conducted similar tests leading to operational implementation of the Videograph. They also conducted tests to compare a wide variety of scattering-type sensors with human observers and transmissometer data. The results were reported by Sheppard (1978). Canadian results, along with testing done by the Air Force (Chisholm and Jacobs, 1975), led to encouragement that "forward-scatter" measurements could provide more representative visibilities in a wide range of atmospheric obscuring phenomena. Forward-scatter sensors measure the light that is scattered at angles of about 10–50° off the axis of the direct beam (in a direction away from the light source) and have no direct beam component. A particular sensor, the EG&G 207 Forward Scatter Meter, was refined and applied by the Air Force in numerous visibility experiments, including an experiment described by Brown (1980) to determine the weather type (fog, rain, snow, etc.) by examining the characteristic outputs of different types of visibility sensors.

5.2 Recent Experiments

In the late 1970s, rapid developments in microcomputers and optoelectronics technology provided encouragement that fairly complete surface aviation observations, including the subjective elements (clouds, visibility, and present weather), could be automated at a reasonable cost. This gave added impetus to the search for reliable field-tested and inexpensive visibility sensors.

The FAA sponsored a number of visibility tests conducted by the Transportation Systems Center (TSC). Test results from Otis Air Force Base, Massachusetts, and Arcata, California, were reported by Burnham and Collins (1982). Additional tests in simulated weather extremes were performed at the large climatic

TABLE 2
Instruments Evaluated at Test and Evaluation Division Visibility Facility

Instrument	Measurement Principle	Path Length or Measurement Volume	Light Source	Detector Characteristics	Scattering Angle	Range of Measurement (miles)
Tasker transmissometer (2)[1]	Extinction	500 ft	Incandescent sealed beam lamp	Silicon photodetector with optics		0.1–4
Videograph (6)	Backscatter	13 ft³	Parabolic reflector and xenon flashlamp	Silicon photodetector with optics	177°	0.3–12
EG&G 207 (3)	Forward scatter	1.7 ft³	Mechanically chopped quartz halogen lamp	Silicon photodetector (focused)	20–50°	0.05–20
Fog 15	Forward scatter	0.9 ft³	Mechanically chopped quartz halogen lamp	Silicon photodetector	16–40°	0.05–12
FS-3	Forward scatter	0.011 ft³	Xenon flashlamp	Silicon photodetector with optics	40°	0.05–3
EV-1000	Forward scatter and extinction	0.03 ft³ and 4 ft	Xenon flashlamp	Silicon photodetector (focused)	20°	0.05–2
HSS VR-301	Forward scatter	0.014 ft³	Pulsed LED (980 nm)	Hybrid silicon photodetector (focused)	27–42°	0.05–12
Skopograph transmissometer	Extinction	500 ft	Parabolic reflector and pulsed xenon flashlamp	Silicon photodetector with optics		0.1–4
ASL-FSM	Forward scatter	0.34 ft³	Xenon flashlamp	Silicon photodetector	20–50°	0.3–30

[1]Numbers in parentheses following the sensor name indicate that more than one sensor was evaluated.

chamber at Eglin Air Force Base, Florida, described by Burnham (1983). This led to the identification of a number of sensors that were deemed suitable for measuring extremely low visibilities. Some of the available state-of-the-art sensors covered a wide enough range of visibilities that they were also candidates for use at automated stations.

A number of these sensors were subsequently made available to NWS for installation and evaluation at the Test and Evaluation Division (T&ED) in Sterling, Virginia.

5.3 Summary Description of Sensors

Table 2 is a summary of instrument characteristics. Note that the range-of-measurement column is not the manufacturers' stated range but is usable sensor range that might be expected in daytime. This was determined by comparing sensor-derived visibilities (based upon the manufacturers' calibrations and conversions) with concurrent daytime human observations. Enough comparisons were made to justify confidence that sensors were capable of covering the range given. Sufficient data were not available for nighttime determinations.

5.3.1 Tasker Transmissometer RVR-500

The two 500-ft base-line transmissometers used in the evaluations were updated versions of the original NBS transmissometer. A dramatic improvement in the overall sensor performance was noted when the old tube-type electronics were replaced with solid-state modification kits purchased from Tasker Systems. Calibration stability was greatly improved, providing 1% or less difference in output between the two systems.

5.3.2 FF Impulsphysik Videograph

Throughout the evaluations, at least three Videographs were installed in the T&ED visibility facility in Sterling, Virginia. However, the most reliable were two that were specially built for AUTOB field sites. These were designed to perform in Arctic environments and therefore had been specified to stringent environmental tolerances, and had special temperature-compensated printed circuit boards to minimize output drift at low temperatures.

5.3.3 EG&G 207 Forward-Scatter Meter

The EG&G 207 sensor evolved as the Air Force gained experience with it in numerous field experiments. Two sensors were supplied by the Air Force. Earlier versions tested at T&ED showed significant susceptibility to sunlight noise and lamp burnouts. The Air Force units and a third unit of later design were found to be less sensitive to noise.

5.3.4 Wright and Wright Fog 15

The Fog 15 has many similarities to the EG&G 207 since it is an outgrowth of the EG&G design. However, its mechanical chopping rate is faster than the EG&G 207, decreasing its sensitivity to background light. Other modifications have lengthened lamp life, improved maintainability, and reduced cost.

5.3.5 FF Impulsphysik FS-3

The FS-3 forward-scatter sensor, like other sensors that use a xenon flashlamp and associated electronic filtering, is relatively insensitive to sunlight. Although it does not cover as wide a range of visibilities, it was found to be generally in close agreement with other forward-scatter sensors for visibilities below 3 mi. The sampling volume is shielded somewhat by the support joining the projector and detector. This tends to reduce the unit's sensitivity to precipitation.

5.3.6 Fairchild Weston EV-1000

This unit was found to cover a smaller range of visibilities than the other sensors, making it least suitable for automated applications. However, it has two unique features that deserve special mention. It switches alternately between a transmissometer-type measurement and a forward-scatter measurement and utilizes a "smart" processor to change the time between visibility updates. When visibility is below 3 mi, the lamp is flashed continuously 28 times per minute. A mirror switches every 30 s between the extinction and forward-scatter measurement modes. When visibility is above 3 mi, there is a 3-min "off cycle" between visibility updates. A serious problem with the unit was its exterior shielding with hoods covering portions of the sampling volume, making the sensor response, as with the FS-3, relatively insensitive to precipitation. Additionally, good agreement with other sensors was found only for visibility measurements below 2 mi in fog.

5.3.7 HSS VR-301

The VR-301 is different from other forward-scatter sensors in that it measures "off-axis" forward scattering. This means that the projector and detector do not point toward each other, and thus light blocks are not needed to prevent the direct beam from reaching the detector. It is also unique in that it has a pulsed infrared LED project source and infrared filter for the detector to minimize the susceptibility to background light. State-of-the-art electronics are used to minimize noise.

5.3.8 FF Impulsphysik Skopograph Transmissometer

The Skopograph was installed on a 500-ft base line near one Tasker modified transmissometer. It has design characteristics similar to the Videograph and FS-3, the other visibility sensors built by the German manufacturer, Impulsphysik. The amplifier filters out everything but the short-duration xenon pulse and is thus insensitive to background light. It therefore can have a wider field of view, and is less sensitive to slight alignment shifts that are a problem with the NBS-type transmissometer. The Skopograph held calibration with no maintenance other than window cleaning for over a year. It agreed well with the Tasker transmissometers, which were realigned and recalibrated about once a month.

5.3.9 NWS Advanced Systems Laboratory Forward-Scatter Meter (ASL-FSM)

In response to the limited availability of commercial forward-scatter sensors, the NWS Advanced Systems Laboratory (ASL) built such a sensor in-house. The instrument incorporates state-of-the-art electronics in a low-cost design. It also uses time-tested technology such as a xenon flashlamp for long lamp life and a silicon photodiode. The photodiode is pointed downward to minimize its exposure to background light. The amplifier electronically filters the scatter signal for the short-duration xenon flash, further reducing the sensor's susceptibility to electronic and background noise. A serious problem with the sensor was its degraded performance in precipitation when water drops would accumulate on the exterior cover glasses and thus block the transmitted and scattered light. Heating the cover glasses would correct this problem.

5.4 Recalibration of the Videograph

When the ASL-FSM was completed and calibrated, it was installed in the T&ED visibility facility at Sterling, Virginia. After a period of evaluation it became the standard of comparison in a special task to improve the Videograph calibration, since the Videograph tended to indicate significantly lower visibility in light fog and haze. When the recalibration effort was under way it became clear that the Videograph response could come close to covering the dynamic range of the ASL-FSM. It was also determined that the previous calibration was seriously underreporting visibilities with respect to the ASL-FSM and other forward-scatter sensors. A linear regression equation of the ASL-FSM output to the Videograph photodiode response was then derived for cases of fog and haze, resulting in a new calibration. With the installation of the HSS VR-301 and Wright and Wright Fog 15, Videograph comparisons using the new calibration were extended to include these sensors. The results of these intercomparisons are discussed in Section 5.6. Comparisons with the EG&G 207 and transmissometer are also included in that section.

5.5 Data Collection and Analysis

In 1976 NWS established a visibility test facility at the T&ED. The specific purpose was to maintain a "standard" Videograph to be used to calibrate other Videographs before they were shipped to operational AUTOB sites. The facility evolved to include the capability of automatically collecting data from up to eight Videographs and two 500-ft base-line transmissometers with additional capacity to read other sensors. It has been used for the evaluation of off-the-shelf visibility sensors and sensor measurement techniques (extinction, forward scatter, and backscatter).

The test layout is shown in Fig. 8. Transmissometers at the 14-ft level are on each side of the central sensor pads. Videographs are located on the large center pad to facilitate calibration comparisons. Other sensors are arrayed out from this pad so as to avoid beam interference. Two EG&G 207s are located at each end of the central instrument array to provide an additional check on visibility variations near the ground.

All sensors are fed to a programmable calculator that scans analog outputs and samples the transmissometer pulse counts. A hygrothermometer, day/night indicator, and heated tipping bucket are also sampled with the visibility sensors. Data are stored on a nine-track magnetic tape, and real-time processing is included to allow immediate intercomparisons of sensor-derived visibilities and comparisons with human visibility.

5.6 Results of Analysis

Figures 9, 10, and 11 are matrices showing the number of joint occurrences of visibility values within the specified

ranges for the given instrument pair (KV = ASL-FSM, EP/EN1 = EG&G 207, SV1 = Videograph). These are 1-min averages of hourly data taken during the months of October and December 1982. Because of its limitations in precipitation, the ASL-FSM data are restricted to fog and haze cases. Figure 9 shows the good agreement between the two different forward-scatter sensors, the ASL-FSM and EG&G 207. Figure 10 shows similarly good agreement between the Videograph with the new calibration and the ASL-FSM. Finally, Fig. 11 shows the overall good agreement between the Videograph and EG&G 207. This particular figure also includes rain and snow data from December 1982. The points where the Videograph indicates lower and higher visibility were generally the result of selective backscatter responses to snow and rain, respectively, which will be discussed later. The Videograph in fog and haze showed excellent agreement with the EG&G 207, further verifying the new equation.

A breakdown of visibilities occurring with different obstructions to vision reveals areas of agreement/disagreement that are dependent upon the method of sensor measurement. Comparisons between the transmissometer (TW),

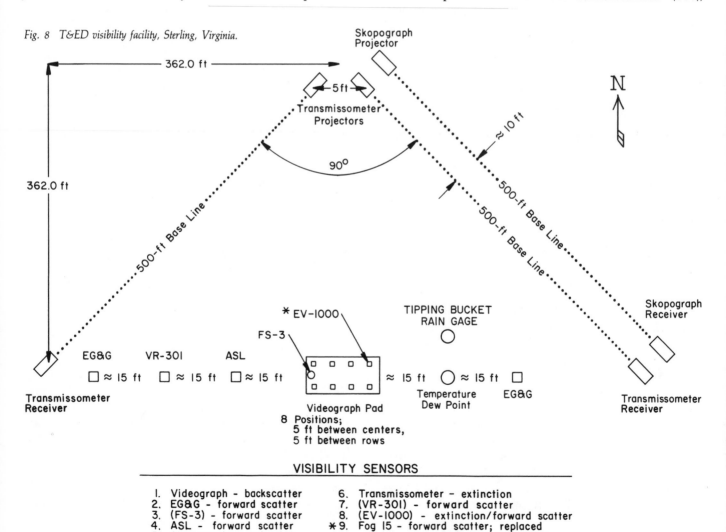

Fig. 8   T&ED visibility facility, Sterling, Virginia.

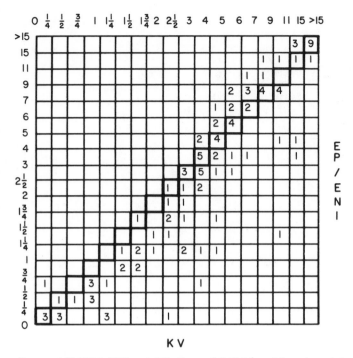

Fig. 9 ASL-FSM (KV) and EG&G 207 (EP/EN1) visibilities (in miles) within specific ranges.

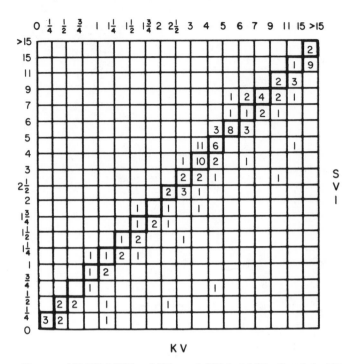

Fig. 10 ASL-FSM (KV) and Videograph (SV1) visibilities (in miles) within specific ranges.

Videograph (SV1), and Fog 15 (F15) during fog are given in Fig. 12 from data collected in December 1982. Each plotted point represents a 10-min running average as do the remaining data in subsequent figures. Good agreement between the sensors is apparent. Little difference is thus seen between the different types of sensors (extinction, backscatter, forward scatter) in fog.

Figure 13 shows a different relationship during a rain event. Rain was identified by the tipping bucket rain gage. Here the Videograph indicates higher visibility than the transmissometer, by about a factor of two in some instances. On the other hand, the forward-scatter sensor, the Fog 15, indicates about half the transmissometer visibility. The Videograph visibilities are usually at least twice as great as those of the Fog 15.

A snow event in February 1983 is illustrated in Fig. 14. Plotted again are the transmissometer, Videograph, and Fog 15. The general indication is that forward-scatter measurements are more optimistic (higher visibility) than either the extinction (transmissometer) or the backscatter (Videograph) measurements, a reversal of the situation in rain. This is especially noticeable when comparing the forward-scatter and the backscatter sensor responses. The Videograph shows the typical backscatter response (lower visibility) in snow.

Similar comparisons in fog, rain, and snow were performed on the VR-301. The general relationships vs the transmissometer were the same as with the Fog 15. Figure 15 compares three forward-scatter sensors under a variety of weather conditions (rain, snow, fog) from data taken on three different days in December 1982. Besides the Fog 15 and

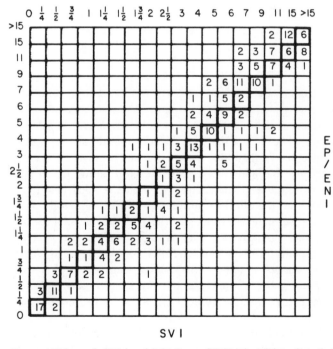

Fig. 11 Videograph (SV1) and EG&G 207 (EP/EN1) visibilities (in miles) within specific ranges.

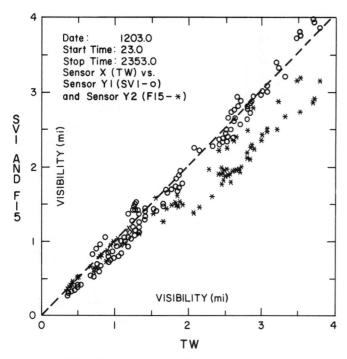

Fig. 12 Videograph (SV1), Fog 15 (F15), and transmissometer (TW) visibilities in fog.

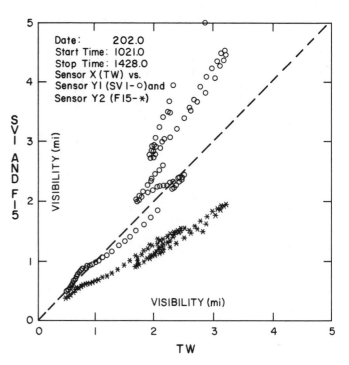

Fig. 13 Videograph (SV1), Fog 15 (F15), and transmissometer (TW) visibilities in rain.

VR-301 (VRH), an EG&G 207 (EP3) is included for reference. Excellent agreement among the sensors is noted.

5.7 Discussion of Results

A relatively simple atmospheric model accounts for all the observed phenomena. The transmissometer, measuring extinction of light across a path, will be responsive to a broad range of scattering particles. Raindrop-sized particles scatter substantial light into the direct beam and thus will not greatly decrease the light reaching the detector. Snow, which produces significant backscatter, will scatter a great deal more light from the direct beam for a given particle density. Fog, having substantially more particles with greater total scatter, will also significantly reduce the light reaching the detector.

Similarly, certain characteristics can be deduced about other classes of sensors. Backscatter sensors will obviously respond more strongly to the preferential backscattering from snow. A similar effect is observed for dry aerosol particles.

Forward-scatter sensors, on the other hand, can be quite optimistic (higher visibilities) in terms of their visibility indication when snow is falling. In the case of an optically thick ensemble of weakly absorbing particles such as snow, the collective backscatter predominates at the expense of forward scatter. Dry aerosols similarly reduce the relative forward scatter. The relationship between forward scatter and backscatter is approximately linear for fog droplets and hydrated aerosol particles, resulting in similar degrees of scat-

tering. Rain will increase the relative predominance of forward scatter, causing forward-scatter sensors to indicate visibility somewhat lower than the transmissometer and much lower than a backscatter sensor.

Data from the VR-301 provide an important additional insight into the problem of scattering measurements. Since the VR-301 uses an infrared source, its projected light is scattered very little by small particles (below 1 μm) that influence the other sensors. Its visibility indications are therefore somewhat optimistic in haze. Although the variation in the number of these particles is important to the other visibility sensors, it does not have much effect on the VR-301. The VR-301 relationship to the other sensors is thus a good indicator of the submicrometer particle content of the atmosphere.

6 SUMMARY AND CONCLUSIONS

Atmospheric visibility depends upon the light-scattering and absorption characteristics of the path between an observer and target, upon the scattering properties of the ambient atmosphere (including clouds), and upon the reflective properties of surrounding and viewed terrain. Furthermore, the process by which an observer determines visibility involves the spatial viewing or scanning of visibility targets with subsequent spatial integration of contrast information.

Although progress is being made toward developing an instrument that closely parallels the human observer's assessment of contrast and visibility, most active research has been

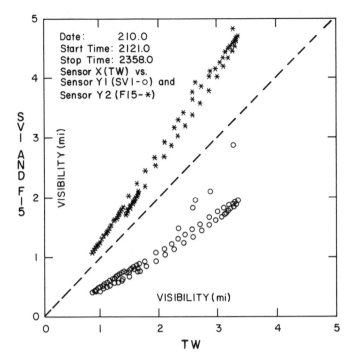

Fig. 14 Videograph (SV1), Fog 15 (F15), and transmissometer (TW) visibilities in snow.

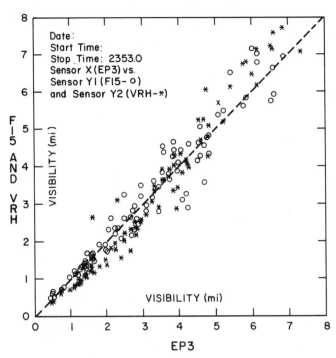

Fig. 15 Fog 15 (F15), VR-301 (VRH), and EG&G 207 (EP3) visibilities in rain, snow, and fog.

in the area of testing field instruments that can routinely measure the atmospheric scattering characteristics. It must be pointed out, however, that such measurements have fundamental limitations. For example, a measurement of the forward scattering, backward scattering, or extinction at one point along the path is useful, but not sufficient to determine the visibility. More importantly, there is not a unique relationship between visibility and backscattering and there is not a unique relationship between visibility and extinction coefficient. It is therefore unreasonable to expect that measuring a single quantity will, under all circumstances, adequately determine the visibility.

A comparison of visibility sensors based on scattering principles has been performed by the NWS T&ED. New-generation forward-scatter sensors were compared with older-generation systems. Solid-state modifications to the widely used NBS-type transmissometer provided a reliable standard for comparisons at visibilities below 4 mi. A third transmissometer, the Impulsphysik Skopograph, was well correlated with the NBS transmissometers and required minimal maintenance. Two EG&G 207 sensors that were cross-checked to ensure stable calibration were used as standards at higher visibilities. Videographs provided a backscatter visibility measurement for comparison with the forward-scatter and transmission measurement devices.

A state-of-the-art forward-scatter sensor built in-house by NWS provided another "standard" of comparison. By reducing electronic noise, the system designers were able to achieve a dynamic range equivalent to visibilities from 0.3 to 30 mi. This sensor was used to derive a new calibration for the Videograph to improve its representativeness in fog and haze. The dynamic range covered by the Videograph was found to be at least from 0.25 to 12 mi, with good agreement with the other forward-scatter sensors and transmissometers when precipitation was not occurring.

The results of the comparisons show that each sensor has unique characteristics dependent not only on its measurement method (forward scatter, backscatter, or extinction), but also on design characteristics such as transmitter light-source wavelength and optics. However, all the sensors under consideration for automated station applications, such as the VR-301, Fog 15, Videograph, and transmissometer, are electronically stable and provide reproducible readings for given weather conditions. This means that if the atmospheric conditions are characterized as to temperature, dew point, precipitation rate, and obscuration type, the sensor's output could be used to reliably and accurately determine visual range. Obviously, one problem for automation is the need to identify the particular obstruction(s) to vision.

A conceivable scenario would (through algorithms) involve developing modifications to sensor calibrations dependent upon the weather conditions identified. Otherwise, considerations of the intended uses and ranges of visibility measurements required and attainable would strongly govern the selection and application of sensors for automated observations. Since no sensor is "perfect," some trade-offs between requirements and the extent to which they can be satisfied may be appropriate.

Gaseous Tracer Technology and Applications

Walter F. Dabberdt, SRI International, and Russell N. Dietz, Brookhaven National Laboratory[1]

1 INTRODUCTION

Current interest in the use of gaseous tracers for probing the atmospheric boundary layer has resulted largely from increasing concern for the quality of the air. In the past, tracers were used primarily to quantify atmospheric characteristics such as transport and diffusion, but in recent years their use has been extended to study emissions from complex pollutant sources and chemical reactions in the atmosphere.

Rather than review specific studies, we attempt only to summarize the role and scope of tracer methods. Our focus is on the controlled application of artificial gaseous tracers to the study of micro- and mesoscale phenomena; we will not treat natural tracers and macroscale applications. Radioactive tracers (gases and particles) also will not be considered because of the real and perceived hazards associated with their use in atmospheric experiments. However, radioactive materials from nuclear detonations and power plant emissions have been used as tracers of "opportunity" in atmospheric studies; Reiter (1978) has given an extensive review of the use of radioactive tracers to study atmospheric transport processes. Additionally, the reader is referred to Johnson (1983) for a broad survey of nonradioactive gaseous and particulate tracers, and remote and in situ measurement techniques.

2 AVAILABLE GASEOUS TRACERS

2.1 Properties of Gaseous Conservative Tracers

A conservative tracer is generally a gaseous constituent that, when released into a parcel of air, provides a measure of the transport and dilution of that component as a function of distance from the source. For model verification it represents an upper bound for the concentration of a trace species from an emission source at a downwind receptor site.

The essential properties of a conservative gaseous tracer are that it is (or has):

• Nondepositing—during the transport time of interest (perhaps up to two to four days), the tracer must not be adsorbed onto surfaces of airborne particulate matter nor dissolve in water (to prevent rainout).

• Nonreactive—neither photochemical, aqueous phase, heterogeneous, nor other homogeneous reactions must involve the tracer. The half-life of the tracer in the troposphere should be greater than two months, and it should be several years for global experiments.

• Low atmospheric background—the lower the ambient concentration the lower the amount of tracer that needs to be released to detect a meaningful signal.

• Limited industrial use—if the source of the tracer being measured in an experiment is not unique to the release, then measured concentrations may not unambiguously identify the source-receptor relationship.

• Readily and sensitively detectable—the tracer should be sampled easily and inexpensively and, to make full use of its capability, be detectable down to the current ambient levels at low cost.

• Harmless—neither the users nor the general public must be affected in any adverse way. The materials used must not be toxic or contribute detrimentally to the environment.

To keep the cost of a tracer to a minimum, it should have a low background concentration, have limited industrial use, and be easily sampled and detected. Thus, conservative tracers with concentrations at or below the part-per-trillion (ppt, vol./vol. basis) level have been sought.

As a result of the development of the electron capture detector and the subsequent studies of its use for the detection of halogenated compounds (Lovelock and Gregory, 1962; Clemons and Altshuller, 1966), it became clear that these materials might make suitable conservative gaseous tracers (Saltzman et al., 1966). However, the atmosphere contains many such anthropogenic and natural halogenated compounds, some of which are present in the hundreds-of-parts-per-trillion range (see, for example, Singh et al., 1983, and Rasmussen et al., 1982).

To put these concentrations into perspective, Table 1 lists some constituents of the atmosphere. Although the sum of all the halocarbons in the atmosphere is equivalent to a concentration of less than 5,000 ppt, they are in substantial excess of suitable tracer compounds, i.e., those with concentrations less than 1 ppt. How these tracers are sampled, separated from the large number of other halocarbon compounds, and quantitatively determined is discussed later. Put simply, however, an air sample, either collected whole (e.g., in a plastic bag or evacuated bottle) or passed through a tracer-adsorbing medium (e.g., charcoal), is presented to a gas chromatograph column, which separates the constituents prior to detection by the electron capture detector.

One other type of compound has been demonstrated to be a suitable gaseous tracer, especially for long-range transport.

---

[1]Under Contract No. DE-AC02-76CH00016 with the U.S. Department of Energy.

TABLE 1
Some Atmospheric Constituents

Gas	Rural Concentration (pp10¹²)	Formula	Concentration Range
Nitrogen	780,900,000,000	N_2	%
Oxygen	209,500,000,000	O_2	
Argon	9,300,000,000	A	
Carbon dioxide	335,000,000	CO_2	ppm
Methane	1,480,000	CH_4	
Nitrous oxide	315,000	N_2O	ppb
Ozone	35,000	O_3	
Nitrogen oxides	3,000	NO, NO_2	
Methyl chloride	630	CH_3Cl	ppt
Halocarbon 12	305	CCl_2F_2	
Halocarbon 11	186	CCl_3F	
Carbon tetrachloride	135	CCl_4	
Chloroform	20	$CHCl_3$	
Sulfur hexafluoride	0.85	SF_6	ppq
Bromotrifluoromethane	0.75	$CBrF_3$	
Perfluorodimethylcyclohexane	0.026	C_8F_{16}	
Perfluoromethylcyclohexane	0.0036	C_7F_{14}	
Perfluorodimethylcyclobutane	0.00035	C_6F_{12}	sub-ppq
Deuterated methane	0.00030[1]	CD_4	
Deuterated methane-13	0.00001[1]	$^{13}CD_4$	

[1]Current limits of detection; actual backgrounds estimated at > 0.000001 pp10¹².

The deuterated methanes of mass 20 and 21 as shown in Table 1 have the lowest background levels, less than one part per quintillion (pp10¹⁸). They have been used in several long-range (500- to 1,500-km) tracer experiments (Cowan et al., 1976; Fowler, 1979) but are sampled and analyzed differently from the halocarbons. Generally, cryogenic whole-air samplers are used. The methane fraction, including both normal CH_4 (present at 1.5 parts per million in air) and the isotopic methanes, is separated by preparative gas chromatography and the ratios of isotopic to normal methane determined on a high-resolution mass spectrometer.

Since potential candidate tracers are those with background concentrations less than 1 ppt, it is useful to consider the units to be used in reporting those concentrations. Typical units used for parts per trillion are ppt, $\times 10^{-12}$, pp10¹², and pℓ/ℓ (picoliters/liter). For parts per quadrillion, units often used are ppq, $\times 10^{-15}$, pp10¹⁵, and fℓ/ℓ (femtoliters/liter). Current convention would have us use pℓ/ℓ and fℓ/ℓ.

2.2 Selected Conservative Gaseous Tracers

Since for the past 20 years the most sensitive device for determining sub-part-per-trillion levels has been the electron capture detector, the prime tracer candidates have been the halocarbons and related compounds.

2.2.1 SF_6 and Halocarbons

The most widely used intentionally released tracers, listed approximately in the order of decreasing ambient background concentration, are shown in Table 2. Sulfur hexafluoride (SF_6) was the first electron-attaching compound to be used for atmospheric tracing (e.g., Saltzman et al., 1966; Dietz and Cote, 1973). The property of SF_6 that promulgated its current standing as the number one tracer, in addition to its high electron capture sensitivity, was its ready availability as a liquefied gas in large steel cylinders at a reasonable cost (~ $10/kg). Its low boiling point (and hence its high vapor pressure) simplified the controlled release from the cylinders.

A unique property, however, was that SF_6 could be eluted ahead of all other atmospheric constituents during its chromatographic determination on a 5A molecular sieve column (Dietz, 1970; Simmonds et al., 1972) as shown in Fig. 1 (Dietz and Cote, 1972). The SF_6 concentration was 3.3 pℓ/ℓ, higher than tropospheric background because of local usage.

SF_6 is used primarily as an electrical insulating gas for high-voltage power transmission equipment. Currently about 1,800 metric tons (2,000 tons) are produced per year and, like the ubiquitous chlorofluoro(halo)carbons, ultimately released to the atmosphere. In September 1975, the background SF_6 was 0.44 ± 0.05 pℓ/ℓ (Dietz et al., 1976a); in November 1976,

TABLE 2
Gaseous Convervative Tracers: Properties and Costs

Distance: 100 km
Desired Concentration: 100 times background at center line
Release Time: 3 h

Tracer	Symbol	Formula	Molecular Weight	Phase at 20°C	Boiling Point (°C)	Supplied Form	Ambient Concentration (fℓ/ℓ)[1]	Cost ($/kg)	Released Quantity (kg)	Relative Tracer Cost (1,000$)
Sulfur hexafluoride	SF$_6$	SF$_6$	146	Gas	−64	Liq./gas	2,000[2]	10	2,320	23.2
Bromotrifluoromethane	F13B1	CBrF$_3$	149	Gas	−58	Liq./gas	750	15	887	13.3
Perfluorodimethylcyclohexane	PDCH	C$_8$F$_{16}$	400	Liquid	102	Liquid	26	120	82	9.8
Dibromodifluoromethane	F12B2	CBr$_2$F$_2$	210	Liquid	25	Liquid	>20	30	>33	>1.0
Perfluorocyclobutane	FC-318	C$_4$F$_8$	200	Gas	−6	Liq./gas	?	250		
Perfluoromethylcyclohexane	PMCH	C$_7$F$_{14}$	350	Liquid	76	Liquid	3.6	100	10	1.0
Perfluorodimethylcyclobutane	PDCB	C$_6$F$_{12}$	300	Liquid	45	Liquid	0.35	500	0.83	0.42
Deuterated methane	CD$_4$	CD$_4$	20	Gas	−160	Gas	0.60[3]	3,000	0.095	0.29
Deuterated methane-13	^{13}CD$_4$	^{13}CD$_4$	21	Gas	−160	Gas	0.02[3]	50,000	0.0033	0.17

[1]1,000 fℓ/ℓ equals pℓ/ℓ or 1 ppt.
[2]Near-urban SF$_6$ is 2,000 fℓ/ℓ or more in many locations because of significant use; tropospheric background is 850 fℓ/ℓ.
[3]Values for deuterated methanes represent current limits of detection (S/N = 2) for a 1-m^3 air sample; actual backgrounds are about 0.0005 fℓ/ℓ.

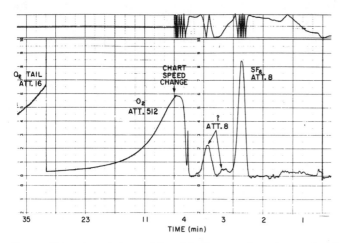

Fig. 1 Chromatogram of 2.5-mℓ Long Island air sample on a 5.2-m (17-ft) NO-treated molecular sieve 5A column with a 9.1-m (3-ft) precut column (same material) at 100°C with an N$_2$ carrier flow rate of 14 mℓ/min.

0.51 ± 0.02 pℓ/ℓ (Dietz et al., 1978); and in December 1981, 0.90 ± 0.10 pℓ/ℓ (Singh et al., 1983). Assuming 1,400 metric tons (1,500 tons) of SF$_6$ was released to the atmosphere each year, the background should have increased by 0.075 pℓ/ℓ per year; this is in excellent agreement with the measured rise in the ambient SF$_6$ levels.

The chemical stability of SF$_6$ makes it a conservative tracer but at the same time causes the background concentration to be ever increasing. In fact, local usage has established backgrounds in many areas that are from 2 to 10 pℓ/ℓ (Dietz et al., 1976c). Because it is readily available in large quantities it is reasonably inexpensive compared with other tracers. But when normalized to the quantity needed to exceed the background level, it is the most expensive of all the tracers listed in Table 2.

The second tracer to be used to some extent, but significantly less than SF$_6$, is halocarbon 13B1, or bromotrifluoromethane (Saltzman et al., 1966; Lamb et al., 1978b). CBrF$_3$ has nearly identical properties to SF$_6$ in terms of boiling point and cost. It is widely available because of its use as a fire-extinguishing agent, and thus its use in long-range transport is also precluded because of its current background level of 0.75 pℓ/ℓ. CBrF$_3$ also can be discretely separated from all the other atmospheric halocarbons using the same molecular sieve column. In fact, the larger unknown in Fig. 1 is CBrF$_3$ (Lamb et al., 1978b; Singh et al., 1983). Global background concentrations of CBrF$_3$ are about 0.8 pℓ/ℓ in the Northern Hemisphere and about 0.7 pℓ/ℓ in the Southern Hemisphere, while the detection limit for routine tracer analyses is typically about 3 pℓ/ℓ by gas chromatography and electron capture detection without preconcentration of the sample. However, the relative response of SF$_6$ is 20-fold to 50-fold greater than that of CBrF$_3$ (Clemons and Altshuller, 1966; Lamb et al., 1978b), making SF$_6$ the preferred choice. Both are used many times when dual-tracer experiments are performed to obtain transport and dispersion information from two sources to the same receptor or from simultaneous ground and elevated releases (C.R. Dickson and G.E. Start, personal communication, 1983).

Another tracer that has been used to a limited extent (C.R. Dickson and G.E. Start, personal communication, 1983; Elias, 1977) is halocarbon 12B2 or dibromodifluoromethane (CBr$_2$F$_2$). Because of its poor chromatographic separation from the numerous other compounds boiling in the same 25°C range, the present background concentration of CBr$_2$F$_2$ has not been determined. The current limit of detection places the background below 0.1 pℓ/ℓ, but, because it is a by-product in the manufacture of CBrF$_3$ (Lovelock and Ferber,

1982) and is used much less, its background concentration is probably less than 0.01 p$\ell\ell$.

As a by-product, it is most certainly present in small quantities in the main products, namely, $CBrF_3$ and bromochlorodifluoromethane ($CBrClF_2$). Thus, inadvertently, as these products are consumed, the background of the CBr_2F_2 will increase, ultimately reducing its utility for long-range tracing studies, even if an acceptable analytical approach were found.

As a single class of tracers, SF_6 and $CBrF_3$ are about comparable in cost, both in terms of the tracer and in terms of sampling and analysis procedures. CBr_2F_2 could potentially be an order of magnitude cheaper to use in terms of tracer cost. The sampling costs should be comparable, but the analysis might be more costly because of the greater time needed to resolve the tracer from other interfering constituents, if, indeed, it can be done at all.

For some limited tracer experiments (i.e., less than 1 km fetch) in which three tracers are needed simultaneously, halocarbon 12 (CCl_2F_2) has been used jointly with SF_6 and CBr_2F_2. In determining air circulation patterns around buildings, the three tracers were released from different locations but sampled and analyzed on the same 5A molecular sieve column (Start et al., 1980). Although its background concentration is 300 p$\ell\ell$, sufficient CCl_2F_2 can be economically released, provided the sampling is performed reasonably close to the source.

In summary, then, the preferred atmosphere tracers for short- (up to 1 km) to moderate- (up to 50 km) range dispersion studies are SF_6 followed closely by $CBrF_3$. CCl_2F_2 is equally useful but only for short-range tests. CBr_2F_2 might prove useful for moderate- to long- (up to 1,000 km) range tests, if a suitable high-resolution column is found.

2.2.2 Perfluorocarbon Compounds (PFCs)

Considerable experience has been demonstrated with three PFCs, namely, PDCH, PMCH, and PDCB. The full chemical names and current ambient concentrations are

shown in Table 2. All three are fully fluorinated, saturated cyclo-organic compounds and have high electron capture sensitivity (Lovelock and Ferber, 1982), nearly comparable to that of SF_6 (Senum, 1981). Branched perfluoroalkanes also have the same high sensitivities, but the normal alkanes are two orders of magnitude less sensitive (Lovelock and Ferber, 1982).

Additional properties of the PFCs that make them extremely attractive as atmospheric tracers are:

• Low ambient background—ranges from 0.3 f$\ell\ell$ for PDCB to 26 f$\ell\ell$ for PDCH (see Tables 2 and 3).

• Chemical stability—resistance to decomposition is useful for separating PFCs from the less stable ambient halocarbons which would interfere with the electron capture detection.

• Essentially nontoxic—the saturated alkanes and cyclo-alkanes have been shown to be completely innocuous (Senum et al., 1980).

Two example chromatograms of the analysis of approximately 25-ℓ background air samples are shown in Fig. 2 (R.N. Dietz, unpublished results, 1982). Specific details on the

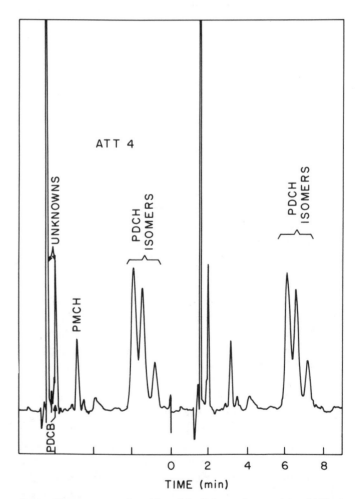

Fig. 2 Chromatograms of 25-ℓ Long Island air samples on a 1.2-m (4-ft) column of Carbopack C with 0.1% SP-1000 at 140°C and a 5% H_2 in N_2 carrier flow rate of 22 mℓ/min.

TABLE 3
PFC Background Air Samples, Yaphank, New York, 4 November 1982

| Sample Volume (ℓ) | Ambient Concentration (f$\ell\ell$) | | | | | |
| | PDCB | PMCH | PDCH | | | |
			Meta	Para	Ortho	Total
24.94	0.373	3.72	12.40	10.30	4.09	26.79
24.36	0.340	3.57	12.57	10.43	4.41	27.41
24.88	0.359	3.53	11.44	9.73	3.74	24.91
24.11	0.329	3.47	10.86	9.34	3.39	23.59
23.37	0.290	3.50	11.68	9.67	4.13	25.47
22.11	0.383	3.70	11.61	9.89	4.00	25.50
Average:	0.346 ±0.034	3.58 ±0.10	11.76 ±0.63	9.89 ±0.41	3.96 ±0.35	25.61 ±1.36

sampling and analysis will be described later; basically the PFCs in the 25-ℓ air sample were collected on a solid, charcoallike adsorbent and thermally desorbed in the gas chromatograph. The three isomers of PDCH (meta, para, and ortho) are clearly shown at their current background levels as is the PMCH that eluted earlier (at about 3.3 min). The PDCB, which eluted at 2 min just ahead of an unknown, was not clearly resolved from the unknown. By reducing the column temperature to 100°C and precutting the PDCH, the PDCB could be clearly resolved from the unknown (Fig. 3). Table 3 gives the results from six background air samples analyzed at a column temperature of 140°C. The reproducibility was within ±10% even for the sub-$f\ell\ell$ PDCB concentration.

The principal use of the PFCs has been as coolants for electronic equipment; because of their chemical inertness and high dielectric strength, they are extremely useful for cooling by convection or boiling. In fact, the chemical inertness of the PFCs assists in their separation from other interfering compounds during chromatographic analysis. The adsorbed sample is thermally desorbed and then passed through a palladium catalyst bed (3% Pd supported on 5A molecular sieve) to destroy interfering halocarbons and molecular oxygen (O_2) as shown in Fig. 4. By maintaining the bed temperature above 220°C, all interfering constituents are reduced by hydrogen in the 5% H_2 in N_2 carrier gas. Only SF_6 and the PFCs survive. However, it should be mentioned that in the presence of O_2 from even a small leak in the plumbing, even the PFCs can be partially reacted, with PDCH being the most susceptible (and meta more than the para or ortho isomers).

This unique chemical stability of the PFCs poses a problem in that once released to the atmosphere, they will persist with a half-life substantially in excess of 100 years. Thus the 26-$f\ell\ell$ background concentration of PDCH arose from the use of more than 900 metric tons (1,000 tons) of commercial grade material as a coolant during the World War II period (G.J. Ferber, personal communication, 1977). Assuming a uniform tropospheric background, 26 $f\ell\ell$ corresponds to 1,434 metric tons of PDCH released to date. Clearly, for the PFCs, what we release today will be with us for many years to come.

In 1976, Brookhaven National Laboratory analyzed the commercial grade of PDCH and found that it contained two other components. The composition was

PDCH: 88.1 vol. %
PMCH: 11.7 vol. %
Unknown: 0.2 vol. %

with the unknown most likely being PMCP (Lovelock and Ferber, 1982) or PCH. Based on this composition and the present-day ambient PDCH concentration, the expected ambient concentrations of PMCH and PMCP are shown in Table 4, which lists current and potential perfluorocarbon tracers (PFTs). The expected PMCH is almost identical to the current measured ambient concentration. Thus, the expected ambient concentration of PMCP is probably no more than 0.06 $f\ell\ell$.

The relative tracer cost for detection at 100 times background is shown in Table 2. Even though the PFCs cost ten to 50 times as much as SF_6, the significantly lower background concentrations mean less material needs to be released. Thus, for PMCH, the relative tracer cost is more than ten times less than if SF_6 or $CBrF_3$ had been used in a

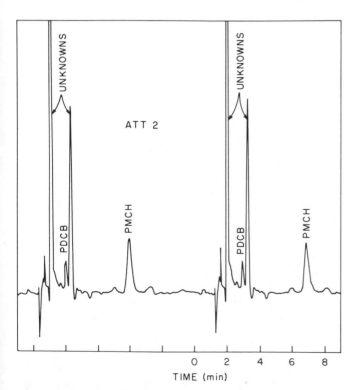

Fig. 3 Same as Fig. 2 but with column at 100°C.

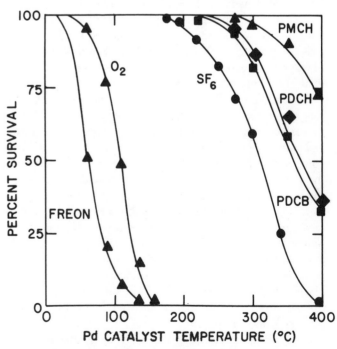

Fig. 4 Effect of catalyst bed (0.32-cm [1/8-in.] lumen by 2.5-cm [1-in.] long) temperature on destruction of PFCs and ambient interferents.

TABLE 4
Current and Potential PFTs

No.[1]	Symbol[2]	Name (perfluoro-)	Ambient Concentration ($f\ell/\ell$)	
			Measured	Expected[3]
1	PDCH (mixture)[4]	-dimethylcyclohexane	25.6	25.6
	PDCH (meta)	-dimethylcyclohexane	11.8	
	PDCH (para)	-dimethylcyclohexane	9.9	
	PDCH (ortho)[5]	-dimethylcyclohexane	4.0	
2	PMCH[4]	-methylcyclohexane	3.6	3.4
3	PDCB	-dimethylcyclobutane	0.35	
4	PMCP[5]	-methylcyclopentane		0.06
5	PCB	-cyclobutane		
6	PCH	-cyclohexane		
7	PDCP	-dimethylcyclopentane		
8	PECP	-ethylcyclopentane		
9	PECH[5]	-ethylcyclohexane		

[1]Only 1, 2, and 5 are currently commercially available in economically large quantities.
[2]All are liquids at room temperature with the exceptions of PCB (supplied as a liquefied gas) and PCH (a subliming solid).
[3]Based on composition of commercial grade PDCH (contained 11.7 vol. % PMCH and 0.2 vol. % PMCP) and present ambient PDCH concentration.
[4]Currently available in large quantity from I.S.C. Chemicals Limited, England.
[5]Has been prepared in the past by I.S.C. Chemicals.

given tracer experiment, for a savings of about $10,000 to $20,000. Of course, for a 1,000-km experiment, the savings would be an order of magnitude higher.

If the ortho isomer of PDCH could be made relatively pure at the same cost as PMCH, then it would be equal in cost-effectiveness to PMCH, since it has essentially the same background concentration as PMCH (see Table 4). At one time I.S.C. Chemicals Limited did produce the pure ortho isomer (D.S.L. Slinn, personal communication, 1983). Similarly, PDCB was at one time manufactured by duPont at a cost comparable to PMCH; since its background is an order of magnitude lower, it would be ten times less expensive than PMCH to employ as a tracer.

Certainly other PFCs listed in Table 4 may prove useful as long-range tracers if they can be supplied at a cost of $500/kg or less, since their background concentrations are potentially less than one-tenth that of PMCH. However, many of these compounds are present in varying degrees as impurities in the principal industrial PFCs. The continued indiscriminate release of these compounds occurs at the rate of ten metric tons (11 tons) or less per year, which corresponds to an increase in the background of no more than 0.2 $f\ell/\ell$. But for the potential long-range PFTs, this could mean a doubling of the ambient concentration every few years or less. Thus at some point in the near future, some restrictions may have to be placed on the commercial use of PFCs in order to preserve the near-zero background of the family of future PFTs (Lovelock and Ferber, 1982).

It should be noted that many of the PFCs that are useful and potentially useful tracers are liquids at room temperature and must be released by either atomization or vaporization followed by dilution to prevent condensation until the vapors are sufficiently diluted below the ambient dew point concentration. Thus the release apparatus is somewhat more complex than for the liquefied gaseous tracers such as SF_6 and $CBrF_3$.

In summary, the PFCs are one to, potentially, two orders of magnitude less expensive to use as moderate- to long-range (up to 1,000 km) atmospheric tracers compared with conventional SF_6 and halocarbon tracers. There is significant potential for developing a useful family of as many as five to ten perfluorocarbon tracers.

2.2.3 Deuterated Methanes

As indicated earlier, another class of tracers is the heavy methanes, $^{13}CD_4$ and $^{12}CD_4$. They have been used in a number of long-range tracer experiments, both alone (Cowan et al., 1976; Mroz, 1983) and in conjunction with other tracers such as SF_6 and PFTs (Fowler, 1979; Fowler and Barr, 1983).

The significant difference between the heavy methanes and the tracers mentioned earlier is that the latter are analyzed by electron capture chromatography, whereas the heavy methanes are determined by mass spectrometry.

Whole-air samples are collected in plastic bags equipped with pumps for a total of 30 to 50 ℓ of ambient air. Alternatively, cryogenic samplers extract methane (both normal methane present at 1.5 ppm and the tracer methane) from the air by adsorption on activated charcoal maintained at liquid nitrogen temperature (Fowler and Barr, 1983). Up to 1 m^3 (1,000 ℓ) of air can be collected in 0.5 h. Since a minimum quantity of 1 mℓ of CH_4 is needed for subsequent sample handling and analysis, about 1 mℓ is added to the bag samples during transfer to storage cylinders. The cryogenic samplers contain sufficient natural ambient methane in a 1-m^3 sample. Thus the limit of detection from bag samples is reduced about 20-fold below that shown in Table 2.

Following sample collection in bags, the sample is analyzed for normal methane concentration; for the cryogenic samples without added methane, the normal ambient concentration of 1.5 ppm is assumed. Then the methane fraction, which is separated and purified by a preparative gas chromatographic procedure, is analyzed by mass spectrometry.

In the mass spectrometer, the heavy methane concentration is measured relative to that of the normal methane. For the $^{13}CD_4$:CH_4 ratio, the instrument noise level is 5×10^{-12}; that for the $^{12}CD_4$:CH_4 is about 30 times poorer (Fowler, 1979; Fowler and Barr, 1983). Thus, at a signal-to-noise ratio of 2 or 3 and assuming only normal ambient methane is present, the limit of detection is 0.60 and 0.02 fℓ/ℓ, respectively, for $^{12}CD_4$ and $^{13}CD_4$.

The costs relative to background of the heavy methane tracers, \$3,000/kg for $^{12}CD_4$ and \$50,000/kg for $^{13}CD_4$, compared with those of the PFTs and other tracers are shown in Table 2. A savings in tracer costs of about a factor of 3 for $^{12}CD_4$ compared with PMCH is realized, and about a factor of 6 for $^{13}CD_4$ compared with PMCH. For $^{12}CD_4$, the cost is associated with the availability of deuterium. But for $^{13}CD_4$, the cost of carbon-13 ($\sim$\$50/g) predominates.

The cost of the tracer alone does not constitute the entire cost of applying a tracer in a transport and dispersion experiment. There are also the costs of the release mechanism, the sampling apparatus, and the analytical system. For heavy methanes, the release is simply controlled by a regulator, metering valve, and flow meter on a small pressurized cylinder, which is weighed before and after the release, not unlike the technique for the controlled release of SF_6 and $CBrF_3$. Sampling is also done with bag samplers, in much the same way as for SF_6, or with cryogenic adsorbent samplers, which are probably not as expensive as PFT programmable samplers. Thus the cost of the release and sampling apparatus would not discourage the use of heavy methanes over the PFT types of tracers.

However, there is a decided difference in the complexity of the analytical procedures for the heavy methanes and those of any of the other tracers that are determined by electron capture chromatography. For the latter, the sample is generally automatically or manually injected into the electron capture detection (ECD) gas chromatograph and the output concentrations reported automatically with a cycle time of from 3 to 10 min, depending on the system. For the heavy methanes, there is a manual sample transfer, followed by analysis of the total methane concentration, followed by preparative gas chromatographic separation of the methane fraction, and finally analysis of the methane fraction by mass spectrometry. Although none of the references indicates the exact time required for each sample analysis, it can be estimated to be substantially longer, perhaps a total of 1 to 2 h per sample.

In conclusion, the use of heavy methanes in long-range atmospheric tracing is less expensive from the standpoint of tracer materials cost than, for example, the perfluorocarbon tracer, PMCH. But for an entire system, which includes release, sampling, and analysis, the PMCH tracer is comparable or cheaper and possibly even easier to implement. When some of the new PFTs proposed in Table 4 become a reality, the PFT system will emerge as the least expensive to implement and the most flexible because there is a family of tracers from which to choose.

2.2.4 Preparation of Gaseous Tracer Standards

During the previous discussion, quite low concentrations of tracers in air are presumed to be routinely sampled and analyzed. Such analytical equipment must, however, be calibrated. Thus preparing standards at or near these sub-part-per-trillion levels may not be routine.

Calibration of the electron capture detector for response to the strong electron-capturing species such as the PFTs, SF_6, and certain halocarbons can be accomplished by:

- Assuming coulometric response or a response factor,
- Injecting various size samples from prepared gas standards or from permeation devices, or
- Using an exponential dilution method.

Although the assumption of coulometric response is attractive in that one electron is presumed to react with one tracer molecule with a probability close to unity (Lovelock, 1974), in practice both the variability in ECD systems and the variability in electron capture cross sections (Senum, 1981) and response (Clemons and Altshuller, 1966) preclude satisfactory performance from this technique.

An alternative is to calibrate using prepared gas standards and various sizes of chromatograph gas sample loops. Typically, a primary standard is prepared in He at a concentration of 100 to 1,000 ppm, either gravimetrically or by pressure-volume techniques, followed by verification with a thermal conductivity gas chromatograph system that has been calibrated by the pure compounds. This primary standard can then be diluted in steps of tenfold or 100-fold in large compressed gas cylinders, using accurate pressure gages. Replication of this process has shown that standards down to 1 pℓ/ℓ can be prepared with a precision of about ±2% and an accuracy of ±5%. Low concentrations of SF_6 can be safely stored for years in conventional carbon steel cylinders. However, PFT and halocarbon standards can deteriorate by wall adsorption in such cylinders. Treated aluminum cylinders, prepared in part by a proprietary process (Wechter, 1976; Airco, 1976), have been used by Brookhaven National Laboratory to prepare part-per-trillion PFT standards that showed no wall losses in more than five years (cf. Table 5).

TABLE 5
Comparison of Gas Cylinder Standards

Tracer in Air	Expected Concentration (pℓ/ℓ)	Measured Concentration (pℓ/ℓ)	
		Spectra-Seal[1] Aluminum	Carbon Steel
SF$_6$	10.2	9.9 ± 0.1	10.1 ± 0.2
PDCB	10.0	10.2 ± 0.1	8.3 ± 0.4
PDCH	10.0	10.1 ± 0.1	3.4 ± 0.2

[1]Airco Industrial Gases proprietary treated aluminum cylinder.

Gas standards of halocarbons (e.g., Vidal-Madjar et al., 1981) and PFTs (Dietz and Cote, 1982) have also been prepared through the use of permeation devices. These devices, which have emission rates in the range of 10 to 100 nℓ/min, will produce concentrations in the range of ten to 100 parts per billion (ppb) when placed in a temperature-controlled gas stream flowing at 1 ℓ/min. Calibration of the gas chromatograph is then accomplished using various sample sizes, dilution flow rates, and/or permeation device temperatures, since the emission rates do increase with temperature. Advantages are that the devices over a period of a few months can be accurately calibrated gravimetrically with a precision of about ±5% or, alternatively, in a few hours by an absolute pressure technique (Dietz and Smith, 1976) with similar precision. The disadvantage is the need for a temperature-controlled bath to contain the device and the need for periodic gravimetric measurements. Brookhaven prefers the flexibility and reliability of the prepared gas standards in aluminum cylinders.

An alternative procedure is the exponential dilution flask method, which allows a complete calibration curve to be generated over the entire calibration range of the chromatograph system (e.g., Lamb et al., 1978b). The advantage is that a single primary standard is all that is needed; measurements can be made consecutively from the upper range of detection down to the lowest limit of detection. The disadvantage is the need to accurately know the starting volume of the primary standards, the dilution flow rate, and the flask volume, as well as the time after the start of the dilution flow.

In calibrating ECD gas chromatographs, a word of caution is needed about assuming linearity in the ECD response when operated in the constant-current mode (the present-day commercially favored approach). This mode provides a dynamic range of about five to six orders of magnitude. It has been shown that for strong electron-capturing species there is a nonlinearity that results in an enhanced signal as the quantity of analyte (tracer) increases (Lovelock, 1974; Knighton and Grimsrud, 1983). This has been observed for SF$_6$ as shown by the twofold enhanced response in Fig. 5 (Dietz et al., 1976a). The enhancement started at an SF$_6$ concentration of 2 × 10^{-10}, which was equivalent to 1 pℓ of SF$_6$. Interestingly, on another "linear" ECD gas chromatograph (GC) system, PDCH response also showed an S-kink commencing at 1 pℓ as shown in Fig. 6 (R.N. Dietz, unpublished results, 1982). In this case, there was a decrease in response of about 30%

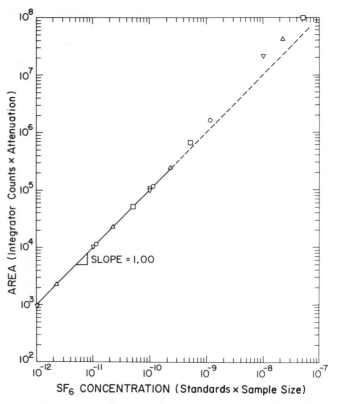

Fig. 5 *Calibration of a constant-current electron capture detector (CC ECD) gas chromatograph (GC) system for response to SF$_6$. (Note: 10^{-9} on abscissa was equivalent to 5 pℓ of SF$_6$.)*

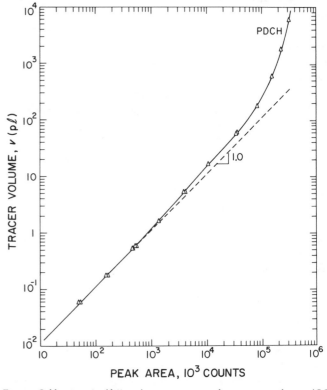

Fig. 6 *Calibration of a (different) constant-current electron capture detector (CC ECD) gas chromatograph (GC) system for response to PDCH.*

rather than an enhancement. Thus, there is a need for care in calibrating these ECD GC systems.

2.2.5 Tracer Lifetimes

The tracers described in the preceding sections (with the exception of the heavy methanes) have atmospheric residence times in excess of 100 years (Lovelock and Ferber, 1982). As we pointed out, the amounts of these tracers released now will, for all practical purposes, be with us forever.

On the other hand, the heavy methanes reside in the atmosphere as methane for only two to seven years (Fowler, 1979). Thus, the long-term consequences of the indiscriminate release of the methanes are markedly less than for PFTs and certain other fluorocarbons.

3 RELEASE CONSIDERATIONS

3.1 Release-Rate Estimation

The rate of tracer release (e.g., in kilograms per hour for a stationary source) is usually determined by a sort of optimization of two basic criteria: cost of the tracer material that is released, and ambient concentration required at any sampling location. In turn, the minimum desired release rate for an atmospheric experiment is a function of the following factors:
- Sampling location (height, range, bearing),
- Wind speed,
- Atmospheric diffusion (surface roughness, atmospheric stability, mixing depth),
- Background concentration (χ_{bkg}),
- Instrument detection threshold (C_{min}) and dynamic range, and
- Sampling duration.

The procedure for estimating the necessary minimum release rate is to first estimate the minimum ambient concentration (χ_{min}) required to provide an acceptable detection signal-to-noise ratio, and then to estimate the atmospheric dilution and subsequently the release rate. Because of the uncertainties in the various estimates, a factor of safety is often applied to the calculated release rate, or it may be incorporated into the calculations by choosing conservative estimates of the individual parameters.

The required minimum ambient concentration, χ_{min}, is given as

$$\chi_{min} = k\, MAX\,(\chi_{bkg},\, C_{min}) \qquad (1)$$

where k is frequently set equal to 10. The definition of χ_{min} will vary according to the nature of the experiment, but for studies of plume dispersion from a single isolated source it can be practically expressed as the concentration two standard deviations from the maximum concentration (χ_{peak}) on the plume center line; for a Gaussian plume profile

$$\chi_{min} \cong 0.1\chi_{peak} \qquad (2)$$

or

$$\chi_{peak} \geq 100\, MAX\,(\chi_{bkg},\, C_{min}) \qquad (3)$$

Atmospheric dilution usually can be estimated within a factor of 2 or 3 from a number of available workbooks; for example, Turner (1970) addresses application of the Gaussian plume formulation to dispersion from continuous point-source releases, and Dabberdt (1981) considers line and area sources. Figure 7 is an example from Turner and illustrates the variation of the normalized ground-level plume center-line concentration as a function of range (X), source height (h), mixing-layer depth (H), and stability (taken as neutral in this illustration). As an example of the procedure, consider a surface source $(h = 0,\ X = 10\ km,\ and\ H = 100\ m)$; therefore $\chi_{peak} U/Q = 7.5 \times 10^{-6}/m^2$ and $Q = 1.33 \times 10^5\ (m^2)\,\chi_{peak} U$. Assuming that SF_6 is the tracer being evaluated, then $\chi_{bkg} \approx 1\ ppt = 6 \times 10^{-6}\ mg/m^3$, and $C_{min} \approx 5\ ppt\ (3 \times 10^{-5}\ mg/m^3)$ for a continuous analyzer of the type described by Dietz and

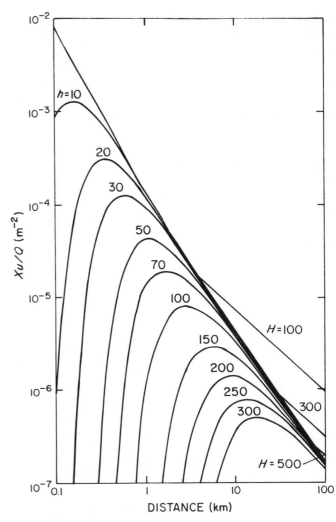

Fig. 7   *Range variation of normalized ground-level, plume-center-line concentration for various source heights (h) and mixing depths (H); neutral stability.*

Goodrich (1980). Recalling Eq. (3) and introducing a design wind speed of 10 m/s, the SF_6 release-rate estimate is

$$Q = [1.33 \times 10^5 \, (m^2)][10^2 \times 3 \times 10^{-5} \, (mg/m^3)][10 \, (m/s)]$$

$$= 4.0 \times 10^3 \, mg/s$$

$$= 14.4 \, kg/h$$

Estimates for atmospheric releases of other tracers can be made analogously using the data on atmospheric background concentrations given in Section 1 and the instrument specifications in Section 4. Estimates of release rates (or amounts) for applications within a controlled atmosphere (e.g., a building) will generally follow the same rationale, except that the dilution is controlled by the size and integrity of the containment structure. For studies of infiltration in buildings, Dietz and Cote (1982) have presented a useful review.

3.2 Release Mechanisms

Tracers are available in pressurized cylinders as liquefied compressed gases (e.g., SF_6, $CBrF_3$, PFCB) or liquids (CBr_2F_2, PMCH). Generic schematics of release systems for gas-phase and liquid-phase tracers[2] are illustrated in Figs. 8 and 9, respectively. Compressed gas systems include the storage cylinder, pressure regulator, heat exchanger, dry test meter, and balance. The latter two components provide a redundant means for monitoring and controlling the emission rate. Electrical heating tape wrapped around the cylinder serves to offset the heat lost through the rapid expansion of the tracer at the regulator. A 15-m coil of 0.64-cm copper refrigerator tubing serves a similar purpose downstream. Where the release rate (and heat loss) is particularly large (e.g., around 25 kg/h SF_6 from a single 1A gas cylinder[3]), the cylinder and coil can be immersed in water in a 208-ℓ (55-gal) drum. Where

high release rates are required for extended periods, the heat loss is more effectively controlled by connecting several cylinders in parallel through a manifold (see Fig. 10) and maintaining a low release rate at each cylinder.

Liquefied tracers require atomization of the liquid and complete evaporation in a gas stream (ambient atmosphere or a controlled flow). Figure 9 illustrates schematically the components of a typical release system. In the illustration, the storage tank is not pressurized, and the liquid is withdrawn by a metering pump; the liquid can also be forced out by pressurizing the tank with compressed air or nitrogen. The atomizer illustrated is a high-pressure hydraulic nozzle that generates droplets with a mean diameter in the range of 20–30 μm. Injecting the spray into a heated air stream assures rapid vaporization of the liquid aerosols. The rate of vaporization is a function of the heat of vaporization of the tracer, droplet size, release rate, and temperature and velocity of the air stream.

4 SAMPLING AND ANALYSIS

Earlier we discussed briefly the manner in which the various types of tracers (i.e., SF_6 and halocarbons, PFTs, and heavy methanes) are sampled and subsequently quantitatively determined. In this section, the various techniques for sampling both on the ground and aloft will be described in further detail. Since some applications also required real-time instrumentation, this will also be discussed. Some examples of the application of each sampling technique will also be provided as well as a brief description of the laboratory analyzers used to determine tracer concentrations from collected samples.

---

[2]The phase or state at atmospheric temperature and pressure.

[3]A single 1A cylinder contains about 45 kg of SF_6.

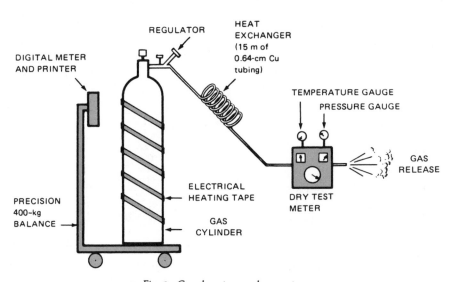

Fig. 8 Gas-phase tracer release system.

4.1 Whole-Air Samplers

The sampling objective is to collect either an instantaneous point sample, a space-integrated sample, or, more typically, a time-integrated point sample, and to return the sample to the laboratory for analysis of the tracer material. Whole-air samplers usually consist of plastic bags equipped with pumps, syringes, plastic squeeze bottles, and evacuated bottles.

This type of sampling requires that the samplers be carefully checked for leaks to maintain sample integrity. They should be purged, sealed, and stored in a location where they will not be contaminated with the tracer of interest. Many experiments have been compromised and even ruined from inattention to this point, especially when sampling occurs near background levels.

In addition, the compatibility of the sampler container and the tracer must be checked. Are there any wall losses? Memory effects? Sample recovery problems? Although most plastic materials are compatible with the gaseous tracers, it has been demonstrated that fluoroplastics (e.g., Teflon) and fluoroelastomers (e.g., Viton) have a high solubility for PFCs and SF_6 (Senum et al., 1980; Dietz et al., 1976c) and thus can cause significant memory effects. Additionally, carbon steel cylinders can cause significant wall losses of collected PFTs (e.g., cf. Table 5).

Bag samplers have been used routinely by many investigators. Some have been homemade, equipped with pumps, and placed in plastic or metal drums (Dietz et al., 1976b; Fowler and Barr, 1983); others have been commercially produced, such as the programmable AOS-II by Environmental Measurements, Inc. (Clements, 1979). Radio-controlled bag sampling techniques have also been implemented (Sehmel, 1982). The results of bag sampling 100 km downwind during

a 7-h collection period for a 3-h multitracer release of SF_6, $^{12}CD_4$, and $^{13}CD_4$ are shown in Fig. 11 (Fowler, 1979). When care is taken, good results can be obtained even when the peak measured concentration is only ten times the background as was the case for SF_6 (Dietz et al., 1976b). Bag sampling can also be performed from aircraft by filling under ram air pressure through a controlled external port (Fowler and Barr, 1983). Automated bag sampling on aircraft has also been implemented (e.g., C.R. Dickson and G.E. Start, personal communication, 1983). The samples should be analyzed immediately or transferred to evacuated containers for shipment and subsequent analysis in the laboratory (Dietz et al., 1976a), since bags can be easily damaged and are difficult to ship.

Syringe sampling is a very simple and effective means of manually collecting nearly instantaneous air samples, and it has been used by numerous investigators (e.g., Lamb et al., 1978a; Sehmel, 1982). Typically, 30-ml plastic syringes are filled in 10 s, at 30- to 60-s intervals while traversing a plume. Stationary time-averaged samples have also been collected by automated syringe samplers (e.g., Developmental Sciences, Inc.), collecting 30-ml samples averaged usually over a 1-h period (e.g., Lamb et al., 1978a; Drivas and Shair, 1975). Aircraft samples have also been collected with a screw-driven syringe sampler, collecting 30 ml at a constant rate over a 1-min period (Drivas and Shair, 1975).

A simple and effective means to collect air samples is to squeeze a 250-ml polyethylene bottle about ten times, and then cap the bottle on the last fill (Drivas et al., 1972). However, the technique does have some memory effects.

A novel technique for collecting three sequential whole-air samples by remote radio-controlled operation relies on the use of three 1-ℓ evacuated steel bottles, each equipped with a

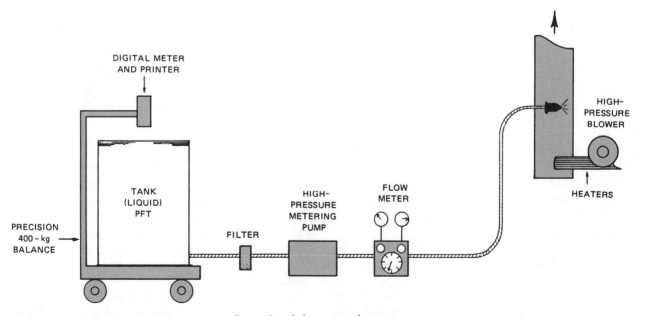

Fig. 9   Liquid-phase tracer release system.

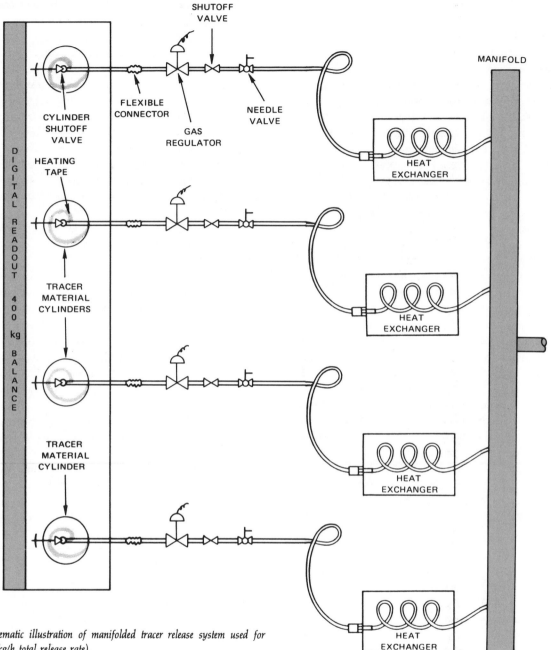

Fig. 10 Schematic illustration of manifolded tracer release system used for CBrF₃ (130 kg/h total release rate).

toggle valve, solenoid valve, and critical orifice (Raynor, 1978). The orifices were designed for constant sampling rates for periods of 0.5–2 h. Each of the three bottles in 20 sampler units could be opened and closed on command from the radio controller, which had a broadcast range of more than 50 km from an airplane.

Bottles have also been used to collect whole-air samples by pumping at a controlled rate (Jaffer et al., 1981). Aluminum bottles treated by the "Summa" electropolishing process (Robinson et al., 1977) are suitable for halocarbon and, likely, PFT sampling as well. Generally, whole-air sampling for SF₆ is the most practical approach.

4.2 Adsorbent Samplers

Occasionally, tracers have been sampled from the air by adsorption processes. The more volatile tracers, e.g., the heavy methanes, can only be adsorbed at cryogenic temperatures. Even for SF₆, which is quite volatile, only small collection volumes can be achieved at room temperature. Some adsorption sampling has been used for determining ambient halocarbons, but essentially none has been used for halocarbon tracers. The most successful application of this approach has been for PFTs; both programmable and passive samplers have been used.

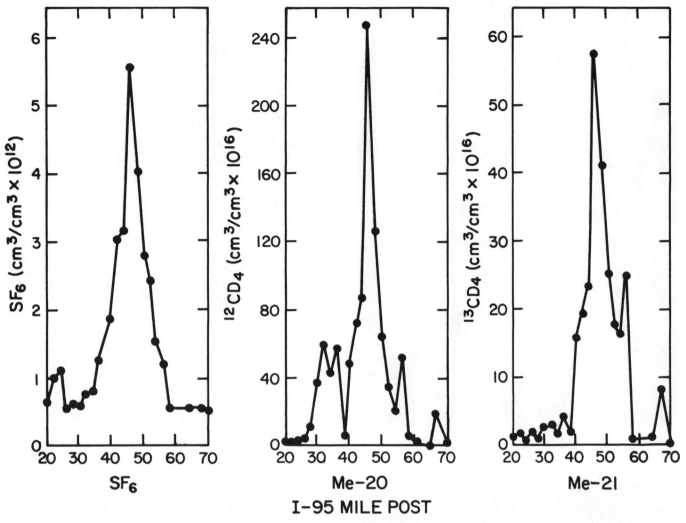

Fig. 11 Bag-sampled 7-h integrated ground-level SF₆ and heavy methane concentrations 100 km downwind.

4.2.1 Cryogenic Samplers

To collect sufficient ambient methane for analysis of the heavy methane tracers, a minimum quantity of air on the order of 1 m³ has been collected on activated charcoal held at liquid nitrogen temperatures in a cryogenic sampler developed at Los Alamos National Laboratory (Fowler and Barr, 1983). Samples were collected over a 2.3-h period.

4.2.2 Adsorbent Samplers for SF₆

Activated coconut charcoal (Clemons et al., 1968) and molecular sieve 13X traps (Clemons et al., 1968; Dietz et al., 1976b) have been used to concentrate whole-air samples collected in bags in order to improve the limit of detection. At room temperature, 0.5-g traps of either material retained most of the SF_6 in air samples up to 50 mℓ. Cooling the molecular sieve trap to the temperature of dry ice allowed

much larger air samples to be concentrated; however, a 40-mℓ air sample was sufficient to determine the ambient SF_6 concentration (about 0.5 pℓ/ℓ in 1975) with a precision of ±3 % (Dietz et al., 1976b).

Field adsorption sampling for SF_6 has been implemented by adsorption onto 0.1 g of activated charcoal at a rate of 3 mℓ/min for up to 30 min. Flow rate was controlled by displacing water using a peristaltic pump (Van Duuren et al., 1974).

4.2.3 Adsorbent Samplers for Halocarbons

Very little work has been done on the adsorption sampling of halocarbons, with whole-air sampling the predominant technique. However, halocarbons from up to 3 ℓ of air have been collected in a 10-cm trap of Porapak N (Russell and Shadoff, 1977) and up to 30 ℓ on a 13-cm Carbosieve B trap (Bruner et al., 1978, 1981).

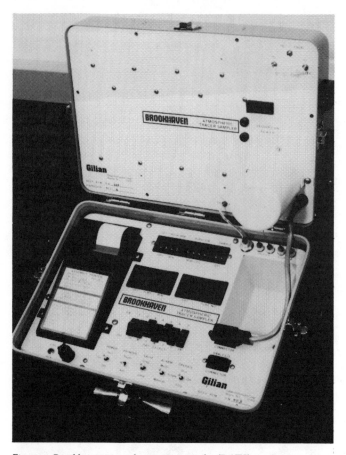

Fig. 12 Brookhaven atmospheric tracer sampler (BATS).

Fig. 13 BATS lid assembly showing the 23 sampling tubes.

4.2.4 Perfluorocarbon Tracer (PFT) Adsorbent Samplers

Two types of adsorbent samplers have been developed for PFTs: a programmable sampler (Ferber et al., 1981) and a passive sampler (Dietz et al., 1981; R.N. Dietz et al., in preparation; Dietz and Cote, 1982).

4.2.4.1 Programmable PFT Sampler

Dietz developed this sampler at Brookhaven. It was commercially manufactured for the National Oceanic and Atmospheric Administration (NOAA) by Gilian Instrument Corporation as the Brookhaven atmospheric tracer sampler (BATS). The entire unit, shown in Fig. 12, measures just $36 \times 26 \times 20$ cm and weighs 7 kg. The lid contains 23 sampling tubes (Fig. 13), each containing 150 mg of 20–50 mesh type 347 Ambersorb (Rohm and Haas Co.), which can retain all of the PFTs in more than 30 ℓ of air. Internal batteries provide power for up to a month of unattended operation of all the automatic sampling and recording features (Ferber et al., 1981).

Sample recovery is accomplished by direct ohmic heating of the adsorption tube to 400 °C. The PFTs are purged from the BATS tube through an automated ECD GC system (Ferber et al., 1981). All 23 tubes are automatically analyzed in about 3 h. As demonstrated in Table 3, the precision of this system is ±10 % at a concentration of 0.35 fℓ of PDCB per liter of air for a 25-ℓ air sample.

This PFT system was tested in a 600-km tracer experiment in which five tracers (PMCH, PDCH, SF_6, $^{12}CD_4$, and $^{13}CD_4$) were released and sampled at 100- and 600-km ground-based stations (Ferber et al., 1981); the $^{13}CD_4$ methane was even detected up to 2,500 km away (Fowler and Barr, 1983). As shown in Fig. 14, the maximum PMCH concentration at 600 km was nearly three orders of magnitude above background. Quite likely PFT samplers at 2,500 km would also have readily detected PMCH if they had been deployed at that distance.

4.2.4.2 Passive PFT Sampler

Originally developed as a means to measure the indoor PFT concentration during the determination of air infiltration and air exchange rates in homes and buildings using miniature PFT permeation sources (Dietz et al., 1981; Dietz and Cote, 1982), the passive sampler has also been used in atmospheric tracer studies (R.N. Dietz et al., in preparation).

In its first configuration, one end of the sampler contained a 1-mm capillary tube and so was named the capillary adsorption tube sampler (CATS). The present configuration of a CATS is shown in Fig. 15. The passive sampler, which is made from glass tubing 6 mm in outside diameter (OD) by 4 mm in inside diameter (ID) and 6.4 cm long, contains 64 mg of Ambersorb 347. A sample is taken by Fickian diffusion when one cap is removed, as shown. From the depth to the bed (2.76 cm), the cross-sectional area (0.126 cm²), and the empirically derived diffusion coefficients of the PFTs in air, it was determined that the CATS sampled at a rate equivalent

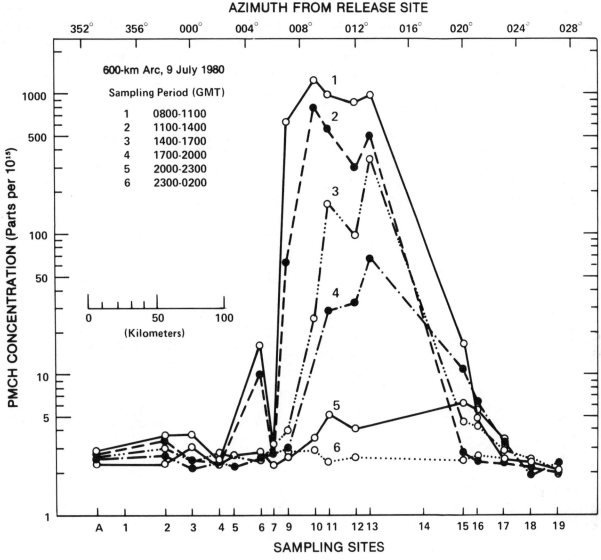

Fig. 14   *Average 3-h PMCH concentrations at a 600-km arc from the release point of 192 kg of PMCH over 3 h.*

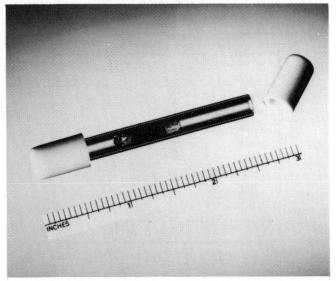

Fig. 15   *Capillary adsorption tube sampler (CATS) shown in its sampling mode with one cap removed.*

to about 200 m*ℓ* of air per day for PMCH. In some recent comparisons between BATS and CATS, the actual sampling rate was found to be equivalent to 232 m*ℓ* of air per day for PMCH and 217 m*ℓ* of air per day for PDCH (R.N. Dietz et al., in preparation). With two caps removed, increasing wind speed causes increasing errors.

In a short-range (10-km) tracer experiment conducted at Brookhaven, PMCH was released at 5.3 kg/h for 1.25 h and sampled at 20 sites about 8 to 11 km downwind as shown in Fig. 16. The PMCH concentration measured by the CATS is shown in Fig. 17; average peak concentrations of about 2 p*ℓ*/*ℓ* were adequately detected during the 135-min sampling period. The limit of detection was about 0.1 p*ℓ*/*ℓ*.

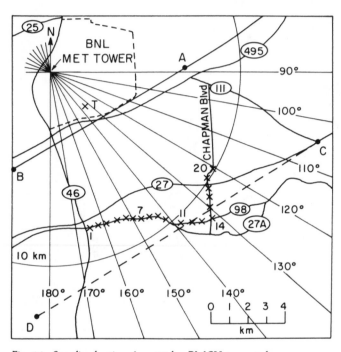

Fig. 16 Sampling locations for a 10-km PMCH tracer study.

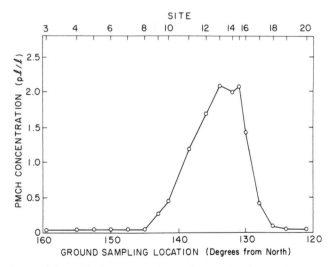

Fig. 17 Ground-level passive sampler results at 10 km downwind.

Ambient concentration measurements of PMCH and PDCH were recently determined with ten CATSs deployed in the Washington, D.C., area for 30 days. As shown in Table 6, the background of both tracers was identical to that obtained from the 25-*ℓ* BATS samples (cf. Table 3).

4.3 Vertical Sampling

For model verification, investigators need to know the fate of the tracer aloft as well as on the ground. For this reason, techniques have been implemented to measure the vertical distribution of tracer but, because of technical and logistical difficulties, not to a great extent.

Tethersondes have been used to measure fluorescent zinc cadmium sulfide particles to heights of 300 m (McElroy and Pooler, 1968) and to 450 m (Hilst and Bowne, 1971). The Norwegian Institute for Air Research (B. Sivertsen, personal communication) has used vacuum-actuated plastic syringes suspended on the cable of a tethered balloon to measure SF_6 to heights of several hundred meters. Numerous investigators have used aircraft to collect gaseous tracer bag samples aloft. Crosswind vertical SF_6 tracer profiles have been obtained from aircraft operating semicontinuous (Dietz et al., 1976a; Dietz and Cote, 1973) and continuous SF_6 analyzers (Dietz and Goodrich, 1980; Dabberdt et al., 1982; Baxter et al., 1983). The programmable PFT sampler (BATS) has also been used in aircraft to make crosswind traverses of a PMCH tracer release in short- (R.N. Dietz et al., in preparation) and long-range (R.N. Dietz, unpublished results) tracer tests. A semi-real-time dual-trap analyzer has been proposed for detection of PFTs in aircraft at concentrations down to 10 f*ℓ*/*ℓ*. Vertical PFT data have been collected with a special 490-m vertical atmospheric sampling cable (VASC) suspended from a 100-m³ tethered balloon (Ferber et al., 1983). Passive

TABLE 6
Ambient PFT Determinations with Passive Samplers (CATS)[1]
Washington, D.C., Area, 30-Day Period, June 1983

No.	CATS No.	PFT Concentration (f*ℓ*/*ℓ*)	
		PMCH	PDCH
1	925	3.52	26.3
2	936[2]	22.33	40.1
3	953	3.85	26.9
4	913	4.00	28.5
5	969	3.91	27.9
6	932	3.39	26.3
7	926	3.07	24.3
8	300[2]	23.40	45.8
9	954	3.14	27.3
10	959	2.90	20.5
	Average:	3.5 ± 0.4	26.0 ± 2.6

[1]Equivalent sampled air volumes were 7.0 and 6.5 *ℓ*, respectively, for PMCH and PDCH.
[2]Excluded from averages; CATS no. 936 had a cracked cap.

samplers on tethersondes (R.N. Dietz et al., in preparation) have also been proposed.

4.3.1 BATS in Aircraft

Using the programmable sampler, but connected to a 500-mℓ/min critical orifice and 28-V DC aircraft vacuum pump, crosswind profiles of the PMCH plume from the release at Brookhaven (R.N. Dietz et al., in preparation) were obtained along line C-D in Fig. 16. The four tracer profiles shown in Fig. 18 had peak concentrations, when fitted with a Gaussian plume (R.N. Dietz et al., in preparation), of 2–5 pℓ/ℓ, comparable to the time-integrated average ground-level concentrations shown earlier.

Similar crosswind profiles were obtained 70 km downwind from a release in a Pacific Northwest Laboratory aircraft using an on-board BATS as shown in Fig. 19. The flight was part of the Department of Energy CAPTEX (Cross-Appalachian Tracer Experiment) study geared toward tracing PFTs for more than 1,000 km (R.N. Dietz, unpublished results) with a 200-kg PMCH release over 3 h.

4.3.2 Vertical Atmospheric Sampling Cable

A cable was designed at Brookhaven to sample four 122-m layers from 0–122 m up to 366–490 m, as shown in Fig. 20. Four separate 0.6-cm OD polyethylene sampling cables were bundled together in a braided Kevlar sheath (Cortland Line Company) and suspended from a 100-m³ balloon tethered 490 m above the ground in the complex mountainous terrain of the Geysers area north of San Francisco (Ferber et al., 1983). Each cable had intake holes at 15-m intervals over the designated sampling span. Pumps on the ground pulled the air through each cable to the programmable samplers, bag samplers, and a real-time dual-trap analyzer.

About 445 g of PDCH were released about 4 km upwind of the tethered balloon site and 380 m higher in elevation. The 1-h release commenced at midnight in order to study the drainage flows from the mountains into the valleys. The vertical cable allowed the concentration distribution aloft to be recorded on an hourly basis with the BATS programmable sampler and the dual-trap analyzer shown in Fig. 21 (Ferber et al., 1983).

4.4 Laboratory and Real-Time Analyzers

4.4.1 Laboratory Analyzers

Specific details of the technique for every gaseous conservative tracer are beyond the scope of this presentation. A brief description of the procedure for the heavy methanes was given in Section 2.2.3. More details are available elsewhere (Fowler and Barr, 1983; Fowler, 1979).

A number of portable gas chromatographs are commercially available (e.g., Systems, Science and Software, La Jolla, California; ITI, Inc., Burlington, Massachusetts) for rapid

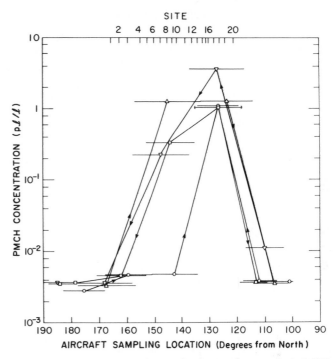

Fig. 18 Aircraft traverses at 300 m in altitude using the 500-mℓ/min BATS. PMCH concentration versus crosswind location at about 11 km downwind.

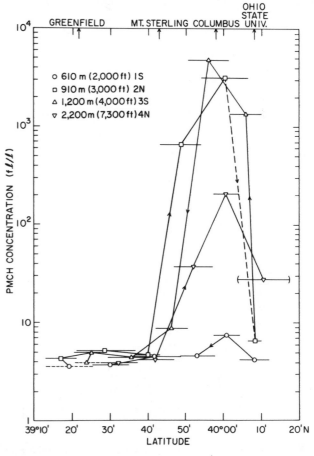

Fig. 19 BATS aircraft traverses at 70 km downwind during a 3-h release of 200 kg of PMCH.

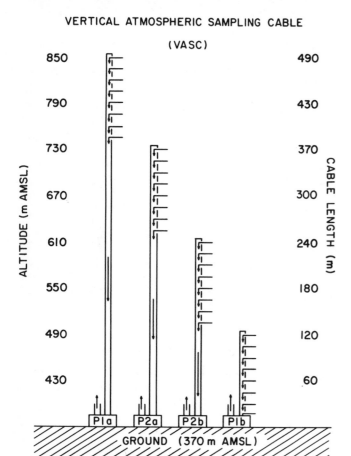

Fig. 20 *Schematic configuration of the tethered cable.*

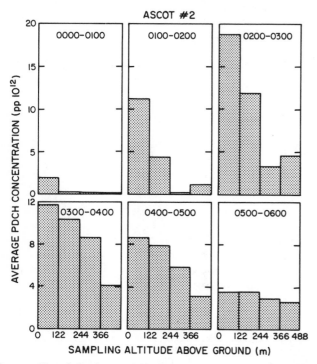

Fig. 21 *Vertical distribution of PDCH tracer measured with the dual-trap analyzer connected to the VASC.*

analysis of SF_6 samples at concentrations of 5 pℓ/ℓ and above in whole air. Greater precision and a lower limit of detection have been achieved using a precut column technique (Dietz and Cote, 1971) and a concentrator (Dietz et al., 1976c).

Ambient halocarbons have been measured by a number of researchers. Many of their techniques would be applicable to the halocarbon tracers (e.g., Singh et al., 1983; Lillian et al., 1976).

The perfluorocarbon tracers (PFTs) have been analyzed from both whole-air and adsorbent samplers using a laboratory gas chromatograph system. Basically the sample is injected from a sample loop or thermally desorbed and passed through a Pd catalyst bed, permeation dryer, and precut column before being reconcentrated on an in situ trap. The trap prevents the collection of unwanted constituents of low molecular weight, the precut column prevents the passage of unwanted constituents of high molecular weight, and the dryer removes moisture from the ambient samples. After the trap is thermally desorbed, the sample is injected into the main column after passing through another Pd catalyst bed and finally through the electron capture detector. Limits of detection range from 0.5–5 fℓ up to a maximum of about 5,000 pℓ or, roughly, six orders of magnitude (Ferber et al., 1981; R.N. Dietz et al., in preparation). Some typical chromatograms are shown in Figs. 2 and 3.

4.4.2 Real-Time Analyzers

In many instances, there is a need for rapid, preferably real-time, continuous determination of tracer profiles. Such profiles, collected both on the ground using vehicles and aloft using aircraft at different altitudes, can provide useful information on the standard deviation of the plume width as well as the vertical standard deviation or height of the mixing layer.

For downwind distances of up to 10–100 km, crosswind aircraft traverses have been made with two types of real-time SF_6 analyzers, a semicontinuous and a continuous version.

The semicontinuous monitor worked on the principle of frontal chromatography using a molecular sieve 5A column. When the SF_6 frontal was eluted and until the oxygen frontal reached the detector (a period of 90 to 180 s), the output followed the input SF_6 profile (Dietz et al., 1972). Then the column was backflushed with carrier gas for twice the forward time to remove oxygen and other interferents before the cycle was repeated. During the SF_6 window, continuous detection of an SF_6 plume at distances of up to 90 km downwind (about 7 pℓ/ℓ) was demonstrated (Dietz et al., 1973).

Attempts to develop a truly continuous SF_6 monitor resulted in the application of the Pd catalyst bed combustor (Simmonds et al., 1976). Air was continuously mixed with half as much hydrogen to catalytically convert all of the oxygen to water, which was removed by a condenser, and to destroy unwanted interferents. Subsequently, numerous improvements have been made including the use of a permeation-type Nafion dryer (Dietz and Goodrich, 1980) and mass flow controllers (Baxter et al., 1983) to reduce altitude effects.

These continuous analyzers have been used in a number of short- (about 10 km) range (e.g., Dabberdt et al., 1982, 1983; Baxter et al., 1983) and moderate- (up to 100 km) range (e.g., R.N. Dietz et al., unpublished results) SF_6 dispersion experiments. Although continuous measurements are desirable, experience has shown that the practical or useful limit of detection with the continuous version, about 10–30 $p\ell/\ell$, is about an order of magnitude poorer than that for the semi-continuous version.

The continuous analyzer can also detect perfluorocarbon tracers and, therefore, was used in a feasibility study of the detection of PFT-tagged clandestine explosives (Senum et al., 1980). However, the limit of detection is about 70% poorer than for SF_6 (i.e., about 15–40 $p\ell/\ell$). Thus, for long-range transport studies, in which the PFT concentration is primarily in the sub-$p\ell/\ell$ range, a different type of real-time approach is necessary.

An instrument originally conceived and developed by J.E. Lovelock (unpublished results) was modified at Brookhaven to concentrate the PFTs in a 5-ℓ air sample and analyze the desorbed sample on an in situ chromatograph column and detector. This version had two Ambersorb traps. While one was sampling at 1 ℓ/min for 5 min, the other was desorbed and analyzed. Since the traps reversed position every 5 min, no tracer was lost (Ferber et al., 1981). The limit of detection is about 10 $f\ell/\ell$, making the instrument quite useful for potential aircraft traverses in long-range (1,000 km) PFT experiments. At the very least the dual-trap analyzer might be used to direct the aircraft for on-board programmable sampler collection of the otherwise invisible plume. Useful data were also obtained in complex terrain experiments, as shown in Fig. 21 (Ferber et al., 1983).

4.4.3 Future PFT Analyzers

For potential detection of PFTs on a continental scale, a capillary chromatograph must be employed. This will provide both the requisite resolution and sensitivity to detect the tracer among the various interfering constituents. It is likely that such an approach will also be necessary if a family of five or more PFTs is released during the same experiment.

5 APPLICATIONS

5.1 Overview

Atmospheric applications of gas tracer technology generally include quantification of one or more of four processes:
• Transport,
• Diffusion,
• Emission rates, and
• Infiltration.
It frequently is important, but difficult, to quantify near-surface or elevated atmospheric trajectories, particularly

when the terrain is complex (e.g., mountainous or near a sizable body of water). Numerous gas tracer studies have been conducted to document flow regimes and their variation as functions of the synoptic weather pattern, time of day, and geographic location. This type of tracer experiment is the most elementary and, frequently, also the most difficult to conduct successfully because of the large area (and long time period) that must be sampled and the uncertainty about the wind flow. A typical experiment involves the point release of a tracer for an intermediate period (e.g., 2–6 h) with subsequent sampling by both fixed-location, time-integrated samplers and continuous or discrete techniques from moving platforms at the ground and aloft. Since continuous real-time SF_6 analysis instruments have been available on an operational basis only since 1981, most of the trajectory studies to date have suffered from a lack of real-time experimental control. Previously, the location and structure of a tracer plume could only be reconstructed after the fact. One modestly successful attempt to overcome this limitation involved the use of constant density-altitude balloons (i.e., tetroons) to physically mark a mass of air into which the tracer was released. A lightweight (135-g) 403-MHz transponder attached to the tetroon has been used to determine its location. The NOAA Air Resources Laboratory Field Office, Idaho Falls, Idaho, uses a modified M-33 X-band radar to track the transponder to a range of 80 km, while SRI International has modified the weather radar on its Beechcraft A-80/8800 Queen Air aircraft to locate the unit up to a range of 6 km. With the tetroon location known, grab samples could then be collected by aircraft and on the ground. Often, however, vertical wind shear and horizontal divergence patterns result in significant differences between tetroon location and surface tracer distribution.

Figure 22 is an example of some results from a dual-tracer transport experiment conducted in north-central California (Dabberdt, 1983). SF_6 and $CBrF_3$ were released at 2-h intervals in the mid- to late morning; release locations and times were different for each test. On the test that is illustrated both tracers were released at the same ground-level location: SF_6 (25 kg/h) from 0900 to 1100 PDT and $CBrF_3$ (34 kg/h) from 1030 to 1227 PDT. The plotted data are hourly values for 1400–1500 PDT and clearly indicate the complex nature of transport conditions throughout the region and also the significance of the 90-min differential in the two release intervals.

With the exception of exploratory tests conducted recently with perfluorocarbon tracers (i.e., Ferber et al., 1981) and a few studies using $CBrF_3$, SF_6 has been used almost exclusively for mesoscale transport studies. Because of the high SF_6 background concentration (~ 1 ppt) and the effects of fugitive sources, it is not practicable to use either SF_6 or $CBrF_3$ beyond a range of about 100 km.[4] PMCH, on the other hand,

---

[4]A notable exception is the study of Smith et al. (1981) in California's Sacramento Valley, where SF_6 was released for 4 h at a rate of 43 kg/h and was detected at levels of 43 and 13 ppt at ranges of 176 and 268 km, respectively.

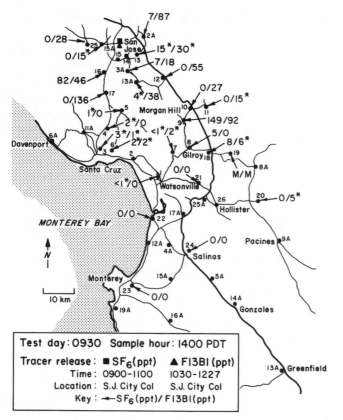

Fig. 22 Hourly surface tracer concentration measurements.

provides a potential practical range of greater than 1,000 km with an hourly cost for tracer materials about twice that of a 100-km SF_6 experiment. However, sampling, logistics, and analysis costs would be significantly greater.

A common application of tracers is to measure the atmospheric diffusion of plumes. The procedure is quite straightforward. Horizontal crosswind profiles of tracer concentration are measured through the tracer plume at different altitudes and ranges. Sampling periods generally vary from 15 to 60 min depending upon the scale of the experiment. Samples typically consist of time-integrated air samples at fixed locations and continuous or discrete (i.e., grab) sampling from vehicles moving through the plume. Figure 23 illustrates schematically the design of a microscale dispersion experiment for quantification of lateral and vertical dispersion in the shoreline transition zone. The tracer measurements enable determination of diffusion "coefficients," such as the eddy diffusivity or the "sigma function" of the Gaussian parameterization. For simplicity of illustration, consider the Gaussian steady-state point source plume model,

$$\chi = \frac{Q_P}{2\pi\sigma_y\sigma_z U} \exp\left(\frac{-y^2}{2\sigma_y{}^2}\right)\left\{\exp\left[\frac{-(z-h)^2}{2\sigma_z{}^2}\right] + \exp\left[\frac{-(z+h)^2}{2\sigma_z{}^2}\right]\right\} \tag{4}$$

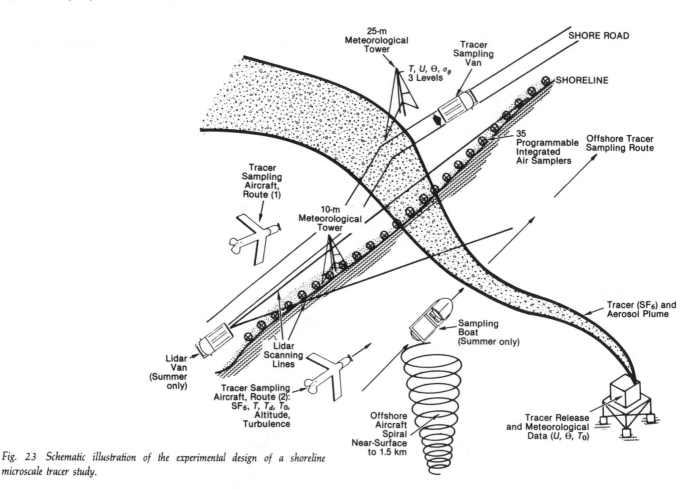

Fig. 23 Schematic illustration of the experimental design of a shoreline microscale tracer study.

where y is the normal horizontal distance from the plume center line, z is receptor height, U is average wind speed, h is the plume source height, and Q_p is the point source release rate (grams per second). The lateral and vertical dispersion functions are given by σ_y and σ_z, respectively, and are assumed to be independent of height. The lateral function, σ_y, is calculated directly from the variance of the crosswind, time-averaged profile at any height. Figure 24, for example, illustrates 14 ground-level crosswind profiles obtained in a 1-h period along the road shown in Fig. 23; the individual profiles were taken over approximately 1-min periods by an instrumented van. With similar profiles obtained aloft by aircraft, the vertical dispersion function can also be determined through use of Eq. (4); the procedure is described later in this section.

Other studies have investigated dispersion from semi-infinite line sources (Q_L), such as highways. The tracer is released from vehicles moving along the highway and is sampled at fixed locations downwind of the highway. At the surface ($z = 0$), Eq. (4) can be integrated in the horizontal crosswind direction to obtain the line source dispersion equation

$$\chi = Q_L/(\sqrt{\pi/2}\, \sigma_z U \sin \phi) \qquad (5)$$

which can be used directly with the tracer data at single points to estimate σ_z. The term $\sin \phi$ accounts for nonorthogonality of the wind and roadway to angles as small as $25°$. Estimating emission rates of complex or uncooperative pollution sources is a potentially powerful application of tracer technology. First, we introduce nomenclature: Q_S is the unknown emission rate of the pollutant source, and Q_T is the tracer release rate; χ_{STOT} is the total atmospheric concentration of the pollutant, χ_{SB} is its ambient background concentration, and χ_S is the contribution by the source of concern; and, similarly, χ_{TTOT} is the total atmospheric concentration of the tracer, χ_{TB} is its background concentration, and χ_T is the contribution by the tracer source. When the tracer is released coincidentally with the pollutants and both are inert gases, then by definition

$$\chi_S/Q_S = \chi_T/Q_T \qquad (6)$$

and

$$(\chi_{STOT} - \chi_{SB})/Q_S = (\chi_{TTOT} - \chi_{TB})/Q_T \qquad (7)$$

Rearranging terms,

$$Q_S = [(\chi_{STOT} - \chi_{SB})/(\chi_{TTOT} - \chi_{TB})]Q_T \qquad (8)$$

Dabberdt et al. (1981) have extended this concept to estimate carbon monoxide emissions from traffic in both directions on a major freeway. Two vehicles traveled continuously around a loop on an east-west freeway. They emitted SF_6 from the central westbound lane and returned in the central eastbound

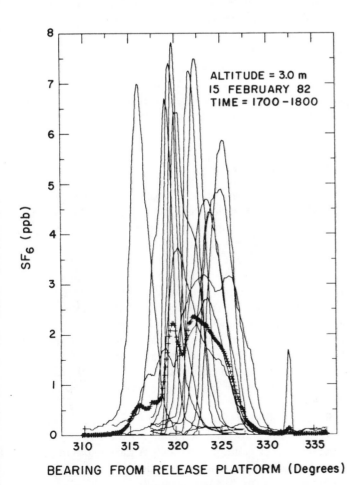

Fig. 24 Example of 14 individual tracer profiles and the hourly average.

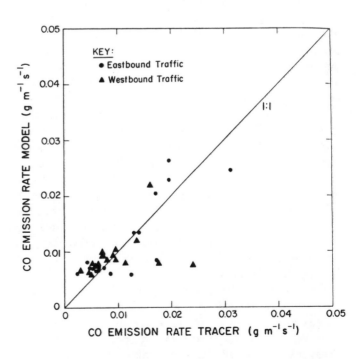

Fig. 25 Comparison of model prediction of CO emission rate with computation based on tracer and ambient CO measurements.

lane, emitting CBrF$_3$. Figure 25 illustrates the agreement between emission rates estimated from the tracer and CO data with estimates made by a speed-dependent vehicle emissions algorithm.

Fugitive emissions from a large petrochemical facility have been estimated by Ludwig et al. (1983) on the basis of a controlled SF$_6$ release along a road upwind of the facility and collocated hourly samples of the tracer and three halogenated hydrocarbons. Concentrations were measured far enough downwind to assume equality of the dispersion from the two sources, yet close enough to assume that the chemical plant could be represented as a semi-infinite area source in the crosswind direction. The 800-m-long tracer line source (i.e., the road) was 1.6 km upwind of the plant, which has dimensions of approximately 225 m (along the road) by 150 m. Samples were collected 1 to 5 km downwind of the plant, with winds within 45° of perpendicular to the road.

Ludwig et al. assumed that σ_z could be approximated by the form ax^b, where x is range and a and b are stability-dependent "constants." Using this relationship and integrating Eq. (5) in the x direction, concentrations from the area source (χ_A) are given by

$$\chi_A = \frac{2Q_A(x_u{}^{1-b} - x_d{}^{1-b})}{\sin \phi \sqrt{2\pi} \, Ua(1 - b)}, \quad b \neq 1 \tag{9}$$

where x_u and x_d are distances from the sampling location to the up- and downwind edge of the area sources (emission rate, Q_A). For neutral and stable conditions, b is equal to about 0.65, and Q_A can then be estimated from the ratio of Eqs. (5) and (9):

$$Q_A \approx 0.35\chi_A Q_L/[\chi_L(x_u{}^{0.35} - x_d{}^{0.35})x^{0.65}] \tag{10}$$

Independent emission estimates were not available in the study for evaluation of the technique. The authors, however, recommend more frequent and denser sampling to account for the large variability in space and time of the fugitive emissions.

The method of Ludwig et al. (1983) is preferred only when it is not possible to obtain access to the source to release the tracer(s). B. Sivertsen (personal communication, 1983) has described a study in which SF$_6$ and CBrF$_3$ were released within a petrochemical complex, and a simple proportionality model was used to estimate leakages of ethylene, propylene, ethane, propane, and isobutane from different parts of the complex. Sivertsen's proportionality model is similar to Eq. (9), except that the tracer-release locations were only an approximation of the (unknown) leakage locations; tracer and hydrocarbon samplers were collocated.

Gas tracer techniques generally provide the only direct means of measuring air infiltration in buildings under actual usage conditions. Several modes of operation have been used (Kronvall, 1981), the most common ones being the tracer decay method (Bassett et al., 1981; Hunt, 1980; Harrje et al., 1979) and the steady-state tracer method (Harrje et al., 1975;

Condon et al., 1980). In the tracer-decay method, a small amount of tracer is released instantaneously in the building; the logarithmic decay of concentration with time is directly proportional to the infiltration rate expressed as air changes per hour (ach). The exact amount of tracer released need not be known, but only periodic concentrations (5–15 min apart) must be measured. The steady-state method requires tracer injections on a periodic basis to maintain a constant tracer concentration within the structure; the infiltration rate is proportional to the tracer injection rate. In a more recent approach, Dietz and Cote (1982) have used multiple, miniature PFT diffusion sources in a house (one per room; tracer mixes uniformly in the room in 3–4 min) to determine infiltration rate from a constant emission source and periodic measurements of the concentration. Samples are integrated over intervals from 20 min to several hours or weeks. By emitting at a known rate, a building tracer mass balance can be expressed as

$$dV(t)/dt = Q_T - R_V(t)[V(t)/V_b] \tag{11}$$

where $V(t)$ is the amount of tracer gas remaining in the building at any time t, Q_T is the tracer release rate, R_V is the time-dependent building infiltration rate, and V_b is the interior building volume. Knowing Q_T and measuring the time-dependent tracer amount within the building enable Eq. (11) to be solved for the infiltration rate. Dietz and Cote have successfully used a PFT diffusion source (described in Section 4) and both passive and active adsorption tube samplers to measure concentrations. Figure 26 illustrates PDCH concentrations measured at various locations in a residential dwelling over a 16-day period and the associated values of the derived infiltration rate.

5.2 Analysis Methods

We describe several specific analysis methods for determining diffusion functions and mass balance estimates; also included are selected methods for estimating and quantifying the uncertainty of the calculations.

In studies of plume dispersion, it is usually necessary to quantify the lateral and vertical spread of the plume in time and range, and to relate these variations to independent parameters such as surface roughness, stability, wind speed, and so forth. The functions σ_y and σ_z are most commonly used for this purpose. Two methods can be used to estimate these functions: direct calculation of the second moment of the measurements, and calculation of the value of each function that provides the best fit to an assumed concentration distribution.

Figure 24 illustrates a series of 14 horizontal crosswind traverses made through a plume in 1 h at a range of 10 km from the release point. It also indicates the composite hourly average profile. The standard deviation of the concentration distribution is then calculated as

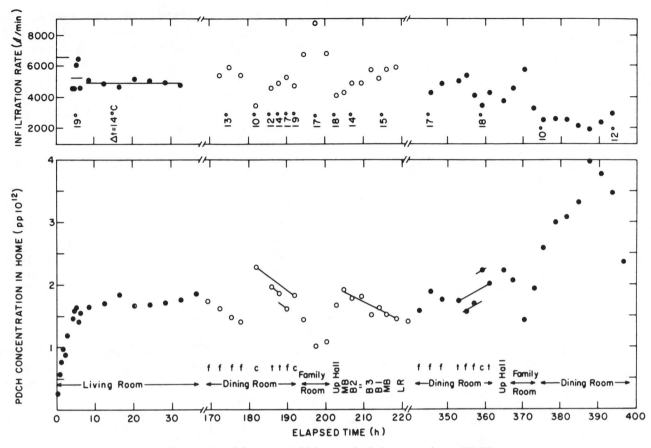

Fig. 26 Extended testing period (about 16 days) showing steady-state PDCH concentration as a function of time (lower plot) and the calculated infiltration rate (upper plot), including the average inside/outside temperature difference (pp10^{12} is equivalent to pℓ/ℓ).

$$\sigma - (\Sigma \chi y^2 / \Sigma \chi)^{1/2} \qquad (12)$$

where y is the departure or distance from the plume center line and χ is the corresponding value of the observed concentration. The extent to which the data are normally distributed can be evaluated by calculating the third and fourth moments (e.g., see Panofsky and Brier, 1958) and comparing the coefficients of skewness and kurtosis with those of a normal distribution (0 and 3, respectively).

When the sampling line is not perpendicular to the plume center line, it is desirable to make a simple adjustment by mathematically rotating the sampling line; the procedure is illustrated in Fig. 27.

The actual sampling line or traverse is denoted by AB and makes an angle with the plume center line of only about 60°; A'B' is the desired orientation of the sampling line. A linear adjustment can be made[5] to the concentration at each sampling point according to the ratio of the actual range (R_i) to the adjusted range ($R_i - \Delta R_i$), where

$$\chi_i(\text{adjusted}) - [R_i/(R_i - \Delta R_i)]\chi_i(\text{actual}) \qquad (13)$$

and

$$R_i - \Delta R_i \equiv R_0 \cos \beta_i \qquad (14)$$

The angle β_i is measured from the plume center line (C_L) to the line connecting the release point and sampling point (i). The measured and adjusted tracer distributions are illustrated in part b of Fig. 27, together with the adjustments made at sampling points 1 and 2 whose locations are shown in Fig. 27a. The magnitude and sign of the correction are consistent with the range dependence of concentrations presented by Turner (1970) for a continuous surface release. Turner's calculations using a Gaussian model and F stability (i.e., stable stratification) agree within 12% of estimates made with Eq. (13), where the range separation between receptor points (6 and 10 km) used by Turner is much greater than those actually used in the adjustment procedure here (typically less than a few hundred meters over a range of 6–15 km).

Dabberdt et al. (1982) have calculated sigma functions by using a linearization-of-function routine to obtain the best fit of the data to a Gaussian distribution. Many hourly tracer

---

[5]Alternatively, a dispersion model such as Eq. (4) or (5) could be used to describe the range dependence of the adjustment, particularly when the ratio $\Delta R/R$ becomes large.

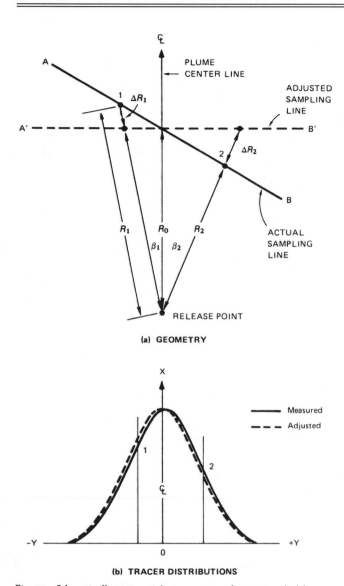

Fig. 27 Schematic illustration of the tracer range-adjustment methodology.

distributions have significant multimodal structure (i.e., well-defined, multiple peaks). Accordingly, their analysis also included determination of the standard deviation and center-line position of up to three individual distributions (or modes). The sequence of steps in the analysis is summarized briefly below and then described more fully in the following discussion:

(1) Represent all hourly profiles as arrays of tracer concentration, range (from release point), and bearing.

(2) Analyze the profiles to obtain a first estimate of the plume center-line position and standard deviation.

(3) Adjust the initial arrays (step 1) by mathematically rotating the orientation of the sampling lines so they are perpendicular to the plume center line (step 2); repeat steps 2 and 3 until there is no significant change in the plume center-line position.

(4) Using the final estimate of plume center-line position and the corresponding adjusted data arrays, analyze the (adjusted) hourly profiles for multiple tracer distribution.

The correction or adjustment described in steps 2 and 3 is only significant when the acute angle between the bearing of the plume center line and the sampling line is small. For the data shown in Fig. 24, the angle was about 62°. Figure 28 shows both the original tracer data and the adjusted data, as well as the fitted unimodal Gaussian distribution and the fitted composite bimodal distribution. The data tabulation to the right of the plot lists

- Altitude;
- Date;
- Hour;
- Range, R_0;
- Center-line bearing, θ_0;
- Crosswind distance from plume center line, Y; and
- Standard deviation, σ.

The latter two parameters (Y and σ) are given first for the composite plume (denoted by the subscript zero) and then for each of the individual modes—in this case, two. Note that by definition, $Y_0 \sim 0$.

The analysis routine used to calculate the standard deviation is a linearization-of-function technique described by Bevington (1969). The technique is based on a Taylor expansion of a nonlinear function, $f(m)$, and uses a gradient search method to minimize chi-square (χ_{fit}^2, the goodness of fit); in generic format,

$$\chi_{fit}^2 \equiv \sum \left\{ \frac{1}{\sigma_i^2} [f_i - f(m_i)]^2 \right\} \quad (15)$$

where f_i is the observed value of the variable (e.g., tracer concentration) and $f(m_i)$ is the value of the variable calculated for m_i from the nonlinear function. The curve-fitting routine requires two subjective choices: the function to be fit, and an "initial guess" of the fit. The data were fitted to a multiple Gaussian distribution with zero background,

$$\chi = \sum_{i=1}^{q} \chi_{peak_i} \exp[-(y - \bar{y})^2/2\sigma_i^2], \quad q = 1, 2, \text{ or } 3 \quad (16)$$

In Eq. (16) χ_{peak_i} is the center-line or peak concentration, y is the crosswind distance from the center-line position ($\bar{y}$), σ is the standard deviation of the Gaussian distribution, and q is the number of peaks. The number was determined subjectively by first reviewing the composite hourly profiles.

The vertical diffusion parameter, σ_z, is more difficult to determine because the plume impinges at the ground surface and elevated inversions. Dabberdt et al. (1983) introduced an approach that uses the linearization-of-function routine described above and the crosswind-integrated form of the Gaussian plume equation (with single reflections at the surface boundary and aloft). Because of multiple peaks in the lateral tracer distributions, the crosswind-integrated concentration (χ_{cwi}) provides a more robust and conservative measure with which to assess the shape of the vertical concentration profile. This approach also is fully independent of σ_y calculations because χ_{cwi} is calculated directly from the observations. Conditions with multiple reflections are ignored because an unambiguous value of σ_z cannot be determined when σ_z approaches H, the mixing depth.

The point-source Gaussian plume equation for the above conditions is

$$\chi = \frac{Q}{2\pi\sigma_y\sigma_z U} \exp\left[-\frac{1}{2}\left(\frac{y}{\sigma_y}\right)^2\right]\left\{\exp\left[-\frac{1}{2}\left(\frac{z-h}{\sigma_z}\right)^2\right]\right.$$

$$+ \exp\left[-\frac{1}{2}\left(\frac{z+h}{\sigma_z}\right)^2\right] \quad (17)$$

$$\left.+ \exp\left[-\frac{1}{2}\left(\frac{z-2H-h}{\sigma_z}\right)^2\right]\right\}$$

Integrating Eq. (17) along y yields the crosswind-integrated form of the plume equation,

$$\chi_{cwi} \equiv \int_{-\infty}^{+\infty} \chi\, dy = \frac{Q}{\sqrt{2\pi}\sigma_y\sigma_z U}\left\{\exp\left[-\frac{1}{2}\left(\frac{z-h}{\sigma_z}\right)^2\right]\right.$$

$$+ \exp\left[-\frac{1}{2}\left(\frac{z+h}{\sigma_z}\right)^2\right] \quad (18)$$

$$\left.+ \exp\left[-\frac{1}{2}\left(\frac{z-2H-h}{\sigma_z}\right)^2\right]\right\}$$

where χ_{cwi} has units of grams per square meter. Equation (18) has been used, first, to solve simultaneously for the terms σ_z and $Q/(\sqrt{2\pi}\,\sigma_z U)$, and, second, to solve simultaneously for σ_z, $Q/(\sqrt{2\pi}\,\sigma_z U)$, and h. The term $Q/(\sqrt{2\pi}\,\sigma_z U)$ is an amplitude term $(\hat{\chi})$ that enables us to evaluate the quality of the measurements and the representativeness of the model by using independent measurements of Q and U to estimate $\hat{\chi}$. Estimated and fitted values of $\hat{\chi}$ can then be compared. Disagreement indicates that either the model form incorrectly approximates the actual dispersion conditions or the wind-speed value is inappropriate. The independent determination of h in the second application cited can be used to evaluate whether the plume rises significantly (e.g., because of mountains).

An arbitrary weighting scheme was used to assign relative weights to the data according to the nature of the measurements. Surface bag samples and mobile measurements with eight or more traverses per flight level per hour were given a relative weight of 4. Profiles from five to seven traverses were given a weight of 3, three and four traverses a weight of 2, and one and two traverses a weight of 1. In the linearization-of-function regression routine, a weight of 4 was simulated by repeating the datum four times, and so on for the other weights. Figure 29 is an example of the σ_z analysis.

An additional evaluation of the model and the data involves the calculation of the tracer mass flux through the y-z plane of the measurements. Given the vertical profile of χ_{cwi}, the mass flux, Q' (in grams per second), is calculated from the height integral of the transport rate.

$$Q' = \int_0^H u\chi_{cwi}\, dz \quad (19)$$

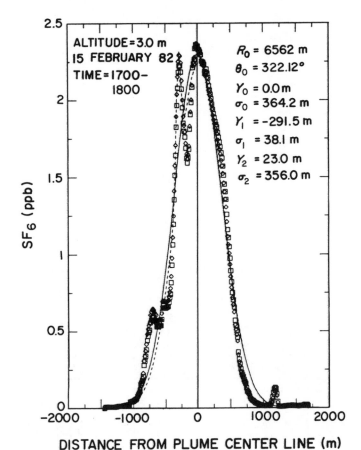

Fig. 28 Example of hourly composite data from mobile sampling platform—both raw (squares) and range-adjusted (diamonds)—and the fitted single (solid line) and multimodal (dashed line) Gaussian distributions.

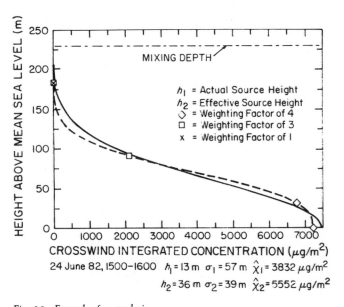

Fig. 29 Example of σ_z analysis.

TABLE 7
Mass Balance and Uncertainty Analysis of
Sackinger et al. (1982)

	Test 1	Test 2
Q (SF$_6$ released, kg)	90.8	236.1
Q' (SF$_6$ mass est., kg)	111.2	211.9
$Q' \div Q$ (%)	123	90
Estimated data uncertainty:		
ϱ, density (%)	5	5
H, inversion height (%)	40	40
χ, SF$_6$ concentration (%)	30	30
U, wind speed (%)	25	25
l, sampling interval (%)	10	10
ϕ, sample line orientation (rad)	0.175	0.175
θ, wind direction (rad)	0.698	0.698
Calculated uncertainty of total mass (%)	25.4	25.5

TABLE 8
Gaussian Plume Parameters and Uncertainty
Analysis of Sackinger et al. (1982)

Traverse Number	1	2	3
Plume parameters:			
y_0 (km)	3.38	1.56	1.54
σ_y (km)	0.38	0.23	0.18
χ_{max} (ppb)	2.73	9.15	9.41
Estimated data uncertainty:			
χ (%)	30	30	30
y (km)	0.16	0.16	0.16
Calculated uncertainty:			
y_0 (km)	0.06	0.08	0.06
σ_y (%)	11.5	18.2	14.2
χ_{max} (%)	21.5	33.4	31.2

The calculated mass flux should equal the tracer release rate, provided the vertical wind and crosswind-integrated concentration profiles are adequately represented by the measurements.

Sackinger et al. (1982) present and apply a methodology for quantitatively estimating the uncertainty of mass balance and diffusion-parameter calculations based on the uncertainty in specifying each of the independent meteorological/geographical variables and in measuring the dependent variable, the tracer concentration. For a function (Q', σ_y, or σ_z) of several variables,

$$m = f(n_1, n_2, n_3, \ldots n_p) \qquad (20)$$

The uncertainty in the dependent variable m is a function of the uncertainties in each of the independent variables n_i (Bevington, 1969), where

$$\sigma_m{}^2 = \sum_{i=1}^{p} \sigma_{n_i}{}^2 \left(\frac{\partial f}{\partial n_i} \right)^2 \qquad (21)$$

The partial derivatives are to be evaluated for each variable that is a source of uncertainty. Assuming that the independent variables are independent of each other and are themselves randomly distributed about their mean value, Sackinger et al. applied Eq. (21) to estimate the uncertainty by considering analogous forms of Eq. (19) and Eq. (4) for Q' and σ_y, respectively. Uncertainty estimates of mass balance were made for two mesoscale tracer tests along a 160-km section of the southern California coast (see Table 7). In test 1, Q' was estimated at 123% of the actual SF$_6$ release (Q), and in test 2 it was 90%. The uncertainty of the mass estimates was calculated to be about 25% for both tests, indicating that both mass estimates agreed with the actual tracer release amounts within the uncertainty of the independent variables. Table 8 summarizes the uncertainty in the calculations of plume center-line location (y_0), σ_y, and χ_{max} for three microscale dispersion tests on the Santa Barbara (California) coastline.

The uncertainty in y_0 is estimated to be very small (60–80 m) in all cases; σ_y uncertainty is 11–18% of the calculated values, and χ_{max} uncertainty is 21–33%. These uncertainties of the dependent variables are less than or approximately equal to the maximum uncertainty of the individual independent parameters, reflecting the effect of statistical cancellation of random errors and demonstrating the validity of complicated mass balance calculations.

A Comparative Overview of Active Remote-Sensing Techniques

R.L. Schwiesow, NCAR

1 WHAT IS REMOTE SENSING IN THE BOUNDARY LAYER?

To answer this question we look at remote sensing as a field with some broadly defined characteristics and as a historical entity with roots from before the turn of the century. Relevant subtopics are: How does remote sensing function? When and where did it start? Who worked on it?

In the boundary layer, considered as a three-dimensional fluid, remote sensing means measuring the characteristics of some region in the fluid with instrumentation that does not have a sensing element in or surrounding the volume of interest. Any such definition is more or less arbitrary, but it is helpful to consider the function and properties of remote sensing under two headings, topology (geometric relationships) and effect on the measured variable. A topological difference between remote sensors and sensors immersed in the region to be probed is evident if we consider remote sensors as having a space of at least the linear dimension of the instrument between the region of interest and the instrument. This form of definition excludes, for example, sonic anemometers and Lyman-alpha hygrometers, which are discussed earlier in the book in the context of in situ sensors. Under the heading of effect on the measured variable, remote sensing can be defined as not disturbing the variable being measured. This noninterference characteristic gives remote sensing a potential advantage in accuracy over immersion sensors in some cases, but exploiting the advantage requires careful experiment design. For example, remote temperature sensors have no error caused by heat flow through the sensor-supporting structure, radiation errors are eliminated, and wetting and dynamic heating have no effect; likewise, remote velocity sensors have no hardware or towers to distort the measured wind field.

These definitions illustrate that it is not possible to place a restrictive or exclusive boundary around remote sensing. Avoiding a limiting approach can encourage development. This is appropriate for a field that is still changing and testing applications for feasibility.

Remote sensing of boundary-layer variables can be done actively or passively. Active measurements involve transmitting acoustic or electromagnetic radiation to the region of interest and measuring the portion of the radiation that is returned from the region to the instrument. This section of the book emphasizes the three "dars": lidar, radar, and sodar. These words for active remote sensing came from acronyms, now in common use. The suffix dar stands for *detection and ranging*, the prefixes, for the type of radiation used: *light*,

radio, and *sound* waves. Some active remote-sensing techniques use a transmitter and receiver on either end of a path and infer properties of the boundary layer from variations in the transmitted radiation, rather than from the scattered radiation as in the case of the three dars. The optical crosswind sensor is an example of a two-ended, path-averaged remote sensor. Precise range information is difficult to obtain using transmitted, rather than scattered, radiation. Passive measurements involve receiving and analyzing radiation naturally emitted from the atmosphere. As we might expect, it is difficult to obtain precise range information from passive data. Visual observations and infrared radiometry are examples of passive remote-sensing techniques.

In contexts other than the boundary layer, remote sensing can have other meanings, particularly if the target is a two-dimensional surface rather than the three-dimensional atmosphere. Aerial photography or Landsat multispectral images are types of remote sensing for land use (e.g., Barrett and Curtis, 1982). Infrared radiation emitted from a surface (e.g., the ocean or a cloud) can be remotely sensed with a radiometer to infer surface temperature. Multispectral images (visible to microwave wavelengths) taken from satellites are used for observation of rain belts and rainfall monitoring (Barrett and Martin, 1981). Infrared radiometric sensors can also measure radiation at short range from molecules in the region of the sensor to infer air temperature, but such sensors are not remote in the sense that they do not have a space between the instrument and the region measured. Lightning strike location systems remotely sense data from an active source. However, none of these systems does active remote sensing of standard meteorological variables in the boundary layer in the sense that we discuss here.

The following four articles discuss lidar, radar, and sodar as mainstream examples of remote-sensing systems for probing the atmospheric boundary layer. Although the three dars do not cover all appropriate techniques for remotely sensing the boundary layer, they do represent a core technology and serve as an introduction to the field.

Remote sensing of the atmosphere has gone on since people began watching the sky for signs of a change of weather, but active remote sensing with intentionally directed radiation is comparatively new. The human eye and ear acting as detectors, together with the brain as an integrated data processor, still provide a benchmark in radiation sensitivity, dynamic range, compactness, and energy efficiency. Only in spectral range and response time are modern techniques now superior. The individual articles on sodar, radar, and lidar mention the history of the development of the techniques as applied to the boundary layer, but it is interesting to note that basic active remote sensing is older than a few decades.

Tyndall (1875) in England investigated acoustic scattering in the atmosphere before the turn of the century. Figure 1 shows an early remote-sensing instrument. Gilman et al. (1946) started the modern era in sodar.

Radar returns from the ionosphere were obtained by Appleton and Barnett (1925), but the development of shorter-wavelength radars with steerable antennas during World War II made boundary-layer measurements practical. Soon after the war, radar measurements from precipitation targets (Marshall et al., 1947; Wexler, 1947) made use of newly developed equipment. More than a decade of radar development was necessary to achieve sufficient transmitter power and receiver sensitivity to obtain reliable clear-air returns (Hardy et al., 1966).

Lidar at first used large, modulated searchlights separated from the receiver location and scanned in elevation angle to intersect a vertically pointing receiver beam at various altitudes up to 60 km (Elterman, 1951). Although many people originally thought that a laser was impossible, after its demonstration in 1960 it quickly replaced other light sources for lidar. Fiocco and Smullin (1963) demonstrated a lidar based on a ruby laser, and since then, many different types of lasers have been used for lidar.

Major advances in remote sensing have been made in the past 20–30 years, but new ideas in sensing techniques and applications appear regularly. The beginner can learn from the established technology but should not limit his or her outlook to present developments. The four articles in this book on remote sensing are offered in that spirit.

2 SHOULD I USE REMOTE SENSING?

A meteorologist with an observational problem that involves probing the atmospheric boundary layer can be intrigued, confused, or both, by the array of remote-sensing techniques available. By considering the general applications and common properties of the remote sensors that are reviewed in detail in the following articles, I hope to provide a basis on which an experimentalist can decide if remote sensing is likely to be useful for a particular problem. If remote sensing appears to be applicable, then one must decide which technique to use. To help make these decisions, this article discusses some meteorological variables and compares the capabilities of acoustic (sodar), radio (radar), and optical (lidar) techniques for measuring each variable.

The goal of this overview is to provide an outline on remote-sensing techniques that can guide you to other, more detailed sources. It should indicate the article in this book that is most closely related to the variables and operational scale

Fig. 1 Transmitter for an early sodar. The receiver was the unaided ear of the observer. (From Tyndall, 1875.)

of your application, but all the articles contain data and insights useful to both beginners and experts in remote-sensing measurements.

2.1 General Applications for Remote Sensing

Remotely sensed meteorological measurements are particularly useful where the application requires a spatial and temporal data density that cannot be obtained economically with a multitude of conventional point sensors, or where the region of the atmosphere to be explored is not accessible to normal instrument platforms. In situ and remote-sensing instruments are complementary rather than competitive in most cases. For example, aircraft-borne immersion instruments can obtain drop-size distributions and temperature cross sections in thunderstorms that are inaccessible using radar, but a radar can provide nearly synoptic data on the flow field in the precipitating region. Such spatial data density from radar on the three-dimensional (3-D) wind field at many levels and over a 10- to 40-km-diameter region cannot be obtained from immersion instruments carried by aircraft before the storm changes character. Furthermore, some storm environments may be inaccessible to aircraft for safety reasons. McCarthy et al. (1983) discuss a set of experiments involving complementary remote and in situ sensors, mounted both at the surface and on aircraft, to obtain both spatially dense and multiparameter data sets.

Regular measurement of profiles of wind, temperature, and humidity is another area where remote sensing is useful. Although aircraft-mounted immersion sensors or dropsondes can measure profiles at altitudes far above the region accessible to towers, only remote sensing can provide profile data of sufficient temporal density to compete with balloon-borne radiosondes on a semi-operational basis. Ecklund et al. (1982) discuss the use of long-wavelength VHF radar for profiling winds up to, and beyond, the tropopause. Hogg et al. (1983a) show that the temporal data density for remote sensing can be greater than that for any reasonable balloon launch schedule. Although in situ sensors on meteorological towers provide excellent temporal data density, they are of course severely limited in the altitude to which they can collect data. Remote sensors can be applied to regions of the boundary layer where regular access by other means is difficult.

In short, remote-sensing techniques are applicable where in situ methods will not work or are uneconomical and where the atmosphere interacts with the transmitted or received radiation in a way that allows the meteorological variable of interest to be measured.

2.2 Common Properties of Different Techniques

Remote sensors can be divided into two geometric classes depending on whether the transmitter and receiver are located in practically the same spot (so that the sensing depends on backscatter), or are separated by a substantial base line but aimed at the same volume in space. Members of the former class, called monostatic sensors, are by far the most common configuration; they are sensitive only to backscattered energy. Examples of the separate receiver-transmitter class, called bistatic sensors, operate at a scattering angle that depends on the base-line separation and location of the sample volume. Because of difficulties in accurately coordinating beam steering, practical bistatic systems are usually wide-beam-width systems and often use acoustic radiation.

Remotely sensed measurements are spatial averages over volumes (or lines as in the case of a crosswind laser anemometer) that depend on the parameters of the instrument; remote sensing is suitable if spatial averages are desired. A single reading from a remote-sensing instrument is equivalent to the average of an ensemble of in situ instruments distributed within the sample volume of the remote sensor. Thus, the remotely sensed value is more representative of a mean quantity than is a single-point measurement, and for some applications a single value from a remote sensor can replace a time average value from a single immersion sensor. Comparison of rain rate measured by radar backscatter with that measured by a rain-gage network is one example; comparison of lidar wind estimates with those from a sonic anemometer is another. The shape of the averaging volume for typical systems can be estimated from the information in Section 2.3.

One characteristic that is common to sodar, radar, and lidar is that all utilize returns from a target consisting of scatterers distributed in space, but traditions associated with the three communities treat the scatterers differently. These differences can lead to confusion when the approaches are compared. One can consider the medium a continuum with uniform scattering characteristics, a model suited for long-wavelength sensors, or one can consider the medium a collection of scattering centers with a particular scattering strength and angular pattern associated with each scatterer. In the discrete representation the scatterers have a spatial density (number per unit volume) in the medium. This density factor introduces a continuum (i.e., statistically uniform) characteristic into the particle picture. The scattering cross section of the target is sometimes defined in terms of total power scattered out of the beam per unit volume of scattering medium divided by the incident flux density (power per unit area in the beam), particularly in radar work. Another definition is the differential scattering cross section, usually just called the cross section, which is the power scattered out of the beam in a given direction per unit solid angle and per unit volume of the medium, divided by the incident flux density. In lidar, the differential cross section is often ascribed to the scatterer as power scattered from a single particle or scattering center in a given direction per unit solid angle, divided by the incident flux density. When multiplied by the number of scatterers per unit volume, this definition becomes dimensionally equivalent to a continuum representation, and it is practically equivalent if the number of scatterers per volume resolution element is very large. Although all definitions have the same basic units if sr^{-1} (the unit per steradian) is dropped, the first assumes that the scattering is isotropic and usually that the medium is continuous, whereas the latter two allow for an angle- and polarization-dependent scattering pattern, and the third applies to a distribution of discrete scatterers. In the remote-sensing literature, one clue to the cross section definition assumed is the presence of a factor 1/4 π in the equation for received signal when a total cross section is used and the absence of that factor when a differential cross section is applied.

All three techniques usually assume that the target fills the sensor beam when applied to the atmosphere. Beam filling means that the spatial extent of the scattering medium is greater than the size of the sample volume that is determined by the transmitter and receiver characteristics, and it assumes that the cross section of the medium is independent of position. For a target that fills the beam in azimuth and elevation, the received power will vary as r^{-2}, where r is the range to the target, because the effective target area increases as r^2, canceling the r^{-2} dependence of the incident flux density on range. The solid angle of the receiver as seen from the scattering region varies as r^{-2}. On the other hand, for a target that does not fill the beam, such as an aircraft in radar work, the received power varies as r^{-4}. Beam filling in the range coordinate implies that the received power increases with increasing length of the sample volume. The assumption that the cross section of the medium is independent of position is rarely strictly true, but it is a useful idealization. In practice, approximate constancy of cross section with position can be thought of as

$$(1/\eta)[r^2(\partial\eta/\partial r)^2 + (\partial\eta/\partial\theta)^2 + (\partial\eta/\partial\phi)^2]^{1/2} \ll 1$$

in spherical (r, θ, ϕ) coordinates, where η is the (total or differential) scattering cross section. The relation simply indicates that the fractional change in cross section is small over distances relevant to the remote-sensor application.

Although remote sensing involves volume averages, not all data on scales smaller than the spatial resolution of the sensor are lost. For example, the width of the Doppler spectrum (i.e., the second moment) measured by acoustic, radio, and optical waves contains information about velocity fluctuations on scales smaller than the sensing volume. The eddy dissipation rate (see the article by Friehe in this volume) is available from volume-averaged spectral data (see Kropfli, in this volume), at least in principle (Frisch and Clifford, 1974).

On the other hand, remote sensing sometimes requires averaging over dimensions larger than the basic spatial resolution of the instrument. The wind vector, for example, is usually obtained from wind component measurements at different azimuths. Multiazimuth data can be obtained by using multiple remote sensors with a spatial separation on the order of the sensing range and aimed at a common volume in space, or by using a single monostatic sensor that is steered to different azimuths in a conical scan, for example. (The collocated, triple monostatic acoustic sounder [Neff and Coulter,

this volume] is equivalent to a scanning sensor as far as spatial averaging is concerned.)

Separated, multisensor setups such as the monostatic dual-Doppler radar can probe a common resolution volume; however, they are expensive because of the duplication of hardware, and coordinating the data from independent systems separated by kilometers can be a problem. Bistatic sodar operates over much shorter base lines, does not duplicate as much hardware, and is less expensive than longer-range multisensor arrangements. Transmitter and receiver beam alignment is easier with wide-beam-width sodars than it is for systems with higher angular resolution. Azimuth-scanning remote sensors are simpler and less expensive by comparison, but they are limited in spatial and temporal resolution. The horizontal extent of the spatial-averaging volume of an azimuth-scanning system is the projection of the region swept out by the scan: for example, a cone with a base diameter equal to twice the range setting of the remote sensor times the cosine of the elevation angle. This averaging dimension is much larger than the resolution of the basic sensor, but under some conditions, spatial averages of quantities such as Reynolds stresses and variances due to small-scale turbulent fluctuations in the flow can be obtained (see the article by Kropfli in this volume). The temporal resolution of values inferred from a scanning system is on the order of the maximum spatial dimension of the scanned volume divided by the mean wind, because this is the time for the atmosphere to change within the scanned volume. This averaging time can be many minutes. Obviously, scanning systems are adversely affected by horizontal inhomogeneities in the vector fields and are best suited to measuring profiles in horizontally homogeneous situations, such as over flat, uniform terrain, and are not as useful for measuring vector fields in horizontally inhomogeneous situations such as in clouds and over complex terrain. In such cases, multiple, spatially separated sensors are generally more useful. For some special problems, for example waterspouts (see my article on lidar measurements in this volume), a single, nonscanning sensor can give useful (but not complete) high-resolution data on a vector field.

The scales of atmospheric structure that interact most strongly with radiation from a remote sensor are approximately one-half the wavelength of the radiation. Much smaller targets are called Rayleigh particles. Scattering from Rayleigh particles is strongly dependent on wavelength, decreasing as wavelength to the minus fourth power. The best sensor for various targets can be estimated from Sections 2.3 and 3.1. Using a cloud with droplets in the 5- to 50-μm-diameter range as an example, we find that lidar measures a strong return but does not penetrate far, short-wavelength radar gives a weaker scattering return than lidar and useful cloud penetration, and long-wavelength radar and sodar largely ignore the cloud and give a signal from raindrops and refractive index fluctuations.

The wavelength of the sensor radiation relative to the antenna aperture diameter primarily determines the angular resolution (i.e., beam width) of a remote-sensing system. High angular resolution results from a very large aperture/wavelength ratio, so to increase the resolution with a given antenna diameter, one should decrease the wavelength. Any antenna transmits and receives some radiation in directions other than the direction in which the main beam is pointing. These response side lobes generally result in unwanted signal (i.e., noise) and can be minimized by proper antenna design. The width of the side-lobe pattern decreases as the aperture/wavelength ratio increases. Another advantage to a narrow beam width is greater spatial resolution. Even though usual antenna sizes are not the same for different sensors, lidars generally have the highest angular resolution and sodars the lowest.

It is fairly straightforward to estimate the performance of a remote-sensing system based on satisfactory performance of a system with similar operating principles but different system parameters. In general for all carefully designed remote sensors, the detected signal power S is a function of transmitted power P, range resolution Δ, antenna area A_e, along-path attenuation coefficient α, and target range r as

$$S \sim P\Delta A_e \exp(-2\alpha r)/r^2$$

Other dimensional variables are also involved in the full signal equations; see the technique articles for details on signals and noise. Actually, the performance of a remote-sensing system is limited by the signal-to-noise ratio. The noise equations are different for sodar, radar, and lidar, and for different operating regimes within a given technique, so we cannot discuss noise in detail in an overview even though it is highly important in evaluating performance. If the noise in the system is reduced to that inherent in the signal, then the achieved signal-to-noise ratio depends only on average power and not on how it is compressed into pulses. As an example of an application of the signal proportionality, in order to maintain the same signal power in a system while doubling the observing range and halving the range resolution, the transmitter power must be increased by a factor of eight even if there is no path-dependent attenuation. Noise is likely to be a different function of these variables and to include additional variables. Knowing the signal proportion-

ality and properly considering noise sources should allow you to project the performance of systems described in the technique articles to your needs.

2.3 Parameters of Active Remote-Sensing Systems

All sodars, radars, and lidars are not the same; wide variations exist in system parameters. Representative values for the parameters, listed in Table 1, give some idea of the range of usefulness of different remote-sensing techniques. The figures in Table 1 are not extreme cases in any sense, but represent typical values. For example, some radars have narrower beam widths than the typical values shown, and other fixed-antenna types operate at wavelengths of more than 600 cm. Consult the individual technique articles for details on extreme values for system parameters and special configurations. Costs listed in the table are order-of-magnitude estimates and vary widely depending on transmitter power, the amount of in-house development done, etc.

Although it is clear that a long-wavelength radar is ideal for long-range measurements, Table 1 gives no help in answering the question, "Long-range measurements of what?" Thus this table is not helpful for selecting a technique except to give some idea of the capabilities of a technique once it has shown promise of usefulness on the basis of other criteria. Other criteria are discussed in the next section.

3 WHICH SENSOR?

There is no objective, clearly correct answer to the question of which sensor is most appropriate for measuring a particular boundary-layer variable. Cost, state of development, portability, dependability, maintenance requirements, and staff familiarity enter a technique judgment, as do range, range resolution, beam width, and averaging volume. As an example of different approaches to profile measurement, compare the well-developed system used by Hogg et al. (1983a), which is capable of measurements throughout the troposphere, with the experimental, small, high-resolution system proposed by Schwiesow (1983), which is suitable only for lower altitudes in the boundary layer.

TABLE 1
Representative Parameters for Remote-Sensing Systems

System	Max. Range (km)	Range Resolution (m)	Beam Width[1] (mrad)	Wavelength	Cost[2]	Development
Sodar	1.5	5–50	150	15–30 cm	Low	Commercial
Long-wavelength radar	300	100–1,000	35	>30 cm	Med.	Semicommercial
Short-wavelength radar	50	1–300	10–30	0.3–30 cm	Med.-high	Semicommercial
CW[3] lidar	1	10–200	0.1	10 μm	Med.	Research
Pulsed lidar	20	3–300	0.1–3	0.3–10 μm	Med.-high	Research

[1]$1° = 17.45$ mrad.
[2]Low $\cong$ \$50K; Med. $\cong$ \$100K; High $\gtrsim$ \$500K.
[3]Continuous wave.

The goal of this section is to give some guidance for choosing among remote-sensing techniques for measurement problems as discussed in Section 2. Quantitative data on performance are given in the technique articles (Neff and Coulter; Chadwick and Gossard; Kropfli; Schwiesow; all this volume).

In any evaluation of this kind, there is an unavoidable element of subjectivity. I have tried to reduce the subjectivity by having the tables reviewed by experts in each of the techniques, but the reader should evaluate cautiously and critically the information on different techniques and seek additional opinions wherever possible. There is still art in the application of remote sensing to the boundary layer, and there is no easy answer to the question, "Which sensor?" Recognizing these uncertainties, Tables 2a and 2b give some guidance in selection. The rest of this section discusses some details of remote-sensing approaches, classified by meteorological variable, to help evaluate remote-sensor techniques for probing the atmospheric boundary layer.

3.1 Concentration and Visualization

To estimate the concentration of natural tracers, the choice of a remote sensor that gives the strongest return for a given transmitted power is fairly easy: for molecules such as water vapor, ozone, and other constituents, and particles $<1\ \mu m$ in diameter, short-wavelength (visible) lidar; for larger particles (1 to 20 μm in diameter), infrared lidar or perhaps very-short-wavelength (<3-cm) radar; for cloud droplets, short-wavelength radar; for raindrops, radar; for temperature and humidity fluctuations, radar; for temperature and velocity fluctuations, sodar; and for insects and birds, radar with a wavelength on the order of the target dimension. It is also possible to generate artificial tracers to study dispersion. Smoke and oil fog are useful for lidar experiments, and chaff (conducting fibers cut to one-half the transmitted wavelength) for radar (Kropfli, this volume).

Table 3 contains remote-sensing applications grouped by technique rather than by target as discussed in the preceding paragraph. Concentration measurements can be quantitative, if appropriate attenuation corrections are made and scattering cross sections are known. Flow visualization results, such as time-height cross sections of signal return, are qualitative in terms of backscatter, but quantities such as cell dimension, wave period, and inversion lifting rate can be inferred.

3.2 Velocity

Remotely sensed winds are usually obtained from the Doppler shift of radiation scattered from tracers that are presumed to move with the flow. Radar and lidar, which can produce 3-D tracer concentration maps, can also be used to estimate wind from the spatial displacement of concentration inhomogeneities with time. In either case, the information on tracers in Section 3.1 is relevant to wind measurement applications. However, velocity estimates from precipitation tracers must be corrected for fall velocity, and insects are not reliable flow markers in some cases because they move independently of the wind. The discussion in Section 2.2 on the determination of vector variables emphasizes that wind measurements by any technique require multiple sensing systems with large base-line separations or an assumption of horizontal homogeneity in the flow field. Application of

TABLE 2a
Remote-Sensing Techniques for Meteorological Variables at Short Range
($\lesssim 1$ km)[1]

System	Constituents	Visualization	Velocity	Temperature	Humidity
Sodar[3]		×	×	×[2]	
Short-wavelength radar[3]	×	×	×	×[2]	
CW lidar			×	×	×
Pulsed lidar	×	×		×	×

[1]Long-wavelength radar inapplicable.
[2]Combined sodar and radar required (see Section 3.3) for RASS.
[3]Also useful for refractive-index gradients.

TABLE 2b
Remote-Sensing Techniques for Meteorological Variables at Long Range
($\gtrsim 1$ km)[1]

System	Constituents	Visualization	Velocity	Temperature	Humidity
Long-wavelength radar[2]	×	×	×		
Short-wavelength radar[2]	×	×	×		
Pulsed lidar	×		×	×	×

[1]Sodar and CW lidar inapplicable.
[2]Also useful for refractive-index gradients.

remote velocity sensors to inhomogeneous flow fields is difficult, although the problems seem to be solved for multiple Doppler radars applied to severe storms.

Table 4 lists some advantages and disadvantages of the remote-sensing techniques for wind. In general, long-range measurements require radar, and short-range applications use lidar or sodar depending on whether spatial resolution or low cost is a primary criterion for selection. For some requirements, such as a low-cost sensor for aircraft use, there may not be at present a suitable remote-sensing technique.

Remote-sensing techniques are still being developed for velocity measurement. One approach is to use separated sensor beams and measure the time it takes for scattering inhomogeneities to move from one beam to another. This measures velocity components perpendicular to the sensing beams rather than along the beam, as is done with Doppler methods. In a rather striking demonstration of how active remote-sensing approaches may be used with different types of probing radiation, this method is being applied on different scales with sodar (Mastrantonio and Fiocco, 1982), with radar (called the spaced-antenna drift [SAD] technique [Röttger, 1980]), and with lidar (the time-of-flight technique [Lading et al., 1978]). When combined with a Doppler analysis for the along-the-beam component, a time-of-flight or inhomogeneity drift remote sensor using a nonscanning sensor system operating from a single location can give the three velocity components at a single point in an inhomogeneous field.

TABLE 3
Techniques for Concentration Measurements

	Advantages	Disadvantages
Sodar	Applicable for time-height profiles of layered structures in the boundary layer	Weight and slow pulse rate limit azimuth and elevation scanning speed, thus
	Traditionally useful for locating height and time of convective breakup of inversions	Not useful for 3-D plots; practically limited to time-height cross sections
	Useful for depicting waves (in time), such as Kelvin-Helmholtz, and for frontal passages	Only usable targets are temperature fluctuations (monostatic systems) or temperature and velocity fluctuations (bistatic systems)
	Comparatively inexpensive	
	Operates unattended for long periods	
Radar	Long-range, 3-D concentration maps of chosen targets available	Systems comparatively large and expensive
	Suitable targets include (depending on wavelength) cloud droplets, precipitation, insects, large particulates in the aerosol, temperature-humidity inhomogeneities (clear air), and chaff	Antenna side lobes limit usefulness close to the ground (low elevation angles)
	Applicable to 2-D and 3-D flow visualization in convective cells, waves, etc.	
	Dual-polarization analysis available	
Lidar	High-resolution, 3-D concentration maps of chosen targets available	Possible danger to eyes
	Suitable targets include molecules (constituent selective) and aerosol particulates (smoke, dust, haze)	High system cost
	Information available on target size and shape by multiwavelength and dual-polarization analysis	Not all-weather, i.e., beam strongly attenuated by clouds and fog
	Applicable to flow visualization for drainage wind, droplet deposition, smoke plume dispersion, etc.	

TABLE 4
Techniques for Velocity Measurement

	Advantages	Disadvantages
Sodar	Bistatic signal strength depends on turbulent microstructure	Flow tracers not uniformly distributed; i.e., sometimes only senses wind in special layers
	Comparatively inexpensive	Sensitive to noise from precipitation, high wind, and vehicles
Radar	Long range with appropriate tracers	Systems comparatively large and expensive
	3-D vector fields available with multiple sensors	Antenna side lobes limit usefulness close to the ground
		Clear-air targets nonconservative (e.g., temperature fluctuations) and require high transmitter power
Lidar	Very narrow beam widths	Possible danger to eyes
	Uses conservative tracers	Beam attenuated by cloud and fog

3.3 Temperature and Humidity

Lidar temperature profiling can use data from at least three different kinds of interaction between incident light and molecules. They are the rotational Raman scattering spectrum, Cabannes (Rayleigh) scattering linewidth, and temperature-dependent molecular absorption. The only other useful active technique for remote temperature measurement involves using radar to track the speed of an acoustic pulse launched vertically upward in the radar acoustic sounding system (RASS) method (Frankel et al., 1977). Passive microwave radiometry for temperature and humidity is not covered in the detailed remote-sensing articles, but is mentioned briefly in Section 3.4.

Humidity is a special case of concentration measurement. Two different kinds of lidar interactions are the only demonstrated ways to measure humidity with an active, single-ended remote sensor, although it is theoretically possible to determine humidity profiles from multiple-frequency sodar backscatter intensity. The optical methods are *differential molecular absorption lidar* (DIAL) and vibrational Raman scattering. (Ozone can also be measured with DIAL.)

Table 5 lists active (backscatter) ways of remotely sensing temperature and humidity. Of those, only the lidar techniques are covered in this volume.

3.4 Other Parameters and Techniques

Boundary-layer variables in addition to the meteorological state variables of concentration, velocity, temperature, and humidity can be sensed remotely using active techniques. Refractive index fluctuations provide signals for sodar (Neff and Coulter, this volume) and radar (Chadwick and Gossard, this volume) investigations of atmospheric structure parameters. These structure parameters for index of refraction fluctuations can depend on temperature, humidity, and velocity fluctuations on space scales smaller than the sensor resolution volume. The fluctuating variable and scale size that cause the strongest return depend on the type of probing radiation, its wavelength, and the sensing geometry (monostatic or bistatic). Other variables that can be conveniently measured by remote sensors include turbulence and wind shear, both directly and deduced from vertical time sections. Tables 2a and 2b are in no sense complete or exhaustive lists of the boundary-layer variables that can be remotely sensed. Rather, they list only a few of the most generally applied variables. Tables 6–8 include some additional commonly measured variables such as depth of the planetary boundary layer, turbulent kinetic energy, heat flux, inversion height, stress, eddy dissipation rate, and the related eddy diffusion coefficient. Two-ended paths instrumented with multiple sensors allow inference of temperature and water vapor fluxes (Wyngaard and Clifford, 1978; Coulter and Wesely, 1980), although this volume does not go into great detail on these special applications. Visibility may be measured by lidar (Werner, 1981), by telephotometer (a passive remote sensor), and by short-path in situ instruments (Viezee and Lewis, this volume).

TABLE 5
Techniques for Temperature and Humidity Measurement

	Advantages	Disadvantages
Sodar/radar (RASS)	Good altitude resolution	Temperature only
		Wind causes the acoustic wave front to drift, requiring radar receiver to be moved to compensate
Lidar	Range to > 1 km	Possible danger to eyes
	Simple geometry	Not all-weather
	Useful for temperature and humidity	
	Choice of physical interaction	

TABLE 6
Sodar Performance

Parameter	Max. Range (m)	Min. Range (m)	Range Resolution (m)	Accuracy	Temporal Resolution (s)
Concentration	1,500	10–50	2–10	Qualitative	2–3
u, v	1,500	10	35–50	±10%	3[3]
w	1,500	10	35–50	±20 cm/s	3
PBL[1] depth	600	10–20	2–10	±10%	3
Turbulent KE[2]	600	10	< 5	±3 dB	3
Heat flux	500	50	35–50	±50 W/m²	20

[1]Planetary boundary layer.
[2]KE means kinetic energy. Also senses C_v^2, structure function for velocity. Similar performance for C_T^2, structure function for temperature.
[3]Approximately the mean wind/scan diameter for triple monostatic sodar.

Although we emphasize sodar, radar, and lidar in the remote-sensing articles, passive microwave radiometry is also used to obtain profiles of temperature and humidity. Westwater et al. (1983) review temperature profiling, and Hogg et al. (1983b) discuss humidity measurements. Radiometry generally has rather low resolution in altitude, so it misses sharp vertical gradients in the boundary layer. It is more applicable to large-scale, low-resolution tropospheric profiles than it is to boundary-layer profiles, but with the aid of auxiliary information from radar or sodar on the location of strong scattering layers, effective resolution of the technique is improved (Westwater et al., 1983; Hogg et al., 1983a). Other passive techniques are useful. Mach and Fraser (1979) give an example of an optical image technique for obtaining temperature profiles in the lower part of the boundary layer, and Brunner (1982) analyzes another optical refraction technique for temperature profiling. Infrared radiometry can be used both for measuring surface temperature (assuming the emissivity is known) and for measuring air temperature adjacent to the radiometer, depending respectively on whether the radiometer operates at a wavelength where the atmosphere is nearly transparent or the atmosphere is strongly absorbing. Satellite-based infrared radiometry, where the pressure broadening of molecular line widths gives altitude information, does not have sufficient vertical resolution to be useful for profiling temperatures in the boundary layer.

The reader who is interested in more detail on current work in remote-sensing techniques than can be found in the following articles should consult the preprint volumes of various meetings sponsored by the American Meteorological Society. In particular, see the preprints of the 21st Conference on Radar Meteorology, Edmonton, Canada, 19–23 September 1983, and 11th International Laser Radar Conference, Madison, Wisconsin, 21–25 June 1982 (NASA Conference Publication 2228).

4 HOW WILL IT PERFORM?

In this section, we combine sensing systems and measurements to give quantitative estimates of system performance in tabular form. The cautions about subjectivity and uncertainty in previous tables apply to this section as well. There

TABLE 7a
Long-Wavelength Radar Performance

Parameter	Max. Altitude (km)	Min. Altitude (km)	Altitude Resolution (km)	Accuracy	Temporal Resolution (s)
$\mathcal{U}, \mathcal{V}$	8–100	1–3	0.1–2.5	±1–2 m/s	0.1–10
$\mathcal{W}$	8–100	1–3	1.5	±0.1 m/s	10–300
Inversion height	20	3	1.5	±750 m	300

TABLE 7b
Short-Wavelength Radar Performance

Parameter	Max. Range (km)	Min. Range (km)	Range Resolution (m)	Accuracy	Temporal Resolution (s)
Concentration	12–50	0.02–10	1.5–150	±20%	1[1]
$\mathcal{U}, \mathcal{V}$	12–50	0.03–10	25–300	±0.2–2 m/s	1–10[1]
$\mathcal{W}$	10	0.03–1	25–300	±0.5 m/s	1–10[1]
Stress	2	0.1	25	±0.1 m²/s²	1–10[1]
EDR[2]	2	0.1	25	±5 cm²/s³	1–10[1]
PBL depth	2	0.05	1.5	±10%	1–10

[1]Per resolution cell.
[2]EDR means eddy dissipation rate.

TABLE 8
Lidar Performance

Parameter	Max. Range (km)	Min. Range (m)	Range Resolution (m)	Accuracy	Temporal Resolution (s)
Concentration	2–20	50	1.5–150	±20%	1[1]
$\mathcal{U}, \mathcal{V}, \mathcal{W}$	1–10	20–1,000	5–300	±0.2–2 m/s	< 1[1]–300
PBL depth	3	50	5	±10%	5
Humidity	1.5–3	50	30–100	±10–20%	10^2–10^3
Temperature	2–5	30–500	5–150	±1°C	10^2
Eddy diffusion coefficient	2	500	300	±6 m²/s	300

[1]Per resolution cell.

are many different systems with different sizes and powers under the general heading of radar, for example. Within the ranges of typical system parameters, it may not be possible to achieve simultaneously a chosen range and accuracy. Typical performance values can be extended in most cases by spending enough money on development and better hardware. For another approach to remote-sensor performance evaluation, see Little (1972).

Tables 6–8 are largely self-explanatory. They show the rather broad ranges of performance within and between technique classes. In most cases, the values in the tables are extracted or estimated from particular systems described in further detail in the sodar, radar, and lidar articles.

Acknowledgments. Leif Kristensen and Donald Lenschow encouraged the development of this chapter by their insightful questions about the applicability of remote-sensing techniques, and they have helped guide the answers toward a user's point of view. I appreciate the efforts of my remote-sensing colleagues in this course, William Neff, Richard Coulter, Russell Chadwick, and Robert Kropfli, who made suggestions toward correctness and balance in the presentation of their particular technical specialties.

Lidar Measurement of Boundary-Layer Variables

R.L. Schwiesow, NCAR

1 WHAT LIDAR CAN DO FOR THE BOUNDARY-LAYER METEOROLOGIST

1.1 Remote Sensing

Lidar is an acronym for *light detecting and ranging*; it is the optical counterpart to the familiar radar technique for remotely sensing information about a distant target. In our case, the target is a region in the boundary layer, and we seek information about the meteorological state of the region. A lidar system transmits a pulse or pulses of light into the atmosphere and analyzes the backscattered signal for intensity as a function of time. Because the pulses travel at the speed of light, it is possible to convert time to range and consider the lidar signal to be backscattered intensity as a function of range. In addition to being range-dependent (i.e., time-dependent), the returned signal may have a spectral distribution that carries information about the atmosphere. As an alternative to a pulsed system, the backscattered signal from a continuous-wave (CW) beam as a function of the distance at which the lidar system is focused gives a typical lidar signal of backscattered intensity as a function of range. It is possible to transmit at multiple wavelengths and measure atmospheric parameters by differences in the range-dependence of the signals at different wavelengths.

Lidar measurements exhibit all the desirable characteristics of remote-sensing techniques. The measured values represent spatial averages over volumes that can be varied by proper design of the lidar system. Such spatially averaged values are usually more representative of the overall state of the turbulent atmosphere than are single-point measurements, and in many cases a few remotely sensed values can replace a single-point time average. Some aspects of the spatial average are easily varied so that, for example, the averaging interval in altitude can increase with altitude. This type of dependence can match the data density to the expected gradients in the boundary layer.

Because the sampling region can be moved about in the boundary layer by steering the lidar beam, a single lidar system can simulate many aspects of a large, three-dimensional array of point sensors. In this way it is often possible to obtain more information about meteorological fields than would be possible with a number of conventional instruments.

Remote-sensing instrumentation is more practical than in situ instrumentation in many difficult measuring environments. For example, measurements above the sea surface in a coastal environment can be done sometimes more easily with lidar than with towers. Lidar measurements can reach to altitudes higher than most towers, especially temporary or mobile installations. Lidars can be used where towers represent a hazard, as they do near airports. In contrast to in situ instruments, which can affect the meteorological parameters being measured, lidar is a noninterfering measurement technique. Noninterference with the flow is important for vertical wind measurements, for example. Airborne lidars can reach beyond aircraft flight altitudes.

1.2 Variables Measured

Some of the basic boundary-layer variables measured by lidar in a chosen sample volume are:
- Aerosol backscatter coefficient and depolarization ratio
- Longitudinal (along-the-beam) wind component
- Transverse (across-the-beam) wind component
- Water vapor density
- Temperature
- Concentrations of some other constituents.

From these basic measurements it is possible to infer many other boundary-layer properties such as mixing depth, wind shear, inversion height, aerosol type, and so on. In addition, boundary-layer processes such as diffusion and dispersion, ice-to-water conversion, and material transport can be studied with the help of lidar measurements. The examples of observational problems in Section 3 help clarify the measurement abilities of lidar both for directly accessible variables and for inference of properties.

1.3 Present State of Research

Lidar techniques are probably the least fully developed of the instrumentation techniques discussed in this volume. While research with the other techniques centers on applications, lidar is still the subject of active research to expand its capabilities. This stage is both exciting and frustrating for the beginner. It is exciting because much definitive work remains to be done and because new applications wait to be developed. It is frustrating because there is no consensus on the best method for measuring a particular variable and because each researcher must develop and verify, in person or by proxy, each lidar system for each application.

With the exception of two European laser ceilographs (for example, model LD-WHL by Impulsphysik GmbH, Hamburg, Federal Republic of Germany) for measurement of cloud-base height, there is no commercially available, series-built lidar, to my knowledge. Commercial organizations do, however, produce lidar systems to order for particular applications. One reason for individual development of lidar systems is that there is no such thing as an all-purpose lidar. As shown in Section 2, the variety of measurement interactions, types of system operation, and applications is reflected in a wide variety of lidars.

1.4 Overview of the Article

Readers interested in a quick answer to the question of the best type of lidar for their needs will be disappointed with this article. The aim here is to provide readers with the tools to generate their *own* answers to questions such as "Is lidar likely to work in my application?" and "What type of lidar is best suited to my observational problem?"

In order to understand the capabilities of lidar and the results in sample applications, we need to discuss first some topics relevant to the physical principles underlying lidar sensing:

- Scattering and absorption interactions
- Lidar equation, signal-to-noise ratio, and system components
- Spatial and spectral resolution, refraction, and coherence.

The many different sorts of possible interactions, optical configurations, and resolutions emphasize that the term meteorological lidar covers a range of instrumentation rather than a specific instrument.

After an introduction to the physics of lidar, we are prepared to consider some sample applications:

- Aerosol loading and dispersion
- Wind from ground-based and airborne systems
- Humidity and other constituents
- Temperature.

We conclude with mention of lidar limitations and possible directions of future development.

This article on lidar measurements of boundary-layer variables is in no way a complete review of the present state of lidar. For one thing, lidar has applications outside the boundary layer such as oceanographic measurements (e.g., depth, constituents, and temperature) and stratospheric measurements (e.g., wind, sodium [Na] molecular density, and temperature), which we do not discuss. For another, we emphasize recent work without an attempt at historical completeness. Readers interested in the background of a particular application can refer to the references for guidance back into the literature. Although this article discusses many different types of lidar, it covers only a limited part of a broad field. A book by Measures (1984) presents an extensive review of lidar fundamentals and applications, which covers the overall field in more breadth and depth than is possible in this article.

2 WHY LIDAR WORKS

2.1 A Look at the Details

Lidar works because molecules and aerosols scatter some of the incident radiation back to the lidar system. The scattered photons carry information about the scatterer, and they can undergo attenuation along the beam path, thus giving information on the intervening atmosphere. It is possible to relate quantitatively some aspects of lidar system design to measurement performance. This allows a lidar designer or potential user to choose the type and scale of lidar components best suited to an application.

In order to streamline the discussion, we list in a table with this volume notation for quantities that appear in our lidar analysis, together with common units for the quantities.

2.2 Measurement Interactions

2.2.1 Absorption

For most lidar purposes it is convenient to divide the absorption experienced by a lidar beam into parts caused by aerosol absorption (absorption coefficient a_a) and by molecular absorption a_m as

$$a(\lambda) = a_a(\lambda) + a_m(\lambda) \tag{1}$$

Usually a_a changes slowly with changing wavelength λ, but a_m has many sharp spectral features both in the ultraviolet (UV) and in the near- and middle infrared (IR). The principal importance of a_a is as part of the overall attenuation coefficient α, although its spectral dependence contributes to a number of atmospheric effects such as pollution-based brown clouds and red sunsets. McCartney (1976) discusses absorption in more detail.

In general, attenuation is a result of both absorption and scattering out of the beam. Broad-band attenuation is dominated by variable aerosol effects. For example, McClatchey et al. (1971) give the attenuation coefficients in Table 1 for atmospheric models characterized by visual range. Values in Table 1 are only typical; particular cases of aerosol attenuation vary widely for different meteorological conditions. Broad-band attenuation changes by approximately 50 % from the blue (higher attenuation) to the red (lower attenuation) ends of the visible (VIS) spectrum. The effect of broad-band attenuation is especially important for spectroscopic lidar techniques, such as vibrational Raman scattering, that cover a significant spectral range. In Raman analysis, differences in atmospheric attenuation at two different wavelengths affect the relative signal intensity at the reference and measurement wavelengths. Broad-band attenuation also reduces the signal-to-noise ratio of a lidar relative to that expected if atmospheric effects along the propagation path were neglected, as we observe later in this section. The problem is especially important in the UV.

The spectral features of a_m show up mostly as narrow-line absorptions and are the basis of lidar differential absorption measurements of a number of molecules in the atmosphere. Table 2 lists some molecules and the spectral regions where they have useful absorptions for lidar purposes. Fredriksson et al. (1979) and Korb and Weng (1982) give further examples of the effect of molecular absorption on lidar propagation.

Natural narrow-line absorption exists in the VIS and IR regions. Figure 1 includes part of the spectral data from Curcio et al. (1964) and emphasizes the existence of spectral features. The line absorption features can have an effect on the signal-to-noise ratio when laser or scattered wavelengths coincide with absorption lines. The 0.540- to 0.852-μm region is discussed in Curcio et al. (1964) and the 0.440- to 0.550-μm interval in Curcio et al. (1955). At wavelengths shorter than 0.440 μm the visible absorption spectrum is generally structureless down to 0.35 μm. McClatchey and O'Agati (1978) explain a technique for calculating high-resolution absorption spectra at selected (laser-line) locations.

2.2.2 Scattering

Scattering can be considered usefully in terms of the differential scattering cross section per scatterer $\sigma(\theta,\lambda)$, which is a function of scattering angle and wavelength, and the backscatter coefficient $\beta(\lambda)$, which includes the density of scatterers. The backscatter coefficient is

$$\beta(\lambda) = \sigma(\pi,\lambda)n \qquad (2)$$

where $\sigma(\pi,\lambda)$ is the differential scattering cross section at backscatter, and n is the density of scatterers. The scattering cross section also depends on other variables, which are not explicitly included in Eq. (2), such as the scatterer's size, shape, and composition. Given a distribution of scatterers, Eq. (2) shows the variables over which an experimenter has some control. To obtain the backscatter coefficient of a distribution of scatterers, one must weight the individual backscatter coefficients by the particle number density distribution, i.e., integrate over the distribution. The overall

scattering contribution to attenuation can be found from integrating $\sigma(\theta)$ over all angles in the form

$$I(\lambda) = \int\int n\sigma(\theta,\lambda)\,d\phi\,d\theta \qquad (3)$$

so that

$$\alpha(\lambda) = a(\lambda) + I(\lambda) \qquad (4)$$

A more complete analysis, including polarization effects and particle-size distributions, is both beyond the scope of our discussion and not immediately relevant to the lidar backscatter problem.

Like absorption, scattering can be divided into an aerosol component σ_a or β_a and a molecular component σ_m or β_m so that

$$\beta(\lambda) = \beta_a(\lambda) + \beta_m(\lambda) \qquad (5)$$

The aerosol component β_a changes more slowly with wavelength than β_m does; the difference in their spectral behavior is the basis of a multiwavelength lidar technique for differentiating between molecular and aerosol scatter (e.g., Uthe et al., 1982). In lidar work the blue-to-red backscatter ratio, for example, characterizes the spectral dependence of scatterers,

TABLE 1
Attenuation at 0.5145-μm Wavelength

Visual Range (km)	α_a (km^{-1})	α_m (km^{-1})
5	0.858	0.015
23	0.176	0.015

TABLE 2
Some Absorbing Gases Measured by Lidar

Wavelength Region:	UV	Near-IR	Middle IR
Molecules:	SO$_2$	H$_2$O	O$_3$
	NO$_2$	O$_2$ (temp)	CO$_2$ (temp)
	O$_3$		Hydrocarbons

Fig. 1 *Absorption spectrum in the boundary layer by Curcio et al. (1964).*

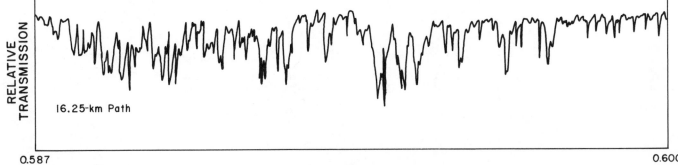

RELATIVE TRANSMISSION

16.25-km Path

0.587

0.600

WAVELENGTH (μm)

being smaller for aerosol particles than for molecular scattering.

In general, the variation of molecular backscatter with wavelength goes as

$$\beta_m \propto \lambda^{-4} \tag{6}$$

but β_a decreases less rapidly with increasing wavelength. This means that the aerosol-to-molecular backscatter coefficient ratio increases with increasing wavelength. Therefore, an aerosol-sensing lidar tends to operate better at longer wavelengths, thus avoiding the masking effects of molecular scatter, and a molecular scattering lidar operates better at shorter wavelengths where the backscatter coefficient is larger.

In addition to spectral dependence, σ_a exhibits an angular pattern that depends on the aerosol particle size parameter, which is given by

$$\zeta = 2\pi\gamma/\lambda \tag{7}$$

where γ is the aerosol particle radius. For large values of ζ, the scattering is peaked in the forward direction as in Fig. 2 on the right.[1] For $\zeta < 1$, the scattering diagram is more dipolar, and in the limit $\zeta \ll 1$ the scattering diagram is completely dipolar with equal forward and backward lobes as on the left in Fig. 2. Molecular scattering shows a dipolar (Fig. 2, left) scattering behavior. When strongly forward-peaked scattering is seen from the atmosphere in the VIS, then the scatterers are a few micrometers in radius or larger.

Most contributions to total scattering are from particles of $\zeta \cong 1$ or larger for typical particle-size distributions, so that the increase of overall σ_a with ζ occurs often enough to overcome the unfavorable back-to-front ratio for backscatter at large values of ζ. The net result is that lidar aerosol backscatter is dominated by particles near $\zeta = 1$ because for $\zeta \ll 1$ the scattering is weak and for $\zeta \gg 1$ the particles are too few. A proper choice of λ allows the lidar experimenter to tailor an aerosol backscatter system to the particle size of interest.

Depolarization of the backscattered signal is important for aerosol identification but not for molecular analysis. If a particle is small with respect to the lidar wavelength (i.e., $\zeta \ll 1$), the particle or molecule acts as a spherical scatterer, and the scattered radiation preserves the state of polarization of the incident radiation. Thus light from a linearly polarized lidar transmitter will be backscattered by a small particle as linearly polarized light. A large, nonspherical scatterer will depolarize the incident light so that the backscattered light contains both a polarized and an unpolarized component. The ratio of unpolarized signal (measured with an analyzer crossed to the incident polarization direction) to the signal polarized parallel to the incident direction for light backscattered from an aerosol target gives information on the shape of the target particles, as we shall see in the applications section. This ratio is called the linear depolarization ratio (Ryan et al., 1979).

Depolarization of the lidar beam as a result of propagation in a normal atmosphere is negligible. Höhn (1969) has observed a maximum depolarization ratio in the forward direction of 5×10^{-5} for a 0.63-μm laser beam over a 4.5-km atmospheric path. This small depolarization is attributable to atmospheric scattering rather than refracting mechanisms.

At least three types of molecular scattering are useful for lidar purposes. The most important is Rayleigh scattering, which occurs with a near-zero frequency shift. The dependence of the Rayleigh backscatter coefficient on scattering wavelength is given in Eq. (6). Its magnitude can be estimated from Table 3.[2] At VIS and shorter wavelengths there are regions of the upper troposphere and middle stratosphere where the lidar return is commonly assumed to be purely Rayleigh scattering. Rayleigh scattering has a number of components, including Brillouin scattering and the central Cabannes peak, and there is some uncertainty in nomenclature (Young, 1981).

---

[1]Figure 2 is for the case where the polarizations of both incident and scattered beams are parallel to the plane defined by the beams. The patterns for perpendicular polarization are different, but in the case of backscatter both patterns must be identical, of course. The conclusions on front-to-back scattering ratio as a function of ζ are independent of polarization at $\theta = \pi$.

[2]The backscattering from the aerosol can vary over a much larger range than molecular backscattering can because the particle loading in the aerosol varies widely for different meteorological conditions. Aerosol values in Table 3 are only representative. Values an order of magnitude larger or smaller can be observed (e.g., Schwiesow et al., 1981a).

Fig. 2 Polar scattering diagram for aerosol scattering from very small particles (left) and large particles (right).

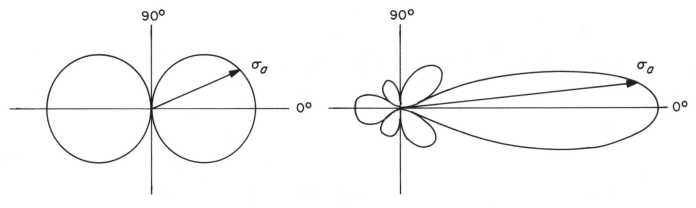

TABLE 3
Representative Backscatter Values

Backscatter $(m^{-1} \, sr^{-1})$	Target	Wavelength (μm)	Altitude (km)
1.3×10^{-6}	Molecules	0.5	5
2.3×10^{-6}	Molecules	0.5	0
0.9×10^{-6}	Aerosol	0.5	5
4.2×10^{-7}	Molecules	0.69	0
8×10^{-9}	Aerosol	10.6	0
1×10^{-10}	Aerosol	10.6	5

TABLE 4
Some Applications of Different Types of Scattering

Scattering Mechanism	Application
Aerosol	Plume tracking, dispersion
	Inversion height measurement
	Cloud identification, water/ice ratio
	Wind by tracking concentration inhomogeneities
	Wind by Doppler shift of scattered radiation
	Target for differential absorption measurements
	Visibility, slant visual range
Rayleigh	Density (and also temperature from gas law)
	Temperature (by line width)
	Reference level for aerosol backscatter
	Atmospheric transmission
Raman	Humidity
	Temperature (energy level population)
	Pollutant concentration
	Density (N_2)
Fluorescence	Dispersion (fluorescent tracer concentration)

Raman scattering follows the wavelength dependence of Eq. (6), but the scattering cross section is at most 2% of the cross section for Rayleigh scattering. The usefulness of Raman scattering is that the scattered light is frequency-shifted by an amount that is characteristic of the scattering molecule. The Raman spectrum can thus be used to identify the scatterer, measure its number density, and measure its energy state by means of the populations of different energy levels.

There are two types of Raman scattering, rotational and vibrational. In rotational Raman scattering, the energy difference between incident and scattered photons corresponds to a transition between rotational energy levels in a molecule. This difference is typically a few hundred wave numbers (in units of per centimeter), i.e., a few nanometers in wavelength at a 0.5-μm exciting frequency. Both Stokes (scattered photon energy lower than incident) and anti-Stokes (scattered photon energy higher than incident) rotational Raman scattering are possible. In vibrational Raman scattering, the energy difference corresponds to a vibration-rotation transition with an energy change of a few thousand wave numbers, i.e., 10 to 100 nm at a 0.5-μm exciting frequency. Only Stokes vibrational Raman scattering is useful in the boundary layer. Naturally, with the wider spectral separation in vibrational Raman than rotational, there is less chance of spectral interference.

Resonance scattering occurs without substantial energy or wavelength change, but the scattering cross section is orders of magnitude larger than the Rayleigh cross section for incident wavelengths matching an absorbing (and reemitting) transition in the molecule or atom. Fluorescent scattering and resonance scattering are similar in that a match between the transmitting laser wavelength and an absorbing transition is required, but the fluorescent scattering is at a longer wavelength than the incident and is characteristic of the scattering material. Fluorescence from individual molecules or atoms is usually quenched in the boundary layer, although it can be useful in the stratosphere and above. Fluorescent scattering from dye particles in the aerosol is possible in the boundary layer and can be used to differentiate lidar return from tracer particles and lidar return from ambient aerosol particles and molecules.

Some types of scattering and their applications are summarized in Table 4.

In multiple scattering, received photons have been scattered more than once. Multiple scattering is important only in clouds, heavy dust, and other targets with a high concentration of scatterers. The theory of multiple scattering is complex; readers interested in more details should consult Carswell (1983). For lidar design purposes it is sufficient to note that multiple scattering delays parts of the return signal, so that range resolution of a lidar in a multiple-scattering situation is generally degraded. Range to the edge of a dense cloud, as discussed under applications, can still be measured with the basic lidar accuracy (see Section 2.4.1) if the leading edge of the multiply scattered return pulse is used for range determination.

The effects of multiple scattering can be reduced by making the receiver field of view (also discussed in 2.4.1) as small as possible. In fact, studying lidar signal strength as a function of receiver field of view is one way of determining the contribution of multiple scattering to a lidar return (e.g., Weinman, 1976). Another way to estimate the amount of multiple scattering is to reduce the backscatter coefficient by increasing the incident wavelength. In this case the probability of multiple scattering is reduced by an exponential power (greater than 1) of the reduction in backscatter coefficient. The exponent is related to the amount of multiple scattering. In general, the aerosol backscatter coefficient, and therefore the multiple scattering, are reduced by transmitting longer lidar wavelengths. Cloud droplet size and water droplet concentration studies make use of multiple-scattering returns.

2.3 System Operation

2.3.1 Lidar Equation and Signal-to-Noise Ratio

In simplified form, the equation for the detected signal energy from a transmitted light pulse of energy E as a function of range r is given by

$$S(r) = E\sigma n(r)lA_e T\eta_e \exp(-2\alpha r)/r^2 \tag{8}$$

where A_e is the effective telescope area, T the filter transmission, and η_e the optical efficiency. Some of the geometric variables can be seen in the sketch in Fig. 3, which shows a common telescope (transceiver) for the transmitter and receiver. If the atmospheric attenuation coefficient is not constant with range, then the exponential factor must be replaced by

$$atten = \exp[-2\int_0^r \alpha(x)\,dx] \tag{9}$$

In the case of a CW lidar in the near field (where the lidar telescope can still focus the beam), l is dependent on A_e and r, as we shall see in the spatial resolution subsection (2.4.1), so that the lidar return signal can be independent of range for this special condition. Another point in Eq. (8) worth noting is that $S(r)$ is proportional to r^{-2} even though the signal per scatterer is proportional to r^{-4}. The transmitted flux density goes as r^{-2}, and the received signal per scatterer for a constant flux density goes as r^{-2} because the solid angle subtended by the receiver as viewed from the scatterer goes as r^{-2}, but the number of scatterers in the sample volume goes as r^2. Recall from Eq. (2) that $\sigma \cdot n(r) = \beta(r)$ is a characteristic of the scattering medium.

The parameters in Eq. (8) can be divided into those characteristic of the atmosphere and system parameters. Although the experimenter cannot control the atmospheric parameters β and α directly, a proper choice of λ selects the relative importance of various components of β or α. The absorption coefficient is generally smallest in the 0.3- to 1.0-μm and 8- to 13-μm spectral windows of the atmosphere. There are many strong, spectrally sharp absorption features that can be selected or bypassed by a proper choice of λ.

System parameters in Eq. (8) are often limited by component cost. From Eq. (8) it is obvious that l should be as large as possible, consistent with the required spatial resolution for the application, in order to maximize $S(r)$. Parameters E, A_e, T, and η_e should also be as large as economically possible. The linear dependence of $S(r)$ on these parameters in Eq. (8) makes it easy to do cost-benefit studies on the relative advantages of increasing laser power or telescope area, for example.

The primary sources of noise in a proper lidar system are shot noise on the photon arrival rate, detector noise (dark current), and sky background noise. The lower limit on noise in the received signal is given by Poisson statistics on the number of detected photons. If the number of detected signal photons is $S/(hf)$, where $f\lambda = c$, then this limit is

$$noise = [S(r)/(hf)]^{1/2} \tag{10}$$

The best possible signal-to-noise ratio is then just

$$S/N_s = [S(r)/(hf)]^{1/2} \tag{11}$$

For molecular scatter, for example, Eq. (11) with Eq. (4) predicts that

$$S/N_s(molecular) \propto f^{3/2} \tag{12}$$

so that a molecular-scattering lidar should operate at as short a wavelength as practical. The detector noise can be made smaller than signal noise in most cases by using a photomultiplier with photon-counting electronics or an infrared heterodyne detection arrangement, depending on the operating wavelength.

Sky background energy in the receiver goes as

$$N_b = 2RlA_e\eta_e w\Omega T/c \tag{13}$$

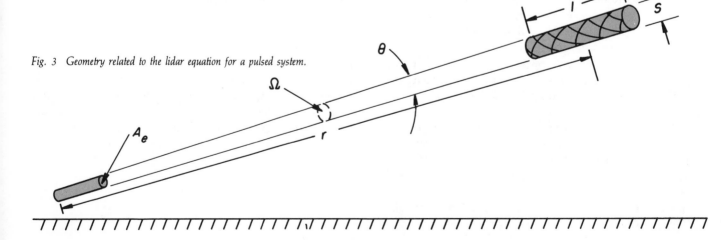

Fig. 3   Geometry related to the lidar equation for a pulsed system.

where R is sky radiance, Ω the solid-angle field of view, w the receiver spectral bandwidth, and c the speed of light. This includes the fact that the time interval the receiver is turned on (i.e., open) for a given resolution length l is

$$\Delta t = 2l/c \qquad (14)$$

R varies widely around 2.5×10^{-1} W/(m² sr nm) for a daytime sky at a wavelength of 0.5 μm (Schwiesow and Lading, 1981). The most effective way to reduce N_b is to reduce Ω and w because these parameters do not appear in Eq. (8), which assumes that transmitter and receiver fields of view are properly matched. Sky background is not usually a problem in heterodyne infrared lidars because w is very small (on the order of 10 to 100 kHz, i.e., 4 to 40 fm [1 fm = 10^{-15} m] at a 10.6-μm wavelength) and because heterodyne Ω is very small. Sky background is smaller in the middle UV and shorter wavelengths ($\lambda < 300$ nm) because of the falloff of the sun's energy in the UV and because of absorption of UV by ozone in the upper atmosphere.

Overall, the signal-to-noise ratio is

$$\text{S/N} = [S(r)/(hf)]/\{[S(r)/(hf)] + [N_b/(hf)] + [N_d/(hf)]\}^{1/2} \quad (15)$$

where N_d is the detector noise energy in time Δt. This expression is simply an application of Poisson statistics to the sum of photons detected. In general, S/N is maximized by making $S(r)$, S/N$_b$, and S/N$_d$ as large as possible. An experimenter interested in maximizing S/N can substitute from Eqs. (8) and (13) into Eq. (15) to determine the dependence of S/N on parameters such as E, A_e, l, and f for different operating conditions.

2.3.2 Components

Basic lidar system components are lasers, telescopes, and detectors. (We discuss data processing in Section 2.5.)

Lasers are now universally used for lidar transmitter sources. Common types are optically pumped, solid-state lasers (e.g., Nd:YAG and ruby), electrically pumped gas discharge lasers (e.g., CO_2, excimer, and Ar⁺), laser- or flash-lamp–pumped dye lasers, and electrically pumped (injection) diode lasers. Table 5 gives general properties of some types of lasers that make them suitable for different lidar applications. With the help of Table 5 it will be easier to understand why certain lasers have been chosen for the applications in the next section. Peak powers for the solid-state lasers are high enough to allow frequency "doubling" (conversion of the output to a half, a third, or even a quarter of the basic laser wavelength), which is why multiple spectral regions are listed in the table. Sometimes the wavelength conversion efficiency is large enough that with Eq. (6) the lidar return energy is larger at the shorter wavelength than at the laser fundamental.

Different types of telescopes are suited to the various types of lidar. Table 6 compares the features of some common types. In the table, limited spectral range means difficulty in covering the 0.3- to 10.6-μm interval that is most used in

TABLE 5
Lasers for Lidar

Laser	Peak Power	Average Power	Pulse Width	Spectral Width	Tunability	Spectral Region
Nd:YAG	High	High	Short	Med.	Slight	IR-VIS-UV
Ruby	High	Med.	Short	Med.	Slight	VIS-UV
CO_2	Med.	High	Med.-CW	Sharp	Discrete	IR
Excimer	Med.	Med.	Short	Narrow	Discrete	UV-VIS
Ar⁺	Low	High	Short-CW	Sharp	Discrete	VIS-UV
Laser-dye	Low	Low	Short	Narrow	Yes	VIS-IR
Lamp-dye	Med.	Med.	Med.	Narrow	Yes	VIS-IR
Diode	Low	Low	Short	Broad	Slight	Near-IR

TABLE 6
Telescopes for Lidar

Type	Open/Closed Tube	f/#	Advantages	Disadvantages
Newtonian reflector	O	Med.	Moderate cost, diffraction-limited	Size, central obscuration
Cassegrain	O	High	Compact, near diffraction-limited	Cost, central obscuration
Catadioptic (Schmidt, etc.)	C	Low	Near diffraction-limited, compact	Cost, central obscuration, limited spectral range
Fresnel lens refractor	C	Low	Light, compact, inexpensive	Low spatial resolution, limited spectral range
Prime-focus reflector	O	Low	Simple, inexpensive	Central obscuration, limited flexibility
Off-axis parabola	O	High	No central obscuration, diffraction-limited	Cost, size
Refractor	C	High	Near diffraction-limited	Limited spectral range, size, cost

lidar. Open-tube telescopes are subject to thermal in-homogeneities in the optical path and dust on the optics, but the optical element at the aperture of a closed-tube system can reflect enough energy to destroy detectors in an arrangement where the same telescope is used to transmit and receive. Central obscurations introduce diffraction effects and, in a transmitter, block some of the most intense part of the output beam. More sophisticated compound optical systems are available, but are not widely used in current lidar work. Newtonian reflectors are widely used. Compared with telescopes used for imaging applications, lidar telescopes require a much smaller field of view, so that off-axis aberrations such as curvature of field and distortion do not present difficulties for lidar work. Often, chromatic corrections are not needed for lidar, and coma is not a problem.

Usually, telescopes are used both on the transmitter beam, to reduce beam divergence, and as part of the receiver, to collect adequate backscattered light. The two telescopes can be arranged side by side (i.e., biaxially) or with the smaller centered in the aperture of the larger (i.e., coaxially). The biaxial arrangement achieves full transmitter-receiver beam overlap only beyond some range limit, depending on geometry, whereas the coaxial arrangement achieves overlap much closer to the lidar system. Therefore, for short-range applications a coaxial geometry is favored. Not all ranges can be simultaneously in focus, however, so the time (range) gate must be set for the optical focus of the telescope(s). One telescope can be used for both the transmit and receive functions (transceiver) if some sort of beam switch is provided. A transceiver is a special case of a coaxial layout. Suitable beam switches include mechanical or electrooptical choppers for pulsed work and polarization coding for CW work. A transceiver is particularly appropriate for lidars with high spatial resolution (see Section 2.4).

The proper detector for a lidar system depends on the spectral region. In the middle IR near 10.6 μm, liquid-nitrogen–cooled HgCdTe photodetectors are widely used. Although costly, such detectors have good quantum efficiency,[3] good heterodyne efficiency, and comparatively low detector noise levels. They can be used to frequencies of 100 MHz if suitable preamplifiers are included. Because of detector noise in the IR, heterodyne operation is needed to achieve the best signal-to-noise ratio in Eq. (11). Solid-state photodiodes are useful in the near-IR because they have good quantum efficiency and are comparatively inexpensive. Sometimes photomultipliers are used in the near-IR because they have lower detector noise than photodiodes, but the quantum efficiency of photomultipliers is low in this spectral region. In the VIS, photomultipliers are widely used because of their low detector noise and reasonable quantum efficiency. It is possible to achieve signal-limited signal-to-noise ratios (Eq. 11) in the VIS when the background energy is low. However, in background-noise–limited applications a solid-

state photodiode is competitive with a photomultiplier because of the greater quantum efficiency and simplicity of the photodiode. Thus a lidar system that will be used in daylight with low spatial resolution may operate better with a photodiode than with a photomultiplier. This is another illustration of the fact that there is no best, all-purpose lidar; systems must be specifically designed for the application, and there are many system variables. Photomultipliers are generally used for UV lidar detectors because the quantum efficiency of solid-state photodiodes is low in the UV. Table 7 summarizes typical performance parameters for some popular lidar detectors.

The two detection methods used in boundary-layer lidar systems are called direct and heterodyne detection. Direct detection, where the backscattered photons are collected in a "light bucket," is by far the most common. The wave nature of the light is not important for the detection process; if one uses the electromagnetic field representation of the return signal, the phase of the signal drops out in the calculation of the output of the photodetector. Because of its independence of the phase, direct detection is referred to as measurement of the *intensity* of the return signal. The intensity is proportional to the photon arrival rate at the photodetector. In contrast is heterodyne (also called reference beam) detection, where the backscattered light is optically mixed with a locally generated reference optical beam. In this case, when the electromagnetic fields of the signal and reference beams are added and used to calculate the output of the photodetector, the phase of the signal is important. Many scatterers contribute depending on their phase. Heterodyne detection is referred to as measurement of the electromagnetic *field* of the return signal. Because of its phase dependence, the heterodyne signal from an ensemble of scatterers can change rapidly as the scatterers undergo relative motion, even though the total number of scatterers (and the backscattered intensity) remain relatively constant. This fluctuation in detected field is called speckle noise. The added complication of heterodyne detection is useful to overcome detector noise and for applications requiring very high spectral resolution. If these advantages are not important for a particular application, then direct detection is usually preferred.

2.4 Propagation Effects

2.4.1 Spatial and Spectral Resolution

Figure 4 shows a schematic view of the r^{-2} behavior of the signal from Eq. (8) applied to a homogeneous atmosphere. The transformation between range and time is simply

$$r = ct/2 \qquad (16)$$

where t is round-trip travel time, so that these variables may be used interchangeably. The question of spatial resolution along the line of sight (range resolution) can be expressed as

---

[3]Quantum efficiency is the probability that an incident photon will generate a charge carrier in the detector. A large quantum efficiency maximizes the available signal S and signal-to-noise ratio in Eq. (15).

TABLE 7
Popular Lidar Detectors

Type	Spectral Range (μm)	Typical Gain	Peak Quantum Efficiency (%)	Detector Noise	Photon Counting
Photomultiplier	0.3–0.8	10^6	20	Very low	Yes
Silicon diode	0.4–1	1	50	Medium	No
Silicon avalanche	0.4–1	10^2	50	Medium	No
Germanium diode	0.5–2	1	50	High	No
HgCdTe diode	8–13	1	50	High	No[1]

[1]Signal-noise-limited detection (equivalent to photon counting) possible with heterodyne operation.

the determination of what regions of the atmosphere contribute to the signal at time t_1.

For range resolution, consider a scatterer at range r and a pulse of length l as in Fig. 5, where

$$l = c\tau \qquad (17)$$

From Eq. (16), the range corresponding to a round-trip travel time t_1 is

$$r_1 = ct_1/2 \qquad (18)$$

where we measure time from the center of the pulse. Continuously distributed scatterers at a range r_a will scatter from the leading edge of the pulse, which departs at a time $-\tau/2$, and will contribute to signal at t_1. This range is

$$r_a = c[t_1 + (\tau/2)]/2 \qquad (19)$$

Contributions to t_1 can also come from range r_b, where particles scatter from the trailing edge of the pulse. Solving for the range resolution Δr between r_a and r_b, we find

$$\Delta r = r_a - r_b = c\tau/2 \qquad (20)$$

The best possible range resolution for a pulsed lidar can be remembered as

150 meters per microsecond.

If the detector time resolution is not shorter than τ, then the range resolution is degraded. In fact, any reasonable value for $l > c\tau$ can be chosen for a range gate by using appropriate time integration at the receiver. The range resolution in Eq. (20) also applies approximately to any pulse shape if τ is the full width of the pulse at half maximum.

A CW lidar achieves range resolution in a manner similar to the mechanism for limiting the depth of field in a camera. The length of the focal volume (i.e., the region of highest incident optical flux density) is approximately

$$\Delta r(CW) \cong 8.5r^2\lambda/D^2 \qquad (21)$$

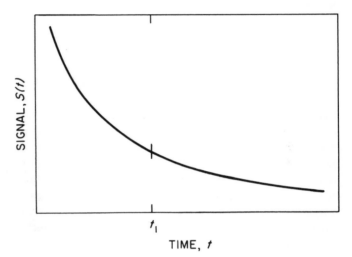

Fig. 4   Lidar signal vs range.

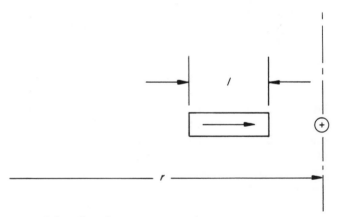

Fig. 5   Lidar pulse and scatterer.

in the near field[4] of a diffraction-limited optical system with a well-illuminated transceiver aperture D. Dickson (1970) gives methods for more precise calculations. This CW range resolution can be remembered as

10 meters at 100 for CO_2,

[4]The near-field limit on range is approximately r(near field) $\lesssim D^2/4\lambda$.

which applies to a 30-cm transceiver operating at $10.6~\mu m$. The squared dependence of Δr(CW) on range in Eq. (21) is a severe restriction on the range resolution of a CW lidar used at long ranges.

We can see now how Eq. (8) is independent of range for a focused CW lidar. From Eq. (21), l for such a system depends on r^2, so that the overall r dependence in Eq. (8) is r^0. Similarly, the telescope area A is proportional to D^2, so the overall A dependence in Eq. (8) is A^0 because l goes as A^{-1}. This independence of range and telescope area is valid only in the near field.

To determine angular resolution, we note that the angular spot size of a diffraction-limited lidar is approximately

$$\theta(min) \cong 2.44\lambda/D \qquad (22)$$

where θ is the full width of the first minimum of the Airy diffraction pattern. To achieve this resolution, both the transmitter and receiver must have telescope apertures of size D, or a transceiver must be used. Many lasers used for lidar are not diffraction-limited (i.e., they do not operate in a single transverse mode of the optical cavity), so that the resolution in Eq. (22) cannot be achieved. In the case of a non–diffraction-limited transmitter, it is necessary to assure that the angular spot size of the receiver, which is determined by the focal length of the receiver telescope and the diameter of the field stop, is adequate to include the image of the entire illuminated volume.

If the angular resolutions of the transmitter and receiver are the same and the telescope focal volumes completely overlap, then the lidar system is matched in spatial mode. We can think of each diffraction-limited focal volume as a spatial mode of the optical system. A spatial-mode-matched system provides a better signal-to-noise ratio than an unmatched system. As an example of a system with a single spatial mode, consider a diffraction-limited lidar at 0.5-μm wavelength with a 30-cm transceiver. The limiting angular resolution of this lidar is approximately 4 μrad, which is an illuminated target diameter of only 4 mm at a range of 1 km. Such high resolution requires high-quality optics. The resolution achieved in practice is usually less than that permitted by diffraction limitations because of refractive index inhomogeneities in the atmosphere (see Section 2.4.2).

One reason for working at the highest angular resolution possible is illustrated by Eq. (13). The sky background energy is directly proportional to Ω, the solid angle field of view, which is given by

$$\Omega \cong 0.8\theta^2 \qquad (23)$$

for circular field stops. A reduction in Ω reduces background noise without reducing signal for a properly matched transmitter and receiver. Another illustration of the importance of spatial resolution concerns the number of scatterers that are simultaneously in the sample volume of a focused CW lidar system. From Eqs. (21) and (22) in the linear rather than

angular version, the sample volume V of a focused CW system is approximately

$$V \sim 45\lambda^3(r/D)^4 \qquad (24)$$

If we compare a VIS-wavelength system having parameters $\lambda \sim 0.5~\mu m$, $r \sim 50$ m, and $D \sim 15$ cm with an IR system with parameters $\lambda \sim 10.6~\mu m$, $r \sim 500$ m, and $D \sim 30$ cm, then the volumes are approximately in the ratio $V(\text{IR})/V(\text{VIS}) \sim 5 \times 10^6$. If the density of scatterers is the same in both cases, then the statistical fluctuations in the return signal will be vastly greater in the VIS case than in the IR case. Such differences have important consequences in data processing and in the entire measurement principle employed in different lidars.

A variation in the number of scatterers in the sample volume causes a statistical fluctuation in the backscattered signal power. If we model the total number of scatterers in the sample volume as a random variable with an expectation value nV to which Poisson statistics apply, then the fractional fluctuation in signal power should be proportional to $(nV)^{-1/2}$. This is analogous to the photon statistics in Eq. (10). Thus, in the example in the previous paragraph, the VIS system will have signal fluctuations approximately 2×10^3 larger than the IR system. One possible measurement technique in the case of large signal fluctuations, say $nV \sim 1$, is to detect the presence or absence of a particle in a sample volume. Appropriate arrangement of sample volumes in this case allows the time of flight of a particle between volumes to be determined. On the other hand, if nV is very large, then it is appropriate to use lidar to measure mean or ensemble-averaged properties of the scattering target such as n or mean velocity.

The spectral detail in the received signal (related to wavelength selectivity or optical frequency resolution) can be no sharper than the spectral width of the transmitter laser. For some applications, such as differential absorption measurements and Doppler lidar anemometry, the laser line width is a critical parameter.

Even if the laser operates in a single longitudinal cavity mode, its line width is limited by the pulse length τ. Essentially, the line width from pulsing is given by the Fourier transform from time to frequency space. For a working approximation, the pulse length and frequency spread are related as

$$\tau~\Delta f \cong 1 \qquad (25)$$

One example of this limitation is for pulsed Doppler lidar, where a radial velocity change δv gives a frequency shift of

$$\delta f = 2\delta v/\lambda \qquad (26)$$

Combining Eqs. (20) and (26) in Eq. (25), we obtain the resolution product

$$\Delta r~\delta v \cong c\lambda/4 \qquad (27)$$

which shows that there is a fundamental limitation on the range and radial velocity resolution of a pulsed Doppler lidar. The limitation is wavelength-dependent, so shorter-wavelength pulsed Doppler lidars can operate at higher-resolution products of Δr times δv. The frequency broadening from the pulse length in Eq. (25) is approximately

1 megahertz per microsecond

for a working order of magnitude, and for a CO_2-laser–based Doppler system the frequency shift from a radial velocity is

200 kilohertz per meter-per-second.

Although it is possible to locate the peak of a frequency spectrum to better accuracy than $\pm\Delta f/2$ if the signal-to-noise ratio is high enough, the width of a spectral line is a good working limit for the uncertainty in line position, as it has always been for classical optical spectroscopy.

Some velocity measurement techniques (e.g., in Section 3.2.1), such as the time-of-flight approach and incoherent differential Doppler (also called the real fringe method), do not rely on an analysis of the width of the optical spectrum. In these cases Eq. (27) does not apply because Eq. (26) is not appropriate; velocity information comes from intensity modulation of backscattered power as a scatterer passes through a spatially varying intensity pattern rather than from a frequency shift of the backscattered radiation. The bandwidth of the intensity modulation can be orders of magnitude smaller than the spectral width of the incident optical carrier in these types of systems.

Many lasers have a line width considerably wider than that given by the pulse limit (Eq. 25). In every case, the laser line width is the lower limit of the achievable spectral resolution of the lidar system. One reason for a comparatively wide laser line width can be optical inhomogeneities in the resonant optical cavity. This is a problem particularly for solid-state lasers, where thermal gradients cause refractive index inhomogeneities, and for dye lasers, where turbulence in the flowing liquid dye causes inhomogeneities. Another reason for a wide laser line width is the inherent spectral width of the lasing transition. If the excited molecules or atoms are inhomogeneously broadened (i.e., the excited centers have different energy-level structures), as in the hyperfine spectra of a copper-vapor laser, then narrowing the laser line width by means that are internal or external to the optical cavity causes a loss in output energy proportional to the fractional narrowing. On the other hand, if the excited molecules are homogeneously broadened (i.e., each excited center has an inherently broad spectrum), as in dye lasers, then in-cavity narrowing of the laser output spectrum can be done with much less output energy loss than in the case of inhomogeneously broadened lasers.

The requirements on spectral resolution are widely different for different types of boundary-layer lidars, as we shall see in the section on applications.

2.4.2 Refraction and Coherence

The same sort of atmospheric refractive index inhomogeneities that cause stars to twinkle affect the path of a lidar beam. The inhomogeneities cause beam broadening, which is a degradation of the angular resolution given in Eq. (22). Such degradation is usually characterized by a lateral coherence length given by (see Yura, 1979)

$$\varrho_0 \cong (3.4 C_n{}^2 r/\lambda^2)^{-3/5} \qquad (28)$$

where $C_n{}^2$, the refractive index structure constant, is usually within a factor of 10 of 1×10^{-15} m$^{-2/3}$ in the lower atmosphere (Lawrence and Strohbehn, 1970). The angular resolution for a lidar system in the inhomogeneous atmosphere is then given by Eq. (22), where D is the smaller of ϱ_0 or the telescope aperture diameter. Recall that loss of angular resolution can mean increased background noise and require a larger telescope field.

For lidars that operate with a transceiver rather than two separate telescopes, Eq. (28) provides an underestimate of ϱ_0; that is, the angular resolution with refractive inhomogeneities is better than would be expected from the ϱ_0 in Eq. (28). The performance of a coaxial lidar in the presence of refractive inhomogeneities is an active area of research. Schwiesow and Calfee (1979) discuss a related experiment, and Clifford and Wandzura (1981) and Clifford and Lading (1983) discuss the theory for a monostatic (transceiver) system.

Another manifestation of refractive effects is the distortion of plane or spherical waves propagating from or to the lidar telescopes. Originally smooth wave fronts (surfaces of constant optical phase) are broken up into patches of scale diameter ϱ_0. Such phase-front distortions reduce the efficiency of optical heterodyning because nonplane received wave fronts will not mix efficiently with a plane-wave local oscillator. The effect on a heterodyne lidar of breaking the wave front into n individually coherent patches is approximately to reduce the heterodyne signal by $n^{1/2}$. This is essentially a random walk problem, i.e., the coherent addition of n signals with random relative phases. Such atmospheric decohering effects are beyond the lidar user's control, but they provide an upper limit on the achievable signal-to-noise ratio.

In summary, we may think of the phase-front distortions introduced by the atmosphere to exist on three relative scales. On a scale much larger than the telescope aperture, the atmosphere acts as a weak, variable prism that causes wave-front tilt. On a scale approximately equal to the telescope aperture, the atmosphere acts as a weak, variable, generally astigmatic lens that changes the effective focal distance. On smaller scales, the atmosphere acts as an inhomogeneous phase screen that causes the optical path length between the lidar system and the scatterers to be different for different regions of the telescope aperture. Biaxial lidar systems, with separate transmitter and receiver telescopes, are most strongly affected by wave-front tilt. Coaxial lidar systems are insensitive to tilt; lens effects dominate. Coaxial transceiver heterodyne lidars are most strongly affected by fluctuations in optical path length.

Both spatial and temporal coherence are important in lidar applications. Some lidars are basically spatially incoherent in the sense that the illuminated target volume is much larger than that corresponding to the diffraction limit of the largest telescope in the system, i.e., larger than a single spatial mode. The van Cittert-Zernike theorem (Born and Wolf, 1965) relates the diameter s of the coherently usable source (the illuminated target spot) and the receiver diameter D over which the illumination is approximately coherent in the form

$$s \cong 0.84\lambda r/D \qquad (29)$$

This is a slightly more restrictive criterion than Eq. (22) for coherence and so illustrates the degree of approximation in various ways of expressing coherence. Other lidars, particularly those with very high spectral resolution, operate in a single spatial mode.

High-resolution lidars are temporally coherent in the sense that the spectral width of the return is a very small fraction of the optical frequency. The spectral resolution in a Doppler lidar, for example, is on the order of

$$f/\Delta f \cong 10^{10} \qquad (30)$$

No lidar can be temporally coherent in the sense that the optical phase of the return signal is in a fixed relationship to the phase of the transmitter. This is because each scatterer of the distributed source in the extended scattering volume will scatter independently, and the phase of the scattered wave will be determined by the exact range on a wavelength scale. Scattering from all the centers adds coherently, resulting in a fluctuating amplitude and phase at the receiver. This vector addition is considered more completely by Hardesty et al. (1981) and gives a result often called a Rayleigh phasor.

2.5 Data Processing Approaches

The basic pulsed lidar signal is similar to that sketched in Fig. 4. It occurs on time scales of a few microseconds, so rapid data recording and time gates suited to the range resolution are necessary. Earlier, oscilloscope photographs were used for data recording, but now high-speed transient digitizers with digitizing rates up to 100 MHz are available and are commonly used. Lidars with high repetition rates present data storage problems. The large amount of data available in a lidar return of 1,024 words of eight bits and a return repetition rate of 10 or 20 Hz require careful experiment planning. Various approaches to handling the data storage problem are being tried, but no method is yet preferred. One obvious option is to average a large number of lidar returns before recording. This is workable only for a nonscanning system and stationary boundary-layer variables. Another approach is to consider signal levels at a few (e.g., two) different ranges as

is done by Werner (1981). Simple analog recording on an intensity-modulated chart is used for cloud ceilographs.

Lidars for which spectral data are also important present an additional dimension to the data-processing problem. If only a few spectral channels are important, then separate signal-vs-range traces can be recorded for each channel. An alternative is to record a complete spectrum (at some suitable repetition rate) for a single range cell, which is determined by a time gate or telescope focus. The spectrum is often digitized before recording if a time-sequential spectrum is available.

Calibration of the return is related to the digitization and storage procedure. Special targets with known (ideally, Lambertian) scattering characteristics are often used to provide a known signal for intensity calibration. In addition to requiring knowledge of overall optical quantum efficiency of the lidar system for calibration, most lidars require correction for geometrical optical effects such as transmitter-receiver beam overlap and optical distortions. One way to reduce calibration problems is to concentrate on lidar measurements that can be reduced to ratio data, such as relative concentration, backscatter at two different wavelengths, or spectral line width.

Microcomputers, sometimes in conjunction with hard-wired processors, are used in on-line data-processing applications. Scanning and control of a basic aerosol lidar, for example, constitute one such use, as is the averaging of signal traces. Another application is the combination of a hard-wired digital spectral peak finder on a CW Doppler lidar to give a time-series output of velocity and a microcomputer for scanning, analyzing the velocity-vs-scan data, and recording the resultant wind vector.

Part of the data-processing problem is the production of multiparameter displays. Even the simplest lidar usually measures both return signal and range at different times. A scanning lidar can be more complex, giving, for example, signal and scatter depolarization ratio for a three-dimensional array as a function of time. Because lidar applications vary widely, we cannot give general rules on display technique. Data displays must be considered early in the design of a lidar system because the display format often demands certain scan capabilities, spatial averages, repetition rates, and other characteristics.

Techniques for reducing the dimensionality of lidar data displays include signal strength contours on a two-dimensional spatial plot, time averaging to a single signal-vs-range plot, and spatial sectioning such as time-height profiles and horizontal cuts. Previous work summarized in Section 3 can serve as a guide for developing appropriate analysis and processing techniques for a particular application.

3 HOW LIDAR IS BEING USED IN THE BOUNDARY LAYER

This section mentions a number of lidar applications as an aid in understanding the techniques and in evaluating the

suitability of lidar methods to the reader's observational problem. Because we are interested in an overview rather than in
detailed results, the applications are illustrated with simplified
schematic diagrams. Readers can find more detailed figures
and information in the referenced papers.

In practical applications of lidar, the experimenter must be
concerned with eye safety. Ways of insuring eye safety include (a) limited transmitter pulse energy, (b) transmitter
beam expansion, (c) operation in the infrared or ultraviolet,
and (d) scanning above the horizon coupled with a simple
radar or closed-circuit TV camera and observer to avoid aircraft. Sliney and Freasier (1972) provide detailed data on eye
safety, and Eberhard (1983) gives an example of a lidar that
will not harm the eyes.

Although this section is organized around individual
boundary-layer parameters such as wind, constituent concentration, and temperature, lidar is also useful for measuring
fluxes in the atmosphere. The volume averaging inherent in
remote sensing is useful for meaningful flux measurements
when the fluctuations of a quantity about its mean can be
derived from lidar data. Most lidar studies so far have concentrated on the measurement of single parameters rather
than the cross correlation of fluctuations, but Benedetti-
Michelangeli et al. (1974) give an example of flux
measurements using lidar. More lidar flux experiments can
be expected as experimenters develop more confidence in
lidar measurement of mean values and fluctuations of single
parameters.

3.1 Aerosol Loading and Dispersion

Aerosol backscatter measurements were among the earli-
est lidar applications. Most aerosol experiments give maps of
received backscatter intensity for inferring various source
strength and dispersion parameters.

3.1.1 Qualitative Patterns

Collis (1969) summarizes early work on lidar applications
to meteorology. For example, the path of a cloud of insec-
ticide released from a low-flying aircraft into drainage flow
showed the velocity of the flow, the settling velocity, and the
qualitative dispersion rate. To make these measurements, an
Nd:YAG lidar was scanned in elevation at a fixed (approximately upwind) azimuth to produce range-height indicator
(RHI) diagrams similar to those in Fig. 6. The spatial resolution in such patterns is typically a few meters in range and as
small as 10 mrad in elevation (corresponding to a vertical
resolution of 10 m at 1 km). The contours show relative
backscatter coefficient at constant decibel intervals. Similar
work is reviewed by Collis and Russell (1976).

It is possible to produce quantitative data on dispersion
from the geometry of qualitative backscatter measurements,
such as those sketched in Fig. 6. Hoff and Froude (1979) obtained Eulerian average cross sections of a plume and derived

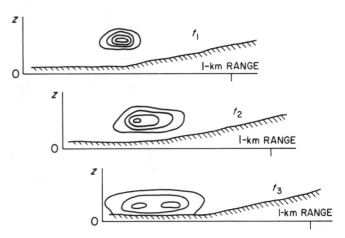

Fig. 6 Backscatter intensity contours showing dispersion of an aerosol plume.
(After Collis, 1969.)

horizontal and vertical dispersion coefficients and plume rise.
For these measurements they neglected attenuation in the
plume (see Section 3.1.2). One important finding from the
lidar measurements is that Gaussian plume formulations lead
to incorrect conclusions about dispersion under some conditions. The lidar technique allows derivation of moments of
the plume concentration distribution that do not rely on any
assumptions about plume shape. Johnson (1983) also gives examples of determining dispersion coefficients from qualitative lidar data. He contrasts values obtained with and without
fitting to a Gaussian assumption.

An alternative way of producing RHI lidar data like those
shown above is to use a downward-looking lidar in an aircraft. The lidar produces the backscatter-vs-height data, and
the aircraft translation produces the horizontal scan. While
the reliable horizontal extent of the RHI plot from a fixed
lidar is a few kilometers, the horizontal extent of an aircraft
scan can be 100 km or more. An example of boundary-layer
development from Uthe et al. (1980) is shown in Fig. 7. Such
data indicate the growth of low-level aerosol layers, pollution
sources, and clouds. Particle plumes can be tracked long
distances downwind with an airborne lidar. The instrumentation used by Uthe et al. included an Nd:YAG lidar operating
at both 1.06- and 0.53-μm wavelengths.

Another example of lidar instrumentation for boundary-
layer research is included in the study of convection cells by
Kunkel et al. (1977), which used a ruby lidar. When a convective plume carries air upward that has a backscatter coefficient larger than the ambient air at some altitude, then the
plume becomes visible to a lidar. From the shapes of the
aerosol-marked plumes and their changes with time over a
range of 1 to 7.5 km horizontally and 0 to 3.5 km vertically,
Kunkel et al. concluded that the cells are plumes rather than
bubbles, the aspect ratio of most plumes is between 0.5 and
1.5, and the circulation patterns show maximum rising motion on the upwind side of the cells and sinking on the downwind side.

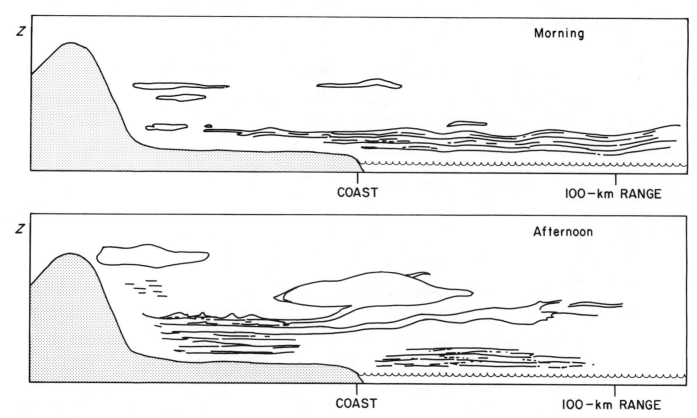

Fig. 7 Aerosol backscatter cross sections using intensity modulation for backscatter levels. (After Uthe et al., 1980.)

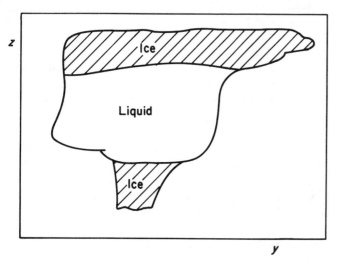

Fig. 8 Side view of spatial extent and water phase in a cloud.

It is possible to infer the freezing level in a cloud, or identify precipitation as ice or water, by measuring the depolarization ratio of lidar backscatter. Field measurements in the boundary layer have been reported by Derr et al. (1976) and Sassen (1980), among others. This application is an example of lidar data for which the range variation of the lidar signal is relatively unimportant; the variation of depolarization ratio with azimuth and elevation is more im-

portant. An example of the type of data obtainable is indicated schematically in Fig. 8. If the cloud is quite dense (i.e., optically thick), then the lidar data characterize the water phase at the surface of the cloud rather than in some interior section. Lidar data with an azimuth-elevation scan show the target as viewed in the direction of the lidar beam rather than as viewed perpendicular to a plane containing the vertical and the lidar beam as in an RHI display. Other cloud data such as drop size and number density can come from lidar multiple-scattering information (Pal and Carswell, 1976).

It is possible to infer approximate aerosol particle size from two-wavelength lidar returns. Uthe et al. (1982) used an airborne lidar operating at 1.06- and 0.53-μm wavelengths to emphasize aerosol scattering compared with molecular (using 1.06 μm) and to derive information related to visibility (using 0.53 μm). Strong attenuation of the surface backscatter at 0.53 μm compared with that at 1.06 μm was interpreted as evidence for submicrometer particle sizes.

The depth of the daytime convective boundary layer can be inferred from lidar measurements of the backscatter coefficient. The backscatter profile expected by Kaimal et al. (1982) under conditions of low particle concentration in the aerosol and fairly high relative humidity is shown in Fig. 9. The height resolution in such a lidar trace can be a few meters. Results from a ruby-based aerosol lidar compare well with results from other techniques, although lidar estimates of the top of the boundary layer sometimes tend to be high (Coulter, 1979). Kaimal et al. concluded that a properly representative measurement of the temperature lapse rate is the best way to infer the depth of the boundary layer. Thus, a temperature-profiling lidar (see Section 3.4) may be better than an aerosol-backscatter–profiling lidar for measuring the boundary-layer depth. This type of comparison illustrates that different lidar possibilities can be considered for a particular observational problem. A thorough understanding of the physics of lidar interaction with scatterers is the best basis for a decision about the applicability of a lidar system to a desired measurement.

These examples show only a few of the uses of an aerosol lidar to measure qualitative patterns and only a few of the inferences that can be drawn from such patterns. The references cited in this article can lead the reader to other useful ideas and stimulate the imagination.

3.1.2 Quantitative Aspects

There are two basic concerns in deriving quantitative aerosol values from correctly calibrated backscatter measurements. The first problem is the conversion between particle concentration and backscatter. If the particle shape is spherical and the size distribution and refractive index of the particles are completely known, then it is possible to calculate the backscatter from Mie theory. Generally the properties of the particles are not well known and assumptions must be made. The second problem is the attenuation of backscattered signal from distant particles by the closer aerosol between the distant particles and the lidar. Methods of handling these problems are subjects of active research and are not yet standard procedures.

For information on measurement of aerosol properties, the reader should consult the article on aerosol measurements in the boundary layer (Pueschel, this volume). One example of the use of assumed aerosol properties to interpret lidar backscatter results is the measurements by Post (1978). Post used Mie calculations and a measured particle-size distribution to show in one case that a 10.6-μm-wavelength lidar was most sensitive to particles with a radius $\gamma \sim 1.5$ μm. For larger radii the particle density decreases rapidly, and for smaller radii the lidar sensitivity is sharply reduced. The generation of fluorescent tracer particles having known fluorescent scattering properties is one way to reduce the aerosol characterization problem in the boundary layer. Lidar can then be used to track the concentration of known tracers, at least for short distances (Schuster and Kyle, 1980; Kyle et al., 1982). Generating a sufficient fluorescent-particle number density for long-range tracking has been a problem.

The use of lidar with artificial tracers (either Mie or fluorescent scatterers) is the remote-sensing analog to methods outlined in the article on gaseous tracer technology and application (Dabberdt and Dietz, this volume). Eberhard (1983) gives another example of tracking tracer plumes. The lidar used in this experiment, which was safe for eyes, could track plumes to a range of 5.5 km when the tracer concentration was high enough. It took 2.5 min to obtain a vertical cross section through the plume with $\sim$ 7.5-m range resolution and $\sim$ 100

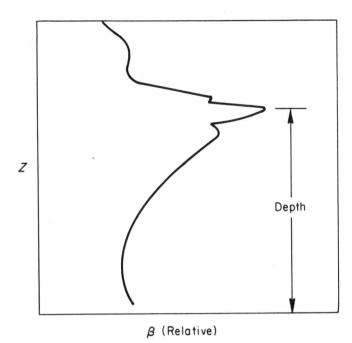

Fig. 9   Backscatter profile showing boundary-layer depth.

resolution elements vertically, but subsequent improvements in the laser have reduced the measurement time for a cross section to less than 1 min.

If we integrate the aerosol scattering coefficient over all angles, we obtain the total aerosol scattering coefficient given in Eq. (3). The value of this coefficient represents an attenuation by scattering out of the beam. If we know the aerosol backscattering coefficient, we can also determine a backscattering-to-attenuation ratio β/I. In deriving and using β/I, especially for plume work, one usually assumes that all attenuation in the plume is caused by aerosol scattering. The ratio β/I varies widely with the type of aerosol and the lidar wavelength. Representative values are $\beta/I \sim 0.025$ sr^{-1} $\pm$ 3 dB at 10.6 μm (Shettle and Fenn, 1976) and $\beta/I \sim 0.04 \pm 0.03$ sr^{-1} at visible wavelengths (Collis and Russell, 1976). Measurement of visibility (see the article by Viezee and Lewis on visibility measurement techniques, this volume) requires knowledge of β/I if lidar is to be used. Werner (1981) discusses a simplified lidar method for determining visibility that involves the ratio of backscatter to extinction, for example.

Some quantitative lidar experiments are interested only in backscatter. Schwiesow et al. (1981a) used an airborne lidar to measure backscatter profiles at 10.6-μm wavelength. The lidar used ranging only to move the sample volume away from possible aircraft contamination of the measurements; altitude scanning was by aircraft position. Some of the average backscatter values obtained are in Table 3. Individual measurements were up to ten times larger or smaller than the average.

At the present state of lidar development, the complications of backscatter-to-concentration conversion and backscatter-to-attenuation ratios are usually removed by assuming some fixed factor for the first and assuming negligible attenuation for the second. As these complications become more clearly understood, it is likely that lidar will become more useful for quantitative aerosol measurements than it is now.

3.2 Wind

3.2.1 Profiles

It is possible to measure wind profiles by tracking inhomogeneities in the atmospheric backscatter coefficient. In general, such measurements determine only the horizontal wind. Figure 10 represents lidar tracking of an idealized spherical inhomogeneity. As shown, the distance Δxy in some time Δt determines the wind. Actually, sophisticated pattern recognition and correlation routines have been used by Sroga et al. (1980) with a ruby lidar and by Sasano et al. (1982) with an Nd:YAG lidar. Sroga et al. measured winds over an averaging volume approximately 1 km in diameter and 100 m in altitude and with a 5-min averaging time to obtain wind values up to 600 m in altitude that compared with tower measurements to within root-mean-squared dif-

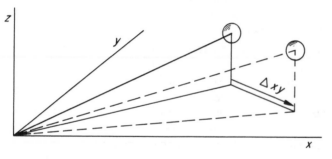

Fig. 10 Measuring wind by measuring displacement of backscatter inhomogeneities (here a sphere).

Fig. 11 Geometry and velocity-vs-azimuth plots for a conical-scan wind-vector-measuring scheme.

ferences of 1 m/s and 10° at a single tower altitude. Sasano et al. averaged measurements over a 4- by 4-km square on a slightly sloped scanning plane varying between 60 and 95 m in altitude and with a 10-min averaging time to obtain wind estimates that compared with tower-mounted anemometer values to ±15% on the average. It is important to note that the pattern-recognition technique for wind measurement relies on *variations* of the backscatter coefficient on the scale of the spatial resolution of the lidar scan rather than on the backscatter value itself. The method can suffer from signal loss in comparatively homogeneous, but strongly scattering, environments. It is possible that larger backscatter fluctuations would be observed at mid-IR wavelengths.

Another wind-profile measurement technique uses the Doppler shift of backscattered light to measure the radial (along the lidar beam) component of the wind directly. By scanning the lidar sample volume on a circle at the base of an inverted cone, one can infer all three (but in practice only the horizontal) components of the wind. The geometry of such a measurement and a typical data element are shown in Fig. 11. Such a scan is often referred to as the velocity-azimuth display (VAD) method. Velocity estimates at different altitudes are obtained by changing the lidar range, elevation angle, or both. Köpp et al. (1984) have used a conically scanned Doppler lidar operating at 10.6 μm (CO_2 laser) to measure wind profiles over spatial averaging dimensions varying from 400 to 1,200 m in diameter and from 4 to 700 m in altitude and with a 50-s time average at each altitude point. For profiles up to 750 m in altitude, the lidar and balloon-sonde wind estimates compare with root-mean-squared differences of 1.3 m/s and 12°. In all volume-averaging methods, uncertainties in the wind estimate are introduced by horizontal inhomogeneities in the wind field.

An example of the effect of horizontal inhomogeneities is given by DiMarzio et al. (1979), who show Doppler lidar results that demonstrate large velocity changes on scales of 2 to 6 km. In such wind fields, average wind values are of questionable utility, and wind measurement techniques that operate with smaller spatial averaging are more useful. Partial

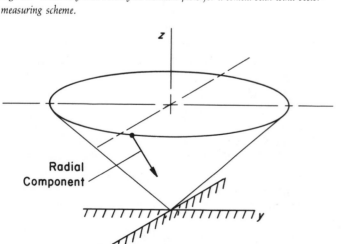

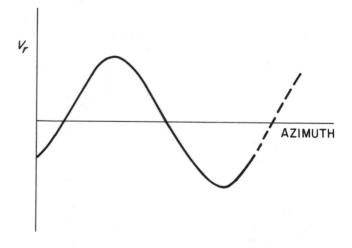

(sector) conical scanning can be used, and other lidar wind-measuring techniques are available.

Congeduti et al. (1981) have measured profiles of the vertical wind by using a vertically pointing, pulsed lidar operating at VIS wavelengths (Ar⁺ laser) and measuring the Doppler shift of backscattered light. Their measurements use an averaging length of 75 m in the vertical and only a few centimeters in the horizontal, and show an internally consistent precision of 0.1 m/s for a 25-s averaging time. Measurements were made over a 160- to 520-m altitude range at night to avoid problems with sky background (see Eq. 13). Schwiesow and Cupp (1981) made similar vertical velocity measurements at 100-m altitude in the daytime (no profiles) with a CW Doppler lidar operating in the IR (CO_2 laser). The averaging length was 12 m in altitude and a few centimeters horizontally, and velocity spectra were averaged for 5 s to produce velocity estimates with a ±10% uncertainty and 0.25-m/s threshold. Compared with the pulsed VIS lidar, the CW IR lidar has the advantage of easy daytime operation and fast response, and the disadvantage of altitude resolution that degrades as the square of the altitude (range) as in Eq. (21). A vertically pointing, pulsed IR Doppler lidar is not appropriate for profiles of vertical velocity in the boundary layer because of the resolution limitations in Eq. (27).

The time-of-flight lidar technique is well suited for measurement of atmospheric velocity components perpendicular to the lidar line of sight. Although boundary-layer profiles using this method have not been reported in the open literature, single-altitude results to a 100-m range show much promise for measurement of vertical profiles of the horizontal wind in the boundary layer with good spatial resolution. The lidar system for time-of-flight measurements in a simple form projects two beams, shown shaded in Fig. 12, from a common telescope aperture. Because the two beams are focused to separate, diffraction-limited spots at the altitude of interest, an airborne particle or aerosol inhomogeneity that passes through both focal volumes will give a double-pulse backscattered signal. The transverse velocity component (V_H) is determined from the time delay between the two pulses and the spacing of the beam foci. Practical time-of-flight lidars for wind measurement at VIS wavelengths (Ar⁺ laser) have demonstrated useful ranges of 100 m (Bartlett and She, 1977) and operation in daylight conditions (Lading et al., 1978). The spatial resolution for such systems is a few millimeters transverse to the beam (along the

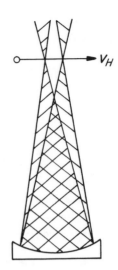

Fig. 12   *Geometry for time-of-flight lidar method for measuring the velocity component perpendicular to the beams.*

direction of the wind component measured) and approximately 1 m along the beam at a range (altitude) of 100 m. The temporal resolution is shorter than 1 s; it depends on aerosol particle number density and range. Lading (1980) found a correlation coefficient of 0.97 between simultaneous wind estimates made from a time-of-flight lidar and a cup anemometer with a 20-cm spacing between the lidar measurement volume and the cup anemometer. A 10-s averaging time was used for both systems. Note that this technique depends on aerosol inhomogeneities on the scale of the two spots, which is the order of a few millimeters (see Eq. 22). Thus, molecular scattering will not give a useful signal, and operation in the IR to emphasize large-particle scattering may give better results at long range than in the VIS.

Lidar wind measurements provide an example of the rich variety of atmospheric characteristics and scales that may be exploited by lidar techniques. Table 8 serves as a summary of this section and shows some of the differences and similarities between the lidar wind measurement methods that have been applied to the boundary layer. Methods that rely on the Doppler shift depend on the mean particle density. Methods that rely on the variation in particle density rely on different scales and mechanisms to produce the variation. The type of detection (direct detection of scattered energy or heterodyne detection of scattered electric fields, including phase values)

TABLE 8
Comparison of Lidar Wind-Measurement Techniques

Signal Source	Scale	Method	System	Velocity Component
Large-scale turbulence modulation of backscatter coefficient	Kilometers	Correlation of patterns	Direct detection	Transverse and radial to beam
Number fluctuations of particles, and single particle passage	Centimeters to meters	Time-of-flight, differential Doppler (fringe)	Direct detection	Transverse
Changes in particle position in the measuring volume	Millimeters to meters	Doppler shift	Heterodyne or direct detection	Radial

also is important in evaluating the performance of various methods. The statistics of the particle density relevant to the various lidar techniques is an area of active research (Lading et al., 1980).

3.2.2 Special Observational Problems

In addition to providing wind profiles, lidar wind measurements have been useful for some special problems, for example, where the advantages of lidar remote sensing outweigh the limitations of single-component sensing. DiMarzio et al. (1979) used a pulsed IR Doppler lidar to produce plan-position-indicator maps of the radial component of the wind as a function of azimuth and range to a distance of 5 km from the lidar. Their data showed the effect of a sea breeze front and horizontal wind shear associated with a thunderstorm gust front. Spatial resolution (300 m) and velocity resolution (~ 2.5 m/s) were limited by Eq. (27), but were adequate for the intended purpose. One might consider also the pattern-recognition technique of Sroga et al. (1980) for this application. Brown et al. (1978) used a CW IR Doppler lidar to measure mean speed and turbulence in a smoke plume. Turbulence decreased with distance downwind from the stack, although, very close to and vertically above the mouth of the stack, the flow was more laminar. The observations show that the (horizontal) plume velocity is greater than the ambient wind. Although the averaging time was shorter than 0.5 s, the spatial averaging interval of the lidar was larger than the plume diameter, so that only spatial averages over the plume diameter were measured. Neither of the reports referenced above gave a wind component accuracy estimate for the measurements.

A VIS Doppler lidar has given simultaneous measurements of aerosol concentration and vertical velocity to allow determination of the vertical eddy flux of aerosol particles in work by Benedetti-Michelangeli et al. (1974). They used a height resolution of 300 m and an integration time of 300 s to achieve an estimated error of 0.03 m/s in measurement of vertical velocity and 6 m²/s in the eddy diffusion coefficient at three different altitudes. Eberhard and Schotland (1980) tested a different type of VIS Doppler lidar using two different transmitted optical frequencies. Their measurements of the wind at a range of 20 m, a range resolution of 2.2 m from beam overlap (noncoaxial transmitter and receiver), and an integration time of 1 to 2 s showed a root-mean-squared difference between lidar and propeller anemometer values of approximately 0.6 m/s.

Schwiesow et al. (1981b) used the radial-component velocity-sensing capability of an airborne, CW, IR Doppler lidar to measure the velocity structure of waterspouts. Although the range resolution of up to 300 m was much larger than the diameter of the vortices, the resolution of less than 10 cm transverse to the beam allowed adequate detail to be observed in a velocity-vs-radius plot for the vortex. Because an aircraft scans a waterspout rapidly, Schwiesow et al. used a 14-ms integration time to achieve a velocity uncer-

tainty of less than 1 m/s, which was dominated by the turbulent width of the velocity spectra. Although it was used on a number of waterspouts the lidar provided no data on vertical velocity in the waterspouts because of the limitation of the system to radial component only. In another application of a CW IR Doppler lidar, Schwiesow and Lawrence (1982) measured profiles of an onshore breeze at altitudes from 4.7 to 66.5 m above sea level. Vertical resolution varied from less than 1 m at low elevation angles to 12 m at the highest altitude measured. A 1-min integration time led to absolute velocity uncertainties of ±10 %, but the relative uncertainty between two different velocity values was only 10 % of the velocity difference. Even though the geometry of the flow restricted the variation to two dimensions, the lidar used was unable to make unambiguous vector wind measurements because of its limitation to measuring only the radial component. This application illustrates the need for an ability to measure remotely three components of the wind at a single point.

3.3 Profiles of Constituents

Most work measuring constituents in the atmosphere has been based on Raman scattering or differential absorption. In the Raman technique (see Section 2.2.2), the intensity of backscatter at the wavelength corresponding to the Raman-shifted scattering from the constituent of interest is compared with the intensity from a reference material, which is usually atmospheric nitrogen. Figure 13 shows the relevant spectrum. From the backscatter intensity ratio it is easy to calculate the number density of unknown constituent molecules relative to reference molecules if the ratio of scattering cross sections of the molecules and instrumental response are known from previous calibration. Usually, vibrational Raman spectra are used, as sketched in Fig. 13, rather than rotational Raman spectra because the multiconstituent spectra are more separated and because much lower spectral resolution is required in the receiver. Although ratio measurements are direct and simple, one major drawback of the Raman technique is that signal strength is very low because the scattering cross section is small. On the other hand, the constituent mixing ratio is independent of lidar system variables such as laser power and alignment, and it is

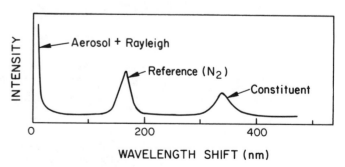

Fig. 13 Backscatter spectrum showing Raman signals from reference and constituent molecules.

independent of wavelength-independent atmospheric attenuation and refractive index inhomogeneities.

In the *differential absorption lidar* (DIAL) technique, the difference in atmospheric attenuation between two closely spaced transmitter wavelengths is used to infer the amount of absorbing material in the beam. Figure 14 shows idealized signal behavior for a simplified case where all of the measured constituent is in a localized plume. By comparing the return at the absorbing wavelength (double line) with that at the unabsorbed wavelength (single line) a measure of the total molecular absorption can be made from the change in return intensity ΔI_r and constituent-absorbed intensity ΔI_c. In general, the unabsorbed wavelength will still be attenuated, so that the unabsorbed attenuation can be used to correct the results from the absorbed wavelength. This procedure assumes that the broad-band attenuation is the same at both wavelengths. If the constituent is more or less uniformly distributed, rather than contained in a plume, then, by determining the difference in absorption between two different ranges, molecular absorption and therefore concentration in the region between the two range points can be estimated. From these data, one obtains a concentration profile.

A simplified analysis of the DIAL method provides an illustration of the use of the lidar equation and of signal ratios. We rewrite the attenuation coefficient (Eq. 4) in the form

$$\alpha(\lambda_1) = a'(\lambda_1)e + \overline{\alpha} \tag{31}$$

where $\overline{\alpha}$ is a broad-band attenuation coefficient *assumed* to be independent of wavelength, $a'(\lambda_1)$ is a specific absorption coefficient measured on a per-unit-absorber-density basis, and $e(r)$ is the density of the constituent of interest; i.e., $a'e =$ the absorption coefficient a. If we can choose a nearby λ_2 where $a'(\lambda_1) \gg a'(\lambda_2)$, or for a simple illustration $a'(\lambda_2) = 0$, then we write

$$\alpha(\lambda_2) = \overline{\alpha} \tag{32}$$

Note the critical importance of the wavelength independence of $\overline{\alpha}$, which results from all scatterers and absorbers except the constituent of interest. We presume $a'(\lambda_1)$ is known from laboratory measurements. From Eq. (8) and Eq. (13) we write

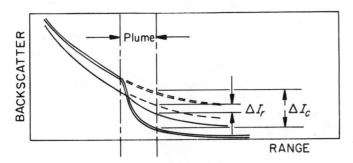

Fig. 14   Lidar returns through an absorbing plume at two different wavelengths shown in single (reference) and double (absorbing) lines.

$$S^*(\lambda_1, r_1) = C(\lambda_1)n(r_1) \exp\left\{-2 \int_0^{r_1} [a'(\lambda_1)e(x) + \overline{\alpha}]\, dx\right\} + N_b(\lambda_1) + N_d(\lambda_1) \tag{33}$$

where S^* includes the backscattered signal plus background and detector energy, and $C(\lambda_1)$ lumps together all range-independent factors in Eq. (8). Equations similar to Eq. (33) can be written for the variables (λ_1, r_2), (λ_2, r_1), and (λ_2, r_2). Additional complications can be introduced if β is a function of r, for instance. It is possible to form the ratio

$$
\begin{aligned}
Q(\lambda_1) &= \{[S^*(\lambda_1, r_2) - N_b - N_d]/[S^*(\lambda_1, r_1) - N_b \\
&\quad - N_d]\} [n(r_1)/n(r_2)] \\
&= \exp\left\{-2 \int_{r_1}^{r_2} [a'(\lambda_1)e(x) + \overline{\alpha}]\, dx\right\}
\end{aligned} \tag{34}
$$

where the S^*, $n(r)$, N_b, and N_d are determined from lidar measurements. $Q(\lambda_1)$ and $Q(\lambda_2)$ are experimentally determined quantities; an analysis of their measurement uncertainty is beyond the scope of this article. To simplify the integral in Eq. (34), we set $r_{12} = (r_1 + r_2)/2$, $\Delta r = r_2 - r_1$, and assume $e(r_{12})$ is constant over the interval (r_1, r_2). Then from Eq. (34) and a similar equation at λ_2 we have

$$\ln[Q(\lambda_1)] = -2a'(\lambda_1)e(r_{12})\, \Delta r - 2\overline{\alpha}\, \Delta r$$
$$\ln[Q(\lambda_2)] = -2\overline{\alpha}\, \Delta r \tag{35}$$

from which the concentration of the constituent of interest can be found as

$$e(r_{12}) = \ln[Q(\lambda_2)/Q(\lambda_1)]/[2a'(\lambda_1)\, \Delta r] \tag{36}$$

The reader can modify Eq. (36) to include other variables if desired, but the constancy of $\overline{\alpha}$, or its known wavelength dependence, remains basic to the method. Browell et al. (1983) present a slightly different form of Eq. (36) that does not consider N_b and N_d explicitly. The problems of measuring ratios near unity from two large (noisy) values restrict the range resolution and concentration accuracy obtainable with the technique. Differences in the aerosol attenuation at the two different wavelengths can also contribute to constituent concentration errors.

3.3.1 Humidity

A number of atmospheric water vapor measurements have been made using vibrational Raman spectra. Among the experiments are those of Melfi et al. (1969), Cooney (1970, 1971), Strauch et al. (1972), and Pourny et al. (1979). Recent experiments measured the water vapor profile to 1,800 m in altitude with 30-m vertical resolution and a 15% uncertainty in the mixing ratio at 1,000 m. To achieve these results, Pourny et al. used 30 shots of a 0.2-J, frequency-doubled ruby laser at a wavelength of 347 nm and a collecting telescope with a 25-cm aperture.

Although these water vapor results are useful, most experiments using the vibrational Raman spectrum of water vapor have been limited to night operation because of skylight interference in the daytime. One solution to the background interference problem is to operate farther into the ultraviolet, as analyzed by Cooney et al. (1980). Renaut et al. (1980) made daytime measurements of the water vapor mixing ratio to 1,000 m in altitude with 30-m vertical resolution and a 10% uncertainty in the mixing ratio at 500 m. The experiment needed 50 shots of a quadrupled Nd:YAG laser (in approximately 30 min), operating with 100 mJ per pulse at a 266-nm wavelength, and a 60-cm telescope aperture. One complication in operating at these wavelengths is the correction for differential absorption at the N_2 and H_2O Raman shifts. Renaut et al. estimate that improved daytime performance is possible, based on a better, but available, laser system than the one used in their experiment.

Werner and Herrmann (1981) report daytime results using the DIAL method for water vapor partial-pressure profiles up to an altitude of 1,500 m with a 100-m range resolution. Their well-developed system uses two independent ruby lasers with 1-J output pulse energy, one tuned to a water vapor absorption line and one tuned off the line. With a 40-cm telescope for the absorbed wavelength, approximately 20 pulses are averaged over 30 min to achieve 1-mb (~10%) uncertainty for each resolution element.

Alternative approaches are possible. Browell et al. (1980) suggest the use of laser-pumped dye lasers, but the wavelength stabilization problem is at least as difficult as that for the ruby lasers. The lidar system of Browell et al. is designed for airborne use. Werner and Herrmann (1981) recommend the use of smaller, more rapidly pulsed ruby lasers as one way of improving existing instrumentation.

Recent DIAL humidity measurements by Cahen et al. (1982) in the near-IR show mixing ratios to 3 km in altitude with a vertical resolution of 30 m and an accuracy of 10% when a 4-min integration time is used for each profile. The large lidar system uses an Nd:YAG-pumped dye laser and three dye amplifier stages to achieve 70 mJ per pulse at a 10-Hz rate, and it uses a 60-cm telescope aperture. This is approximately the same size system as the Raman system of Renaut et al. (1980), but it involves in addition a wavelength stabilization loop to center the dye laser line on a water vapor absorption line. The two different wavelengths are emitted sequentially with a 100-ms delay; the analysis neglects any temporal variations in the atmospheric attenuation, refraction, or scattering properties during this time period. Baker (1983) discusses humidity profiles obtained with a CO_2-laser lidar operating near a wavelength of 10.25 μm. The lidar operated with a 30-cm telescope and direct (not heterodyne) detection to give profiles to an accuracy of ±2 to 4 mb at an average water vapor partial pressure of ~10 mb. Profiles were measured with 300-m height resolution to an altitude of

1,300 km, but the integration time for the measurements was not reported. Height resolution of 30 m was available at short ranges. One possible weakness with the technique used by Baker is that there was a time delay of 13 s between measurements at λ_1 and λ_2. Optical properties of the sample volume and path can change during this time.

3.3.2 Other Gases

The vibrational Raman lidar technique that is used for water vapor is also applicable to measurement of other atmospheric constituents. Because the Raman backscattering cross section is smaller than that for many other kinds of scattering, Raman measurements of constituent concentrations are useful only for concentrations of a few hundred parts per million or more at ranges of a few hundred meters. This restricts such measurements to stack plumes and similar sources. Hirschfeld et al. (1973) reported one of the more sensitive demonstrations of Raman sensing of boundary-layer constituents. Inaba (1976) gives a valuable review of Raman measurements of pollutant concentrations.

The DIAL technique is at present more widely used than Raman scattering for measuring gas concentrations in the boundary layer. Grant and Hake (1975) have measured SO_2 and O_3 concentrations. Fredriksson et al. (1979 and 1981) and Grant and Menzies (1983) provide summaries of the instrumentation for, and results of, boundary-layer DIAL measurements. When the constituents of interest are localized in a plume, the concentration integrated over the plume can be obtained as in Fig. 14, and the plume diameter can be measured by short-pulse lidar. For example, Fredriksson et al. measured an integrated NO_2 concentration of 190 ± 50 mg/m³ • m through a 6-m plume at a range of 2 km. For NO_2 the DIAL system worked in the blue part of the spectrum. For example, using UV light, they made plume measurements of SO_2 at 760 ± 150 mg/m³ • m. Range-resolved measurements of ambient SO_2 had a range resolution of 500 m and a concentration uncertainty of approximately ±10 ppb to a range of 3 km in one case and a range resolution of 100 m and concentration uncertainty of ±15 μg/m³ to a range of 1.5 km in another case. The compact, mobile lidar used for these measurements had a 30-cm telescope aperture and used 0.5-mJ laser pulses at a 10- to 30-Hz repetition rate; however, integration times for the measurements were not reported. Browell et al. (1983) describe a DIAL system and report ozone profile measurements. The lidar they used includes a 35-cm telescope aperture and a complex laser system with an output energy of 30 mJ at a 10-Hz repetition rate. Profiles were measured with a vertical resolution of 210 m, a maximum range of ~2.5 km, and an integration time of 10 s. The lidar ozone values were within ±10% of values measured by in situ instruments.

3.4 Temperature

Rotational Raman scattering, Cabannes scattering (nearly inelastic molecular scattering), DIAL measurements, and vibrational Raman density measurements are some of the lidar techniques that can be used to determine temperature profiles in the boundary layer.

The rotational Raman-scattering spectrum of atmospheric molecules is temperature-dependent because the shape of the envelope of the many spectral lines corresponding to transitions between rotational energy levels in the ground-state manifold reflects the temperature-dependent populations of the energy levels. By measuring the intensity ratio between two parts of the rotational Raman band, one can infer temperature. Cooney (1972) and coworkers (Gill et al., 1979) measured temperature profiles to 2.3 km in altitude with a lidar based on a ruby laser transmitter with a 694-nm wavelength and interference filters to select spectral regions. The height resolution was 75 m, and a profile was measured in 4 min. The average departure from a corresponding radiosonde profile was ±0.85 °C. One limitation of the temperature-profiling lidar used by this group is that measurements were possible only at night because daytime sky background obscured the Raman return. The spatial resolution in the horizontal was a few meters, which is typical for conventional vertically pointing lidars.

One possible improvement in the lidar is to change from the biaxial arrangement used by Gill et al. to a coaxial optical geometry to allow data to be taken at many ranges with each transmitter pulse. Another improvement would be to use a transmitter wavelength in the ultraviolet to reduce sky background interference. The facts that the Raman-scattering cross section is approximately 0.02 times the Cabannes and that only a part of the total rotational Raman band is used constitute a fundamental limitation of the Raman technique.

Molecules in thermal equilibrium move with a translational velocity distribution described by Maxwell-Boltzmann statistics. Light backscattered from the moving molecules will be spectrally broadened by a direct Doppler shift. Thus the Cabannes-scattering (often given the more inclusive name Rayleigh scattering) line width is dependent on temperature because the translational velocity distribution of the molecules depends on temperature. By measuring the scattering line width as a function of altitude, one determines a temperature profile. Figure 15 is a schematic representation of the relevant spectrum. The spectrum in Fig. 15 is strictly applicable only for atmospheric densities characteristic of 2–3 km and above. At higher densities, collective phenomena become detectable because the mean free path of the scattering molecules is on the order of the wavelength of the incident light. Brillouin scattering (see Section 2.2.2) appears as a symmetric doublet with a separation similar to the Cabannes line width. Although the Brillouin doublet is quite weak in backscatter even at surface atmospheric density and is often not apparent (Fiocco et al., 1971), it can influence the

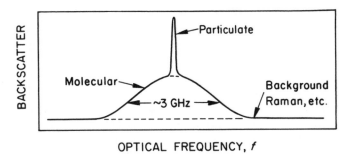

Fig. 15   *Atmospheric scattering spectrum in high resolution. The molecular line width is temperature-dependent.*

accurate line-width measurements required to infer temperature and must be corrected for.

Fiocco et al. (1971) and Benedetti-Michelangeli and Fiocco (1974) used the line-width technique to measure a profile to approximately 5 km in altitude with an accuracy estimated at a few degrees Celsius. They used a CW argon laser transmitter at a 488-nm wavelength and a chopper to pulse the transmitter output, resulting in a range gate of a few hundred meters. With a scanning Fabry-Perot interferometer it took approximately 1 h to obtain a full spectrum and thus the temperature at a single altitude. As in the case of the rotational Raman experiments, measurements were made at night to reduce background light.

Improvements in technology since 1971 allow us to base estimates of the performance of a Cabannes-scattering lidar for temperature profiling on more powerful pulsed lasers and parallel-processing spectral analysis. Schwiesow and Lading (1981) have calculated that a lidar at a wavelength of 488 nm with 1 W of average laser power and a 30-cm telescope aperture can measure temperature at 5 km with 50-m altitude resolution to ±1 K in an integration time of 75 s under daylight conditions. Measurements at lower altitudes can be made more rapidly.

Compared with rotational Raman scattering, the narrower-bandwidth technique that is based on Cabannes line width has the advantage of a greater backscatter cross section and the disadvantage of requiring a narrow-bandwidth transmitter. Both molecular scattering techniques are conceptually direct and can be clearly analyzed. The ability to make daytime temperature measurements with the line-width technique is a natural consequence of the use of a narrow-bandwidth laser transmitter because appropriate lasers operate in single longitudinal and transverse modes, thus making near-diffraction–limited operation an attractive way to reduce the amount of background light detected.

DIAL can also be used for temperature measurements by probing the population of ground-state energy levels in atmospheric molecules of known concentration by means of changes in the absorption spectra of the molecules (Mason, 1975). The absorption at two different energy transitions is a measure of temperature along the absorbing path because the absorptions depend on the ground-state population distribution (assuming the molecular species is well mixed in the atmosphere), which in turn depends on temperature. Range differences in the differential absorption yield temperature profile information.

Choosing an appropriate transmitter laser is difficult (Barton and Le Marshall, 1979), but differential absorption measurements have been made with the help of a retroreflector and two dye lasers (Kalshoven et al., 1981) and with scatter from a hillside and a CO_2 laser (Murray et al., 1980). These measurements were not lidar measurements because the paths were not single-ended, but lidar measurements should be possible when aerosol backscatter, rather than a reflector, is used as a target. Kalshoven et al. measured average temperature over a 1-km path to $\pm 1\,°C$ using integration times of less than 1 min with approximately 50-mW CW laser power in each beam. Murray et al. used a 5-km, double-passed path and determined the path-averaged temperature of CO_2 molecules with an uncertainty of approximately $\pm 1.1\,°C$ using 1.5-J laser pulses and an unstated integration time.

Range-resolved differential absorption data are required to determine a temperature profile rather than a line average. Range-resolved absorption differences are inherently difficult to measure because one is dealing with small differences in comparatively large absorption values at two different ranges. For the absorption differences to be sufficiently large to be above noise, substantial range differences must be used. The goal of one program (not yet achieved) is a range resolution of only 2 km (Kalshoven et al., 1981). This is not appropriate for boundary-layer work.

Another method for inferring temperature is to measure the molecular density profile by means of the profile of vibrational Raman scattering return from N_2. Strauch et al. (1971) determined relative temperature at a single point with a few degrees uncertainty in 3 s of integration time. Although the measurements were at a range of 30 m and a resolution of 5 m in altitude, Strauch et al. estimated that lidar equipment available in 1971 could measure temperature profiles to several kilometers. One complication in the density profile approach is that a statistical model of the pressure profile (or height integration of density) is required to infer temperature. On the other hand, density data are directly available in a lidar mode without dependence on aerosol targets or signal differences.

At the present state of development, molecular-scattering approaches to lidar temperature profiling have the advantages over differential absorption of better height resolution; freedom from restrictions on, and need for control of, laser wavelength; field demonstration in a lidar mode; and simple, direct analysis. Temperature profiling by temperature-dependent spectroscopy has the advantages over the density profile approach of using intensity ratios rather than calibrated intensity and of being free of the need for auxiliary data such as pressure and optical attenuation profiles.

3.5 Path-Averaged Measurements

There are optical remote-sensing techniques for the boundary layer that are not lidar because ranging to the target is not involved. Such path-averaged methods use a double-ended path with a transmitter at one end and a receiver at another, or a collocated transmitter and receiver at one end and a target (retroreflector, hillside, or building) at the other. Although nonlidar techniques are outside the scope of this article, a brief mention for completeness is in order because optical path-averaged instrumentation for the boundary layer is not included elsewhere in this volume and because lidar systems are often easily adapted to path-averaged methods.

Path-averaged crosswind sensing (e.g., Lawrence et al., 1972) is a well-developed technique that has been used over path lengths of up to approximately 10 km to measure such boundary-layer phenomena as drainage flow and convergence near the surface. The technique is based on the correlation of scintillation patterns caused by refractive index inhomogeneities that drift with the wind across the optical path. It is not necessary to use a laser source; incoherent light sources and passive scene illumination are adequate for average crosswind measurements.

DIAL-type concentration measurements can also be made in a path-averaged mode. Endemann and Byer (1981) give examples of humidity measurements with a two-ended path, and the path-averaged temperature data of Murray et al. (1980) and Kalshoven et al. (1981) were mentioned earlier. Fredriksson et al. (1979) report on NO_2 and SO_2 path-averaged concentrations over a 5-km path that have uncertainties of approximately 2 ppb. The structure constant for refractive index inhomogeneities, $C_n{}^2$, can also be determined with a path-averaging optical technique (Wang et al., 1978).

These techniques are all more or less closely related to lidar. Study of path-averaged methods can help in understanding lidar applications to the boundary layer, and study of lidar interactions helps in using path-averaging methods.

4 WHAT WE CAN EXPECT IN THE FUTURE

4.1 Variety and Limitations

The many types of lidar measurements discussed in this article (and there are many more types of lidar for use outside the boundary layer) demonstrate that lidar is not a single instrumentation technique, but is instead a class of optical remote-sensing techniques. These techniques all involve atmospheric backscatter of some sort or another, but the data analysis can focus on absorption or on different types of scattering that involve aerosol particles or gaseous molecules. Some of the lidar-accessible variables are aerosol particle distribution, wind, temperature, water vapor, other constituents, and combinations of these variables. Depending on the variable of interest and the chosen measurement interaction, the lidar transmitter wavelength can range from the middle IR to the middle UV. Lidar systems range from compact, mobile, and airborne systems to large, fixed installations for longer-range sensing. For an appreciation of the wider lidar context beyond the boundary layer, the reader can consult reviews of lidar from space platforms (Atlas and Korb, 1981), lidar for temperature and wind to an altitude of 100 km (Chanin, 1982), lidar for atmospheric sciences (Grams, 1978), and lasers (including in situ applications) for monitoring the atmosphere (Hinkley, 1976).

In spite of this variety, there is enough similarity in different lidar approaches that parts of the same lidar instrument can be used to measure different boundary-layer variables, if the physical interactions on which the measurements are based are chosen correctly. For example, a single, reflective transceiver with a 30- to 50-cm aperture can be used with a dye laser transmitter for DIAL humidity measurements and for time-of-flight wind data, although separate receiver packages are appropriate for each application. The same transceiver with a CO_2 laser can give radial wind components by Doppler analysis and transverse wind components by time-of-flight analysis. In all cases, boundary-layer lidars have in common the fact that they are single-ended remote-sensing devices that provide range-resolved data along the line of sight.

Lidars can operate only as far as there is adequate optical transmission. Thus they do not penetrate far into dense clouds, and they are not completely all-weather devices. Although it is often possible to obtain lidar (especially IR lidar) data at ranges beyond the standard visual range, there is no evidence that lidars can obtain meaningful data from regions at a distance of many times the visual range, e.g., that they can compete with radar for the detection of wind at long ranges under low-visibility conditions. The strengths of lidar compared with remote sensors based on other types of radiation include:

• High spatial resolution
• Operation close to obstructions (no ground clutter)
• A variety of measurement interactions and accessible variables
• Freedom from many types of interference.

Lidar systems are complex instruments with many parts and sometimes sensitive mechanical alignments. Lasers themselves usually have lifetimes of a few thousand hours, so that lidar is less suited on a cost basis than is conventional instrumentation to continuous monitoring for periods of many months. On the other hand, lidar is suited to intensive research campaigns where its advantages of portability, range, area coverage, and noninterference with the variables of interest are important. Although lidarlike systems have been made suitable for military use without regular operator attention or alignment, the lidars developed so far for boundary-layer purposes are usually operated by an interested group of scientists who maintain the laser and optical elements at peak operating efficiency.

4.2 New Methods and Systems

By comparing the variety of physical interactions for lidar in Section 2 with the realizations in Section 3, one can see that many combinations of scattering, absorption, and wavelength remain to be explored. Two examples, for those of us interested in wind measurements, are direct-Doppler heterodyne instrumentation in the VIS and time-of-flight work in the IR. Lidar specialists in other research areas could suggest many other lidar methods that could be tested and developed. It is difficult to predict the directions of such developments because they depend on the perceived needs for boundary-layer data, on improving laser and optical technology, and on the particular skills and interests of individual lidar researchers. It is unlikely that the basic measurement interactions, system operation, and propagation effects discussed in Section 2 will change much in the next few years.

Developments in reducing size and increasing reliability of presently used types of lidars are easier to predict. Tripod-mounted minilidars for aerosol work that are based on battery-operated military range-finder lasers are already being used by Werner (1981). The minilidar optics head weighs less than 17 kg and is less than 40 cm in each dimension. Vaughan (1979) has developed a compact Doppler lidar that weighs only 44 kg (plus electronics) for mounting in the nose of a business jet aircraft. This lidar is used for data on true airspeed and wind shear at a range of a few hundred meters in front of the aircraft. Lidars for other applications summarized in Section 3 will follow this trend as improved lasers and other components are developed and the needs for lidars become more clearly defined, thus allowing systems to be specifically and compactly designed for a particular observational problem. Wave-guide, radio-frequency–excited CO_2 lasers that are more compact and efficient than their low-pressure predecessors are an example of improved components that lead to new and better lidars. A lidar simulation of a tall meteorological tower (Schwiesow, 1983) is an example of a defined problem leading to a specific lidar design. These kinds of system improvements will continue.

Overall, lidar has demonstrated a capability to measure a number of boundary-layer variables such as aerosol backscatter, wind, humidity, and temperature. We have not yet exhausted its potential. Researchers who understand why a lidar works and understand boundary-layer observational problems are in a good position to make substantial progress in experimental boundary-layer meteorology.

Acknowledgments. I am indebted to L. Lading for many exciting and helpful discussions over a period of years. His insights in data processing, turbulence effects, and system principles and interrelationships are reflected in this article. Vernon Derr provided long-term guidance and challenge in exploring the different applications of lidar. My colleagues in the international lidar community have been friendly and helpful in discussing their work and so making possible the synergism that leads to enjoyable progress in the field, the fruits of which I hope you can see in the applications we have talked about.

Radar Probing and Measurement of the Planetary Boundary Layer: Part I. Scattering from Refractive Index Irregularities

R.B. Chadwick and E.E. Gossard, Wave Propagation Laboratory, NOAA

1 BASIC RADAR CONCEPTS

1.1 Introduction

Radar probes the atmosphere by directing electromagnetic energy at the region of interest and receiving the relatively weak signal that is scattered back. This signal is processed to extract various signal parameters such as delay, amplitude, and frequency difference between transmitted and received signals; then these basic signal parameters are related to various meteorological phenomena. The two problems involved are: first, to estimate the basic signal parameters from the return signal and second, to relate these signal parameters to meteorological quantities of interest. Estimating the parameters of a noisy signal is a well-developed procedure, having a firm basis in statistical communication theory. For many mathematical models of signals and noise, optimal estimators for various signal parameters have been derived and their performance studied. These concepts are well covered in the radar literature and so they will only be mentioned briefly here. The meteorological variables alluded to above, on the other hand, are not well covered in conventional radar literature because returns from the atmosphere are generally termed "weather clutter" by radar designers and operators interested in hard targets. They would prefer that their radar displays were uncluttered by weather return. Their only interest in atmospheric return is in determining characteristics that allow them to eliminate it. By contrast, one of the main goals of the radar meteorologist is to try to understand atmospheric processes by interpreting the return from weather "clutter." This is the special province of radar meteorology and will comprise the bulk of our discussion. This article covers the scattering of radar signals from optically clear air where the signals are reflected from refractive index inhomogeneities caused by temperature and humidity fluctuations on scales of half a radar wavelength. The following article by Kropfli will cover the use of radar to study atmospheric processes using man-made chaff and natural particulates as a tracer of atmospheric motions.

1.2 Different Types of Radars

While there are many different types of radars used for a wide range of applications, they can be broken into two broad categories based on the transmitted signal: low-duty cycle and high-duty cycle. In the simplest form, low-duty-cycle radars can measure range but not velocity, while high-duty-cycle radars can measure velocity but not range. However, there are more complicated types of both high-and low-duty-cycle radars that can measure both range and velocity.

Perhaps the most widely used radar in our modern society is the police speed-measuring radar, a high-duty-cycle radar. This device sends and receives a continuous sinusoid. If the target is moving, the received sinusoid is shifted in frequency from the transmitted sinusoid by the Doppler effect and a frequency difference or Doppler signal can be generated by comparing the phase of the transmitted and received signals as a function of time. Every time the target moves half a radar wavelength radially, the Doppler signal goes through a full period. The mean frequency of the Doppler signal is proportional to the mean outward radial velocity V_R through the well-known Doppler equation, $\delta f = -2V_R/\lambda$, where λ is the wavelength of the signal. The length of the signal (hereafter called the coherent integration time) is the time that it takes to make the frequency measurement. However, since the transmitted signal repeats itself every wavelength, any attempt at range measurement is ambiguous. The result is that this type of radar cannot measure range.

At the other end of the duty-cycle scale is a radar that transmits a single short pulse of radio frequency (RF) energy. The pulse is reflected from the target and the radar can estimate the range by measuring the round-trip travel time. This measurement is easy for a short pulse because the pulse has a distinctive beginning and end. On the other hand, a short-pulse radar cannot measure velocity. Since the target moves much less than half a wavelength during the time the pulse interacts with it, much less than one cycle of Doppler frequency is generated at the output of the radar. Thus, the time derivative of phase, i.e., frequency, is not measurable. To quantify this, assume these typical values: pulse length of 1 μs (range cell length of 150 m), wavelength of 10 cm (transmitter frequency of 3,000 MHz), and radial velocity of 10 m/s. Then the Doppler frequency is 20 Hz and one period of the Doppler signal is 1/20 s, or 50,000 μs. Since the signal interacts with the target for 1 μs, only 1/50,000 of a complete cycle of the Doppler signal can be generated and a frequency measurement is impossible. In this case the coherent integration time is the length of the pulse and we might guess that to measure velocity the coherent integration time must be greater than a few periods of the Doppler signal.

It is possible to change or enhance the sinusoid signal or the single short-pulse signal to improve the measurement capabilities. As might be expected, the basic idea is to make the signal either longer or more complex. The short-pulse signal is made longer by transmitting a sequence or train of short pulses that are coherent from pulse to pulse or can be processed in a coherent manner; i.e., the phase as well as the amplitude of the RF signal is used in the processing. For a sequence of pulses the coherent integration time is the number of pulses times the pulse period. Such a pulsed system for Doppler velocity measurements is generally called a pulse-Doppler radar. The range is obtained from the round-trip travel time, and the velocity is obtained from the mean frequency of the Doppler signal. The well-known Nyquist sampling theorem states that the highest frequency to be resolved in the signal must be sampled at least two times per period. If the signal contains energy at frequencies higher than the Nyquist frequency (i.e., half the sampling rate), the Doppler signal can be aliased by higher frequencies. This means that the energy at higher frequencies is "folded back" to frequencies less than the Nyquist frequency (McGillen and Cooper, 1974, pp. 167–170). For the case considered above, the Doppler signal has a period of 50 ms, and to measure its frequency at the output of the radar would require a sample every 25 ms. Therefore, the transmitter must be pulsed every 25 ms. Furthermore, we cannot determine within which 10-m/s interval the Doppler signal is contained; thus, the velocities are ambiguous. The maximum unambiguous velocity has a Doppler signal with a period equal to two transmitter pulse periods. In terms of pulse repetition frequency f_p, the maximum unambiguous velocity V_m is shown to be $f_p \lambda/4$.

The ambiguity in velocity comes directly from the sampling theorem and the sampling theorem would imply that it is desirable to have a high pulse repetition frequency. However, there is also an ambiguity in range, unrelated to the sampling theorem, that argues for a low pulse repetition frequency. A pulse that is reflected from a target and returns to the radar antenna just before the next pulse is transmitted is said to come from the maximum unambiguous range. Pulses returning from slightly greater ranges (beyond the maximum unambiguous range) are called second-trip echoes and are indistinguishable from weak return pulses at ranges slightly greater than zero. The maximum unambiguous range r_m is given by $r_m = c/2f_p$, where c is the speed of propagation of the radar signal. Clearly, maximum unambiguous range and velocity are related, as shown in Fig. 1.

The transmitted sinusoid signal can also be changed to improve some of the shortcomings pointed out above, but in this case we do not need to make the signal longer as in the pulse case. Rather we need to add more structure to the signal. There are a number of different ways of doing this. The two approaches that have been used in meteorological radar are linear frequency modulation and pseudo-random phase modulation. These approaches change the frequency or the phase of the transmitted signal in a way that can be easily recognized in the return signal. This allows ranging measurements to be made even with high-duty-cycle wave forms. The signal processing is more involved than for the pulse-Doppler technique but the same type of ambiguity problems arise. Range and velocity ambiguities that exist for pulse-Doppler systems apply also to high-duty-cycle frequency or phase modulation systems with just one difference: the pulse repetition frequency f_p now becomes the modulation repetition frequency. This is the frequency at which the frequency- or phase-modulation function or code is repeated. So we see that there is a close parallel between a sequence of pulses and a sequence of modulation functions or codes.

The ambiguity characteristics of radar signals can be illustrated as a two-dimensional surface that represents an antenna pattern. This is called the ambiguity function and shows how a particular radar signal weights the target in range-velocity space just as an antenna pattern weights the target in the two cross-beam dimensions. The ambiguity function is discussed in detail by Deley (1970) and many graphic examples for different types of radar signals are given by Rihaczek (1969).

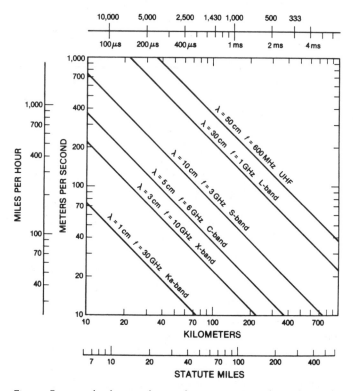

Fig. 1 Range and velocity ambiguity characteristics for radar with periodic transmitter modulation of T. The horizontal axis is the maximum unambiguous range and the vertical axis is the maximum unambiguous velocity.

1.3 Radar Equation

The radar equation, which relates the received power to the target radar cross section or reflectivity, is the basic equation of radar performance. Assume that P_T (watts) is transmitted and radiated isotropically over a large sphere centered at the radar location. If the sphere has radius r, then $4\pi r^2$ is the surface area over which the power is distributed. So the power density (watts per square meter) at range r from an isotropic radiator is $P_T/4\pi r^2$. However, radar antennas are not isotropic radiators; rather they are designed to concentrate or "beam" the energy in a desired direction. The amount by which they concentrate the energy in a given direction is called the antenna gain over isotropic, or simply gain. Suppose that the transmitting antenna has gain G_T in a given direction; then the power density at range r in that direction is $G_T P_T/4\pi r^2$. The target scatters some power back to the receiver. To quantify that scattering, assume that at the target an area σ, in square meters (which may or may not be related to the physical cross-section area of the target), collects the incident power $\sigma G_T P_T/4\pi r^2$. This power is reradiated *isotropically*; thus the power density back at the antenna is $(G_T P_T/4\pi r^2)(\sigma/4\pi r^2)$. This power is then collected by an antenna with effective aperture area A_e so the power at the receiver is

$$P_r = P_T A_e G_T \sigma/[(4\pi)^2 r^4] \qquad (1)$$

This is one form of the basic radar equation relating transmitted and received power. All of the quantities are quite straightforward except for the radar cross section σ, which is a characteristic of the target not necessarily simply related to its physical cross section. For meteorological radar we are more interested in beam-filling targets (see the overview article by Schwiesow in this volume), so we generally replace radar cross section σ, in square meters, with radar cross section per unit volume η, in square meters per cubic meter, or per meter. So σ is $V\eta$, where V is the volume of the range cell under consideration. If the antenna beam is rectangular, with beam widths of ϕ and θ in the two orthogonal directions, the beam area at range r is a rectangle $r\phi$ by $r\theta$. If the range cell length is Δ, then the resolution volume at range r is

$$V = \Delta\, r\phi\, r\theta = \Delta\, r^2\, \phi\theta \qquad (2)$$

We can simplify the radar equation to that used for distributed targets if the same (or identical) antenna is used for transmitting and receiving. The antenna gain is replaced by using an expression relating gain to beam width, i.e., $G_T = 4\pi/(\phi\theta)$. Then the radar equation for distributed targets becomes

$$P_r = (P_T A_e/4\pi)(\Delta/r^2)\eta \qquad (3)$$

This equation is frequently adjusted slightly because antenna beams are not rectangular. The most widely used adjustment is due to Probert-Jones (1962), who assumed that the beam pattern was Gaussian rather than sharp-edged. The effect of this assumption is to reduce the right side of the radar equation by the factor $2\ln 2$.

A second correction, especially for shorter wavelengths, is to compensate for the propagation signal loss. This loss is a complicated function of wavelength, atmospheric constituents, and elevation angle and it is generally available from graphs in units of decibels per kilometer [dB $\equiv$ $10\log_{10}(P_r/P_T)$], although in Eq. (3), the loss is expressed in terms of an atmospheric loss $L_{at} \geq 1$. With these two corrections the radar equation becomes

$$P_r = 0.0574 P_T A_e(\Delta/r^2)(\eta/L_{at}) \qquad (4)$$

Note that the received power is a product of three different factors involving, respectively, radar, geometry, and atmosphere. The radar factor is the power-aperture product $P_T A_e$, generally expressed in watts times square meters. The effective aperture area is related to the physical aperture by aperture efficiency ϱ_e, so that if A is aperture area, then $A_e = \varrho_e A$. For most antennas the aperture efficiency is in the range 0.4 to 0.7. Generally, the power-aperture product is given in terms of physical aperture even though effective aperture is the more important quantity. The second factor in the radar equation is geometry expressed as range cell size over range squared. Note that range has the greater effect so that a suitably designed radar can operate at short ranges in the atmospheric boundary layer and detect very small values of η. The final factor depends on the condition of the atmosphere in the antenna beam. Later sections will relate this radar cross-section term to appropriate meteorological terms.

1.4 Signal-to-Noise Ratio

The signal-to-noise ratio (S/N) is useful because it is an indication of how accurately a radar can measure various parameters of the target under investigation. Also it gives a measure by which the minimum detectable signal, and hence minimum detectable reflectivity, can be estimated. Signal-to-noise ratios exist in different forms and can be specified at different points in a radar system. Because S/N is widely used (and sometimes misused), it is well to always be sure of its precise definition. Here, S/N is the peak signal power at the output of the receiver (before detection) divided by the average noise power at the same point. This is the most widely used definition of output S/N.

In the normal radar case, two competing wave forms are at the input of the receiver: one is the return signal $s(t)$, and the other is unwanted noise from various sources. The noise is generally wide-band with spectral height N_0 (watts per hertz), and the customary assumption is that N_0 is a constant, i.e., "white" noise. While this is not exactly correct (because it implies infinite noise power), over any reasonable band it is usually a very good assumption. Thus we will assume that the noise is white and Gaussian. In this case, the optimum receiver, i.e., the receiver that maximizes the signal-to-noise ratio, is widely known to be a matched filter. The form of a matched filter is given in radar textbooks, and most microwave radar receivers closely approach matched filters within 1 or 2 dB. So we will assume that the receiver is optimum. The signal-to-noise ratio from an optimum receiver for the assumptions given above is well known to be

$$S/N = 2E_s/N_0 \qquad (5)$$

where E_s is the energy in the signal, defined by

$$E_s = \int_0^T s^2(t)\, dt = TP_r \qquad (6)$$

where P_r is the received power averaged over the coherent integration time T. In addition to affecting signal-to-noise ratio, T also affects the velocity resolution of Doppler radars. The signal-to-noise ratio then becomes

$$S/N = 2TP_r/N_0 = (2T/N_0)(P_T A_e/17.42)(\Delta/R^2)(\eta/L_{at}) \qquad (7)$$

1.5 System Noise

If we are able to determine N_0 in the above equation, we will have enough information to calculate S/N and get an idea of radar performance. Most low-noise receivers used in radars are best characterized in terms of noise temperatures. If the receiver were noise-free (which it is not) and if the input to the receiver were terminated (loaded with matching impedance) then

$$N_0 = kT_0 \qquad (8)$$

where k is Boltzman's constant (1.38×10^{-23} J/K) and T_0 is the physical temperature of the termination (kelvins). For room temperature, $kT_0 = 4 \times 10^{-21}$ J and the noise temperature is 290 K. However, the receiver is not noise-free and introduces noise according to its noise figure F, which effectively gives a new temperature, called the receiver effective noise temperature, according to

$$T_{re} = (F - 1)T_0 \qquad (9)$$

This is a fictitious temperature at which the input termination would produce enough noise to account for all of the receiver noise. Manufacturers may give either the noise figure or the effective noise temperature to characterize the receiver.

The system noise has contributions from the receiver, from the RF input hardware, from the atmosphere within the beam, and from space beyond. These are called, respectively, receiver noise, RF hardware noise, atmospheric noise, and cosmic noise. Like the noise figure, these are treated by defining an equivalent temperature called the system noise temperature, which is given by

$$\begin{aligned}T_e &= T_c + (L_{at} - 1)T_{at} \\ &+ (L_{rf} - 1)T_{rf}L_{at} + T_{re}L_{rf}L_{at} = k^{-1}N_0\end{aligned} \qquad (10)$$

where T_c = cosmic noise temperature
T_{at} = temperature of the atmosphere
T_{rf} = temperature of RF input hardware
L_{at} = loss due to atmospheric absorption ($L_{at} > 1$)
L_{rf} = loss due to RF input hardware ($L_{rf} > 1$)
T_{re} = effective noise temperature of receiver.

Note that if there is no atmospheric loss (i.e., $L_{at} = 1$) and no RF hardware loss ($L_{rf} = 1$), then $T_e = T_c + T_{re}$.

The first two terms in Eq. (10) represent the cosmic noise and the atmospheric noise and the combination of these two terms is called the antenna temperature. For microwave frequencies the antenna temperature ranges from about 200 K at zero elevation angle to about 50 K at elevation angles of 20° or more when directed toward space. It is about 290–320 K when directed toward the earth except over water.

1.6 Resolution and Accuracy

In addition to measuring radar cross section, radars can also measure target parameters such as range, velocity, etc., with a certain resolution and accuracy. Resolution is the ability to separate closely spaced targets in space or velocity while accuracy is the root-mean-square (rms) error of estimates of the given parameter (it is assumed that the estimate is unbiased). Resolution is independent of signal-to-noise ratio while accuracy is not.

Range resolution is the ability to separate return signals in time. This is essentially the width of the autocorrelation function for that signal. For any signal of bandwidth B, the range resolution Δ is given by

$$\Delta = c/(2B) \qquad (11)$$

This expression is true for any radar signal, but if the signal is a rectangular pulse of width τ then an equivalent expression is

$$\Delta = c\tau/2 \qquad (12)$$

Velocity resolution is the ability to separate return signals that are closely spaced in Doppler frequency. Better frequency resolution δ is related to an increase in the coherent integration time T, according to the relation

$$\delta = \lambda/(2T) \qquad (13)$$

Angular resolution is the ability to separate return signals that are only slightly separated in space. For the types of radars under discussion here the angular resolution depends on the size of the antenna aperture in wavelength. For a circular aperture, the angular resolution or beam width is given by

$$\theta = 70\lambda/D \text{ (in degrees)} = 1.22\lambda/D \text{ (in radians)} \tag{14}$$

where D is the antenna diameter. Note that the angular resolution is related to gain. Larger antennas have narrower beam widths and higher gain.

The accuracy of estimates made by radars is more complicated than the resolution because accuracy depends on signal-to-noise ratio and on the wave form in a more complicated way. For a complete discussion see Swerling (1970). Here we present expressions for a lower bound on the accuracy of measurements of return signal amplitude A_r, time delay τ, and Doppler frequency δf. These are called Cramer-Rao lower bounds and even though they are derived for point targets, they still represent lower bounds for atmospheric targets. We assume that the estimators of these quantities are unbiased so that the accuracy or rms error is the standard deviation, given by

$$\sigma_{A_r} \geq A_r/\sqrt{S/N} \tag{15}$$

$$\sigma_\tau \geq 1/(2\pi\beta\sqrt{S/N}) \tag{16}$$

$$\sigma_{\delta f} \geq 1/(2\pi t_0\sqrt{S/N}) \tag{17}$$

The terms β and t_0 are the effective bandwidth and the effective time duration, respectively, and can be considerably different from conventional values of bandwidth and time duration. For a rectangular pulse of duration T, the value of t_0 is $T/\sqrt{12}$. For precise definitions of β and t_0 see Swerling (1970).

Finally, consider a rectangular amplitude distribution on an antenna aperture of diameter D. The rms error of angle estimates in radians is bounded by

$$\sigma_\theta \geq (\lambda/\pi D)(3/\sqrt{S/N}) \tag{18}$$

The main point here is that the lower bounds on the accuracy (or rms error) for reflectivity, range, velocity, and angle are inversely related to the square root of the signal-to-noise ratio. Similar results hold for higher derivatives of range. So, for most measurement situations, it is important to keep the signal-to-noise ratio high.

1.7 Coherent and Incoherent Integration

Most radars have two types of integration to improve the output signal-to-noise ratio. The first of these is coherent integration (using a matched filter as discussed in the previous section), in which the receiver uses both phase and amplitude information. This type of integration is generally harder to implement than incoherent integration but is more efficient in that it gives a relatively geater signal-to-noise ratio. It is inversely related to the velocity resolution, so for velocity-distributed targets where the velocity resolution is less than the velocity spread of the target, S/N enhancement does not occur linearly with increased coherent integration time.

Incoherent integration is that which is done after the signal is detected, i.e., after the phase information is removed. The detection can be analog, with a diode and capacitor, or numerical, for example, by summing squares of amplitudes or other such algorithms. After detection, a phase reference is not necessary for signal processing, with the result that incoherent integration is easier than coherent integration. However, it is less efficient because the nonlinear detection process destroys some of the signal. The exact process is discussed by Brookner (1977). However, a simplified and useful rule of thumb for S/N enhancement by incoherent integration exists. For fluctuating targets, if M is the number of signals incoherently integrated then the increase in signal-to-noise ratio due to incoherent integration is approximately $M^{0.6}$ (Barton, 1964). For example, if the signal is a sequence of pulses and these are processed coherently by a Fourier transform, detected by summing amplitude squares, and then averaged, M would be the number of spectra averaged.

Incoherent integration can also increase the accuracy associated with measurements discussed in the previous section. For M signals incoherently integrated, the lower bound on the rms error is reduced by $M^{-0.5}$.

1.8 The Doppler Spectrum

The Doppler signal at the output of a Doppler radar is a single sinusoid for a single point target in the beam. If there are a large number of independent scatterers in the beam, as is the case for meteorological radar, then the Doppler signal is a random signal with a Gaussian distribution. Because of the nature of the receiver, the mean value of this Gaussian process is zero and so it is entirely characterized by its second-moment behavior. If $s(t)$ is the Doppler signal, then if we assume that it is a stationary random process, its autocorrelation function $R(\tau)$ is

$$R(\tau) = \overline{s(t)s(t + \tau)} \tag{19}$$

All of the information relating to the velocity distribution of the target is contained in the autocorrelation function. One approach to estimating mean velocity or spectral width is to estimate values of the autocorrelation function and use certain algorithms to estimate desired parameters. This technique is called pulse-pair processing because adjacent pairs of pulses are used to estimate values of the autocorrelation function. The main advantage of this technique is that it is amenable to real-time estimation of Doppler mean velocities. An analysis of the behavior of pulse-pair estimators is given by Zrnic (1977).

A second way of obtaining velocity is to estimate the power spectral density of the Doppler signal. Since the power spectral density is the Fourier transform of the autocorrelation function, no information is lost by the transformation. The power spectral density of the Doppler signal can be interpreted directly as the velocity distribution of the radar cross section within the range cell. Thus the output of a Doppler radar viewing a meteorological target is a sequence of range-ordered Doppler spectra.

In most applications, one curve (the Doppler spectrum) for each range bin is simply too much information to display and assimilate effectively. The general approach to this data reduction problem is to derive a single number or parameter from each spectrum and display that number as a function of range. Three parameters of the Doppler spectrum have been widely used: the area under the spectrum, the location or center of the spectrum, and the width of the spectrum. These are sometimes called zeroth, first, and second moments or area, mean velocity, and velocity variance, respectively.

If the target is beam-filling, the area under the Doppler spectrum is proportional to the total radar cross section of the illuminated target,

$$\sigma = \eta V = C \int_{-\infty}^{\infty} S(\mathscr{V}_R)\, d\mathscr{V}_R \qquad (20)$$

where $S(\mathscr{V}_R)$ is the spectral density of the Doppler signal and C is a constant which depends on the particular radar and how it is operated. This is the parameter used when a calibrated measure of radar cross section is needed. This parameter does not contain Doppler information and is the same as the output obtained from a conventional non-Doppler weather radar.

The second parameter derived from the Doppler spectrum is the mean of the spectrum, defined as

$$V_R = \int_{-\infty}^{\infty} \mathscr{V}_R S(\mathscr{V}_R)\, d\mathscr{V}_R \Big/ \int_{-\infty}^{\infty} S(\mathscr{V}_R)\, d\mathscr{V}_R \qquad (21)$$

This quantity is roughly the location of the center of the velocity spectrum, and is generally taken as the radial wind component. Very often this parameter is displayed on a PPI in the same manner as the radar cross section.

The third parameter taken from the Doppler spectrum is the spectral width, or velocity variance, defined by

$$\sigma_{V_R}^2 = \int_{-\infty}^{\infty} (\mathscr{V}_R - V_R)^2 S(\mathscr{V}_R)\, d\mathscr{V}_R \Big/ \int_{-\infty}^{\infty} S(\mathscr{V}_R)\, d\mathscr{V}_R \qquad (22)$$

The velocity variance is important for two reasons: First, it is important to know expected spectral widths when designing meteorological radars, and second, the velocity variance contains information on the rate of turbulence-related kinetic energy dissipation. An important feature of both V_R and $\sigma_{V_R}^2$ is that they are calculated from normalized spectra so radar received-power calibration is not important.

A number of different effects can cause broadening of the Doppler spectrum. Four such effects will be covered here and combined to yield a total spectral width. The first two effects result from the finite extent of antenna beam width and are generally undesirable. Even if the wind is uniform everywhere, a spectrum broadening effect arises because the radial velocity is different at the edges of the beam, which results from the difference of the angle with the wind at the beam edges. Nathanson (1969) gives this expression for beam broadening

$$\sigma_{beam}^2 = (0.42\, \mathscr{V}\theta \sin\beta)^2 \qquad (23)$$

where $\mathscr{V}$ is the wind velocity, θ is the two-way half-power azimuth beam width in radians, and β is the azimuth angle relative to the wind. For most narrow-beam radars, this effect is usually small.

The second effect is called shear broadening and is caused by the difference in radial velocity due to vertical wind shear across the beam. Stoss and Atlas (1968) show that for Gaussian beam shape the shear broadening is given by

$$\sigma_{shear}^2 = (S_r R\phi)^2 / 2.76 \qquad (24)$$

where S_r is the shear (per second), R is range, and ϕ is the half beam width to the half-power point in the elevation beam pattern. Nathanson (1969) gives a discussion of expected values for shear broadening.

The third spectrum broadening effect is caused by small-scale atmospheric turbulence imparting random velocities to the radar cross section. This effect, called turbulence broadening, is related to the turbulence dissipation rate ϵ by

$$\sigma_{turb}^2 = K_1 \epsilon^{2/3} \qquad (25)$$

where K_1 is a function whose form depends on the geometry of the resolution cell. This spectral broadening effect allows an estimate of a small-scale turbulence parameter. The analysis of turbulence broadening is reviewed by Gossard and Strauch (1983).

If each of these effects is assumed to be independent, then the broadening terms combine to produce the overall spectral broadening

$$\sigma_{V_R}^2 = \sigma_{beam}^2 + \sigma_{shear}^2 + \sigma_{turb}^2 \qquad (26)$$

There are other spectral broadening effects which add in a similar way; see Nathanson (1969) and Gossard and Strauch (1983).

2 CLEAR-AIR RADAR RESEARCH PROGRAMS

2.1 Past Programs and Results

The use of radar to remotely sense the clear atmosphere is not new; electromagnetic signals verified the existence of the ionosphere in the 1920s before the "invention" of higher-frequency radar in the late 1930s (Appleton and Barnett, 1925; Breit and Tuve, 1926). However, radars for sensing the troposphere and stratosphere have become widespread only in the last decade. Many specific aspects of radar remote sensing of the non-ionized, clear atmosphere are in a recent special issue of *Radio Science*, edited by Gossard and Yeh (1980).

The radio propagation community first became aware of the importance of atmospheric Bragg scatter when experimenters found that one-way VHF signals beyond the horizon decreased much more slowly than the diffraction field around a spherical earth. The effect was sometimes caused by ducting, or by partial reflection from layers of sharp refractive index gradient (Friis et al., 1957) but the explanation for most of the observations came with the publication of the Booker-Gordon theory (Booker and Gordon, 1950) which identified Bragg scatter (from atmospheric volumes high enough in the atmosphere to be common to the antenna beams of both the transmitter and the receiver) as the important contributor to over-the-horizon fields. This kind of scatter is very angle-dependent; the power decreases approximately as $\theta^{-14/3}$, where θ is the scattering angle off the incident direction. Therefore, recognition of its importance in radar backscatter was rather slow to come. Since the advent of short-wavelength radars in World War II, reports of mysterious returns from regions where there were apparently no targets had generated a substantial amount of literature on "angels," "ghosts," and "pixies." There was an ongoing controversy on whether these returns resulted from special features, such as refractive index bubbles, or from birds and insects, and this is well summarized by Plank (1956). In some cases layer structures in the troposphere were present (Lane and Meadows, 1963), and the tropopause was sometimes seen (Atlas et al., 1966). Insects and birds caused some of the return (Glover and Hardy, 1966). However, it was not until the development of an atmospheric radar capable of observing targets at very short range with very high range resolution, the frequency-modulated continuous-wave (FM-CW) radar reported by Richter (1969), that the situation was really clarified. This radar made it clear that both Bragg scatter from the clear air and return from insects are important, depending on climatic regime, time of day, season, and altitude. In Fig. 2 the point returns from the insects are clearly distinguishable from the diffuse return from the layers.

In the late 1960s special sensitive radar systems became available and were used for clear-air observations. A series of radar experiments at Wallops Island, Virginia, begun in the mid-sixties, clearly demonstrated the potential of high-powered radars for boundary-layer research. Three wavelengths, $\lambda = 3.2$ cm (X-band), 10.7 cm (S-band), and 71.5 cm (UHF), were used in these experiments; the simultaneous use of three wavelengths was extremely useful in identifying the various scattering mechanisms. Because of its sensitivity to refractive index fluctuations, its transmitted power, and its spatial resolution, the S-band radar at Wallops Island was most useful for revealing the structure of the boundary layer.

One of the first experiments at Wallops Island was performed during the summer of 1966. A microwave refractometer was suspended from a helicopter and tracked with the S-band radar, a monopulse tracking radar. A data gate, separate from the tracking gate, permitted the reflectivity or cross section per unit volume to be measured without contamination by the echo from the aircraft. In this way, microwave refractive index and S-band reflectivity were measured from nearly the same air parcels at nearly the same time.

Kropfli et al. (1968) utilized the expression of Saxton et al. (1964),

$$\eta = (\pi/8)\chi_r{}^2\sigma_n{}^2F_n(\chi_r) \qquad (27)$$

and the spectral form for refractivity in the presence of homogeneous, isotropic turbulence given by Tatarskii (1961),

$$F_n(\chi) = (2/3)\chi_o{}^{2/3}\chi^{-5/3} \qquad (28)$$

to relate the two measurements. In these expressions, χ_r is the radian wave number corresponding to half the radar wavelength, $\sigma_n{}^2$ is the variance of the refractive index, $F_n(\chi_r)$ is the normalized one-dimensional wave-number spectrum evaluated at $\chi = \chi_r = 4\pi/\lambda$, and χ_o is the wave number of the outer scale of the inertial subrange. The result of this work suggested that S-band radar reflectivity in the planetary boundary layer (PBL) is in fact determined by the refractive index fluctuations at a scale of one-half the radar wavelength as implied by Eqs. (1) and (2). In other words, the radar acts as a narrow-band filter; it senses only a narrow band of the refractive index spectrum centered at $\chi = \chi_r$.

Another program at Wallops Island studied the structure of the convective boundary layer. Figure 3 shows convective cell structure in plan view (top frame) and in vertical cross section (bottom frame). It shows clearly a dome-like or "hummock" structure capping the convective features of the boundary layer. A clear-air layer between 2 and 3 km is also evident in the bottom frame. Readings et al. (1973) proposed a model for this kind of convective hummock (Fig. 4) based on radar observations accompanied by captive balloon soundings. Such observations were quite common at Wallops Island in the summer months during undisturbed daytime conditions. These photographs show how the individual convective elements are strongly reflecting near their tops, where the largest refractive index fluctuations exist, because of entrainment of warm, dry air from above the boundary layer (Wyngaard and LeMone, 1980) and mixing with cool, moist boundary-layer air (Grant, 1965; Konrad, 1970).

Scattering from the clear atmosphere is generally proportional to the variance of fluctuations in radio refractive index N produced by fluctuations in temperature θ and partial pressure of water vapor e (Gossard, 1960):

$$\overline{n^2} = a\overline{\theta^2} + b\overline{e^2} - c\overline{\theta e} \qquad (29)$$

where a, b, and c are constants which depend on atmospheric

16 APRIL 1972

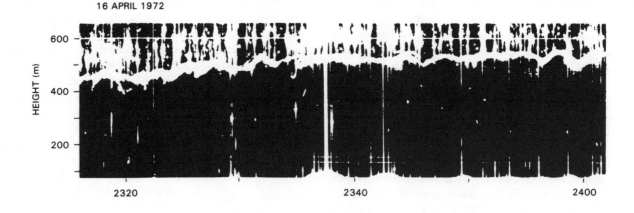

9 FEBRUARY 1973

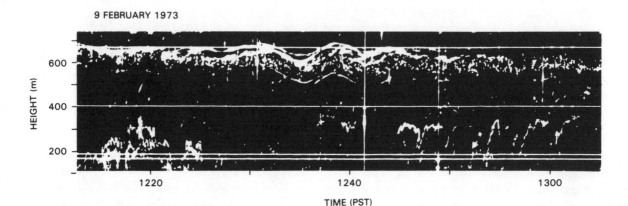

TIME (PST)

Fig. 2 Clear-air backscattered power observed with a vertically pointing FM-CW radar at San Diego. Bragg backscatter from convective features in the boundary layer and from an elevated inversion illustrated by the diffuse, laminar type of return. The dots and vertical streaks result from particulate targets, probably insects and birds. The horizontal straight lines are range markers. (From Richter et al., 1973.)

Fig. 3 Top frame: Plan view of features in bottom frame obtained with PPI scans of the radar, showing doughnut-like appearance of the convective features in cross section. Bottom frame: Dome-like echoes at Wallops Island reported by Hardy and Ottersten (1969). They are height-vs-range plots obtained from RHI radar scans.

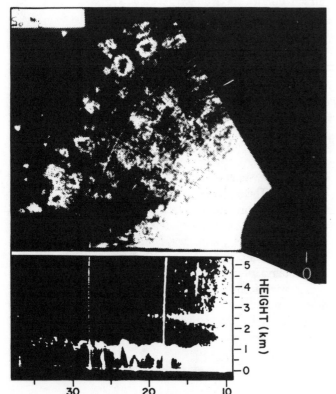

RANGE FROM RADAR (km)

conditions. The correlation term can be either positive or negative and thus either decrease or increase the backscattered power. Wyngaard et al. (1978) show that in the lower boundary layer this correlation is positive and in the upper boundary layer it is negative. As a consequence, the refractive index fluctuations, and therefore the reflectivity, are enhanced near the top of the PBL, as the Wallops Island observations bear out. Furthermore, evidence given by Rowland (1973) indicates that the cool, moist regions of the convective field were the *only* regions visible to the Wallops Island radars.

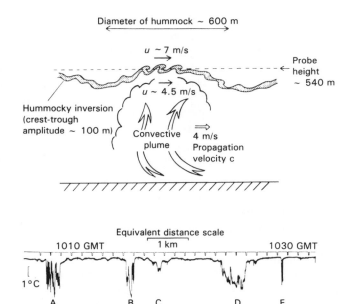

Fig. 4   *Fine-scale structure and mixing within an inversion. Top frame: Model of a hummock as derived by Readings et al. (1973). Lower frame is temperature trace showing the locations of bursts of temperature fluctuations.*

Figure 5 is a schematic diagram showing the variations of η through a convective cell at the indicated levels. Although no ratio of cell diameter to PBL height was reported in the Wallops Island results, cell diameters were described as ranging from 0.5 to 3 km, which were generally comparable with z_i, the PBL height. Although most published figures from the Wallops Island experiments indicate that the central core of a convective cell is relatively echo-free, most of these scope photographs at Wallops Island were taken with the signals deliberately attenuated by 10 or 15 dB in order to clearly show the reflectivity structure near the cell boundary. It was not unusual to observe reflectivity values as much as 15 to 20 dB above the minimum detectable level during these experiments. Table 1 summarizes physical properties of convection cells observed at Wallops Island.

The Wallops Island experiments detected convection cells by changes in reflectivity, but they can also be detected by changes in velocity. Chadwick et al. (1976a) show a case where an FM-CW Doppler radar is measuring radial wind speed at 45° elevation angle. The periodic changes in radial wind occur about every 8–10 min and are the result of convection cells being advected through the beam. In the center of the convection cell, the velocity is upward and at the edge it is downward. Because the cell is being advected toward the radar, the center of the cell shows up as nearly zero radial velocity and the edge shows an inward radial velocity. Frisch et al. (1976) give examples of convection cell detection by other methods.

Although it was generally difficult to identify specific sources of convection, on one occasion new cells repeatedly appeared in the clear air over an airfield at about 4-min intervals (Arnold, 1976). Also, the alignment of cells was usually

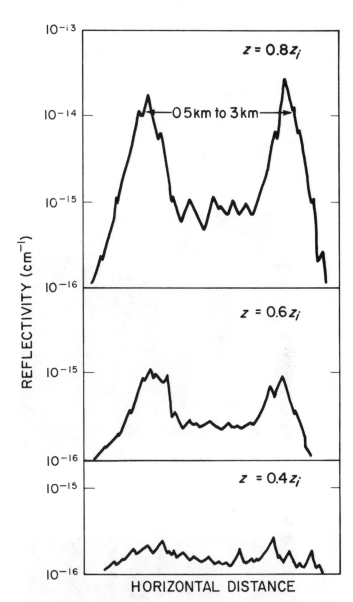

Fig. 5   *Values of radar cross section through convection cell.*

TABLE 1
Physical Properties of Convective Cells Observed at Wallops Island

Property	Value	Source
Size	0.5–3 km	Konrad (1970)
Height	2.5 km	Konrad (1970)
Field growth rate	0.8 m/s	Konrad (1970)
Cell growth rate	0.5–1.5 m/s	Hardy and Ottersten (1969)
% area covered by cells	50% at $z/z_i = 0.8$	Konrad and Robison (1972)
Typical reflectivity	10^{-15} cm^{-1}	Kropfli et al. (1968)

observed to be random, unless there was a decided curvature of the PBL wind speed profile and no change in wind direction. In such cases, thermal "streets" were observed (Konrad, 1968). A detailed example of thermal streeting is also described by Doviak and Berger (1980), in which the PBL wind curvature was large ($|\partial^2 \mathbf{V}/\partial z^2| \cong -4 \times 10^{-7} \text{ m}^{-1} \text{ s}^{-1}$). These observations are consistent with the theory of Kuettner (1971), who deduced that high curvature is necessary to suppress circulation in the plane along the wind and to amplify the cross-wind circulation.

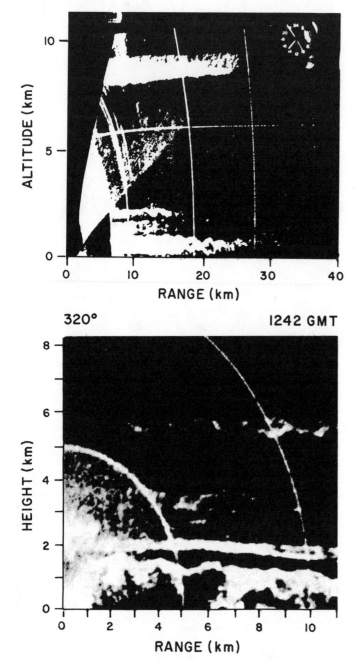

Fig. 6 Range-height display showing braided structure resembling the Kelvin "cat's eye" believed to result from Kelvin-Helmholtz instability. Top frame: Wallops Island radar reported by Katz (1972). Bottom frame: Defford radar reported by Browning (1971).

Perhaps the features most dramatically revealed by clear-air radars are internal waves, breaking billows, and vortices. Figures 6 and 7 show examples. In the top frame of Fig. 6 the layer at about 8 km is a cirrus deck; the bottom frame is from the Royal Air Force radar at Defford, United Kingdom. Figure 7 shows a record of a mountain lee wave from the same radar (Browning, 1971). The FM-CW radar (Richter, 1969) displayed the fine-scale structure of waves and vortices in remarkable detail, as shown in Figs. 2 and 8. Figure 9 (Gossard et al., 1982) shows, with 6-m height resolution, the fine structure of the stable layer capping the convective boundary layer along with height profiles of temperature, wind, and potential refractive index (refractive index calculated with potential temperature instead of temperature) obtained by balloon sounding. The hummocks, Kelvin-Helmholtz instabilities, and fine structure of the capping layer are revealed in great detail by this radar. Atmospheric flow patterns have also been visualized (Bean et al., 1971). This radar has several advantages over pulse radars for boundary-layer investigation: (a) it achieves very high range resolution easily and fairly inexpensively; (b) it has a very short minimum range; (c) it has high average power without a large transmitter; (d) it can operate in a high-clutter environment more effectively than a radar with high peak power.

These characteristics clearly revealed the potential of this kind of radar for sounding the lower atmosphere; however, the potential was not fully exploited because of the success of the corresponding acoustic "radar" systems. The acoustic systems had two very attractive features: they were much less expensive than radars, and it was easy to incorporate a Doppler capability into the acoustic system, whereas it was widely believed to be impossible to achieve Doppler in an FM-CW radar operating on a distributed target such as the atmosphere. This belief prevailed because this type of radar

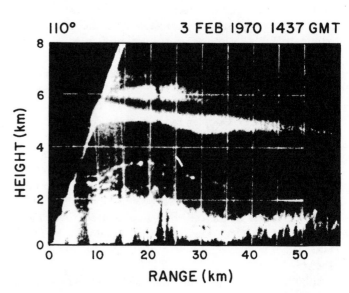

Fig. 7 Range-height display of backscattered power recorded by Defford radar showing large-amplitude lee wave. Two overlapping sets of range markers represent horizontal and slant range.

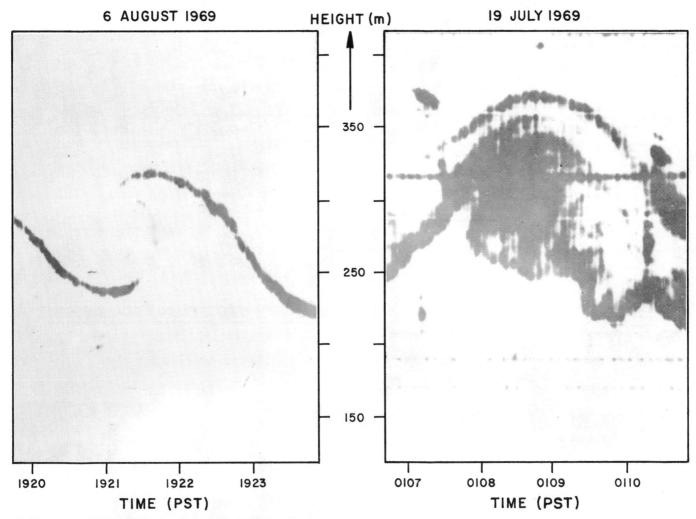

6 AUGUST 1969   HEIGHT (m)   19 JULY 1969

350

250

150

1920   1921   1922   1923   0107   0108   0109   0110

TIME (PST)   TIME (PST)

Fig. 8   Blowup of vortex and breaking wave, illustrating the value of the fine resolution in height (1.5 m) of which the FM-CW radar is capable. (From Gossard et al., 1970.)

already uses the difference between the frequency of the returning signal and that of the transmitted signal to measure the range of the target. However, in 1976 a successful Doppler FM-CW system was reported by Strauch et al. (1976) and Chadwick et al. (1976a), and one advantage of the acoustic systems vanished. Furthermore, it was becoming clear that the acoustic systems had an important limitation; they were fundamentally limited in range (and therefore altitude) by atmospheric absorption (see the article by Neff and Coulter in this volume).

In the meantime a development was occurring that would broaden the whole concept of radar sounding of the atmosphere. At the suggestion of C.G. Little, the large ionospheric research radar at Jicamarca, Peru, operating at a frequency of up to 50 MHz with an antenna aperture of 8.5×10^4 m², was used to detect returns from the non-ionized stratosphere. Success was reported by Woodman and Guillen (1974) and major programs in radar investigation of the non-ionized mesosphere, stratosphere, and troposphere (MST) became

global in scope as ionospheric physicists using facilities such as the radio/radar observatory at Arecibo, Puerto Rico, redirected their attention toward the "lower atmosphere," i.e., altitudes below 40 km.

2.2 Current Clear-Air Research Programs

Various pulse-Doppler radar systems designed for the study of storms (where the targets are large hydrometeors) are useful in clear-air research. Some of these are shown in the top portion of Table 2. Some of these radars have been used in pairs or multiple-radar complexes, greatly adding to the value of the information obtained. In the United States such dual-Doppler systems are operated by the National Severe Storms Laboratory of the National Oceanic and Atmospheric Administration (NOAA) at Norman, Oklahoma; the National Center for Atmospheric Research at Boulder, Colorado; the NOAA Wave Propagation Laboratory at Boulder, Colorado; and the University of Miami in Florida. The technique has been applied by Doviak and Jobson (1979) to natural returns from the clear boundary layer; they attribute their signals to Bragg backscatter from

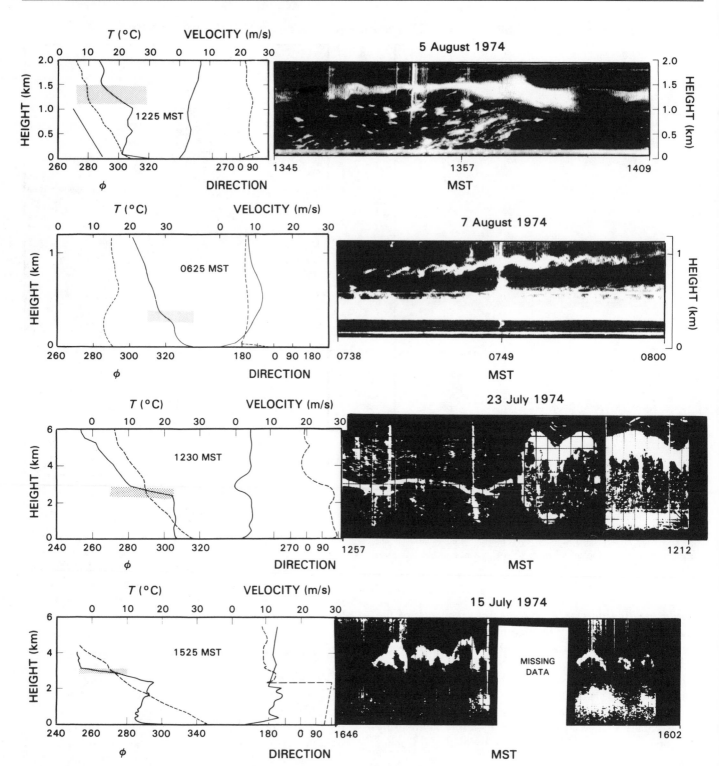

Fig. 9 Clear-air backscattered power from layers capping the convective boundary layer. Profiles are rawin temperature (dashed), radio potential refractive index (solid), wind direction (dashed), and wind speed (solid). In frame 3 the range scale was changed mid-way through the record, after which it matches the height scale of the profiles. The point targets in the top frame are insects being entrained into a convective updraft. If the wavy layer in frame 2 is extrapolated back to 0625 (the time of the rawin), its height matches the stippled zone of large gradient of potential refractive index. (From Gossard et al., 1982.)

TABLE 2a
Cloud- and Storm-Sensing Radars

	Pulse Doppler WPL	Pulse Doppler NSSL	Pulse Doppler Defford	Pulse Doppler Wallops Is	WSR-57 (Hook Model)	WSR-57	K-Band Doppler WPL	TPQ 11	Chatanika Radar
P_T (W)	0.9×10^5	4.7×10^5	3×10^5	2×10^6	4.1×10^5	4.1×10^5	1.5×10^5	1.2×10^5	3.2×10^6
P_r (dBm)	-106	-111.2	-110	-110	-100	-108	-106	-99	-128
λ (cm)	3.22	10.7	10.7	10.7	10.3	10.3	0.86	0.86	23.25
A_e (m^2)	4.02	31	270	146	5.8	5.8	.64	1.8	310
Δ (m)	75	150	30	195	75	600	37.5	75	1500
R (km)	10	10	10	10	10	10	10	10	10
η (m^{-1})	2.7×10^{-12}	9.8×10^{-15}	1.1×10^{-14}	5×10^{-16}	1.6×10^{-12}	3.1×10^{-14}	2.2×10^{-11}	2.2×10^{-11}	3.0×10^{-19}
$C_n{}^2$ (cm$^{-2/3}$)	1.0×10^{-13}	5.6×10^{-16}	6.8×10^{-16}	2.9×10^{-17}	9.0×10^{-14}	1.8×10^{-15}	5.5×10^{-13}	5.5×10^{-13}	2.2×10^{-19}
$C_n{}^2$ (m$^{-2/3}$)	2.2×10^{-12}	1.2×10^{-14}	1.5×15^{-14}	6.2×10^{-16}	1.9×10^{-12}	3.85×10^{-14}	1.2×10^{-11}	1.2×10^{-11}	4.9×10^{-19}
Z (mm^{6}m^{-3})	1.0×10^{-2}	4.5×10^{-3}	5.4×10^{-3}	2.3×10^{-4}	0.625	1.2×10^{-2}	4.2×10^{-4}	4.2×10^{-4}	3.1×10^{-6}
τ (μs)	0.5	1	1.25	1.3	0.5	4	0.25	0.5	10
T (μs)	512	768	1250	960	1520	6097	500	1000	3400
Antenna (m)	3.05	8.5	25	18.4	3.65	3.65	1.12	2.14	26.8
Beam width (deg)	0.86	1.1	0.33	0.50	0.50	2.0	0.5	0.25	0.6

(The Min bracket groups the rows η, $C_n{}^2$ (cm$^{-2/3}$), $C_n{}^2$ (m$^{-2/3}$), Z.)

TABLE 2b
Clear-Air Radar Wind Sounders

	FM-CW WPL	Meteorological Profiler WPL	Sunset, Colorado	SOUSY W. Germany	Poker Flat, Alaska
P_T (W)	200 (av)	1.5×10^3 (av)	1.25×10^5	6×10^5	0.15×10^5
λ (cm)	10.2	32.8	741	561	600
Δ (m)	100*	900**	1000	50	750
R (km)	10	10	10	10	10
η (m^{-1})	1.8×10^{-14}	4.1×10^{-17}	5.9×10^{-19}	2.1×10^{-20}	1.0×10^{-18}
$C_n{}^2$ (cm$^{-2/3}$)	9.9×10^{-16}	3.5×10^{-18}	1.4×10^{-19}	4.6×10^{-21}	2.3×10^{-19}
$C_n{}^2$ (m$^{-2/3}$)	2.1×10^{-14}	7.5×10^{-17}	3×10^{-18}	1×10^{-19}	5×10^{-18}
Z (mm^{6}m^{-3})	6.2×10^{-3}	1.7×10^{-3}	6.2	7.5×10^{-2}	4.8
Antenna (m)	2.44	10	60 x 30	62	100 x 50
Beam width (deg)	2.7	2.3	5 x 9	5	2 x 4
Remarks	Sweep length 50 ms	Integration time 4.6 ms	Integration time 60 ms	Integration time 100 ms	Integration time 60 ms

(The Min bracket groups the rows η, $C_n{}^2$ (cm$^{-2/3}$), $C_n{}^2$ (m$^{-2/3}$), Z.)

*Typical. **Maximum resolution is 1 m.**
**High altitude mode. Low altitude, 100 m resolution.

the turbulent fluctuations in refractive index. Figure 10 shows (top frame) a PPI display of aligned convective rolls and (bottom frame) examples of calculated velocity fields; note the high correlation between successive patterns about 2.5 min apart, indicating that these clear-air features are persistent. Other results using the dual-Doppler technique in the clear boundary layer are given by Doviak and Berger (1980).

FM-CW radars are now used for atmospheric research at Delft University of Technology in The Netherlands and at NOAA's Wave Propagation Laboratory (WPL). The Delft system has been described by Ligthart (1980); the WPL

system (with Doppler capability) has been described by Chadwick et al. (1976a, b). Doppler spectra are shown with a time-intensity display in Fig. 11 for a weather front at a height of about 1,500 m over Boulder. The width of the velocity spectrum is an indicator of turbulence intensity. The peaks of the spectra in the different range cells show no displacement from zero because the mean vertical velocity was zero. A further discussion of techniques for measuring gradients and fluxes in the boundary layer with clear-air radar is presented by Gossard et al. (1982).

Two other FM-CW radars are presently being constructed, one in West Germany and another in New Zealand. However, results from these radars have yet to be published.

The Joint Airport Weather Studies (JAWS) Program was an NCAR project (in conjunction with the University of Chicago) to investigate the wind-shear hazard near airports. Five- and 10-cm pulse-Doppler radars were used to characterize the convective boundary layer near Stapleton Airport, Denver, during the summer of 1982. This was an unusually

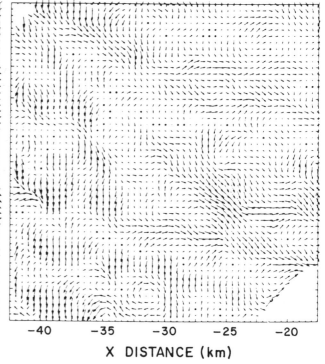

Fig. 10   Top frame: PPI contour display of reflectivity xR⁻² (R is range) from Doppler radar (27 April 1977, 1350 CST). Bright areas of higher reflectivity are aligned parallel to the mean wind and the bands are spaced about 4 km apart. Range marks are 20 km apart. Elevation angle is 0.8°. Bottom frame: Horizontal wind fields in the clear air measured by dual-Doppler radar without chaff. Note the high correlation between successive patterns about 2.5 min apart, indicating that the atmospheric features are persistent convective rolls. (From Doviak and Jobson, 1979.)

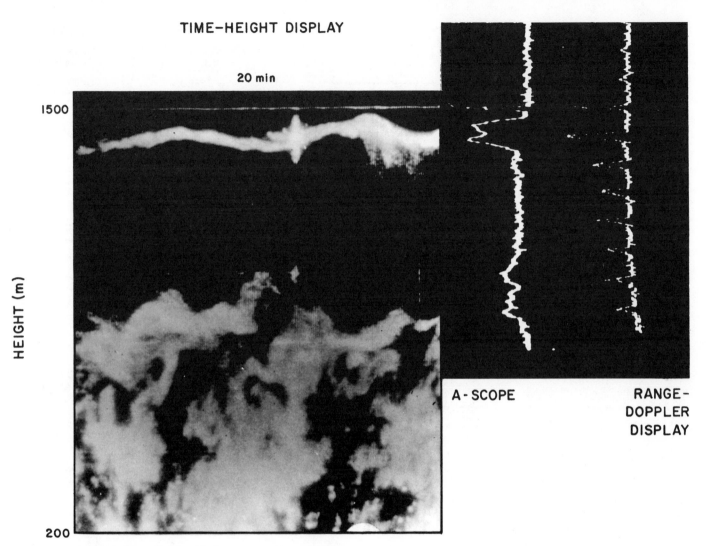

Fig. 11 Left frame: Height-time display of backscatter power from vertically pointing FM-CW radar. Right frame: A-scope display of backscattered power for a single scan compared with the Doppler spectra, in height increments of 87 m. The horizontal line at 1,500 m is a range reference signal.

productive experiment as nearly 200 shear events were detected by ground-based and remote sensors. The results demonstrate that radar *can* detect hazardous wind shear, although it was not clear whether the scattering was from refractive index fluctuations or insects. The shear events occurred mostly in the afternoon and seemed to be associated with, but distinct from, gust fronts. Some preliminary results were presented at the 21st Radar Meteorology Conference by McCarthy et al. (1983) and Wilson and Roberts (1983).

Several VHF radar facilities are now used in global programs to study upper-level winds and even though they do not often have minimum ranges in the boundary layer, they are briefly discussed here because a few can measure at heights within the boundary layer. Such systems are sometimes called MST or ST (mesosphere, stratosphere, and troposphere) radars because of their ability to sense wind and other

fields across several regions of the atmosphere. Radars used in studies of this kind are located at Arecibo, Puerto Rico; Lindau, Federal Republic of Germany; Jicamarca, Peru; Urbana, Illinois; Eniwetok, Western Pacific; Lincoln Laboratory, Massachusetts; Poker Flat, Alaska; Sunset, Colorado; and Platteville, Colorado. Some of the systems designed specifically for atmospheric wind sounding are described in the lower portion of Table 2.

Figure 12 shows wind profiles into the stratosphere, as measured by radar, compared with rawin balloon soundings from (a) Arecibo; (b) Chatanika, Alaska (L-band); (c) Sunset; and (d) Lindau. The ability of these radars to measure winds is now well established. The Poker Flat facility has been described by Balsley et al. (1980). The Jicamarca facility has been described by Woodman and Guillen (1974). The Arecibo facility and its use in the present context have been described by Woodman (1980a, b). The Lindau system has been described by Röttger and Schmidt (1979), and the Sunset facility has been described by Green et al. (1979).

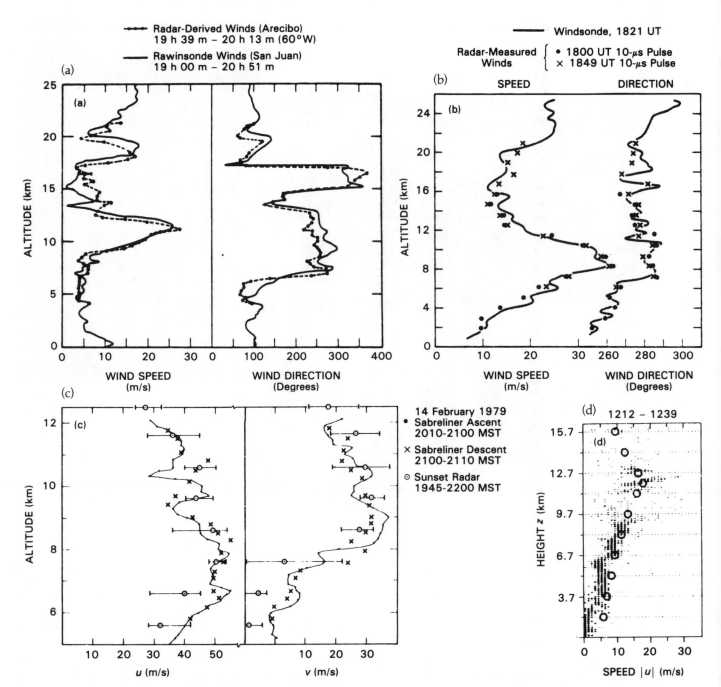

Fig. 12 Wind profiles into the stratosphere measured by radar compared with rawin soundings from (a) Arecibo, Puerto Rico; (b) Chatanika, Alaska; (c) Sunset, Colorado; and (d) Lindau, Federal Republic of Germany. In (d) the circles are balloon sounding observations. Comparisons such as these verify that the radar systems are measuring true winds.

3 FREQUENCY DEPENDENCE OF CLEAR-AIR RADARS

Two important parameters of clear-air radars are power-aperture product and frequency. The power-aperture product for a clear-air radar can vary from $10\,W \cdot m^2$ to $10^{10}\,W \cdot m^2$, and the frequencies are generally in the VHF and UHF regions of the spectrum. The radar reflectivity (radar cross section per unit volume) of the clear atmosphere is given by Ottersten (1969) as

$$\eta = 0.38 C_n^2 \lambda^{-1/3} \qquad (30)$$

where the structure parameter C_n^2 represents a measure of the variability of the refractive index field within the inertial subrange of turbulence. The important thing to note is that the clear-air reflectivity is predicted to be only weakly dependent on wavelength; this is quite different from the hydrometeor case where reflectivity is strongly dependent on wavelength. Because of the weak dependence on wavelength, other factors, such as sky noise, have a significant effect on system sensitivity at different wavelengths. This dependence can be visualized by noting the power-aperture product as a function of frequency required to attain a given output S/N for a given clear-air radar requirement. The requirement we have picked is to achieve an output S/N = 1 for a clear-air Doppler radar operating at 5-km range with 300-m range cell and 2-m/s velocity cell. We assume that the refractive index structure constant, C_n^2, is 3×10^{-16} $m^{-2/3}$ and that the radar takes an averaged velocity spectrum every 10 s. Further, we assume the electronic receiver noise temperature to be 300 K and independent of frequency. These characteristics might be required for a short-range radar designed to monitor a runway approach zone at an airport for hazardous wind shear (Chadwick et al., 1979). A plot of the power-aperture product to meet the requirement is shown in Fig. 13 as a function of frequency. Three different frequency effects are inherent in this plot: (1) the frequency dependence of sky noise, both cosmic and atmospheric; (2) the weak frequency dependence of the radar cross section of the clear air; (3) detector loss or the trade-off between coherent and incoherent integration.

Three additional points should be made concerning Fig. 13. First, although the initial reaction is that the VHF band would find little use for clear-air radar because of the large power-aperture product required, antenna apertures are much less expensive in the VHF band than at microwave frequencies. For example, an antenna with a 50-MHz aperture can be built for about $6/m^2 while it costs about $1,000/m^2 for 4,000 MHz. So, even though a larger power-aperture product is required at VHF than UHF, the VHF radar generally has a cost advantage.

Second, we have considered only the inertial subrange of turbulence, or turbulent scales for which kinetic energy dissipation by viscosity is negligible compared with energy transfer between scales. At the small-scale limit of the inertial

subrange, viscous forces eventually dissipate the turbulent kinetic energy as heat. The scale size where this happens is termed the Kolmogoroff scale of turbulence η_K (see the article by Friehe in this volume), and the range of smaller sizes is called the viscous subrange. Radars whose half-wavelength lies in the viscous subrange need additional power-aperture product beyond that predicted in Fig. 13 because some turbulent energy has been dissipated and is not available to cause refractive index fluctuations. The viscous cutoff is generally on the order of a few millimeters at ground level and increases with height because of an increase in the kinematic viscosity of the atmosphere. Figure 14, from Crane (1980), is a graph of the viscous cutoff as a function of height. The important point is that VHF should be used to sound to stratospheric heights, but UHF is a better choice for boundary-layer applications because of the greater spatial resolution available.

Third, a vertical VHF radar beam will sometimes experience enhanced backscatter from smooth horizontal layers. This is a specular, partial-reflection mechanism rather than Bragg scatter (Gage and Green, 1979). The specular reflection can be more than 10 dB above the usual Bragg backscatter. It generally decreases quickly with angle off of zenith and decreases with frequency because the layers do not appear smooth over a Fresnel zone at shorter wavelengths. There is no evidence for specular reflection at UHF and it is believed to be important only at frequencies in the bottom part of the VHF band and lower.

4 APPLICATION OF CLEAR-AIR RADAR FOR WIND MEASUREMENT

A system of clear-air radars measuring winds through the boundary layer and troposphere could find wide use in weather prediction, pollution transport, and aircraft safety. If

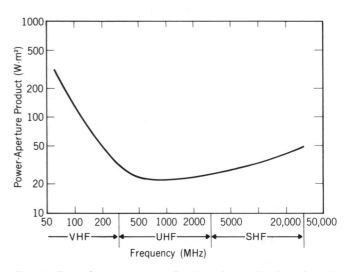

Fig. 13 Required power-aperture product for a clear-air Doppler radar with 300-m range cell and 2-m/s velocity cell to achieve an output signal-to-noise ratio of unity from clear air with structure parameter of 3×10^{-16} $m^{-2/3}$ at 5 km, assuming one output spectrum every 10 s.

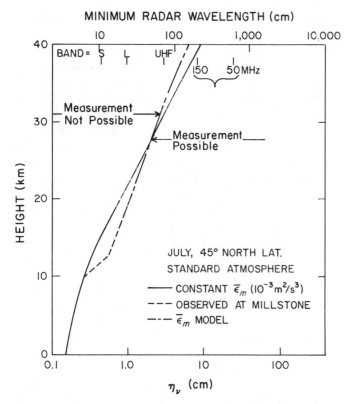

Fig. 14   Height dependence of the inner scale for the inertial convective subrange of turbulence, from Crane (1980). The solid curve is for a constant eddy dissipation rate while the broken curve is from measurements.

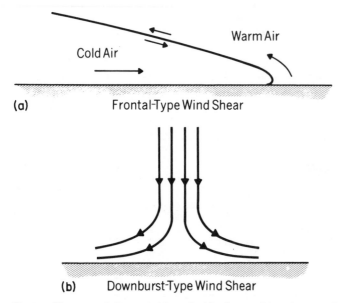

Fig. 15   Two types of airport wind hazards: (a) side view of frontal-type wind shear and (b) cross section of downburst-type wind shear. Both can be detected by clear-air radar.

the system were properly designed it could effectively and cheaply do all three of these jobs.

4.1 Pollution Transport

A widespread wind-profiling system for real-time wind measurement in the boundary layer and troposphere would be invaluable in the event of a large-scale release of polluting material from a contamination source. The wind profiles could serve as input data for plume dispersion algorithms to give a three-dimensional display of the plume in real time. For this reason, it might be desirable to site individual wind-profiling radars near sources of potential pollution.

A major ecological problem, especially in the northeastern United States and eastern Canada, is rainfall that has sufficient acidity to adversely affect aquatic life in lakes. A major source of the acid is probably the large coal-burning power plants of the Midwest. A system of properly placed clear-air radars measuring horizontal and vertical winds could shed considerable light on air trajectories followed by storms showing high acidity. This system of radars could monitor winds through the boundary layer and well up into the troposphere.

4.2 Wind-Shear Hazard

An important use of clear-air radars will be to monitor approach zones at major airports for hazardous wind phenomena. During takeoff and landing operations, aircraft are highly vulnerable to wind shear or wingtip vortices from preceding aircraft. In recent years, the hazards of low-level wind shear have become apparent, and some accidents that previously would have been attributed to pilot error are now considered to be wind-related. Fujita and Caracena (1977) analyzed three commercial airline crashes which occurred in 1975 and 1976 and determined that they were caused by wind shear. Many researchers feel that the solution is a remote sensor scanning the approach zones in such a way as to provide hazard detection and hazard prediction capability.

Several remote-sensing techniques can be used to measure airport wind shear. The most appropriate technique depends on the requirements of an operational wind-shear detection system, which in turn depends on the structure of the meteorological phenomenon that causes the hazardous wind shear. Three different meteorological phenomena are generally recognized as capable of creating hazardous wind shear. The first is a weather front with wind shear across the frontal interface (Fig. 15a). This is a large- or synoptic-scale phenomenon that can be monitored by a clear-air radar so that aircraft may be held during the hazardous period of a frontal passage. The second phenomenon is the gust front, or cold air outflow, of thunderstorms; this is a smaller-scale phenomenon than a front, and is also shorter-lived. Data in Hall et al. (1976) show that cold air outflow from distant thunderstorms produces gust fronts with wind shear that can be hazardous to aircraft. The third phenomenon, now widely believed to be the most hazardous of the three, is the "downburst" or "microburst" (Fig. 15b) described by Fujita and Caracena (1977). This is a small localized region of cold, rapidly descending air, which may be only a few hundred meters across and may last for only a few minutes. The downward-moving air moves outward in all directions when it encounters the ground and a plane flying through this structure would first encounter a headwind and then a tail-wind regardless of the direction of flight. If the aircraft is too near the ground this increased lift followed by a loss of lift could result in a crash. Models for this "microburst" phenomenon are being developed (Mueller and Hildebrand, 1983).

4.3 Wake Vortex Hazard

Wingtip vortices trailing behind large jet aircraft in landing and takeoff configurations constitute a major hazard to smaller aircraft at large airports. During 1964–1978 approximately 12 accidents per year were classified as related to wake vortex. These accidents usually occur when a smaller craft is following a larger one in a landing approach. Deregulation of the airline industry has increased the traffic at large airports and has changed the mix of aircraft so that large and small planes regularly use the same runways.

During the 1970s a great deal of work was done on the wake vortex problem, and a large amount of literature exists on the subject. However, even with all of the effort directed at this problem, a generally accepted, operationally oriented solution has not emerged. It appears that to be widely accepted, a solution must include direct measurement of the vortices in the approach zone of the runway. A major problem over the past decade has been the lack of an all-weather sensor that can detect wake vortices to significant ranges. Experiments with a short-range clear-air radar (Chadwick et al.,

1983) have shown that 10-cm radars can detect and locate vortices of large commercial jet aircraft (Boeing 737 and larger) to ranges of 3 km under clear-air conditions, and longer ranges should be possible under rain or snow.

Figure 16 shows the wake vortex of a Douglas DC-8 as measured by a 10-cm FM-CW radar. As with any clear-air radar, the scattering is from regions with enhanced refractive index fluctuations; probably the heat and moisture from the jet exhaust become entrained in the vortex system and cause increased refractive index fluctuations. The backscattered power from a vortex is typically about 10–12 dB above the background return from the clear air (Chadwick et al., 1983).

It may be possible to use the same clear-air radar to warn of wind shear and wake vortices. Here the technology of clear-air radar is not the problem; it has been proved and is widely accepted that clear-air radar can detect wind shear and wake turbulence. The problem is how the radar or radars should be sited with respect to the runways. A radar sited a long distance from a runway will have two problems. First, a vortex is not a beam-filling target at long ranges and the sensitivity fall-off with range is much faster than the inverse range squared that applies for most meteorological targets. So, long-range detection of vortices may be difficult. Second, at long ranges the radar will have to operate at quite low elevation angles to detect low-level wind shear or wake turbulence. At these low angles the ground clutter will be severe and may preclude detection of the hazard. The alternative to using one large radar located far from the runways is to use a smaller radar at each approach zone. The cost-benefit comparisons of these two approaches are not well known.

Fig. 16   Wake vortex of DC-8 as measured by FM-CW radar. The horizontal lines are range markers and the dark vertical area to the right is the DC-8 saturating the radar receiver. The vortex is shown by an increased spectral width after the airplane passes the beam.

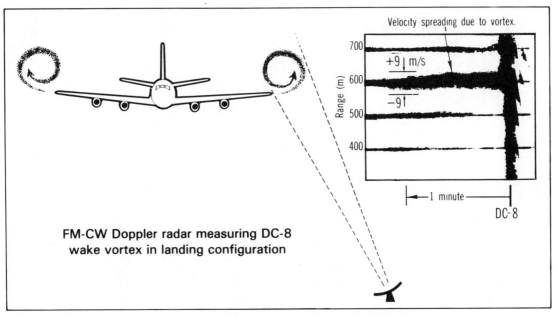

FM-CW Doppler radar measuring DC-8
wake vortex in landing configuration

5 CONCLUSION

Sensitive VHF, UHF, and SHF radars can detect return from refractive index fluctuations in the optically clear air. The best choice of frequency to measure winds and other clear-air phenomena in the upper troposphere and stratosphere is at VHF because large apertures can be realized easily. For clear-air applications in the boundary layer, the UHF band of frequencies is the better choice because of the narrower beam widths and higher spatial resolution available. The power-aperture product is a fairly good measure of the relative sensitivities of two clear-air radars.

One of the unique aspects of scanning Doppler clear-air radar is its ability to obtain almost four-dimensionally continuous "sky-filling" data. It is probably the only observing system that sees the atmosphere as a continuum in the way that the fluid dynamics equations do. For this reason it will continue to be a powerful research tool which is not effectively paralleled by any technique in laboratory fluid dynamics or atmospheric science.

Some of the atmospheric properties that have been measured by clear-air radar are inversion layer height and thickness, gravity waves on layers, Kelvin-Helmholtz breaking waves, depth and breakup of the boundary layer, convection cells, and mountain lee waves. Clear-air radars are capable of measuring numerous properties of wind fields in the boundary layer, including wind profiles, wind shear, vertical winds and divergence, large-scale circulations, density currents, gust fronts, and natural and man-made vortices.

There are important applications for which the clear-air technology is well developed and where there are urgent needs: (1) weather predictions; (2) aircraft routing; and (3) airport-area hazard detection. Of these, the use of clear-air radar for weather prediction is receiving the most attention, and a demonstration system of upper-air sounders is being developed in Colorado by NOAA. The Federal Aviation Administration is supporting work on the use of clear-air radar for aircraft routing and terminal-area hazard detection but operational systems have yet to be defined.

Radar Probing and Measurement of the Planetary Boundary Layer: Part II. Scattering from Particulates

R.A. Kropfli, Wave Propagation Laboratory, NOAA

1 INTRODUCTION

Part I describes how "clear-air" radars have been used to study the planetary boundary layer (PBL). The small-scale structure of the microwave refractive index in the PBL allows these longer-wavelength, usually 10-cm, radars to make such observations to distances approaching 100 km. Velocities can therefore be measured over large areas, with these centimeter-scale refractivity variations being used as tracers of the larger-scale air motion.

This clear-air scattering mechanism is usually not significant at the shorter radar wavelengths of 3 cm and 0.86 cm because the refractivity scales contributing to backscattered power can be within the viscous dissipation range rather than the inertial subrange. This is more likely to be true for the shorter of these two wavelengths. In addition, the transmitted power is usually considerably less at these wavelengths. Nevertheless, the power received from small (compared to the radar wavelength) particles varies inversely with the fourth power of the wavelength, in contrast with the inverse one-third power dependence for refractivity scatter. Thus at these shorter wavelengths particulate scatter dominates the refractivity scatter. As a consequence, shorter-wavelength radars can also be used in the "clear air" although at much shorter range.

Considerations such as ground clutter effects, resolution, and the time required "on target" for an accurate measurement all suggest that short-wavelength radars can also play a useful role in PBL research, especially when high-resolution measurements near the surface are required. For these reasons this part will emphasize the close-range, high-resolution measurements possible with 3- and 0.86-cm-wavelength pulse-Doppler radars.

Preliminary data sets from Colorado near mid-day indicate that natural particulates are lifted from the surface and are well mixed by buoyant plumes during the months of May through September. While the exact nature of these particulate scatterers is uncertain, we strongly suspect they include seeds, insects, and other small particles. When stronger echoes and more distant ranges are desired, the use of highly reflecting tracers of air motion called chaff filaments are artificially introduced into the PBL. High-resolution measurements that utilize the natural particulates in addition to those that are deliberately introduced will be discussed here.

Following a brief discussion of basic radar concepts not covered earlier, Part II is divided into two main subtopics:

- The kinematic structure: How is Doppler radar used to map the detailed three-dimensional velocity structure of the PBL?
- Turbulence and diffusion measurements: How can high-resolution profiles of relevant quantities such as velocity, velocity variances and covariances, shear, and turbulence dissipation rate be measured? What does a 20-km plume from a continuous point source look like when averaged over 30 min? Can convective scaling be applied to such a plume?

We emphasize the recent advances. Some techniques are presently being developed and have only preliminary data sets as indications of their validity. They have not yet stood the test of time as has much of our tower and aircraft technology. Nevertheless, if these techniques can be verified, we may soon be able to obtain measurements of "tower quality" (in some instances, even better) throughout the depth of the PBL and not just within several hundred meters of the surface. We begin with a brief summary of a few radar concepts which are relevant to these topics and which have not been covered in Part I.

2 BASIC RADAR CONCEPTS

2.1 Particulate Scatter of Electromagnetic Radiation

Particles that are small relative to the radar wavelength have a backscattering cross section given by

$$\sigma_i = (\pi^5/\lambda^4)|K_i|^2 D_i^6 \qquad (1)$$

where σ_i is the radar cross section (see Part I) of a single particle i, K_i is a function of the particle index of refraction, D_i is the diameter of particle i, and λ is the radar wavelength (Battan, 1973). The factor K_i is a weak function of temperature and radar wavelength but is usually treated as a constant: 0.93 for water and 0.197 for ice. Equation (1) was derived with the "small-particle" or Rayleigh assumption, i.e., that $2\pi a/\lambda \ll 1$, where a is the particle radius. Strictly speaking, this equation applies to small spherical particles of uniform refractive index. Complications arising when the Rayleigh scattering assumption is no longer valid are discussed in detail by Van de Hulst (1957).

When many small particles are within the radar sample volume the total radar cross section is given by

$$\sigma = \sum_v \sigma_i = \frac{\pi^5}{\lambda^4} |K|^2 \sum_v D_i^6 \qquad (2)$$

where the summation includes all particles within the radar sample or pulse volume. One should be aware that the *radar* cross section defined in Part I and used here varies as D^6 and is not the same as the *physical* cross section. We have dropped the subscript from the refractive index factor since a distinction is never made between particles having different refractive indices within the pulse volume.

Since most targets of interest in radar meteorology are distributed, or "beam filling" (in contrast to a single point target), the cross section per unit volume or reflectivity (expressed as the reciprocal of meters) is often used:

$$\eta = \frac{\sum_v \sigma_i}{V} = \frac{\pi^5 |K|^2}{\lambda^4} Z \qquad (3)$$

In this equation Z is called the reflectivity *factor* and is given by

$$Z = \sum_v D_i^6/V \qquad (4)$$

where V is the radar pulse volume. A reasonable approximation to the cylindrical radar pulse volume is

$$V \cong (c\tau/2)(r\phi/2)^2 \pi \qquad (5)$$

where τ is the radar transmitted pulse length in microseconds, c is the speed of light, r is the range to the target, and ϕ is the beam width. The radar pulse volume is given in Part I for the more general case of an elliptical beam rather than the conical beam discussed here and used most often in radar meteorology. The factor of 1/2 is necessary in the pulse length factor to account for the round-trip distance traveled by the pulse.

When Eq. (2) is substituted into the radar equation for distributed targets discussed in Part I, we have an expression that relates Z to the average received power:

$$P_r = C|K|^2 Z/r^2 \qquad (6)$$

where C is a radar-dependent constant. While the dimensions of Z in Eq. (4) are meters cubed, more commonly used units are $(mm)^6/m^3$, and Z is usually expressed in decibels as

$$Z_e = 10 \log Z \qquad (7)$$

It is referred to as reflectivity factor in "dBZ," where Z is assumed to be in units of $(mm)^6/m^3$. Even when the exact nature, composition, and size of the scatterers are unknown, as is frequently the case, most radars are calibrated to provide

a reflectivity factor appropriate for small water spheres, and the more correct terminology, "effective" reflectivity factor, is sometimes used. A comprehensive discussion of these concepts is given by Battan (1973).

To review what is often done with the received signal to provide a common basis for comparing measurements, the radar is first calibrated to yield an absolute measure of received power. By means of Eq. (6) or Eq. (7), the reflectivity factor, Z or Z_e, is determined from this power measurement, the known range to the target, and other measurable radar parameters, such as the transmitted pulse width, peak power, system losses, antenna size, and radar wavelength. Finally, if desired, one can compute the radar cross section per unit volume, η, from Eq. (3).

It is often useful to deliberately inject strongly reflecting scatterers called chaff into the PBL to be used as tracers of the air motion. Chaff consists of aluminum-coated fiber glass filaments approximately 25 μm in diameter and cut to $\lambda/2$ so that each filament behaves as a resonant dipole and has a very large radar cross section (Cassedy and Fainberg, 1960). Since the fall speed of chaff is about 30 cm/s in still air, these filaments remain in the convective boundary layer for several hours after release from an aircraft because the convective velocity w_* is usually several times larger. The usual method for dispersal of such particles is by means of an airborne chaff cutter, a device consisting of a motor-driven cutter wheel through which a 2,000-strand continuous ribbon of chaff is fed and cut to one-half the radar wavelength.

The equation derived by Schlesinger (1961),

$$N = \eta(0.18\lambda^2)^{-1} \qquad (8)$$

gives the chaff concentration N in filaments per cubic meter as a function of reflectivity (per meter) and the radar wavelength. This relationship assumes that chaff is cut to $\lambda/2$ in length and that the filaments are randomly oriented. The material itself is not biologically harmful and the density of filaments is quite small. For example, a signal-to-noise ratio (S/N) in excess of 20 dB is achieved even when the chaff density is on the order of one chaff filament per half cubic football field!

2.2 The VAD Method for Profiling Wind and Stress Components

In the velocity-azimuth-display (VAD) method for wind profiling, the radar antenna is rotated through 360° in azimuth, with the elevation held fixed to measure the azimuthal variation of the Doppler velocity, V_R. The periodic azimuth variation of the measured radial velocity is discussed by Browning and Wexler (1968). Figure 1 indicates the scan geometry.

If at a particular height the local large-scale wind can be represented as a constant mean plus a linear variation in x and y, the mean Doppler velocity is written as

$$
\begin{aligned}
V_R(\beta) = \ &(R/2)\cos\theta\ DIV + W_0\sin\theta \\
&+ V_0\cos\theta\cos\beta + U_0\cos\theta\sin\beta \\
&- (R/2)\cos\theta\,(<\partial U/\partial x> - <\partial V/\partial y>)\cos 2\beta \\
&+ (R/2)\cos\theta\,(<\partial U/\partial y> + <\partial V/\partial x>)\sin 2\beta
\end{aligned}
\tag{9}
$$

where R is the radius of the measurement circle, θ is the elevation angle, β is the azimuth, $<\ >$ signifies an average over the circle, and $DIV = <\partial U/\partial x> + <\partial V/\partial y>$. Note that the more usual sign convention, i.e., V_R being positive for targets moving away from the radar, is used here in contrast to that found in Browning and Wexler (1968). The velocity components U_0, V_0, and W_0 are averages over the measurement circle of radius R.

A harmonic analysis of $V_R(\beta)$ can yield the following information about the large-scale kinematic properties of the wind:

- A weighted sum of divergence and vertical motion from the zeroth-order harmonic
- The horizontal wind components U_0 and V_0 from the first-order harmonics
- Stretching deformation $<\partial U/\partial x> - <\partial V/\partial y>$ and shearing deformation $<\partial V/\partial x> + <\partial U/\partial y>$ from the second-order harmonics.

The usual procedure is to perform a least-squares fit of the data to an equation of the form

$$
\begin{aligned}
V_R(\beta) = \ &A_0 + A_1\sin\beta + A_2\cos\beta \\
&+ A_3\sin 2\beta + A_4\cos 2\beta
\end{aligned}
\tag{10}
$$

The desired kinematic terms are found from the coefficients of this equation along with the known range and elevation angle.

In practice when we examine a VAD, we observe a noisy sinusoidal variation of V_R with azimuth. The sinusoidal portion is the result of large-scale components of U_0, V_0, W_0, and DIV, and the superimposed noiselike signal is the result of small-scale turbulence and random errors in the estimate of V_R.

An example of this is shown in Fig. 2. These data were taken in the PBL on 27 September 1982, with the 3.2-cm-wavelength (X-band) radar operated by the National Oceanic

and Atmospheric Administration (NOAA) Wave Propagation Laboratory (WPL). They represent the data from one range gate during a scan at an elevation of 45°. Fitting a 360° sine wave by least squares, as shown in the figure, provides an excellent measure of the mean wind components, U_0 and V_0, and the sum of the first two terms in Eq. (9), which we represent as $f(W,DIV)$. In this analysis we set A_3 and A_4 in Eq. (10) to zero since we were not interested in deformation. The two terms comprising $f(W,DIV)$ can provide a measure of W_0 when the elevation angle is high ($\cos^2\theta \ll 1$) and divergence occurs at low elevation ($\sin\theta \ll 1$). Forty-five degrees is an intermediate case, and a separation of W_0 and DIV cannot be made unless other elevations or assumptions are used. (As we shall see later, 45° was chosen to minimize errors in the estimates of $<uw>$ and $<vw>$.)

As shown in Fig. 2, there is a significant amount of scatter about the best-fit sine curve. The scatter here is not measurement noise but is actually the turbulent signal so important in many PBL studies. Because of the possible masking of this signal by random errors in the estimate of V_R, a scan procedure was used to reduce these errors as much as possible. An estimate of this measurement uncertainty in V_R is given by Dennenberg (1971):

$$
\sigma_v{}^2 = \sigma_d\lambda/(8\sqrt{\pi}\,T)
\tag{11}
$$

where $\sigma_v{}^2$ is the error variance of the estimate of V_R, σ_d is the width of the Doppler spectrum, and T is the dwell time or time to acquire a velocity sample. This equation applies when the S/N is large (>10) and is appropriate for many PBL measurements from particulate scatter at close range. Since typical values for σ_d in the PBL are about 0.9 m/s, the 0.3-s dwell time used in the X-band radar measurements of Fig. 2 yielded an uncertainty of 0.08 m/s in V_R. This error estimate is a small fraction of the fluctuations apparent in the figure.

Wilson (1970) describes a straightforward technique in

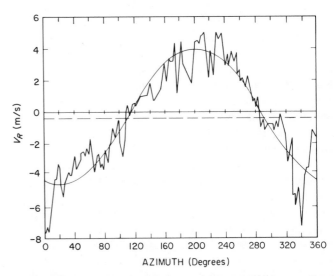

Fig. 2 Radial velocity vs azimuth showing the effects of the mean wind and small-scale turbulence. The least-squares-fit sine curve is indicated.

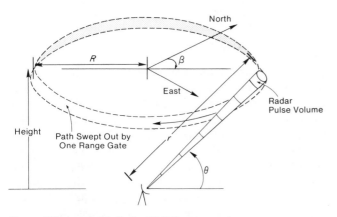

Fig. 1 Velocity-azimuth-display (VAD) scan geometry.

which these turbulent fluctuations can be used to estimate velocity variances and covariances. (This paper by Wilson is apparently the only reference to this technique, which for over 12 years remained ignored.) The following discussion outlines an extension of this technique to include larger scales not included by Wilson.

We start by decomposing the total Doppler velocity in terms of the following average and fluctuating quantities:

$$\mathscr{V}_R = V_R + v_R$$
$$= (U + u)A + (V + v)B + (W + w)C \qquad (12)$$

In this equation, we have written the angle-dependent factors in Eq. (9) as A, B, and C; V_R, U, V, and W represent ensemble averages of many profiles during a period 20 min or longer so as to include the passage of several convective elements or plumes in the average. The velocities v_R, u, v, and w represent deviations from the ensemble average. The variance of V_R is therefore written as

$$\sigma_{V_R}^2 \equiv \overline{(\mathscr{V}_R - V_R)^2}$$
$$= \overline{u^2}A^2 + \overline{v^2}B^2 + \overline{w^2}C^2 \qquad (13)$$
$$+ 2\overline{uv}AB + 2\overline{uw}AC + 2\overline{vw}BC$$

This equation, then, takes on the same form as that of Wilson (1970), except that the variances are now computed about the ensemble average rather than the single-scan measurement of U_0, V_0, and W_0.

$$\sigma_{V_R}^2 = \overline{u^2}\cos^2\theta \, \sin^2\beta + \overline{v^2}\cos^2\theta \, \cos^2\beta + \overline{w^2}\sin^2\theta$$
$$+ \overline{uv} \cos^2\theta \, \sin 2\beta + \overline{uw} \sin 2\theta \, \sin \beta \qquad (14)$$
$$+ \overline{vw} \sin 2\theta \, \cos \beta$$

When computed this way, the variances and covariances include contributions from plume-scale as well as the pulse-volume-scale fluctuations shown in Fig. 2.

From inspection of Eq. (14) we note the following: An integration from $\beta = 0$ to 2π results in no net contribution from the last three terms that involve the covariances. The resulting integral is a function of only θ and the three variances. We also note that if we first multiply Eq. (14) by $\sin 2\beta$ and then perform the integration, all terms are zero except for the term containing $\overline{uv}$ because of the orthogonality of sines and cosines. Similarly, the integration can be performed after multiplying Eq. (14) by $\sin \beta$ and $\cos \beta$ to obtain expressions for $\overline{uw}$ and $\overline{vw}$, respectively, in terms of an azimuth-weighted average of $\sigma_{V_R}^2$. Thus the appropriately weighted average of the measured $\sigma_{V_R}^2$ separates the desired variance from the others, and the choice of the weighting function determines which desired quantity is computed.

A mathematically equivalent but computationally simpler way of performing this calculation, in which $\sigma_{V_R}^2$ is weighted uniformly over each of the four quadrants, was described by Wilson (1970). Averages over the four quadrants are combined in different ways to achieve the above results. Examples given in Section 4.2 follow this method.

In applying this method, four integrals are computed from the fluctuations about the long-term (~ 20 min) least-squares-fit sine function that represents the ensemble mean of u, v, and $f(W,DIV)$:

$$I_1 = \int_0^{\pi/2} \sigma_{V_R}^2 \, d\beta$$

$$I_2 = \int_{\pi/2}^{\pi} \sigma_{V_R}^2 \, d\beta$$

$$I_3 = \int_{\pi}^{3\pi/2} \sigma_{V_R}^2 \, d\beta \qquad (15)$$

$$I_4 = \int_{3\pi/2}^{2\pi} \sigma_{V_R}^2 \, d\beta$$

We assume now that the flow is horizontally homogeneous, i.e., that the variances and covariances of u, v, and w are independent of position and therefore azimuth. This is expected to be true for large ensembles of profiles except when there are features in the wind field that are fixed to the terrain. If we also assume stationarity, so that an ensemble of measurements can be obtained in a short enough time for the real variances and covariances to be constant, we arrive at the following expressions:

$$I_1 + I_2 + I_3 + I_4 = \pi \cos^2\theta(\overline{u^2} + \overline{v^2}) + 2\pi \sin^2\theta \, \overline{w^2}$$

$$(I_1 + I_2) - (I_3 + I_4) = 4 \sin 2\theta \, \overline{uw}$$

$$(I_4 + I_1) - (I_2 + I_3) = 4 \sin 2\theta \, \overline{vw} \qquad (16)$$

$$(I_1 + I_3) - (I_2 + I_4) = 4 \cos^2\theta \, \overline{uv}$$

An error analysis of these equations indicates that optimum estimates are obtained for $\overline{uw}$ and $\overline{vw}$ at $\theta = 45°$, for $\overline{uv}$ and $(\overline{u^2} + \overline{v^2})$ at $\theta = 0°$, and for $\overline{w^2}$ at $\theta = 90°$.

2.3 Radar Resolution

Since a meteorological radar transmits most of its energy along a conical beam, the cross-beam dimension of the radar sample volume increases linearly with range while the dimension of the radar pulse volume along the beam is independent of range. This sample volume, as used in Eq. (5), can be considered cylindrical, with length determined by the radar pulse length, τ, and as having a diameter of $r\phi$, where r is the distance to the cell and ϕ is the radar beam width measured at the half-power points on the beam pattern. Table 1 lists these parameters for the NOAA 3-cm-wavelength (X-band) and

TABLE 1
Resolution Parameters for NOAA WPL Pulsed-Doppler Radars

Wavelength (cm)	Beam Width (deg)	Minimum Pulse Width (Range Resolution) (m)	Pulse Volume Diameter at 1,000-m Range (m)	Pulse Volume at 1,000-m Range (m³)
3.22	0.8	75	14	11.5×10^3
0.86	0.5	45	9	1.9×10^3

8-mm-wavelength (K-band) radars, which have been used in many of the examples that follow.

We examine the horizontal filter response characteristics of the averaging process inherent in conical scanning described earlier for velocity profile measurement. Conical (VAD) scanning results in a uniform weighting of measurements over a circle with radius R; i.e.,

$$R = h/\tan \theta \qquad (17)$$

where h is the height. The horizontal filter response for this averaging process is given in Fig. 3, where the abscissa is the normalized wave number $kR = 2\pi R/\lambda$ and the ordinate is the power transfer function (Stevens, 1967). This function is interpreted as the fraction of power in the horizontal wave number k (or wavelength λ) that is passed by the filter. For example, the normalized wave number $kR = 1$ has about 77% of its power represented in the filtered signal, or, equivalently, its amplitude is reduced to 88% of its true (unfiltered) value. Thus a conical scan that generated a circle of 500-m radius at some height would reduce the amplitude of a wave with a 3.1-km wavelength by 12% at that height. Vertical resolution determined by the radar pulse-volume dimensions as given in Table 1 as well as elevation angle is approximately 50 m between independent samples.

3 BOUNDARY-LAYER KINEMATICS AND STRUCTURE FROM DUAL-DOPPLER RADAR

3.1 Synopsis of the Dual-Doppler Radar Method

Most of the multidimensional velocity measurements in the PBL have followed the coplane method (or modifications of this method) first described by Lhermitte and Miller (1970) and later, in more detail, by Miller and Strauch (1975). This method requires two Doppler radars to scan simultaneously in a series of planes tilted about an axis along their base line as in Fig. 4. The tilt angle is increased until the planes no longer pass through the echo. The two radial velocity measurements are then geometrically combined to generate two-dimensional wind fields within each plane. Finally, the third component is computed by an integration of the equation of mass continuity in cylindrical coordinates, with the radar base line being the axis of the coordinate system.

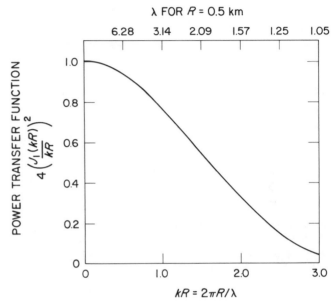

Fig. 3 *Power transfer function for VAD filtering.*

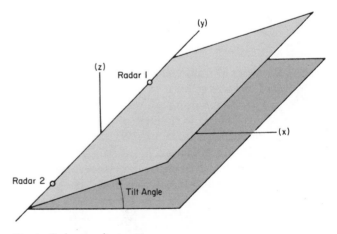

Fig. 4 *Coplane coordinate system.*

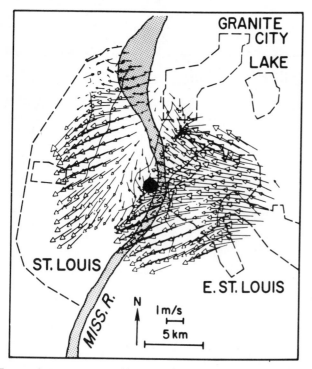

Fig. 5 A 70-min average eddy wind field at 0.3 km above ground level on 28 July 1975. The volume mean of 4 m/s from the northeast has been removed. The location of the St. Louis arch is indicated by the dot. Radars were located off the figure 25 km to the southeast of the arch and 40 km to the northwest.

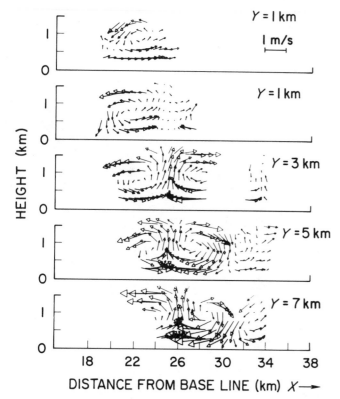

Fig. 6 Vertical cross sections across the convergence line looking upwind on 4 August 1975. The y axis is parallel to the radar base line, i.e., along an azimuth of 25°.

In many observations of this kind, aircraft-dispensed chaff has been used to provide tracers of air motion. The usual method of chaff distribution is to mount a chaff cutter on a small aircraft to provide a continuous source. The aircraft is then flown crosswind near the mid-point of the PBL. When horizontal legs of about 15 km are flown at a distance equivalent to about 30 to 45 min (advection time) upstream of the target area, a uniform echo over hundreds of square kilometers is produced through the depth of the PBL. The following examples show that wind fields measured this way are similar to the large eddy simulations familiar to boundary-layer numerical modelers.

3.2 Results from METROMEX

During the last full-scale field season (July 1975) of METROMEX (Metropolitan Meteorological Experiment), areas of the urban PBL at St. Louis as large as 300 km² were observed (Kropfli and Kohn, 1978) with two NOAA WPL X-band radars. One radar was located 25 km to the southeast of the St. Louis arch and the other about 40 km to the northeast. These observations revealed the existence of persistent horizontal rolls that were occasionally generated by the urban heat island. These rolls were associated with moderate northeasterly winds, and an example of the radar-derived perturbation wind field at the 300-m level is shown in Fig. 5. The 4-m/s mean wind from the northeast has been removed to illustrate the convergence pattern. The figure represents an average of 17 scans over a 70-min period and demonstrates stationarity of the convergence zone fixed to the most heavily industrialized portion of the St. Louis area. Figure 6 is a vertical cross section across the convergence line on a different day with similar wind conditions, and it again shows the existence of well-formed horizontal rolls that filled the PBL. Urban wind fields from these two days, along with five others, are documented by Kropfli (1977).

3.3 Results from the Phoenix Program

In a more recent multiple-Doppler radar experiment involving the use of chaff, Kropfli and Hildebrand (1980a, b) examined the experimental limits of accuracy and resolution that are possible with radar in the PBL. This experiment, called Phoenix (Hooke, 1979), took place at the Boulder Atmospheric Observatory (BAO) and included a network of three Doppler radars, an NCAR C-band (5-cm-wavelength) radar, and the two NOAA X-band radars. The relatively short radar separations of 13 km allowed scales as small as 0.6 km to be resolved. In the radar analysis, efforts were made to include the effect of terrain in the lower boundary conditions, to remove velocity estimates contaminated by ground clutter, and to account for the underestimation of divergence magnitude near the surface. Good agreement was generally found between radar data and data from in situ devices.

The results of a rather severe test of the vertical wind field are shown in Fig. 7. This figure compares normalized profiles of radar-derived vertical velocity variance with an average profile given by Caughey and Palmer (1979) that was derived from in situ measurements in various parts of the world. The radar data follow the Caughey-Palmer profile quite well, including the reduction in variance near the top of the PBL.

Temporal and spatial correlations of the radar-derived divergence field were also used in the Phoenix analysis to study some of the statistical properties of the PBL. The correlation $C_{z,t}(l_x, l_y, \tau)$ was computed from the divergence $DIV(x,y,z,t)$ in the following way:

$$C_{z,t}(l_x, l_y, \tau) = \frac{\overline{DIV(x,y,z,t) \, DIV(x - l_x, y - l_y, z, t - \tau)}}{\overline{DIV(x,y,z,t)^2}} \quad (18)$$

where the overbar represents a horizontal average, z is the height of the measurement, and DIV is obtained from wind fields similar to that shown in Fig. 5. An example of $C_{z,t}(l_x, l_y, \tau)$ with $\tau = 0$ is shown in Fig. 8a and in Fig. 8b with $\tau = 2.5$ min. The indicated values in the figure have been

multiplied by 100 so that the actual correlation value for $l_x = l_y = 0$ and $\tau = 0$ is normalized to 100%, as can be seen by the value at the origin of Fig. 8a. Thus Fig. 8a shows the spatial correlation and provides information about the shape and orientation of an average convective element or eddy. The major and minor axes of the contour at the 40% level in plots similar to Fig. 8a were estimated by eye, and the resulting axial ratios are plotted as a function of the wind

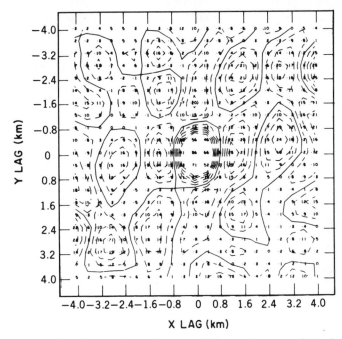

Fig. 8a   Two-dimensional spatial autocorrelation function computed for the radar-derived divergence field at a height of 150 m. Values are given in percent.

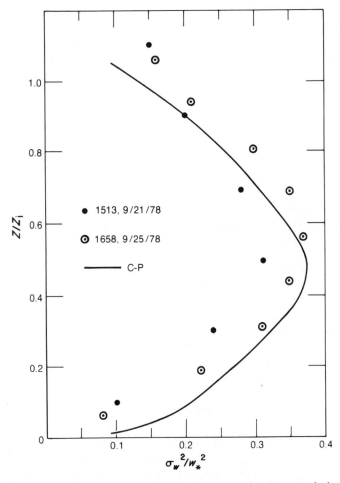

Fig. 7   Normalized profiles of vertical velocity variance, σ_w^2/w_*^2, from multiple-Doppler radar measurements during Phoenix along with an independently derived curve from Caughey and Palmer (1979). (From Kropfli and Hildebrand, 1980a.)

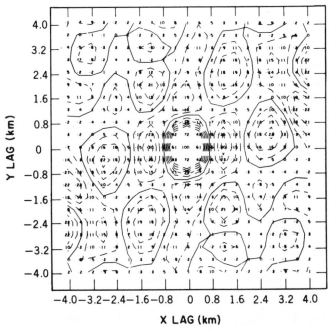

Fig. 8b   Two-dimensional spatial cross-correlation field computed for radar-derived divergence field at a height of 150 m with a temporal lag of 2.5 min.

speed at $z = 150$ m in Fig. 9. This figure shows the tendency for the axial ratio to increase with wind speed; i.e., the average shape of convective elements near the surface tends to be "stretched out" as the wind speed increases. Figure 10 shows that there is also a tendency for the major axis to be oriented along the wind direction at that level.

Figure 8b shows that the maximum value of the correlation $C_{z,t}$ has moved from the origin and has been reduced to 0.86 at $\tau = 2.5$ min. The actual location of the peak $C_{z,t}$ value was observed to track very closely with the mean wind. If one measures the peak correlation as a function of τ, it is possible to estimate the time scale in which the convective field loses its identity. Such a plot of peak correlation versus time is shown in Fig. 11. These data suggest an exponential decay, with an average decay to a value of e^{-1} in 1,116 s and with a standard deviation of 366 s about this mean. The ratio of standard deviation to mean is reduced from 0.33 to 0.22 if

the lag is scaled by z_i/U. This scaling produces a normalized eddy lifetime of 1.68 and suggests that the lifetime can be estimated if z_i and the low-level mean wind, U, are known. These observations were taken away from any significant localized source of heat or large-scale roughness inhomogeneities.

3.4 Measurements at the National Severe Storms Laboratory

The more powerful 10-cm radars at the National Severe Storms Laboratory (NSSL) near Norman, Oklahoma, have been used to measure PBL wind fields without the use of chaff (O'Bannon, 1978; Doviak and Jobson, 1979). This capability has great potential as a forecast tool in identifying regions of convergence in the PBL where convective storms are likely to develop. These observations at NSSL covered areas as large as 600 km². They revealed a variety of PBL structures, including periodic structures (Doviak and Berger, 1980), convergent and divergent areas, vortices, and obstacle flow (O'Bannon, 1978).

Uncertainty remains about the nature of the scattering mechanism, with some evidence that it may be refractive index fluctuations (Doviak and Berger, 1980) as discussed in Part I. Not only do the reflectivity values fall into the range of values from previous measurements at Wallops Island (Kropfli et al., 1968), but there is also good agreement in the probability density functions for the structure constant, C_n^2, derived from radar and from a nearby airborne microwave refractometer. On the other hand, some observers have suggested that these same radars are detecting clouds of insects (Rabin, 1983). These radars are certainly sensitive enough to detect echoes from either kind of scattering mechanism, and the natural echoes observed in Oklahoma are probably caused by both.

3.5 PBL Measurements during the Joint Airport Weather Studies

During the summer of 1982 an experiment designed to investigate low-level wind shear hazardous to aircraft was conducted near Denver, Colorado (McCarthy et al., 1983). A Doppler radar network that included the two NCAR 5-cm-wavelength radars and the NCAR CP-2 10-cm-wavelength radar was the major observational tool in this program. A 3-cm-wavelength Doppler radar was also used on board the NOAA P-3 aircraft (Mueller and Hildebrand, 1983). Analyses of these multiple-Doppler radar data produced new insights into the structure of small-scale downdrafts that penetrate the PBL and produce strongly divergent flow near the surface. (See Part I for a schematic depiction of this.) These phenomena are difficult to detect because of their scale (about 1 km), their transience, and their proximity to the surface. The large velocity differential, averaging 24 m/s within the divergent flow at the surface, makes them particularly

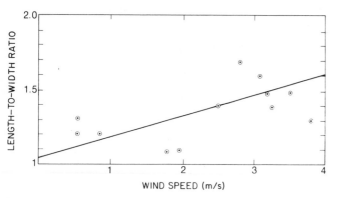

Fig. 9 Length-to-width ratio of contour enclosing the peak at the 40% contour level vs wind speed.

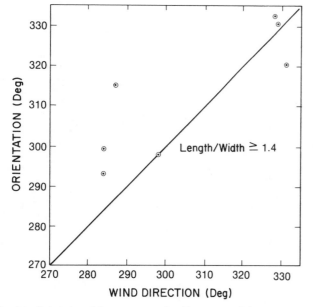

Fig. 10 Orientation of the contours used in Fig. 9 vs wind direction.

hazardous to aircraft (Wilson and Roberts, 1983). Excellent boundary-layer flow patterns from the resulting dual-Doppler radar analysis are found in Kessinger et al. (1983), Wilson and Roberts (1983), and Mueller and Hildebrand (1983).

3.6 Thermodynamic Retrieval from Dual-Doppler Radar Measurements

If the three-dimensional wind field and its temporal and spatial derivatives are measured accurately enough, the momentum equations can be used to recover the pressure and temperature fields (Gal-Chen, 1978).

As an outline of the method, the momentum equations are written in the following way:

$$\partial p/\partial x = -\varrho_0(du/dt) + f_1 \equiv F \tag{19}$$

$$\partial p/\partial y = -\varrho_0(dv/dt) + f_2 \equiv G \tag{20}$$

$$\varrho g + (\partial p/\partial z) = \varrho_0(dw/dt) + f_3 \equiv H \tag{21}$$

where d/dt is the substantial derivative; f_i $(i = 1,2,3)$ represents other forces that are specified or parameterized in terms of the observed kinematics, e.g., turbulent friction and Coriolis force; and the rest of the notation is standard. Thus the quantities F, G, and H are completely specified by the kinematical (Doppler radar) observations. Solutions to Eqs. (19) and (20) exist if and only if

$$\partial F/\partial y = \partial G/\partial x \tag{23}$$

which would be exactly satisfied if the radar-derived quantities F and G could be determined exactly. Since measurement and numerical errors are present, we can solve the system in the least-squares sense:

$$\iint \{[(\partial p/\partial x) - F]^2 + [(\partial p/\partial y) - G]^2\} \, dx \, dy = \text{minimum} \tag{24}$$

This is a standard variational analysis problem (Courant and Hilbert, 1953). The resulting Euler equation is a Poisson equation for the pressure fluctuations

$$(\partial^2 p/\partial x^2) + (\partial^2 p/\partial y^2) = (\partial F/\partial x) + (\partial G/\partial y) \tag{25}$$

This equation can be numerically solved for the pressure deviations from the horizontal mean at each level where observations of the kinematics are available. Equation (21), which at this point has not been used, is then brought in to calculate the deviations from the horizontal average of the density ϱ.

Preliminary tests of this method performed with the Doppler radar data available from the Phoenix experiment are encouraging. For example, the derived pressure field does

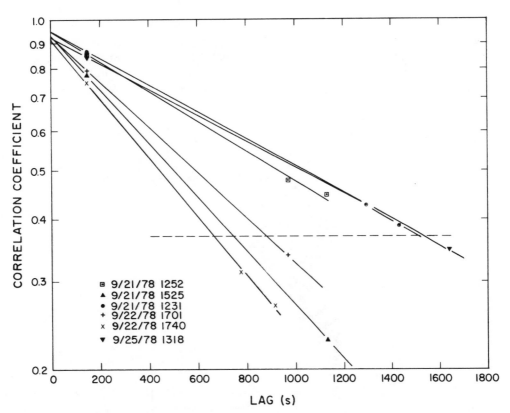

Fig. 11 Peak correlation coefficient vs time for divergence at 150 m on six different occasions. The dashed line is at a value of e^{-1}.

balance the momentum equations, and there is good temporal continuity in the result. The derived profile of heat flux has some of the expected classical features: large positive values near the surface, a gradual decrease with height, and small negative values near z_i. The derived temperature variance profile also seems reasonable since it has a single minimum near $z_i/2$ and values three to four times greater near the surface and near z_i (Gal-Chen and Kropfli, 1983). A similar Doppler radar experiment, Phoenix Reborn, somewhat more closely designed to the requirements of this thermodynamic retrieval method, was conducted during the spring of 1984.

3.7 Summary of Dual-Doppler Radar Measurements in the PBL

Several conclusions can be drawn from the dual-Doppler radar studies that have been performed. Horizontal wind measurements accurate within 0.2 to 0.4 m/s are realizable with these methods; vertical motion estimates can be accurate to within 0.5 m/s. Such accuracy is sufficient for many purposes as this error is considerably smaller than PBL wind-field perturbations. Spatial resolution varies directly with radar separation, but spatial wavelengths as small as 600 m are resolvable. At a sacrifice of spatial resolution, regions as large as 600 km² can be observed, and this is clearly sufficient to resolve rolls and other periodic structures. The large-scale-eddy lifetime was observed to be about 20 min with these techniques and showed a tendency to scale with z_i/U. Such radar measurements usually showed that identifiable eddies drift with the mean wind in the unstable PBL, except when a significant heat source produces a persistent updraft that is fixed to a surface feature such as the urban heat island (as in flows observed over St. Louis during METROMEX). Very

recently, dual-Doppler radar techniques have been used to study hazardous low-level shear caused by microbursts. This phenomenon is difficult to detect but Doppler radar has shown great promise in this area. In a somewhat more exploratory area, statistics of thermodynamic properties of the PBL are being computed from input data derived totally from dual-Doppler radar measurements. A program, Phoenix Reborn, has been conducted to evaluate such methods.

4 TURBULENCE AND DIFFUSION

Since the work of Hildebrand (1977) on the diffusion of chaff in the lower atmosphere there has been a resurgence in the use of chaff for turbulence and diffusion studies. We will first discuss the radar measurement (non-Doppler) of a plume from a continuous point source. This measurement is performed by releasing a plume of chaff from near the surface or on a tower and by observing the reflectivity (and, therefore, chaff concentration) with a nearby high-resolution radar. The second kind of measurement discussed here is that of boundary-layer profiling without the use of chaff, in which vertical profiles of mean wind and turbulence are measured. Although both plume studies and PBL profiling are new applications of radar, they are based on well-established principles in radar meteorology. They are straightforward both from an experimental and from an analytical point of view.

4.1 Plume Diffusion

After a preliminary experiment at the BAO by Moninger and Kropfli (1982), a more systematic chaff plume experiment was conducted in September 1982 (Moninger et al., 1983), and yet another in September 1983. The latter two series of plume experiments are called CONDORS I and II

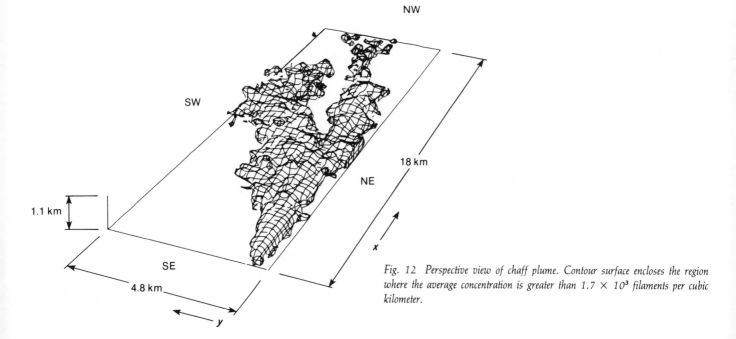

Fig. 12 Perspective view of chaff plume. Contour surface encloses the region where the average concentration is greater than 1.7×10^3 filaments per cubic kilometer.

(for *C*onvection *D*iffusion *O*bserved with *R*emote *S*ensors). The basis for such experiments is the relationship (Eq. 8) between chaff concentration and radar reflectivity. An example of an "instantaneous" concentration contour from Eq. (8) is shown in Fig. 12.

Since hundreds of range samples of reflectivity are acquired instantaneously by a radar, it is much more efficient for the radar to be looking along the plume to minimize the scan sector size. Attenuation of the signal as it passes through the plume is insignificant. An X-band radar can be located as close as 3.5 km upwind of the source of chaff to provide excellent resolution ($\sim$ 70 m) just downwind of the source, with minimal ground clutter contamination. All of these measurements to date have been obtained with the NOAA 3-cm radars, but improved resolution would be possible with a radar of even shorter wavelength (see Table 1). The scan volume time has been typically on the order of a few minutes, and the plume has been easily observed beyond 20 km, although with reduced resolution.

As described by Moninger et al. (1983), such data were interpolated onto a Cartesian grid having a 50-m grid element. During one 29-min period, the radar took nine volume scans of the entire plume. Since the transport distance over 29 min was about 10 km, or about 20 times the mixing-layer depth, a reasonably large "ensemble" of convective eddies passed over any particular point. This assumes that the horizontal eddy scale is about 1.5 times the mixing-layer depth (e.g., Kaimal et al., 1976; Willis and Deardorff, 1976b).

Figure 13 shows crosswind-integrated, normalized concentration, $\overline{C}^y$, compared with similar data from tank model results of Willis and Deardorff (1981). These results are from a 500-m downwind slice. The data were multiplied by $z_i/\Delta z$, where z_i = 500 m and Δz is the data spacing, 50 m. With this "convective scaling," a plume uniformly distributed over the depth of the boundary layer would yield $\overline{C}^y$ = 1 everywhere. This is expressed as

$$\overline{C}^y = z_i \int F \, dy / \int \int F \, dy \, dz \qquad (26)$$

where F is the number of filaments per grid cell. The abscissa in Fig. 13 is the dimensionless downwind distance,

$$X = w_* x/(U z_i) \qquad (27)$$

where w_* is the mixed-layer convective velocity scale and x is the dimensional downwind distance. For a full discussion of X and $\overline{C}^y$, see Willis and Deardorff (1976a). For this case, w_* = 1.44 m/s, z_i = 500 m, and U = 6 m/s. Thus, X = 1 when x = 2.08 km.

The agreement between the radar data and the tank model results of Willis and Deardorff is surprisingly good. Both show the maximum concentration of the time-averaged plume descending to near the surface at $X \cong 0.8$, and then rising. The minimum at the source height near X = 0.75 seen in both data sets suggests that the plume is most often released into an updraft or downdraft, and less frequently finds

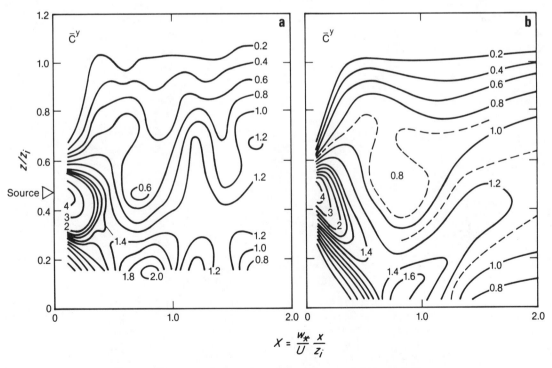

Fig. 13   Contours in the vertical x-z plane of the crosswind-integrated, dimensionless concentration $\overline{C}^y$: (a) radar data; (b) data from Willis and Deardorff (1981).

itself in air with zero vertical velocity. The slowly rising height of the maximum concentration for $X > 1$ in our data may be due to blockage at low radar viewing angles. This effect has not yet been fully investigated.

An important component of CONDORS was to evaluate the effect of the chaff fall speed (30 cm/s) on the result by comparing the radar estimate of chaff concentration with measurements of a simultaneously released oil fog plume observed by the NOAA WPL incoherent lidar. The details of the comparisons can be found in Moninger et al. (1983); Fig. 14 indicates that the effect of chaff fall speed was minimal. The vertical profiles of the normalized-integrated concentration from the radar and lidar in the figure are remarkably similar. The two plumes seem similar enough that they can be confidently used in future experiments comparing surface and elevated releases. The radar resolution is quite adequate, and ground clutter is a solvable problem. Both the lidar and radar show that remote sensors can give a highly detailed quantitative three-dimensional view of tracer plumes.

4.2 Boundary-Layer Profiling

We now review examples of boundary-layer profiling using the VAD technique described in Section 2.2. While the measurements shown here have been obtained in the convective boundary layer over land, they need not be limited to this application. Other situations such as nonprecipitating marine-boundary-layer clouds could also be studied with such techniques, provided that echoes from the cloud droplets are sufficiently strong.

4.2.1 The Scatterers: What Are They?

While the 3- and 0.8-cm-wavelength radars used as examples here are not usually thought of as "clear-air radars" in that they are insensitive to the refractive index fluctuations discussed in Part I, they are, in fact, very sensitive to boundary-layer particulates that are lifted by buoyant parcels rising from the heated surface. This sensitivity to particulates results because the radar cross section for particulate scatterers varies as λ^{-4}, as in Eq. (2). For example, a 0.86-cm-wavelength radar would have a 23-dB advantage over a 3.2-cm-wavelength radar from this wavelength dependency. The scatterers sensed by these radars are larger than aerosols (see the article by Pueschel in this volume), typically in the millimeter size range, and are believed to be insects, seeds, and other small particles. Recent observations made with NOAA X-band radar indicated a reflectivity or radar cross section per unit volume of 10^{-12} cm^{-1}, which is in reasonable agreement with X-band targets identified by Hardy and Katz (1969) as being insects and having a reflectivity of 3×10^{-12} cm^{-1}. These echoes are often found to be strong and uniformly distributed throughout the PBL during May through September in Colorado and have been used as tracers of the air motion. They are usually not found during the winter months because of the weaker surface heating and absence of available particulates.

Evidence obtained so far indicates that the scatterers observed most often in Colorado do not have characteristics of "strong fliers." We have not observed strong lobed patterns in the reflectivity field as observed by Schaefer (1976) for swarms of migrating insects. His observations showed

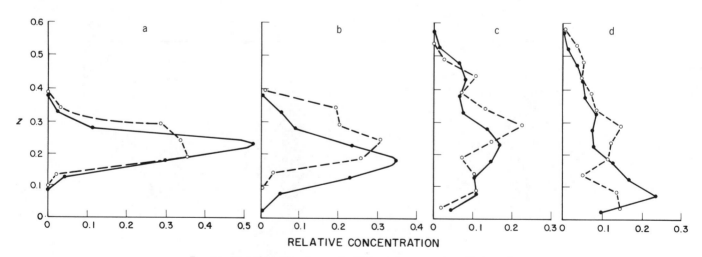

Fig. 14 Vertical profiles of normalized-integrated concentration. Circles and dashed line are the lidar data, and solid dots and line are the radar data. Downwind distances in dimensional and dimensionless coordinates are (a) $x = 280\ m$, $X = 0.135$; (b) $x = 530\ m$, $X = 0.255$; (c) $x = 910\ m$, $X = 0.438$; (d) $x = 1,610\ m$, $X = 0.774$.

two strong reflectivity maxima at azimuths 180° apart, presumably the directions at which the insects were viewed from the side. In addition, the depolarized signal, which is also sensitive to the shape of the scatterers, has been examined and no azimuthal dependence was found (Pasqualucci et al., 1983). Finally, the comparisons of tower and radar measurements of horizontal velocity agree to within 0.2 m/s when the two systems were 3.5 km apart. While there is no evidence so far that the scatterers have significant self-induced motion, this possibility always exists, so efforts are continuing to identify the exact source of the scatterers.

As a first example of radar-derived profiles in the PBL we consider the X-band conical scan data taken 3.5 km east of the BAO on 27 September 1982. Two hours of data were taken starting at about 1200 MST. The antenna was driven at 1 rpm at alternating elevations of 45° and 75°. Scatterers within the PBL on this sunny day provided ample signal strength for the duration of the measurements and S/Ns greater than 10 dB were evident for most of the data at these close ranges. The uncertainty of the estimate of V_R was 0.08 m/s from Eq. (11). Adequate signal was obtained from about 180 m above the surface to more than 1,400 m, which was presumed to be the height of the PBL. To provide the best spatial resolution, range samples were selected at 37.5-m increments, and the transmitted pulse width was 75 m. From this data set, continuous profiles of reflectivity, U, V, W, and the velocity variances and covariances were computed for the 2-h period by the methods discussed in Section 2.2.

Whenever the radar pulse volume averaging is not too great, the reflectivity profile should allow scatter from particulates to be distinguished from refractive index inhomogeneities if one of these mechanisms dominates. The reflectivity factor for particulates would be expected to decrease slowly with height (Rabin, 1983), but the reflectivity profile from the refractive index should show a notable increase just below the capping inversion (Wyngaard and LeMone, 1980). In Fig. 15 the 100-min average reflectivity factor profile derived from the conical scan data shows little variation with height. While the data in Fig. 15 do not show the expected decrease with height, the reflectivity factor magnitude and the absence of an increase near z_i suggest that particulate scattering is the dominant mechanism here.

4.2.2 Mean Velocity Components

Figure 16 is a time-height cross section of the V component derived from a VAD analysis. Each column of numbers was derived from one revolution of the antenna at 45° elevation, and the scan-to-scan continuity is reassuring. The numbers in the figure have been rounded to the nearest meter per second for clarity but the uncertainty of the estimate was much less, as discussed earlier. A northerly component of wind from near the surface to the solid contour just above the mid-point of the PBL is seen here, and above this level the wind reverses and becomes southerly to the top of the echo. The height at which the signal drops out varies by more than

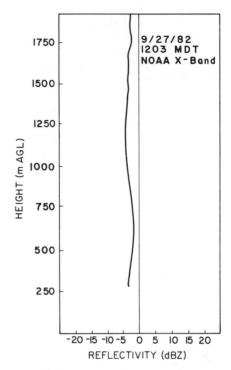

Fig. 15 *Reflectivity profile from particulate scatterers for a 100-min average of data taken with the NOAA 3.2-cm radar on 27 September 1982, starting at 1203 MDT.*

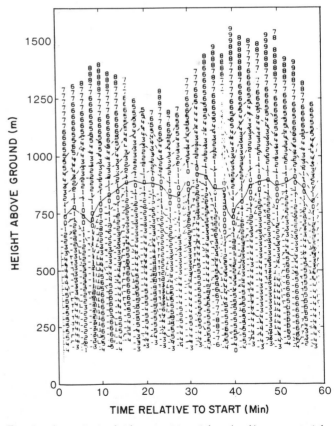

Fig. 16 *A 60-min time-height cross section of the v (north) component of the wind on 27 September 1982, 1203 MDT. Indicated velocities were rounded to the nearest meter per second for clarity. Contours are every 1 m/s.*

TABLE 2
Radar and Tower Mean Wind
Estimates at 200 m

Time	u_R	u_T	v_R	v_T
1200	−1.40	−1.71	−3.70	−3.88
1240	−0.55	−0.92	−4.26	−4.56
1300	−1.40	−1.25	−4.24	−4.34
1320	−0.72	−1.02	−5.27	−4.57
1340	−0.11	−0.08	−5.91	−5.96

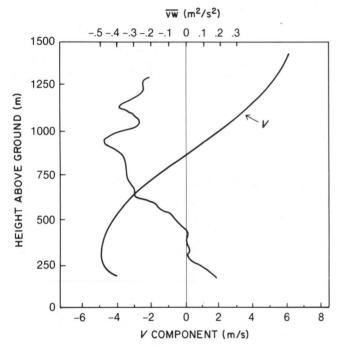

Fig. 17   Profile of V and $\overline{vw}$ for a 100-min period beginning at 1203 MDT on 27 September 1982.

TABLE 3
Eddy Diffusivity Estimates

Time	K_u (m²/s)	K_v (m²/s)
1151		35
1159		50
1211		55
1232	81	52
1240	37	54
1304	235	26
1319	117	−16
1339	68	36

250 m during this period and represents undulations in the height of the mixed layer. Also evident in the figure is the unusually high shear of 8×10^{-3} s⁻¹ over most of the mixed layer.

Mean wind components derived from these data were found to agree well with BAO tower measurements obtained 3.5 km to the west. Table 2 compares the results of corresponding 20-min averages at the 200-m level and suggests that there is only a minimal effect of ground clutter and inherent motion of the scatterers on the radar measurements. The overall 2-h difference between the tower and radar measurements indicates that the wind speed measured by the radar was only 0.16 m/s less than that measured by the tower. This small difference suggests that the data were minimally, if at all, affected by ground clutter or inherent target motion.

4.2.3 Velocity Covariances

Equations (16) were used to compute the velocity covariances $\overline{uw}$ and $\overline{vw}$ for the same data set discussed in the previous paragraph. Profiles of V and $\overline{vw}$ averaged over a 100-min period are presented in Fig. 17. Close examination of this figure suggests that shear and stress are closely related in this case, and one is tempted to investigate the eddy diffusivities:

$$\overline{uw} = -K_u(\partial U/\partial z)$$
$$\overline{vw} = -K_v(\partial V/\partial z)$$

$$(28)$$

Average values of shear and stress were computed for the layer between 210 m and 980 m for successive 20-min samples, and Table 3 shows the resultant eddy diffusivities. Missing values in the table result from very small magnitudes of stress and shear, making the ratio indeterminate. All other values appear reasonable in comparison with the value of ~ 70 m²/s suggested by Draxler (1979) for the daytime PBL with moderate insulation and a wind speed of 4 to 6 m/s. The large variability in K_u reflects the weaker shear in the easterly direction.

The scan-to-scan repeatability of stress measurements is illustrated in Fig. 18. This is a time-height cross section of $\overline{uw}$ from K-band radar data obtained in the PBL on 10 June 1983. Conical scans at 45° elevation were performed every 30 s for a 2-h period, with a 15-min sample being shown in the figure. Plots such as these illustrate that significant variations in wind and stress occur over time scales of 10 min in the undisturbed, convectively driven PBL. It is clear why an averaging time of at least 20 min is necessary to obtain stable estimates of the variances and covariances of the wind components under these conditions.

Other tests have been performed recently to examine the precision of the VAD technique for measuring momentum flux and mean wind profiles. Conical scans at 45° elevation were performed with the NOAA WPL 3.22- and 0.86-cm-wavelength radars while they were collocated and scanning through a nonprecipitating cloud at heights from 1 km to 5 km. Scatter plots of X-band data (ordinate) versus K-band data (abscissa) for U and $\overline{uw}$ are shown in Figs. 19a and 19b, respectively. Root-mean-square error estimates of the wind and the stress components obtained from these measurements were 0.06 m/s and 0.04 m²/s², respectively.

The best opportunity to compare momentum fluxes measured with tower-based anemometers and radar occurs when the mean wind is along the line between the two instruments (Haugen et al., 1975). Such an event occurred on 11 September 1983, when the NOAA X-band radar was located 3.5 km east of the BAO and the mean wind direction was only about 15° off the line of sight between them. Figure 20 shows radar and tower estimates of the vertical flux of horizontal momentum as a function of time. The radar values are near the 300-m level (the level at which the tower measurements were made) and have been advanced by 10 min. It is arguable whether this 10-min shift was necessary to illustrate the agreement. The large increase in momentum flux was the result of a breakdown of a stable layer between 200 m and 250 m at about 1150 MST. Before this time an isothermal layer at this level inhibited vertical mixing. The rapid growth of the unstable boundary layer was marked by a sharp increase in the momentum flux measured by both the tower and the radar.

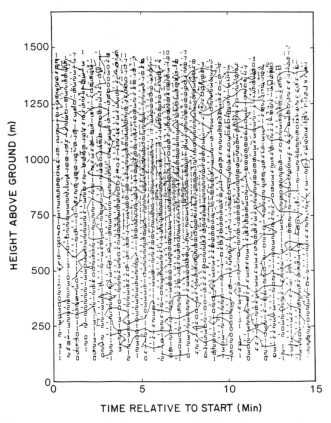

Fig. 18   A 15-min time-height cross section of $\overline{uw}$ for 10 June 1983; data are in m²/s² × 10.

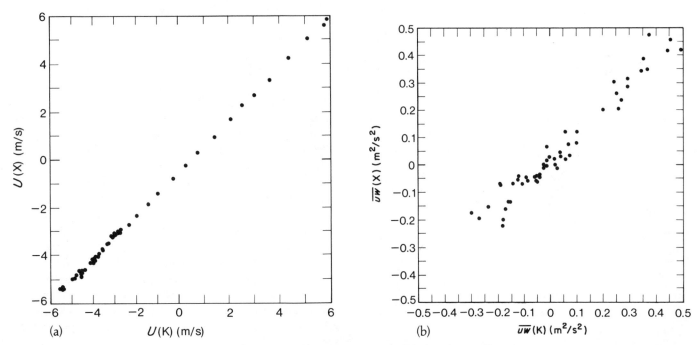

Fig. 19   Scatter plots of (a) U component and (b) $\overline{uw}$, derived from collocated X-band and K-band radars scanning simultaneously in VAD mode at 45° elevation. X-band estimates of U are on the ordinate, and K-band estimates are on the abscissa.

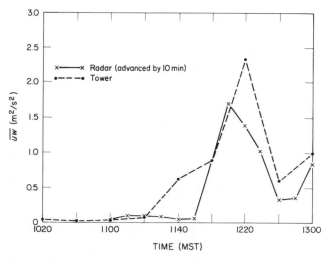

Fig. 20 Tower and radar estimates of vertical flux of horizontal momentum vs time. These are 20-min averages near the 300-m level.

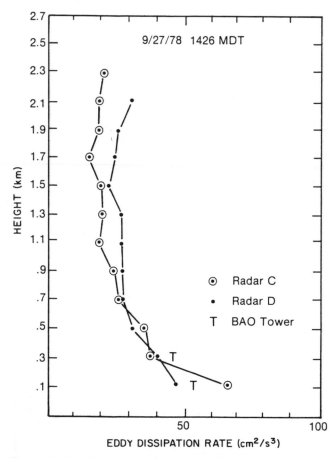

Fig. 21 Profiles of ε computed with volume scan data from two X-band radars spaced 13 km apart during Phoenix. Tower-derived values are indicated at the 200- and 300-m levels.

4.2.4 Radar Estimates of Eddy Dissipation Rate

Part I discusses how the turbulence contribution to the width of the measured Doppler spectrum, σ_T, can be obtained when other factors such as shear, finite beam-width effects, etc., are taken into account. In practice, spectral moment estimators such as the pulse-pair algorithm (Zrnic, 1977; Rummler, 1968) are used to obtain the first and second moments, V_R and $\sigma_T{}^2$, respectively, of the Doppler spectrum rather than computing directly from the Doppler spectrum itself. The quantity σ_T is a measure of the turbulence intensity smaller than the pulse-volume scale and can be related to the eddy dissipation rate ϵ by means of the method developed by Frisch and Clifford (1974). In brief, the method assumes a $k^{-5/3}$ spectral form, and further assumes that the largest scale within this inertial subrange is greater than the maximum pulse-volume dimension. The pulse-volume averaging is treated as a high-pass filter process; i.e., only scales smaller than the largest pulse-volume dimension contribute to σ_T. The result is that ϵ can be expressed in terms of the measured σ_T and known pulse-volume parameters as follows:

$$\epsilon = \phi^{-1}\left[\frac{\sigma_T{}^2}{1.35\alpha(1-\gamma^2/15)}\right]^{3/2} \qquad (29)$$

where $\gamma^2 = 1 - (L/l_w)^2$; l_w and L define the beam width and pulse length, respectively; and $\alpha \cong 1.43$ is the Kolmogoroff constant (Lumley and Panofsky, 1964). The above applies whenever $(L/l_w) \leq 1$. When $(L/l_w) > 1$ the following relation applies:

$$\epsilon = L^{-1}\left[\frac{\sigma_T{}^2}{1.35\alpha(1-4\xi^2/15)}\right]^{3/2} \qquad (30)$$

where $\xi^2 = 1 - (l_w/L)^2$. These equations are given with the corrections to Frisch and Clifford (1974) that were first pointed out by Labitt (1979) and later given by Gossard et al. (1982).

Results of such calculations are indicated by the profiles shown in Fig. 21. These profiles were obtained from volume scans using the two NOAA X-band radars during Project Phoenix in 1978, when they were separated by 13 km. The spectral width estimates were determined when the radars were scanning a common chaff-filled volume in the dual-Doppler radar mode (see Section 3.1). Note that this differs from the conical scanning described above. Values at each height represent horizontal averages over about 30 km² obtained after estimates of $\sigma_T{}^2$ were interpolated to a Cartesian grid. Two levels of BAO tower data are shown for comparison.

4.2.5 The Turbulent Kinetic Energy Equation

From the discussion above and from the horizontally homogeneous form of the turbulent energy equation,

$$\frac{\partial \overline{e}}{\partial t} = \frac{\boldsymbol{\tau}_s}{\varrho} \cdot \frac{\partial \mathbf{V}}{\partial z} - \epsilon + \frac{g}{T}\,\overline{w\theta} + \frac{\overline{\partial(we)}}{\partial z} + \frac{\overline{\partial(wp/\varrho)}}{\partial z} \qquad (31)$$

it is apparent that all but the buoyance and the pressure-velocity correlation terms can be directly measured with radar techniques. Recent work by Gal-Chen (1978) suggests that perhaps even the remaining buoyancy and pressure-velocity terms may be recoverable from radar-derived wind fields by an inversion of the momentum equations.

5 SUMMARY

It is apparent that radar, both conventional and Doppler, can provide information about the PBL that is unobtainable by other means. The longer-wavelength S-band radars give us a detailed look at the overall structure of the convective field. Cell size, shape, and alignment are immediately apparent in these displays. The shorter-wavelength radars, because of their better ground clutter rejection, are able to observe at close range and close to the surface. As a consequence of observing at close range and because shorter pulse widths are available at these wavelengths, the shorter-wavelength radars offer high resolution in the measurement of three-dimensional wind fields by the dual-Doppler method. In addition, the possibility of retrieving thermo-dynamic properties from these data sets shows some promise. The high resolution and reduced clutter at these wavelengths also allow quantitative measurements of narrow chaff plumes. These characteristics, plus the ability to detect small tracers of air motion that occur naturally within the daytime PBL during many months of the year, allow radars of this type to produce continuous profiles of mean wind, Reynolds stress, turbulent kinetic energy, and other turbulence quantities within the PBL. Under certain conditions these data can be considered of "tower" quality.

For several reasons one might argue that some of the statistical quantities directly measured with radar are of even higher quality than what might be obtainable from a tower or aircraft. Under certain conditions a radar can continuously measure a variety of important quantities throughout the depth of the PBL and not just the lowest few hundred meters accessible to a tower or at the particular height flown by an aircraft. More important, however, is the rapid spatial sampling possible with radar. Tens of cubic kilometers of the PBL can be sampled in a few minutes with radar. The requirement for obtaining a good statistical sample of the PBL is much less intimidating. These arguments are made quite convincingly by Wyngaard (1983), who concludes that a "1,000-km line average (by an aircraft) is equivalent to an average over about a 25-km square."

Acoustic Remote Sensing

W.D. Neff, Wave Propagation Laboratory, NOAA, and R.L. Coulter, Argonne National Laboratory

1 INTRODUCTION

The use of conventional instrumentation to measure mean winds, temperature, humidity, and turbulence in the atmospheric boundary layer, coupled with advances in numerical modeling methods, has led to great improvements in our understanding of this region of the atmosphere. Coincident with these advances has been the development of a variety of remote-sensing devices. Exploiting the interaction of acoustic, radio, and light waves with the atmosphere's turbulent microstructure, these devices have provided a new means to measure and visualize the physical processes that govern the structure and evolution of the lower atmosphere. In this article, we describe the development and application of acoustic sounding methods to these studies.

Acoustic sounders, or sodars (an abbreviation for sonic detection and ranging), are unique instruments for the study of the lower atmosphere. Their uniqueness arises from the direct and strong interaction of sound waves with both the thermal and the velocity microstructure of the atmosphere. If the microstructure were uniformly distributed throughout the lower atmosphere, the measurement of turbulence and, through frequency analysis of the returned signal, the mean wind would be easily available for routine use. Because turbulence is intermittent and irregularly distributed, however, the development and application of acoustic techniques are much more difficult. In particular, because the distribution of turbulent microstructure depends on the interrelationship of the mean gradients of temperature and wind, the interpretation of acoustic sounding data is not easy for users conversant only with conventional measurements of wind and temperature. For example, Doppler-derived estimates of the wind are often available only from discrete echo strata under statically stable conditions, and monostatic sodars provide a measure of where turbulence is located within a temperature inversion, not a measure of the inversion itself.

To the engineer, seeking to develop an instrument to measure a continuous wind profile, this might well be disappointing. The atmospheric scientist, however, can gain exciting new insights into the processes of the lower atmosphere.

Progress in the development of acoustic techniques thus reflects two facets:

• Through improvements in engineering, sodars mimic the operation of conventional in situ instruments measuring winds and turbulence.

• By leading to an understanding of the mechanisms underlying the origin of acoustic echoes, sodars provide new insight into the physical processes affecting the lower atmosphere.

The engineering problems associated with the development of sodars have, perhaps, been the easiest to approach, and therefore this facet has received the most attention. The second facet, because it introduces an essentially new view of atmospheric structure, has resisted similar advances. The outline of this chapter reflects this imbalance: Section 2 provides an overview of the classical expressions for the acoustic-scattering cross sections, in terms of inertial subrange turbulence spectra. Included are theoretical expressions that allow the cross sections to be expressed as functions of conventional turbulence quantities and mean gradients of wind and temperature. Section 3 outlines the basic characteristics of sodar instrumentation developed in response to theoretical expectations. Section 4 addresses the integration of scattering theory and instrumentation. In particular, advances made in techniques for the measurement of wind and turbulence are outlined. In large part, Sections 3 and 4 reflect the first facet of sodar development: that of mimicking conventional in situ instruments. Section 5 focuses on the second facet of sodar development: that of relating sodar echo patterns to atmospheric processes. Although this section seems a little out of the general context of this article, the topic is also critical to further development of sodars as quantitative instruments since sodars make measurements only where there are echoes.

2 THE HISTORY AND THEORY OF ACOUSTIC SCATTERING

2.1 Historical Development

The history of acoustic sounder development falls into two parts. Prior to the pioneering papers of McAllister et al. (1969) and Little (1969), a number of essential advances had slowly taken place. Gilman et al. (1946) reported the construction of a device that they called the "sodar." With this device, using time-lapse photography of an oscilloscope trace, they followed the diurnal cycle of acoustic echoes from the atmosphere. Having originally anticipated that echoes would be returned from variations in the mean temperature structure of the atmosphere, Gilman et al. found order-of-magnitude larger returns than expected. In their conclusion (p. 279), they anticipated the later interpretation of the echoes as scattering from turbulent microstructure: "The sodar experimental results are consistent with the concept of a changing congeries of air masses differing slightly in temperature, velocity, vapor, and fog content representing, in effect, a rough finish superimposed on the picture obtained from gross meteorological measurements."

A hiatus in sodar development followed, until further theoretical developments stimulated the scattering experiments of Kallistratova (1959). These results then led to Monin's (1962) correct calculation of the scattering cross section in terms of the spectra of temperature and velocity, which agreed with an earlier, but less-known, calculation of Batchelor (1957). Little further application of sound-scattering techniques to the study of atmospheric structure resulted until McAllister (1968) advanced sodar technology to include a facsimile display similar to that used in sonar devices in oceanography. With this advance, sodar became available as a flow visualization technique.

The first Doppler sodar measurements began with the high-frequency bistatic scattering experiments of Kelton and Bricout (1964). Further development of minicomputers and microcomputers over the last decade allowed the rediscovery and rapid development of Doppler sodars to the point that several units are now available commercially to measure the mean wind.

Our treatment here is of necessity brief. However, many of the details of the early development, particularly advances in scattering theory, and of the experimental work in the decade from 1968 to 1978 can be found in the review paper of Brown and Hall (1978).

2.2 Acoustic Scattering Theory

The theory of sound scattering from locally isotropic and homogeneous turbulence using the Born approximation provides the following expression for the acoustic differential scattering cross-section area per unit volume (per unit solid angle; Batchelor, 1957; Tatarskii, 1961, 1971):

$$\eta(\theta_s) = \frac{1}{8} k^4 \cos^2(\theta_s) \left[\frac{\Phi_T(\varkappa)}{T_0^2} + \frac{\cos^2(\theta_s/2)E(\varkappa)}{\pi C^2 \varkappa^2} \right] \quad (1)$$

where θ_s is the angle at the scattering volume from the transmitter beam axis to the receiver; k, the acoustic wave number; $\varkappa$, the Bragg-scattering wave number, that is, $2k \sin(\theta_s/2)$; T_0, the local temperature; c, the speed of sound; $\Phi_T(\varkappa)$, the isotropic three-dimensional spectral density of temperature; and $E(\varkappa)$, that of turbulent kinetic energy. The principal features of this result are:

- Only $\Phi_T(\varkappa)$ contributes to backscatter.
- Neither $\Phi_T(\varkappa)$ nor $E(\varkappa)$ contributes at $\theta_s = 90°$.
- The spectra are evaluated at $\varkappa = 2k \sin(\theta_s/2)$.

The relation between the scattering cross section represented by Eq. (1) and the spectra of temperature and energy at wave number $\varkappa$, corresponding to the Bragg scale, does not provide the direct information about wind and temperature profiles that one would obtain from conventional in situ sensors. However, as we will see later, theories of the inertial subrange of turbulence, conservation equations for turbulence moments, and experimental results do provide a way to understand and interpret scattering data in at least a qualitative manner.

Equation (1) requires a few caveats: First, it represents the scattering due to small-scale variations in temperature and fluid velocity, both of which affect the phase speed of sound. Similar variations arise from the irregular distribution of humidity, as discussed by Wesely (1976) and in Section 4. Such effects, while generally small, can be significant in special environments such as the marine boundary layer (Burk, 1981). Second, hydrometeors can have a significant Rayleigh-scattering cross section, depending on particle size and sodar operating frequency as calculated theoretically by Little (1972) and observed experimentally by Melling and List (1978a) in the case of falling snow.

2.2.1 Alternative Expressions for Inertial Subrange Spectra

Corrsin (1951) provided the first step necessary to interpret the temperature spectrum required in Eq. (1), proposing an equation of the form

$$F_\theta(\varkappa) = \beta N \epsilon^{-1/3} \varkappa^{-5/3} \quad (2)$$

for the one-dimensional temperature spectrum where β is a constant estimated to be 0.8 (Kaimal, 1973), N is the rate of dissipation of one-half of the temperature variance, ϵ is the energy dissipation rate, and $\varkappa$ is the spectral wave number. The one-dimensional spectrum can be transformed to a three-dimensional isotropic scalar spectrum for use in Eq. (1), by the relation (Ottersten, 1969)

$$\Phi_T(\varkappa) = -(4\pi\varkappa)^{-1}[\partial F_\theta(\varkappa)/\partial\varkappa] \quad (3)$$

This yields

$$\Phi_T(\varkappa) = (12\pi)^{-1}5\beta N\epsilon^{-1/3}\varkappa^{-11/3} \quad (4)$$

Many experimental and theoretical studies utilize the temperature structure function defined by

$$D_T(\mathbf{r}) = <[T(\mathbf{x} + \mathbf{r}) - T(\mathbf{x})]^2> \quad (5)$$

where the temperature is measured at points $\mathbf{x}$ and $\mathbf{x} + \mathbf{r}$ and $< >$ indicates an ensemble average. This, according to Tatarskii (1971), can be related to the three-dimensional spectrum using

$$D_T(r) = 8\pi \int_0^\infty \{1 - [\sin(\varkappa r)/\varkappa r]\}\Phi_T(\varkappa)\varkappa^2\, d\varkappa \qquad (6)$$

Tatarskii shows from Eqs. (4) and (6) that

$$D_T(r) = a^2 N \epsilon^{-1/3} r^{2/3} \qquad (7)$$

where a^2 is an empirical constant equal to 3.2 (Kaimal, 1973). The parameter group $a^2 N \epsilon^{-1/3}$ is normally replaced by the temperature structure parameter C_T^2, which can now be determined experimentally as $C_T^2 = D_T(r)r^{-2/3}$, leading to

$$\Phi_T(\varkappa) = 0.033 C_T^2 \varkappa^{-11/3} \qquad (8)$$

Because of the straightforward differential temperature measurement used to calculate C_T^2, this formulation provides a simple method with which to evaluate the backscatter cross section. The practical application of such a procedure requires the use of a sodar equation relating transmitted and received acoustic energy that also takes into account the loss of energy through spreading and attenuation. Section 4 addresses such applications in more detail.

Similar analyses by Tatarskii (1971) and Tennekes and Lumley (1972) lead to a corresponding expression for the turbulent kinetic energy spectrum

$$E(\varkappa) = 0.76 C_V^2 \varkappa^{-5/3} = 1.5\epsilon^{2/3}\varkappa^{-5/3} \qquad (9)$$

which uses the velocity structure function

$$C_V^2 = <[U(\mathbf{x} + \mathbf{r}) - U(\mathbf{x})]^2> r^{-2/3} \qquad (10)$$

Equations (8) and (9) can then be used in Eq. (1) to rewrite the scattering cross section as

$$\eta(\theta_s) = \frac{1}{8}k^4 \varkappa^{-11/3}\cos^2(\theta_s)\left[\frac{0.106N\epsilon^{-1/3}}{T_0^2} + \frac{\cos^2(\theta_s/2)}{\pi C^2}1.5\epsilon^{2/3}\right] \qquad (11)$$

or alternatively in terms of C_T^2 and C_V^2

$$\eta(\theta_s) = \frac{1}{8}k^4 \varkappa^{-11/3}\cos^2(\theta_s)\left[\frac{0.033C_T^2}{T_0^2} + \frac{\cos^2(\theta_s/2)}{\pi C^2}0.76C_V^2\right] \qquad (12)$$

The first term in the brackets represents the contribution to the scattering at angle θ_s due to temperature inhomogeneities, and the second represents the contributions from turbulent velocity fluctuations. The contribution to the scattering cross section from each term in the equation can be calculated and displayed as in Fig. 1.

2.2.2 Experimental Tests of the Scattering Theory

Kallistratova (1961) reported the first test of the angular dependence of the scatter predicted by Eq. (1), which showed excellent agreement with theory. However, difficulties arise in the measurement of the scattering cross section because of the problems associated with the measurement of system parameters (Section 4). To test the feasibility of such measurements, Neff (1975) and Asimakopoulos et al. (1976) compared backscatter signals with in situ measurements at a single height, finding agreement within a factor-of-two experimental error.

Later measurements by Haugen and Kaimal (1978) showed an attenuation of sound with range in excess of that due to viscous and molecular relaxation effects. Prior to these experimental results, Brown and Clifford (1976) calculated the spreading of the beam due to turbulence that relates to this excess-attenuation problem. Following this, Neff (1978) carried out a multi–beam-width experiment to measure the excess attenuation and compared his result with that predicted from the results of Brown and Clifford, while also evaluating the effect of refraction of a Gaussian-shaped beam by a mean wind. Neff's results showed that application of Brown and Clifford's theory, making the assumptions that the transmitted beam is broader than the receiving beam and that reciprocity does not apply because of the long propagation times involved, underestimates the magnitude of the effect. Calculation of additional turbulence effects on the scattered and received beam (Clifford and Brown, 1979) now appears to provide an adequate theory.

Evaluation of the scattering from the turbulent kinetic energy spectrum is somewhat more complicated because it is

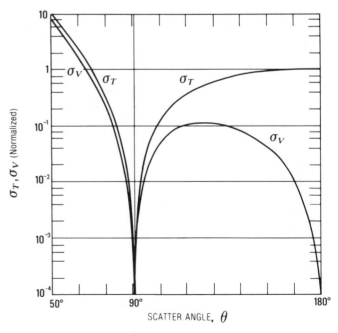

Fig. 1 Contributions to scattering cross section of separate effects of temperature σ_T and velocity fluctuations σ_V as a function of scatter angle.

necessary to account for the geometrical intersection of the two sodar beams used in the bistatic mode. Caughey et al. (1978) and Thomson et al. (1978) evaluated the energy spectrum $E(x)$, at a single height, again finding results within the expected experimental error.

2.3 The Vertical Distribution and Interpretation of the Acoustic-Scattering Cross Section

The theoretical calculations of the acoustic-scattering cross section leading to Eq. (12) followed a decade in which rapid advances had been made in theories of locally isotropic and homogeneous turbulence. Interpretation of the acoustic-scattering cross section by applying these theories to acoustic sounding followed naturally, as reflected in early papers, such as that of McAllister et al. (1969). The useful application of such theoretical results to the development of sodars and to atmospheric studies requires two efforts:

- Increasing our understanding of how the atmosphere arranges to produce the multitude of echo patterns that we observe;
- Developing relations between measurements of $\Phi_T(x)$ and $E(x)$, and other mean and turbulence properties of the atmosphere.

Equation (12) demonstrates the theoretical dependence of the acoustic-scattering cross section on N and ϵ (or equivalently $C_T{}^2$ and $C_V{}^2$). The effectiveness of sodar technology depends primarily on the distribution of these quantities in

the atmosphere combined with sodar system performance (described in Section 4).

2.3.1 Profiles in Simple Boundary Layers

In situ measurements of $C_T{}^2$ and $C_V{}^2$ have been carried out in boundary-layer experiments using towers and balloon tether lines as instrument platforms. For example, Wyngaard (1973) summarized the surface-layer behavior of $C_T{}^2$. Outside of the surface layer, Kaimal et al. (1976) described profiles of $C_T{}^2$ and $C_V{}^2$ in the convective boundary layer, and Caughey et al. (1979) followed with measurement of profiles in the evolving stable boundary layer. Figure 2 summarizes these results. These experiments using in situ sensors have given us some expectations for the behavior of these profiles when the atmosphere is simply structured.

2.3.2 Profiles in the Statically Stable Atmosphere

From the results shown in Fig. 2, one might well expect sodars always to provide a good measure of atmospheric properties in the boundary layer. This is true in the convective boundary layer (as described in detail in Section 4), but the stable atmosphere presents more of a challenge. In particular, the statically stable boundary layer is often shallow, sometimes with a depth less than the minimum range of conventional sodars (20 m to 40 m). To further complicate matters, as pointed out by Brost and Wyngaard (1978) and

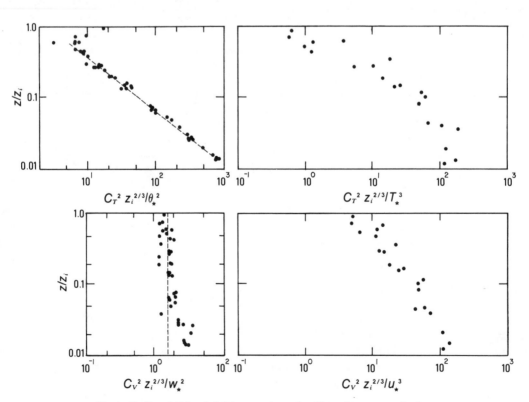

Fig. 2 Profiles of $C_T{}^2$ and $C_V{}^2$ in convective and stable conditions, normalized by conventional surface-layer scaling parameters. (Adapted from Kaimal et al., 1976, and Caughey et al., 1979.)

Nieuwstadt and Driedonks (1979), the low level of turbulence in the stable boundary layer means that its structure is strongly influenced by initial conditions, temporally and spatially changing surface conditions, and large-scale effects such as advection of heat and momentum.

Profiles of wind and temperature measured at 50-m intervals at the Boulder Atmospheric Observatory (BAO; Kaimal and Gaynor, 1983) when compared with digitized sodar returns, show the often-complicated dependence of acoustic backscatter on the vertical distribution of wind and temperature (Fig. 3).

2.3.3 Further Expressions for the Backscatter Cross Section

The turbulence on which sodars depend arises through three principal mechanisms:
• In the statically stable boundary layer, surface friction leads to shear-induced turbulence.
• In the convective boundary layer, solar heating of the surface leads to the upward mixing of air heated at the surface. In the presence of a capping inversion, the impact of buoyant parcels causes entrainment of air into the boundary layer from the overlying free atmosphere.
• In the statically stable free atmosphere, in the absence of latent heat processes, the occurrence of turbulence is usually related to a reduction of Richardson's number, Ri, below 0.25. Ri is defined by $(g/T)(\partial\Theta/\partial z)/|\partial\mathbf{V}/\partial z|^2$, where g is the acceleration due to gravity, T is the local temperature, Θ is mean potential temperature, $\mathbf{V}$ is the wind velocity, and z is the vertical coordinate.

Alternative expressions for the backscatter cross section begin with the equation for the temperature variance as defined, for example, by Wyngaard (1975):

$$(\partial\overline{\theta^2}/\partial t) + 2(\partial\Theta/\partial x_j)\overline{u_j\theta} + U_j(\partial\overline{\theta^2}/\partial x_j) + \partial\overline{u_j\theta^2}/\partial x_j = -2N - N_R \quad (13)$$

where $U_j = (U,V,W)$ is the mean velocity, $u_j = (u,v,w)$ and θ are perturbations from the mean, and repeated indices imply summation. The term N_R represents the rate of radiative destruction of temperature variance following André et al. (1978). Advection by the mean velocity of an inhomogeneous distribution of $\overline{\theta^2}$ is usually neglected. In a steady state under horizontally homogeneous conditions, the production, $2\overline{w\theta}\,\partial\Theta/\partial z$, and divergence of the transport of variance, $\partial(\overline{w\theta^2})/\partial z$, are balanced by the dissipation terms $2N$ and N_R. The temperature-related backscatter term can then be written

$$\eta(180°) = (1/8)k^{1/3}(0.105/T^2)\epsilon^{-1/3}\,[-\overline{w\theta}(\partial\Theta/\partial z) - \partial\overline{w\theta^2}/\partial z + N_R] \quad (14)$$

In the convective boundary layer, all terms are generally significant. In particular, because w is strongly correlated with θ^2 from the heated surface, the divergence term makes a major contribution to the scattering. André et al. (1978) express the radiative term as $N_R = 0.1Q\overline{\theta^2}$, where Q is the mixing ratio of water vapor. This term is normally neglected except in numerical models of the marine boundary layer (Burk, 1981).

In the stable atmosphere, the term $\partial\overline{w\theta^2}/\partial z$ is usually neglected. This approximation has some support from

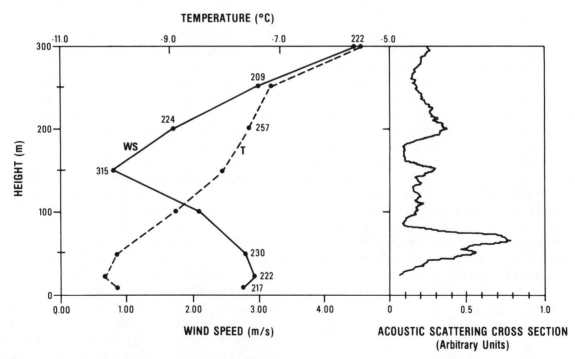

Fig. 3 *Typically complex profiles of acoustic backscatter, temperature, and wind obtained at the BAO.*

surface-layer measurements (Wyngaard and Cote, 1971). Tatarskii (1971) suggests that its neglect is reasonable when the turbulence is locally homogeneous. With these caveats and neglecting N_R, the backscatter cross section takes the form

$$\eta(180°) = (0.105/8)(k^{1/3}/T^2)\epsilon^{-1/3}[-\overline{w\theta}(\partial\Theta/\partial z)] \qquad (15)$$

To provide a simple heuristic interpretation of the backscatter cross section, we introduce the eddy diffusion coefficients for temperature and momentum, K_H and K_M, respectively:

$$K_H = -[\overline{w\theta}/(\partial\Theta/\partial z)]$$
$$K_M = -[\overline{uw}/(\partial U/\partial z)] = -[\overline{vw}/(\partial V/\partial z)] \qquad (16)$$

and the flux Richardson's number, defined by:

$$Rf = Ri(K_H/K_M) \qquad (17)$$

The formulation for K_M follows from assumptions about second-order turbulence closure (Freeman, 1977). The conservation equation for turbulent kinetic energy can be written as (e.g., Brost and Wyngaard, 1978)

$$\epsilon = K_M S^2 - K_H N_B^2 \qquad (18)$$

where $S^2 = |\partial\mathbf{V}/\partial z|^2$ and $N_B^2 = (g/T)(\partial\Theta/\partial z)$. Equation (15) then becomes, for the statically stable, dry atmosphere,

$$\eta(180°) = (0.105/8)(k^{1/3}/T^2)K_H^{2/3}[Rf/(1 - Rf)]^{1/3} \\ \times (\partial\Theta/\partial z)^{5/3} \qquad (19)$$

Inspection of this equation shows that the backscatter under stable conditions depends most strongly on the vertical gradient of potential temperature. However, this equation reveals explicitly the ambiguity in the quantitative interpretation of the scattering: Strong gradients can combine with weak turbulence to produce scattering equal to that associated with strong turbulence interacting with weak gradients. Other expressions for the backscatter cross section are suggested, but usually they involve indirect turbulence parameters such as an "outer scale," as in McAllister (1968). As we shall show later, Eq. (19) has particular utility in second-order closure models that calculate K_H and K_M directly. With higher-order models Eq. (14) becomes useful.

As can be seen from Eq. (19), acoustic scattering depends in a complicated fashion on the distribution of turbulence and mean gradients of wind and temperature in the atmosphere. In the following sections we will address how these and other theoretical results can be used to make practical use of sodar data.

3 INSTRUMENTATION

3.1 Acoustic Sounding Systems

3.1.1 Electronics

Because of the relatively slow speed of sound compared with that of electromagnetic waves, instrumentation for and processing of acoustic data are relatively simple. Therefore, acoustic techniques are relatively inexpensive, and sophisticated analysis of the data is not usually required.

Most components for monostatic sodar are readily available. The parabolic reflector of a searchlight combined with a foam-lined acoustic shield can provide an effective antenna with a conventional compression driver serving as a transmitter/receiver. Display devices can be adapted from standard sonar facsimile recorders. Digital processing for Doppler analysis is now easily available with inexpensive microprocessors.

3.1.2 Facsimile Displays

A major advance in sodar technology was provided by McAllister (1968) when he developed a device to display acoustic echoes from the atmosphere similar to devices used to display sonar echo data from the ocean. In most such devices, a pen is mechanically drawn across a specially treated paper, with each traverse corresponding to a period of interest beginning at the time of the transmitted sound pulse. The electrical current through the paper, from the pen to a conducting roller underneath, is proportional to the signal strength. The burning of the paper then provides a visual record of the return echo strength. After each trace is written, the paper advances in time to draw the subsequent trace adjacent to the previous one. From such a display of transmitted pulses, a time-height record of the atmospheric structure moving through the antenna beam is constructed. The height (or range) of echoes displayed across the paper is deduced by knowing the time from transmission and the average speed of sound. More recently, microprocessor technology, color graphics terminals, and dot matrix printers have provided alternatives to the original analog facsimile displays.

Facsimile records have provided a fascinating picture of atmospheric structure, including such features as thermal plumes, nocturnal inversions, waves, and fronts. Because of the complexity of atmospheric fine structure, the choice of facsimile display is critical to any particular application. In general, the pulse repetition period of the sodar, the pulse length, the filtering of the return signal (narrow bandwidths improve signal-to-noise ratios but reduce temporal resolution, producing smeared recordings), and the dynamical range of the facsimile electronics determine the maximum resolution of such analog recordings. For many routine applications, high-resolution recordings are not necessary. For monitoring diurnal mixing cycles in air quality applications, commercial devices are available with highly compressed records (on the order of 0.015 m/h) that can also be used for developing long-term climatologies. In such devices, dynamic range is not even a critical factor because the vertical extent of the echoes is the major parameter of interest.

For a number of studies, higher resolution is necessary. For example, in air quality tracer studies, release times may be short, and sodar records can provide a useful indication of the type of turbulence regime into which the tracer is released and, hence, an indication of the type of diffusion to be expected. High-resolution records are collected routinely to aid in interpretation of the tower data from the BAO.

3.1.3 Antenna Configuration and Deployment

Sodars can use either of two geometrical arrangements. The first, called monostatic, uses a collocated transmitter and receiver. In this mode, only the temperature spectrum contributes to the backscatter (cf. Eq. 1), and hence researchers use this arrangement to infer the qualitative nature of the temperature profile. The second arrangement, consisting of separated transmitters and receivers, is referred to as a bistatic system. This mode provides signals that are scattered from small-scale velocity fluctuations as well as from temperature inhomogeneities. This mode of acoustic sounding is unique among remote-sensing devices insofar as it is the only remote-sensing method directly sensitive to the small-scale velocity field. Various bistatic arrangements can provide a picture of the vertical distribution of turbulence as well as better signals for Doppler wind estimates when the temperature lapse rate is close to adiabatic.

3.2 Doppler Systems

Doppler sodars are used primarily to obtain estimates of wind speed by analysis of the frequency content of the detected signal (cf. Section 4). Figure 4 is a block diagram of a typical bistatic Doppler system.

Figure 5 reveals two possible Doppler sodar arrangements: the first is referred to as bistatic, and the second is a monostatic configuration.

In the first case the transmitter at T transmits toward and above the receiver point R. The time for the sound to propagate from T to point O and thence, after scattering, to point R is the same for any point on an ellipse defined with T and R located at the foci. The narrow beam of the receiving antenna limits the reception to a narrow section of the ellipsoid near point O. Since only phase shifts associated with phase fronts tangent to the ellipse at point O will be associated with a Doppler shift, the frequency shift measured is that due to the wind along the bisector of the angle between T and R at O. Referring again to Fig. 5, monostatic Doppler sodars use a single narrow-beam transmitter T, which also doubles as a receiver. In this mode, power is concentrated along the beam and the Doppler shift corresponds to motions along the beam.

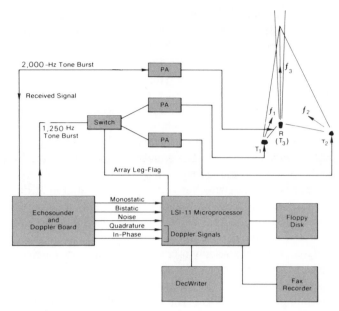

Fig. 4 *Block diagram of microprocessor-controlled bistatic Doppler sodar. (From Neff et al., 1980.)*

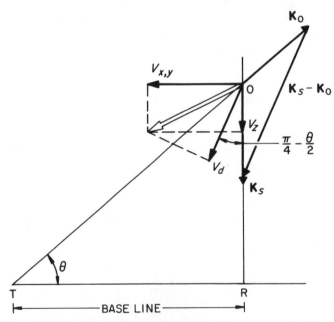

Fig. 5 *Bistatic and monostatic Doppler geometries. The vector component $V_{x,y}$ is the horizontal component of the instantaneous wind velocity in either the xz or the yz plane. The vertical velocity component is denoted V_z. V_d is the projection of the instantaneous wind velocity in the direction defined by the difference between the wave numbers $\mathbf{K_o}$ and $\mathbf{K_s}$. (From Kaimal and Haugen, 1977.)*

Determination of the total wind vector in general requires the use of three transmitter-receiver combinations, each sensing a different component of the wind. This requirement can be accomplished in two basic ways: triple monostatic and bistatic. The triple monostatic system features three antennas, usually collocated, which point at different azimuth and elevation angles in order to obtain the necessary wind components. The bistatic system utilizes two separated receiver-transmitter links along with a single (usually vertically pointing) monostatic transmitter-receiver. These links can be established by using either a single transmitter and two separated receivers, which detect the signal scattered at different directions from the single transmit beam, or a single central receiver (the same as the monostatic receiver) and two separated transmitters, as in Fig. 5.

Signal processing usually proceeds as follows. The received signal is protected from the transmit pulse by a transmit-receive switch, which turns on the receiver only after the transmit pulse has ended. The signal is then preamplified, either at the receiver electronics or at the receiver antenna, to decrease the effects of electrical line noise. This gain usually increases with time to compensate for attenuation and the spherical spreading of the scattered acoustic energy. Within the receiver electronics an envelope detector provides an estimate of the signal amplitude, and a quadrature detector detects the out-of-phase and in-phase components of the signal. The signal may be heterodyned to a lower frequency for digital analysis at reduced sampling rates. The signals are then transmitted to a computer that analyzes and calculates

the Doppler shift (δf), which in turn is related to the wind speed.

The two methods for wind-speed determination have enough differences that a comparison of advantages and disadvantages is important. A significant advantage of triple monostatic systems is that geometrical considerations are minimized: The derived velocity for a given frequency shift is independent of height. The bistatic system requires significant geometrical corrections (Section 4). In addition, the ability to locate the three antennas at one point enhances the convenience and mobility of triple monostatic systems, whereas bistatic systems usually require 100- to 300-m base lines with cabling to match. Also, if there is sufficient return from temperature fluctuations throughout the probing range, the triple monostatic system may be able to achieve greater heights because of the coincidence of the transmit and receive beams. An additional disadvantage of the bistatic system is its increased susceptibility to ground clutter because of refraction and reflection of the broad-beam transmit pulse from low objects. An example of this problem is shown in Fig. 6, where the Doppler shift was biased to zero in the first 200 m of the profile. Reflections from solid objects—in this case, equipment trailers—appear as bands at a constant height. Such effects can be particularly pronounced in very stable conditions when sound can be strongly refracted toward the ground.

On the other hand, the triple monostatic system requires heavy shielding on all three antennas, two of which are not pointing vertically and are thus more susceptible to en-

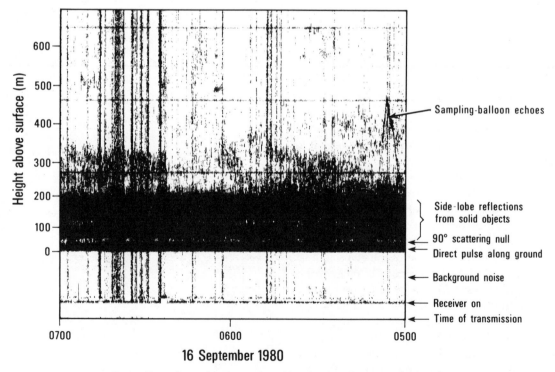

Fig. 6 Facsimile recording showing typical bistatic echoes and an example of contamination from side-lobe echoes reflecting from solid objects.

vironmental noise contamination. More important, perhaps, is the fact that the triple monostatic system depends upon estimates made at three different positions in space to derive the velocity and turbulence estimates. This can be a severe limitation even for mean wind estimates in situations where conditions are not homogeneous—over complex terrain or land-water interfaces, for example. Finally, because the monostatic system is dependent upon C_T^2 alone for its signal source, its signal may be smaller than the bistatic signal, particularly when the atmosphere is adiabatic.

Both systems have a similar, less severe problem in that they transmit alternately among the three antennas. (This means of transmission may not be a necessity for the triple monostatic system if the beam patterns of the three antennas do not interfere with one another.) Thus the individual components of the wind are not derived from the same volume of air. This problem can be overcome for both systems by transmitting at different frequencies from each transmitter and analyzing the signal near each transmit frequency for the different components.

Most commercial and research systems have incorporated some means for estimating the reliability of a given frequency shift, or wind estimate. Details differ in implementation but usually involve some estimate of the signal-to-noise ratio (S/N) and absolute signal level during data collection.

4 QUANTITATIVE ANALYSIS OF SIGNAL AMPLITUDE AND DOPPLER SHIFTS

As we discussed earlier, the potential exists to determine in real time and on a continuous basis, profiles of variables of prime importance for studies of atmospheric transport, dispersion, energy budgets, and turbulence within the lower planetary boundary layer (PBL). These quantities include velocity dissipation rate ϵ, temperature variance dissipation rate N, mean wind, variance of horizontal wind σ_u and vertical wind σ_w, and several other quantities such as heat flux that can be derived from more basic variables. The challenge is to bring the potential closer to reality.

4.1 Sodar Equation

The structure functions of temperature C_T^2 and velocity C_V^2 are measures of the inertial-subrange temperature and velocity spectra, respectively, and are related to the strength of the scattered signal by Eqs. (1), (8), and (9). To estimate their values, we can account for scattering angle, antenna weighting, and geometrical divergence with the dimensionless system functions $S_T = S_T(a)$ and $S_V = S_V(a)$ and the received power dP, written as

$$dP = P_0 A_e d^{-2} e^{-2a\alpha} k^{1/3} (C_T^2 T^{-2} S_T + C_V^2 c^{-2} S_V)\, da \qquad (20)$$

where $2d$ is the distance between the antennas, c is the phase speed of the transmitted energy, α is the molecular attenuation coefficient of sound in air, P_0 is the transmitted power,

and A_e is the effective receiver antenna area. The parameter $a = ct/2$ is one-half the distance the sound pulse has traveled after emission time $t = 0$. Procedures to calculate α are given by Neff (1975), and the calculation of A_e is given by Hall and Wescott (1974). The system functions S_T and S_V are evaluated in an elliptical coordinate system through integration over ellipsoids of revolution defined by the positions of the transmit and receive antennas and by the travel time of the acoustic radiation (for details, see Appendix). These functions account for the finite size and relative weighting by scattering angle effects within the scattering volume as a function of time or distance. In practice, the integrations need not be carried out over the complete half-space because the integral is limited by the angular projection characteristics of the transmit and receive antennas and the length and position of the scattering volume defined by the pulse length and time after sound emission (Fig. 7). Since the most sensitive part of the antenna beam typically subtends an angle of about 10°, the effective size of the volume interrogated at any instant is approximately $0.004\pi c\tau R^2$, where τ is the duration of the pulse and R is the range. In the case of bistatic geometries, this is an upper estimate, because the relative pointing angles of transmitter and receiver sharply limit the region coincident to both beams. Once the directivities (F_t, F_r) of the antennas have been determined, Eq. (20) can be evaluated for a given power output, pulse length $c\tau$, and pointing angle of the antennas as in Fig. 8. The peaked character is created by the finite range over which the two beams intersect.

Because F_t and F_r are functions of both elevation and azimuth angles, there can be significant errors due to misalignment of the transmit and receive antennas' pointing angles (Coulter and Roth, 1979). In particular, although only small errors result from a narrow-beam receiver and broadbeam transmitter, a narrow-beam transmitter-and-receiver combination can lead to errors on the order of 50% for a 5° misalignment. Nevertheless, this method can be used with care under correct conditions to evaluate profiles of C_T^2 and C_V^2.

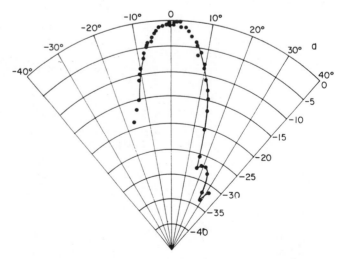

Fig. 7 A measured beam pattern (directivity) for a typical receiver where the radial scale is sensitivity in decibels.

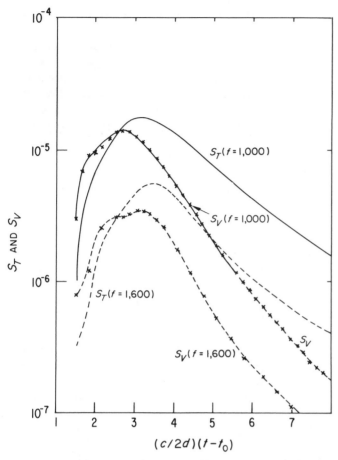

Fig. 8 System functions S_T and S_V for the Argonne sodar at two different frequencies as a function of the time t elapsed after emission of the sound pulse normalized by 2d/c, where d is the length of the base line. Note the sharper character at higher frequencies due to the more peaked character of the antenna directives at higher frequencies.

A much simpler condition results when the transmitter and receiver are collocated in a monostatic system. Then Eq. (20) reduces to

$$\frac{1}{P_0}\frac{dP}{da} = 0.0256 A_e a^{-2} e^{-2a\alpha} k^{1/3} C_T^2 T_0^{-2}$$
$$\times \int_0^{\pi/2} F_r^2(\theta)\sin\theta \; d\theta \qquad (21)$$

Since the integral is not a function of a, i.e., time, it can be evaluated once and for all for a particular antenna configuration. Alignment errors are minimized because the transmit and receive antenna beams coincide. In this configuration, one evaluates only C_T^2 (see Eq. 1); thus the combination of this result with that obtained with Eq. (20) results in the estimation of C_V^2 so long as conditions are homogeneous. In fact, however, the magnitude of C_T^2 is usually one to two orders of magnitude less than C_V^2 (Table 1) so that it can often be neglected altogether in the estimate of C_V^2 (Thomson et al., 1978; Moulsley et al., 1981). As an exception, Coulter and Underwood (1980) found that in the artificially induced turbulence within cooling tower plumes, C_T^2 and C_V^2 are about the same magnitude because of the large amount of thermal turbulence present.

4.1.1 Measurements

To date, there have been a number of sodar measurements of C_T^2 and C_V^2 (e.g., Neff, 1975; Asimakopoulos et al., 1976; Thomson et al., 1978; Moulsley et al., 1981; Coulter and Wesely, 1980). Table 1 summarizes these results. Both parameters vary over several orders of magnitude as a function of meteorological conditions and time. Values range over roughly an order of magnitude on time scales on the order of minutes (e.g., within and outside thermal plumes), and several orders of magnitude diurnally. Thus errors of a factor of 2 may not be as serious as might first appear. In fact,

TABLE 1
Estimates of C_T^2 and C_V^2 Made with Sodar under Varying Conditions

H_T (m)	C_T^2 (°C²/m²/³)	Max.-Min. (Period)	C_V^2 (m⁴/³/s²)	Max.-Min. (Period)	Atmospheric Conditions	Investigator
90	6×10^{-4}	8×10^{-4} (12 min)			Stable	Asimakopoulos et al.
90	1.4×10^{-3}	5×10^{-3} (6 min)			Convective	Asimakopoulos et al.
92	8×10^{-4}	1×10^{-3} (110 min)			Stable	Neff
	1×10^{-3}	6×10^{-2} (80 min)			Convective	Neff
137	3×10^{-4}	5×10^{-4} (60 min)	4×10^{-2}	6×10^{-2} (60 min)	Convective	Moulsley et al.
91	1×10^{-3}		3×10^{-2}		Stable	Moulsley et al.
137	1×10^{-4}		1.5×10^{-2}		Stable	Moulsley et al.
183	4×10^{-4}		2×10^{-2}		Stable	Moulsley et al.
425	2×10^{-4}	1×10^{-4}	4×10^{-3}	1×10^{-4}	Stable	Thomson et al.
250	4×10^{-4}	3×10^{-4} (4 min)	8×10^{-3}	2×10^{-2}	Convective	Thomson et al.
50–150 (from source)	5×10^{-1}*	1.0* (60 min)	8×10^{-1}	6×10^{-1} (60 min)	Stable	Coulter and Underwood

*Source of fluctuations is cooling tower plume rather than atmosphere.

it is probably unrealistic to expect to obtain measurements of structure functions more accurate than this at present.

In addition to calibration errors, there are atmospheric contributions to measurement uncertainties. These include turbulent scattering out of the beam during passage to and from the scatterer; partial filling of the scattering volume—for example, the presence of quasi-laminar layers thinner than the length of the sampling volume; and the contributions to scattering by moisture fluctuations. The structure function of humidity $C_q{}^2$ lends a second-order contribution to acoustic scattering, being relatively unimportant except in very moist environments and near clouds and fog or soon after heavy rainfall. The crossed-structure function of humidity and temperature fluctuations $C_{Tq}{}^2$ is also a potentially important contributor if the correlation of humidity and temperature r_{qT} is large in the size range being interrogated by sodar. Wesely (1976) shows that the modification to measurements of $C_T{}^2$ can be written in terms of the Bowen ratio, $B = H/LE$, where H is the surface sensible heat flux and LE is the surface latent heat flux:

$$C_{na}{}^2 = C_T{}^2(2T)^{-2}[1 + (0.06/B)^2 + 0.5r_{qT}/B] \qquad (22)$$

where $C_{na}{}^2$ is the acoustic refractive index structure function. Thus, corrections to $C_T{}^2$ can be significant at small values of B (< 0.1). It has a more important contribution to sodar determination of surface heat flux, as shown next.

4.1.2 PBL Structure

Measurements of structure parameters can be related directly to PBL structure through the relationships for the energy and temperature spectra in the inertial subrange (Eqs. 7, 8, and 9), whence

$$\epsilon = (C_V{}^2/1.97)^{3/2} \qquad (23)$$

and

$$N = 0.22C_T{}^2C_V \qquad (24)$$

Note that measurement of N requires a knowledge of both $C_T{}^2$ and $C_V{}^2$.

Wyngaard et al. (1971) have shown that the surface heat flux H can be related to measurements of $C_T{}^2$ through

$$C_T{}^2 = 2.68(g/T)^{-2/3}(\varrho c_p z/H)^{-4/3} \qquad (25)$$

where g is the acceleration of gravity, ϱ is the density of air, z is the height, and c_p is the heat capacity of air at constant pressure. The existence of a $-4/3$ dependence upon height in the well-mixed PBL has been observed by several investigators (Tsvang, 1969; Neff, 1975; Coulter and Wesely, 1980). Hence, one may be able to extrapolate mixed-layer measurements of $C_T{}^2$ to surface heat flux with this relationship. The possibility of obtaining surface heat flux estimates

through sodar measurements in the PBL makes it possible to determine surface heat flux values averaged over a larger surface area because of the volume average inherent in every sodar estimate, as discussed above. This technique is particularly relevant to investigations over nonhomogeneous terrain (e.g., variable crop covers). Coulter and Wesely (1980) demonstrated its feasibility (Fig. 9); however, it is necessary to incorporate water vapor effects into the equations both in $C_T{}^2$ and in H:

$$H = 0.48\varrho c_p(g/T)^{1/2}C_{na}{}^2\gamma_{so} \qquad (26)$$

where

$$\gamma_{so} = \{[1 + (0.07/B)^2]/[1 + (0.06/B)^2]\}^{3/4}\{1/[1 + (0.07/B)]\}$$

Even with this correction applied, however, there are still some unexplained large values in the morning hours, possibly due to nonstationary effects of dynamic changes within the growing mixed layer. Because sodar estimates are volume averages, comparison with point measurements in the surface layer is difficult.

The PBL is often bounded by a capping inversion and entrainment layer whose dynamics and characteristics are functions of the temperature, velocity, and humidity differences across the layer. Wyngaard and LeMone (1980) related these differences to the structure functions of temperature and humidity and to temperature-humidity correlations. They showed, for example, that within the capping inversion,

$$C_T{}^2z_i{}^{2/3} = f(H,\Delta\theta_v,\Delta q) \qquad (27)$$

where $\Delta\theta_v$ and Δq are the virtual potential temperature and specific humidity jumps across the interfacial region. If Δq is small, $C_T{}^2$ can be used to predict the temperature jump across the inversion, which is of prime importance to modelers of air pollution concentrations and transport. Simultaneous measurement of $C_q{}^2$ with radar, for example, enables an even better estimate.

4.2 Doppler Analysis

Some of the most useful results from sodar investigations have come recently with the increased reliability of velocity estimates derived from frequency analysis of the returned signal. Measurements of horizontal wind profiles are relatively straightforward; vertical velocities and statistics are more difficult to obtain but quite practicable and useful for PBL investigations and monitoring.

For any form of radiation, the detected frequency is a function not only of the transmitted frequency, but also of the relative motion of the transmitter and receiver. The change in frequency that would be detected at the scatterer (if possible) is proportional to the rate of change of the distance between transmitter and scatterer (V_r) and the unit of frequency (f_0). The energy detected at the receiver is again shifted due to the

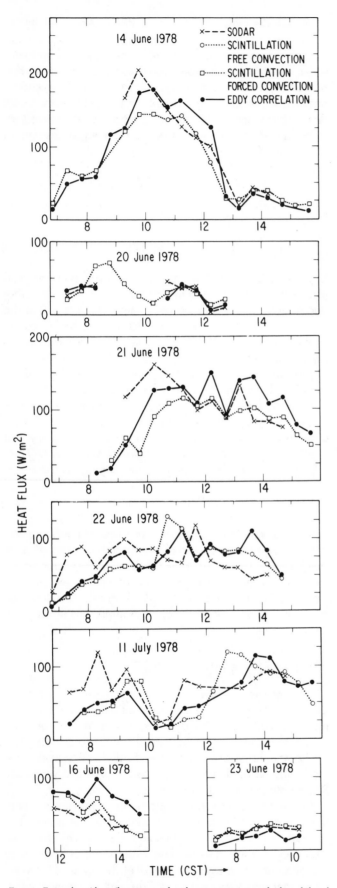

Fig. 9 *Examples of heat flux measured with remote-sensing methods and the eddy correlation method during daytime.*

relative motion. Hence the total frequency shift, to first order, for the simple case of a single scatterer and single antenna is given by

$$\delta f = -2V_r f_0 / c \qquad (28)$$

Thus, if the transmitted frequency is known and the received frequency is measured, the motion of the scatterer relative to the transmitter-receiver geometry is known. It is in principle a simple extension to make three measurements from different directions to determine the three-dimensional velocity vector of the scatterer.

The volume in space interrogated by sodar at any instant is much larger than the acoustic wavelength; hence there are always many individual scattering elements within the volume that contribute to the received signal. Both amplitude and frequency estimates are dependent upon the motions and relative phase of all the scatterers within the volume. That is (Thomson and Coulter, 1974),

$$A_r = \sum_{i=1}^{N} A_i e^{i(\omega t + \phi_i)}$$
$$= e^{i\omega t} \sum_{i=1}^{N} A_i e^{i\phi_i} \qquad (29)$$

where ω is the rotational frequency of the transmitted signal, A_r is the resultant amplitude detected at the receiver at any instant, A_i is the amplitude due to the ith scatterer, and ϕ_i is the phase angle of the ith scatterer, which is related to the Doppler shift for the single scatterer (δf_i) by

$$\delta f_i = -(d\phi_i / dt) \qquad (30)$$

The receiver senses the Doppler shift due to the total ensemble of scatterers, i.e.,

$$\delta f = \frac{d}{dt}\left[\tan^{-1}\frac{\text{Real}(A_r e^{-i\omega t})}{\text{Imag}(A_r e^{-i\omega t})}\right] \qquad (31)$$

Thus the Doppler frequency is a time-dependent estimate derived from the vector addition (in phase space) of the ensemble. In general the wind speed is calculated from a time series that is related to a length in space; since the instantaneous frequency is a fluctuating quantity determined by the motion of scatterers within the range gate, both the mean and statistics on the distribution of values are important.

Acoustic energy travels a million times more slowly than electromagnetic radiation. The response time of thermal or velocity fluctuations on the scale of acoustic wavelengths (10–20 cm) is less than the time between acoustic pulses (3–9 s); thus pulse-to-pulse correlations are not high. As a result, analysis techniques for acoustic waves are somewhat different from those for electromagnetic radiation. Rather than calculate phase changes or derive spectra over several pulses, we make frequency estimates from data within a single pulse, and values from several pulses are averaged together to form the time-averaged estimate.

The general expression for the frequency shift δf due to motion of a scatterer is given by

$$\delta f = (2\pi)^{-1}\mathbf{K} \cdot \mathbf{V}_p \qquad (32)$$

where $\mathbf{V}_p$ is the component of the wind in the plane of the transmitter and receiver boresights and $\mathbf{K}$ is the scattering vector, defined by

$$\mathbf{K} = \mathbf{K}_s - \mathbf{K}_0 \qquad (33)$$

as was illustrated in Fig. 5. Only motions of scatterers along $\mathbf{K}$ (which bisects the angle formed by the incident wave-number vector $\mathbf{K}_0$ and the scattered wave-number vector $\mathbf{K}_s$) lead to frequency shifts. The rate at which scatterers cross the equiphase surfaces (perpendicular to $\mathbf{K}$ and defined as the locus of points such that the total time from transmitter to receiver is constant) determines the magnitude of the Doppler shift. In general, the surfaces are ellipsoids of revolution with foci at the transmitter and receiver, as illustrated in Fig. 7, as long as refraction effects are ignored; when the system is monostatic the foci merge and the surfaces become spheres. Thus, the three-dimensional wind vector can be determined by interrogating the scatterer(s) from three different directions, e.g., a vertically pointing transmitter-receiver to sense vertical motions and two additional transmitter-receiver combinations to sample different horizontal components. The wind profile is obtained through analysis of the time series from each of the receivers at sequential time intervals, or "range gates." Since the transmitter frequency and geometrical orientation of the transmitter-receiver pairs are known, an estimate of δf through analysis and combination of the time series produces the wind vector components. In cases where the vertical velocity is not available, averaging over a suitable time period can reduce the associated error, as discussed by Kaimal and Haugen (1977).

4.3 Frequency Determination Techniques

A number of methods are available for frequency analysis, some of which will be described briefly here. In general, it is necessary to analyze a time series with a duration of approximately 200 ms to make a single estimate, which corresponds to a vertical averaging distance, or range gate, of about 34 m. Usually this time series consists of the in-phase part R (the real part) and out-of-phase part Q (the quadrature, or imaginary part) of the received signal (see Section 3.2 for a block diagram and a description of system operation procedures that supply these inputs), although it is possible to derive Doppler shifts with only the real component of the returned signal. In general, the transmit or carrier frequency must be subtracted from the observed frequency, either before or after analysis.

4.3.1 Tracking Filter

This is an analog technique whereby a variable frequency filter "tracks" the strongest portion of the received echo using a feedback circuit in a manner similar to that in which FM radio receivers are "locked" to the frequency of the signal. Simplicity is its main virtue. However, a tendency to track a single portion of the signal too long, response-time problems, and the advantages of digital analysis have reduced its popularity. Comparative tests of such techniques have been reported by Keeler (1976).

4.3.2 Rotor Techniques

The simplest method of this type is the straightforward analysis of the frequency by counting the number of times the signal crosses zero within the range gate. Problems occur at small signal levels when noise or DC bias can cause spurious zero crossings, as discussed by Melling and List (1978b, 1980).

If one has both real and quadrature components, the technique can be enhanced by dividing the phase space into octants, which enables the detection of smaller frequencies:

$$\delta f = \frac{R_s}{8(R_s P - 1)} \sum_{i=1}^{R_s P} (O_{i+1} - O_i) \qquad (34)$$

$$\cong \frac{1}{8P}(O_{R_s} - O_1) + \frac{M}{P}$$

where R_s is the sample rate, P is the length of the time series, O_i is the phasor octant at the ith sample, and M is the number of complete rotations of the phasor within the period P. For small values of M, this method increases the sensitivity of the zero-crossing technique for small Doppler frequencies. It is easily implemented with three logic tests to determine the octant ($R > 0?$; $Q > 0?$; $R > Q?$) but also is subject to problems at small signal levels. On the other hand, it is possible to control the quality of data within range gates, which is difficult, if not impossible, with other methods. For example, if the ith sample amplitude is too small, it can be eliminated from the sum in Eq. (32) without eliminating the estimate for the entire range gate.

4.3.3 Spectral Estimates

The frequency content of the received sodar signal contains contributions from many sources, some of scientific interest (such as mean and turbulence motions and spatial distributions of scatterers, and volume averaging effects) and some deleterious (such as background noise, electronic noise, antenna "ringing," and reflections). Calculation of the absolute squares of the Fourier amplitudes of the complex sodar signal can give important insights into the relative contributions of these sources and provide an important tool for estimation of mean and turbulence variables. The result is called the rotary spectrum. The spectral coefficients do not "reflect" at the Nyquist frequency. Rather, the second half of the coefficients corresponds to frequencies below the transmit frequency (hence the often-used but misunderstood term of negative frequency) and the first half to positively shifted frequencies. A sampling rate (R_s) of 100 Hz, for example, allows a 100-Hz range of frequencies within the spectrum, namely, $-50 \, \text{Hz} < \delta f < +50 \, \text{Hz}$. The methods listed below are only some of the current methods for utilizing spectral techniques.

4.3.3.1 Complex Covariance

An estimate of the mean frequency δf weighted equally over the entire spectrum S is given by Bello (1965):

$$\delta f = \int_{-\infty}^{\infty} f'S(f') \, df' \bigg/ \int_{-\infty}^{\infty} S(f') \, df' \qquad (35)$$

where $S(f')$ is the spectral intensity at frequency f'. This expression can be readily transformed into the time domain, which results, for a harmonically varying signal, in an expression for δf in terms of the autocovariance $C(\tau)$ at a lag, or delay, time τ:

$$\delta f = \lim_{\tau \to 0} \{(2\pi\tau)^{-1} \text{Arg}[C(\tau)/C(0)]\} \qquad (36)$$

Since for a complex signal $X(t)$,

$$C(\tau) = \overline{X(t)X^*(t + \tau)} \qquad (37)$$

the mean frequency can be estimated by analysis of the time series directly.

In actual applications of this technique, the signal is broken into its cosine and sine components, R and Q, by multiplying by the carrier frequency and its complement obtained by a 90° phase shift. This signal is then low-pass filtered to obtain the signal centered on a range around zero frequency. This is then sampled at period τ_s, and the frequency shift δf is calculated from the equation (Sirmans and Bumgarner, 1975):

$$\delta f = \frac{1}{2\pi\tau_s} \frac{\sum\limits_{i=1}^{N} (R_{i+1}Q_i - R_iQ_{i+1})}{\sum\limits_{i=1}^{N} (R_iR_{i+1} + Q_iQ_{i+1})} \qquad (38)$$

Note that the complex covariance method is equivalent to obtaining the frequency shift from a true spectral estimate weighted over the complete spectrum. Therefore there is no reason for manipulation of the spectrum. This is a very convenient, easily implemented, fast method that can be done in real time for calculation of winds as long as the real and quadrature signals are available and the background noise is white (Sirmans and Bumgarner, 1975). Davey (1976) described early results from this technique, showing its general reliability. Some care must be exercised, however, because the technique depends on the presence of a white noise spec-

trum. Abnormal environmental noise (for example, as reported in Neff et al., 1980) or noise colored by a nonflat frequency response of the acoustic transducer can present significant operational problems. If the quadrature component is not available, the real covariance method can be used but is severely biased by noise (Neff et al., 1980). Other mean-spectral estimators include adaptive linear-predictive filtering (Keeler, 1977).

4.3.3.2 Discrete Transform

The use of discrete frequency estimation techniques has the potential advantage of allowing spectral manipulation (e.g., the removal of noise sources with known spectral characteristics) before the calculation of mean and higher-order moments of the signal. In theory, it also may allow for estimates of turbulence within a single pulse.

Since the time series is usually of short duration, the discrete frequency intervals within the spectrum can be separated by as much as 10 Hz. Portions of the true spectrum not coinciding precisely with the discrete frequency intervals have their energy spread over the remainder of the derived spectrum. The resultant loss, discussed by Harris (1978) and Underwood (1981), can be modified by treating the time series with data windows such as Hamming or Hanning filters before performing the transform. Once the spectral coefficients are obtained, δf can be calculated directly with a discrete, truncated form of Eq. (35).

The spectrum can be manipulated to reduce the effects of noise before calculating the mean frequency estimate, as long as the characteristic to be eliminated is known. For example, ground clutter reflections often occur at low levels, particularly with bistatic systems in stable conditions, because of signal strength enhancement near the surface. That is, the inverted temperature structure causes the sound to be refracted toward the surface and increases the likelihood of reflection from surface objects. In hilly terrain this problem can be exacerbated by land features at higher elevations than the receiver. The result when Eq. (35) is applied is usually a significant underestimate of wind speeds at low elevations. However, because a reflection has a constant Doppler shift near zero, this peak can be detected and eliminated from the calculation. Figure 10 illustrates such a spectrum. When the echo bin is included in the calculation the mean frequency is 22 Hz, but when it is replaced with an interpolation the result is 28 Hz.

A second powerful approach is to limit the integral in Eq. (35) approximately to the expected signal bandwidth rather than to use the full spectrum. In this case, the spectrum is inspected to find the most energetic portion. Calculation of Eq. (35) is then performed over a reduced "signal width" about this portion (Mastrantonio and Fiocco, 1982) to estimate δf. For real-time analysis, the most energetic portion can be defined, with good results, as the peak of the spectrum. When this is done with the spectrum in Fig. 10, for example, the resultant value is 40 Hz; i.e., the echo has no influence on the estimate. The critical choice here is the location of the peak and the signal width over which to integrate. One can then use the previous values to limit the search and increase the stability of the result. That is, portions of the spectrum are eliminated, based upon the previous spectrum. Care must be taken at this point, because a model of atmospheric behavior is being used to investigate the atmosphere.

The means by which the spectral estimates themselves are obtained can vary considerably. Analog techniques, such as a single swept filter, or a bank of discrete filters, such as a comb filter, can be used. Digital techniques include the amplitude or power-weighted histogram of the phasor rotation rate, the discrete Fourier transform, and the fast Fourier transform. Once the spectral estimates are achieved, analysis techniques are similar.

4.3.4 Noise Effects

The limiting factor in velocity determination with sodar is usually the amount of environmental noise included in the signal. Several researchers have investigated the theoretical impact of noise on Doppler signals (Miller and Rochwarger, 1970; Harris, 1978; Underwood, 1981). Here the discussion is limited to measurements of actual signals under varying S/N levels.

Most analyses of the effect of noise upon Doppler signals assume a white noise spectrum; however, this is probably only rarely encountered. Reflections, birds, and insects produce limited-bandwidth signals, which affect the signal in different ways depending upon their type and proximity. Airplanes, a constant nuisance, produce a wider-band signal. Thus, it is important to sample the background noise under varying circumstances and to produce noise spectral estimates.

The effect of noise is usually to bias the measured Doppler frequency toward zero; that is, the effect of white noise on many full-spectrum techniques is

$$\delta f_m = \delta f \left\{ (S/N)/[1 + (S/N)] \right\} \qquad (39)$$

where δf_m is the measured value of δf in the presence of noise.

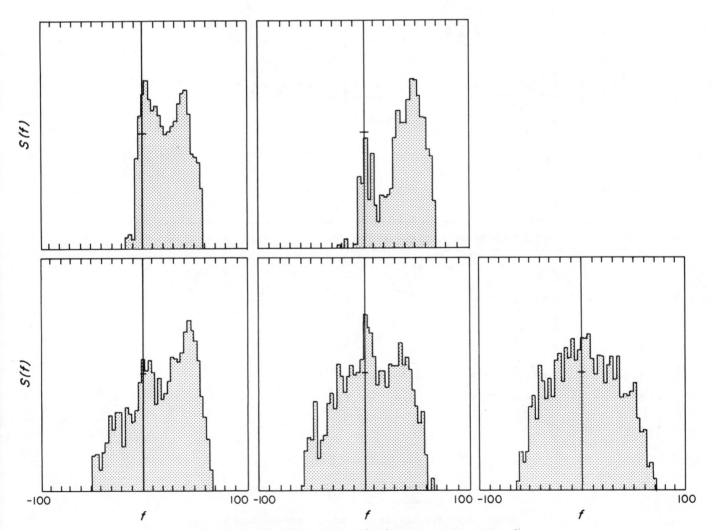

Fig. 10 Examples of bistatic spectra with reflections present, apparent from the second peak at 0 Hz. Spectra are from successive range gates (1–5), approximately 50 m in size.

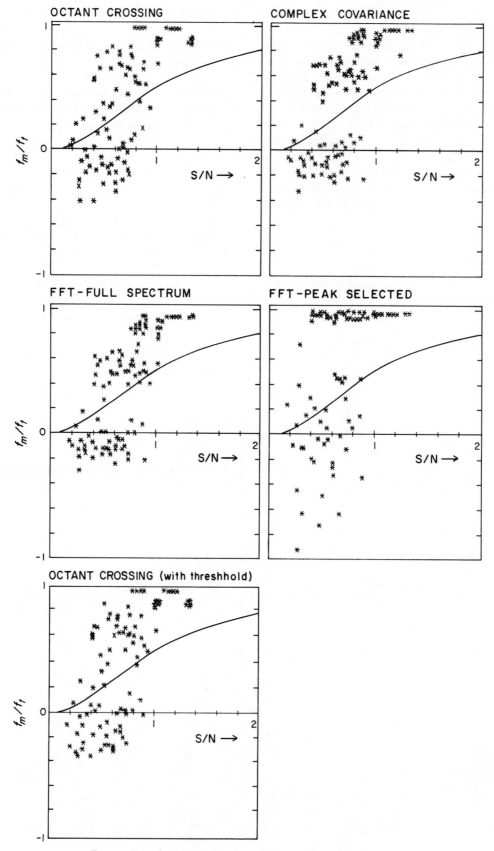

Fig. 11   *Derived frequencies from data created with a constant known frequency imbedded in atmospheric noise surroundings. Measured values f_m are normalized by the true frequency f_t. Note the rather abrupt departure from the true value below S/N = 1, except for FFTP. Solid line is Eq. (39).*

The degree of applicability of Eq. (39) depends upon the analysis technique and the spectral characteristics of the noise. Figure 11 illustrates that this expression is apparently not always valid and that none of the above-mentioned techniques is biased to this extent, at least for that comparison.

4.3.5 Technique Comparison

Figure 12 shows results of intercomparisons of complex-covariance (CXCV), octant crossing (ROT8), FFT full-spectrum (FFTW), and FFT peak-spectrum (FFTP) techniques as a function of S/N. This comparison was carried out on a single data set using sodar data collected under convective conditions so that a range of frequencies was sampled. The figure shows that CXCV and FFTW correlate most closely to one another and have nearly equal values throughout the range of S/Ns, with FFTW showing slightly more bias toward zero at small S/N. The ROT8 method has values much like those of FFTP as long as S/N is above 0.6, although the two are somewhat less well correlated than others. All methods underestimate significantly relative to FFTP for S/N less than 0.7. It is apparent that the full-spectrum estimators are in close agreement and that the differences between amplitude weighting and partial spectrum calculations are not significant above S/N = 1.0.

In order to make an absolute comparison, a known signal mixed with varying amounts of noise is required. Figure 11 was obtained by providing a severely attenuated pure tone at some elevation above the receiver and allowing the atmosphere to provide a realistic spectrum of noise. At low S/N levels it is evident that all techniques are affected: CXCV, ROT8, and FFTW are biased in similar, predictable ways toward zero, whereas the peak spectral technique provides "good" values more often at low S/N. Since the S/N used here is measured over the total receiver bandwidth, the "effective" S/N becomes larger for the FFTP method by the ratio of the bandwidth of the full spectrum to the reduced bandwidth.

For most cases, however, the various methods shown here give comparable results. Examples that have a small S/N but still contain information that can be extracted are relatively few.

4.3.6 Geometry Effects

Estimation of the mean wind from the associated Doppler shift of the scattered signal is based on Eq. (32). For monostatic systems, Eq. (32), relating the wind to the frequency shift, reduces to:

$$\delta f_{x,y} = (2f_0/c)(V_{x,y}\cos\theta - V_z\sin\theta) \qquad (40)$$

and for bistatic systems we have (Kaimal and Haugen, 1977):

$$\delta f_{x,y} = (f_0/c)[V_{x,y}\cos\theta - V_z(1 + \sin\theta)] \qquad (41)$$

where $V_{x,y}$ represents the horizontal components of the wind,

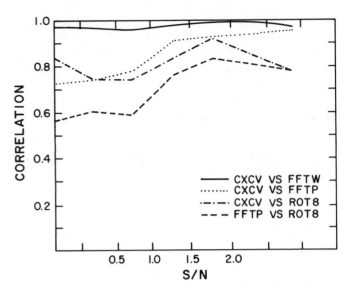

Fig. 12a Correlation coefficient vs S/N for several techniques. Labels are defined in the text.

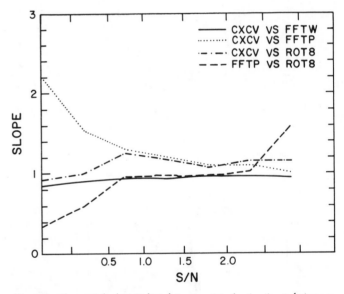

Fig. 12b Slopes of the best-fit lines between pairs of estimation techniques vs S/N for the same techniques as in Fig. 12a. The fit is determined from points derived from data taken in convective conditions.

V_z is the vertical velocity, θ is the elevation angle to the scattering volume, and the center of the scattering volume is assumed to lie vertically above the receiver. If V_z averages to zero, these equations simplify to:

$$\delta f_{x,y} = (2f_0/c)V_{x,y}\cos\theta \qquad (42)$$

$$\delta f_{x,y} = (f_0/c)V_{x,y}\cos\theta \qquad (43)$$

for monostatic and bistatic systems, respectively.

Because the angle θ is known as a function of height, the correction necessary for the estimate of horizontal components can be included straightforwardly as long as one assumes that the scattering center is located directly above the receiver.

An important geometrical correction in bistatic systems derives from the fact that the true center of the scattering

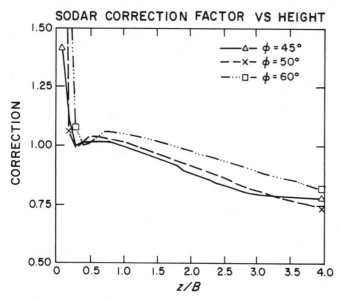

Fig. 13 *Sodar correction factors derived from the directivity patterns of the Argonne system for a frequency of 1,300 Hz, at three different transmitter pointing angles ϕ with a vertically pointing receiver. The term z/B is the ratio of height to sodar bistatic base line.*

volume, resulting from the combination of transmitter and receiver beam patterns, is often not on the center line of the receiver antenna, as assumed above. Normalization with the expected value directly above the receiver leads to correction factors dependent upon height and pointing angle of the transmitter. Figure 13 shows that this correction can be significant for large and small ratios of height to base line, z/B, and small transmitter elevation angles. If the beam patterns are known, however, the correction can be calculated and combined with Eq. (39) to improve wind-profile estimates. It is clear that a good knowledge of the beam pattern is essential for both frequency and amplitude estimation, for the beam pattern interposes a filter between the atmosphere and the receiver, the characteristics of which must be understood.

4.4 Examples

There are certain situations in which sodar-derived wind profiles are particularly useful and realizable. When the scale of the phenomenon being studied is less than 1 km, and continuous (and perhaps unattended), profiling is desired, sodar operation can be very useful (Fig. 14). Complex terrain studies are more reliable if a volume-averaged estimate of winds can be obtained. Because of the enhanced mechanical turbulence often created by topography, sodar wind profiles are attainable to relatively large heights with a bistatic sodar. Lake and sea breeze studies are of a scale appropriate to sodar interrogation because the turbulence associated with sharp temperature differences provides significant sodar signals. Nocturnal boundary-layer development and the low-level jet

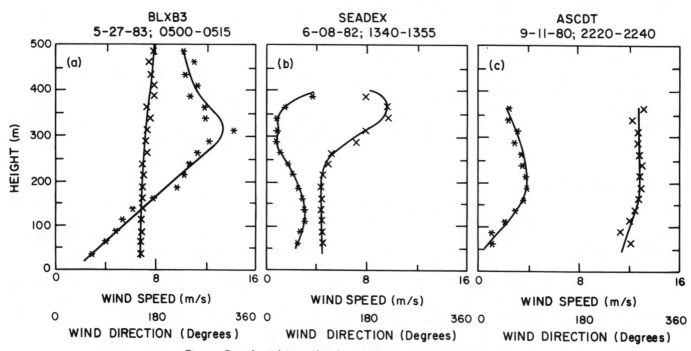

Fig. 14 *Examples of three profiles taken in contrasting conditions: (a) nocturnal jet formation over flat terrain; (b) lake breeze circulation; (c) complex terrain with nocturnal drainage flow. Wind speed is denoted by *; wind direction, by ×.*

are eminently suited to sodar investigation. Wind shear and a strong temperature lapse rate combine to provide strong signals throughout the nocturnal boundary layer; however, the absence of signals above the shear region sometimes makes it difficult to obtain full profiles of the nocturnal jet.

Large-scale vertical gradients of atmospheric variables are normally much greater than horizontal gradients. Hence, vertical velocities can be of particular importance in atmospheric physics, yet they are often the most difficult to measure accurately, because they are usually relatively small and average to zero over large time and space scales. On the other hand, turbulence estimates of the vertical component of motion are often more easily obtained than those of horizontal components.

Measurements of vertical velocities with Doppler sodar are particularly appealing because of the simplicity of using only a single transmitter-receiver and because the statistical properties of the vertical velocity field relevant to pollution dispersion problems may be directly evaluated.

There are, however, some limitations that should be kept in mind. These include the error σ_e of a single estimate of velocity, which results from the broadening of the frequency bandwidth of the transmitted pulse due to its finite duration; the error σ_b, due to the finite beam width of antennas; and errors in vertical alignment and refraction of the beam by horizontal winds during propagation. Detailed discussion of these sources can be found in Spizzichino (1974). Generally, estimates of vertical velocity with sodar have uncertainties near 0.2 m/s, depending upon averaging time and atmospheric conditions. Care must also be taken in the interpretation of results. For example, during unstable conditions, positive velocities may be preferentially selected over negative velocities because they are associated with more turbulent areas of the PBL, particularly with monostatic Doppler sodars. Similarly, the measured vertical velocity variance, σ_{wm}^2, is subject to the same sources of error as the mean velocity, but the error is now additive:

$$\sigma_{wm}^2 = \sigma_w^2 + \sigma_b^2 + \sigma_e^2 + \ldots \qquad (44)$$

Thus, estimates of the actual velocity variance σ_w^2 are subject to overestimation at small values because of system errors. Experience has shown that estimates of σ_w^2 become limited near 0.4 m²/s². Thus, measurements of σ_w^2 at night can be limited by system errors.

A comparison of commercial systems with respect to estimates of w and σ_w^2, carried out at the BAO in 1982, points out this fact. The comparison showed that all systems (which used a variety of techniques) overestimated σ_w^2 at night and somewhat underestimated σ_w^2 during daytime. The applicability of such estimates necessarily depends upon the extent to which the spectrum of sodar-derived vertical velocities extends into the inertial subrange. That is, if such spectra do not extend to frequencies high enough to define the peak in the spectrum, σ_w^2 will be underestimated. The influence of volume averaging becomes important here. The

effect should manifest itself at frequencies determined by the ratio of wind speed to horizontal scale of the antenna. Thus, at a 100-m height and wind speeds near 5 m/s, horizontal volume-averaging effects can be expected to be important near 0.1 Hz. However, because the peak in the spectrum moves to lower frequencies at increasing heights and because the sodar operates at relatively large heights, the derived velocities apparently do reach into the inertial subrange (Asimakopoulos et al., 1978; Underwood and Coulter, 1983), particularly in unstable, convective conditions. Thus it is important to inspect the velocity spectrum before calculating σ_w^2.

Some estimate of σ_w^2 during daytime is routinely available with most commercial systems. In addition, skewness (Taconet and Weill, 1981), which provides information on the rate of kinetic energy transfer (Melling and List, 1980), can be measured. Care must be taken to use periods when the S/N is large throughout the period, or a bias may result (particularly to odd moments of w) because of heavier weighting of strong signal periods. In particular, the velocity and variance fields can be described in detail in convective conditions, enabling examination of individual thermal plumes and entrainment into elevated inversions.

Measurements of the profile of σ_w can be related to the heat flux profile if mechanical production of turbulence and other terms in the turbulence energy equation, such as divergence of fluctuating quantities, are not important. Then in convective conditions (Weill et al., 1980)

$$\sigma_w/u_* \approx M(z/-L)^{1/3} \qquad (45)$$

where L is the Monin-Obukhov length. Thus

$$\sigma_w^3/z \approx M'(H/\varrho c_p)g/T \qquad (46)$$

where M and M' are constants. Figure 15 shows this type of profile. Extrapolation can be used to estimate both the surface heat flux and a first-order approximation of the height of zero heat flux. Thus an estimate of the PBL mixing depth is possible (Weill et al., 1980), particularly if entrainment effects near the capping inversion are included. Figure 16 shows a comparison of sodar-derived surface heat fluxes from C_T^2 and σ_w^3/z with eddy correlation techniques.

Finally, measurement of vertical velocities at separations small enough to remain within the inertial subrange can be used to estimate C_V^2 directly from Eq. (10) (Gaynor, 1977; Weill et al., 1978). Again, care should be taken to obtain values well above the limiting value for σ_w^2 and to include volume and pulse-length averaging effects (Kristensen, 1978).

The convective planetary boundary layer is particularly amenable to investigation with Doppler sodar. Determination of thermal plume structure (Hall et al., 1975; Taconet and Weill, 1981), and of the interaction of plumes with the capping inversion (Kunkel et al., 1977; Weill et al., 1978), and the verification and initialization of models are possible.

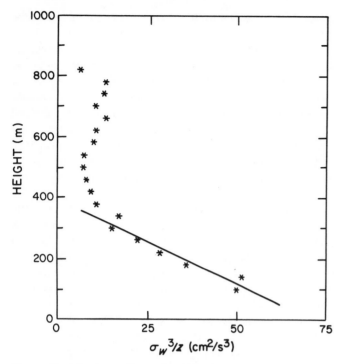

Fig. 15 Example of a profile of $\sigma_w{}^3/z$ obtained in convective conditions (June 1982, in Illinois). Note the linear portion in the lower levels (< 350 m). This portion of the profile can be used to estimate heat flux in the PBL.

5 THE INTERPRETATION OF SODAR FACSIMILE RECORDS

In the history of the physical sciences, advances in observational capabilities have often revealed increasing complexity in physical processes previously believed to be simple. This pattern has been true for the application of sodars to the real atmosphere. For example, following the first sodar measurements, Gilman et al. (1946, p. 274) reported that "the strength of the echoes was such as to lead to the conclusion that a more complicated distribution of boundaries than that measured by ordinary meteorological methods is required in the physical picture of the lower troposphere." Although the last decade has brought understanding of many features of the atmosphere revealed by sodars, many puzzles still remain, particularly in the nocturnal inversion.

The challenge in the interpretation of acoustic sounding data, then, is to use observations of small-scale spectra obtained with high resolution in space and time to deduce information about the atmospheric mean fields and the meteorological processes responsible for the distribution of such echoes. Because such data are of a new character and because they are available in overwhelming detail, such a task is difficult.

In particular, the plethora of detail provided by sodar facsimile recordings often encourages one to focus on and analyze specialized aspects of the records. For example, inversion heights, thermal plume structure, and local wave-

turbulence interactions have been the subjects of many studies appearing in the literature. The continuity of acoustic sounder records in space and time, however, can also lead to a broader interpretation of the boundary-layer response to larger-scale and longer-time-scale forcing. For example, records compressed in time reveal the repeated response of the boundary layer to the diurnal heating-cooling cycle. Variations from day to day in the synoptic weather pattern often are revealed in the growth, rise, and persistence of the inversion that caps the convective layer each morning: stabler air masses produce a slowly growing, persistent capping inversion; less stable ones lead to a rapid growth and disappearance. In the polar regions, long-term records reveal the boundary-layer/inversion-layer response to the alternation between cyclonic and anticyclonic weather systems under perpetually stable conditions. In our discussion we will describe results taken from a number of such experimental programs. Tools that were used in such analyses include tower mean field and turbulence data, rawinsondes, tethered balloons, and other remote-sensing devices.

We will first describe simple boundary layers where adjustment to surface boundary conditions is the dominant factor producing the echoes seen on the acoustic sounder. In this situation, we will consider both convectively unstable and stable conditions. We will then consider the echoes produced within the stably stratified part of the atmosphere that is decoupled from direct frictional influence. Following this, we consider the practical implications of these results for the interpretation of temperature and wind structure from acoustic sounder records. We close this section with a description of a variety of special events that can be recognized from monostatic sodar records.

5.1 The Origin of Acoustic Echoes

5.1.1 Echoes That Reflect Adjustment to Surface Boundary Conditions

5.1.1.1 Convective Conditions

Figure 17 shows a typical convective boundary layer as visualized by scattering of sound from small-scale temperature fluctuations. Since the sodar is sensitive to the small-scale temperature structure, strong echoes are usually observed to considerable heights where the larger organized eddies have carried turbulent air originating near the surface. Additionally, these larger structures, as they carry warmer fluid aloft, are bounded by regions of horizontal shear of the vertical wind. Since there can also be a mean temperature difference between the fluid from the heated surface and the ambient flow, these shear regions can produce additional small-scale thermal fluctuations. A detailed analysis of plume structure as seen by multiple monostatic sounders, and a Doppler-sodar wind-field streamline analysis, can be found in Hall et al. (1975).

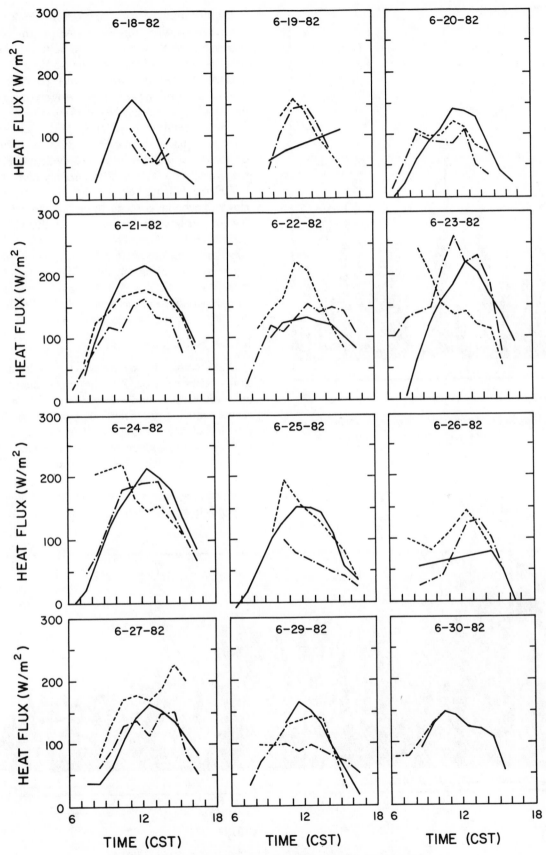

Fig. 16   *Surface heat flux derived from sodar amplitude data (dotted lines), vertical velocity statistics (solid lines), and surface-layer eddy correlation (dashed lines) for 12 days in June 1982.*

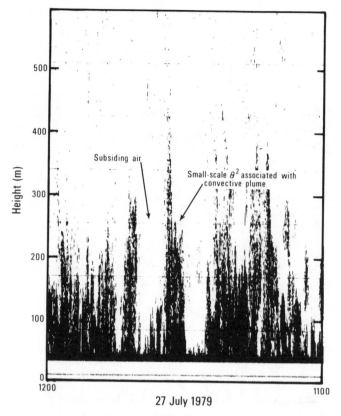

Identification of such convective activity from sounder records is usually quite easy. Because the plumes organize themselves in a variety of sizes, both with height and with time as they pass through the sodar antenna beam, and because the temperature structure parameter characteristically decays with height, an easy-to-recognize echo pattern is produced, even on highly compressed records.

The temperature inversion capping a convective boundary layer is probably the most recognizable atmospheric feature on acoustic sounder records, as is evident in Fig. 18. Whereas the strong echo layer is often strongly perturbed by convective elements from below, a consistent interpretation of the elevated echo layer seems to be that the bottom of the echo region corresponds to the base of the elevated inversion (Kaimal et al., 1982). Although this appears to be a reasonable conclusion, further work with higher-resolution tower data together with measurements obtained using other remote sensors (e.g., Coulter, 1979) may be warranted. Such an analysis would be of particular value for the interpretation of digitally averaged backscatter profiles.

5.1.1.2 Stable Conditions

Under convective conditions, even with a capping inversion, sodars provide a fairly uncomplicated and easy-to-interpret picture; under stably stratified conditions, sodars reveal a much more complicated picture. With a stable lapse rate, turbulence becomes weak or nonexistent, and much of the flow decouples from any direct frictional interaction with the surface. In such cases, particularly with weak synoptic-scale pressure gradients, the frictionally driven boundary layer may be only a few meters deep. Above this the flow responds in a rather sensitive fashion to a variety of external

Fig. 17 Facsimile recording showing monostatic echoes under convective conditions. A pulse length of 0.040 s and a repetition period of 4 s were used.

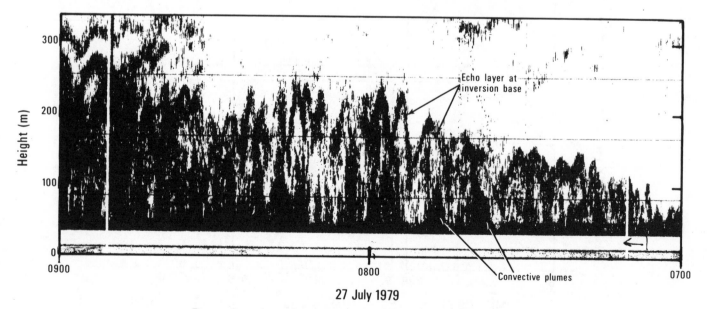

27 July 1979

Fig. 18 Facsimile recording obtained in the early morning as convection develops and a capping inversion forms and moves upward, marking the erosion of stable air aloft.

forcings, such as horizontal pressure and density gradients, inertial oscillations, and momentum advections.

It is useful to distinguish between the nocturnal boundary layer and the nocturnal inversion layer in discussing acoustic scattering in the statically stable atmosphere:

• The nocturnal boundary layer is defined as that part of the stable atmosphere that is influenced directly by surface friction. This layer can range in depth from a few meters to several hundred meters.

• The nocturnal inversion layer is the stable part of the lower atmosphere not directly coupled via turbulence with the surface. Its formation depends on many factors. These include radiative cooling (Garratt and Brost, 1981), the history of the turbulent heat-flux divergence (Brost and Wyngaard, 1978), and advective effects (such as the drainage of cooler air off elevated slopes, as described in Hootman and Blumen,

1983). Its depth, as observed in most sodar records, ranges from a few tens of meters to a kilometer or more, depending on the effective range of the sodar being used.

In a few cases, the nocturnal boundary layer is well defined, and its properties reflect both the turbulent mixing produced at the surface and the adjustment of the surface energy budget. Figures 19 and 20 show two examples of frictionally produced boundary layers under stable conditions. The first (Fig. 19) is a typical echo layer that develops at night with constant surface cooling. The second (Fig. 20) is the type of boundary layer that results when the surface energy budget is balanced and the surface cooling rate is zero. In each case, turbulence near the lower boundary acts to equalize the temperature difference between the air and the surface. As the surface cooling rate decreases, turbulence mixes the boundary-layer air and thus decreases the lapse rate and,

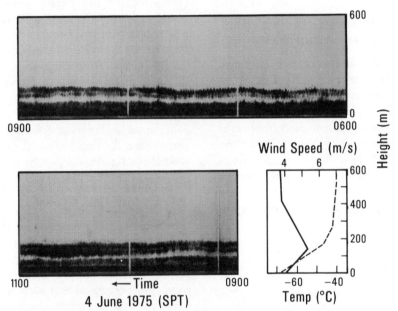

Fig. 19 Facsimile recording obtained under strong inversion, low-wind-speed conditions. The rawinsonde-derived wind profile lacks resolution but suggests a low-level jet. (From Neff, 1980.)

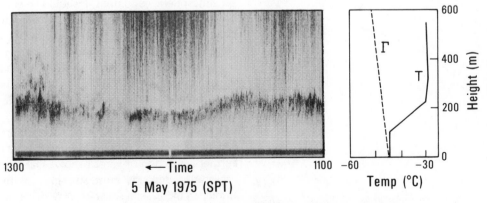

Fig. 20 Facsimile recording obtained under high wind conditions when the surface temperature is much less than the air temperature. A nearly neutral lapse rate forms near the ground in such cases. (From Neff, 1980.)

hence, the temperature fluctuations within the bulk of the boundary layer. At the same time, the temperature inversion near the upper part of the boundary layer increases in strength. A strong elevated echo layer then appears, as in Fig. 20.

The character of the stable boundary layer as seen by sodars often thus reflects the local behavior of the surface energy budget. However, a stable boundary layer with a capping inversion can also form in response to advective processes. For example, the advection of warm air over a colder surface can produce echoes like those seen in Fig. 20. In this case, the sodar echoes again reflect the evolution of the surface energy budget. Such echo patterns are especially prevalent if the advection of warm air is accompanied by low clouds and moderate to strong winds. In such cases, surface radiative losses cease, and the temperature structure of the boundary layer reaches equilibrium.

5.1.2 Echoes That Reflect Mesoscale and Synoptic-Scale Processes in the Free Atmosphere

In addition to the typical features of the nocturnal boundary layer, there is a wide variety of other echo structures in a stably stratified atmosphere that have long intrigued scientists involved in remote sensing. Mid-latitude observations show a highly variable distribution from night to night. Observations from the polar regions, reflecting the absence of a diurnal insolation cycle, show a similar variability, but one that is strongly tied to synoptic-scale changes in the atmosphere. Although much progress has been made in understanding the local details of such layers, both for the

case of acoustic returns and for that of radar echoes, their origin and maintenance remain puzzling. Understanding the origin of such echoes is important for two reasons. First, the presence of such echoes is essential for Doppler sensing systems; sound cannot be scattered from a laminar flow. The choice of remote-sensing devices thus depends on the meteorological phenomena of interest. Second, if one can establish a one-to-one relation between the origin of a particular echo type and a particular fluid dynamical process, the interpretative value of acoustic sounder records is greatly enhanced.

Although echo classification schemes have been proposed (e.g., Clark et al., 1977), few analytical tools have been available with which to examine the meteorological processes that underlie the origin of isolated echo structures. However, numerical modeling techniques developed by Burk (1980) for the moist marine boundary layer and Neff (1980) for the dry polar inversion layer now provide a framework for understanding these echo structures. In the application of these techniques, the importance of Richardson's number (Ri) as a parameter becomes apparent when considering that much of the stable atmosphere maintains a value of Ri near or above its critical value: only when external mechanisms reduce Ri below Ri_{cr} does turbulence result, which leads to echo layers.

In the analysis of sodar echoes from the stable region of the atmosphere, the first, but not always obvious, question to ask is what meteorological process has reduced Ri.

5.1.2.1 Dependence of the Scattering on Mean Field Gradients

In Section 2, we expressed the acoustic scattering cross section in terms of simple turbulence parameters, such as eddy diffusion coefficients and dissipation rates, together with the vertical gradients of the mean potential temperature and wind. We postulated that such relations would hold under most but not all stable atmospheric conditions. Because second-order closure parameterizations, such as that of Brost and Wyngaard (1978), express turbulence quantities in terms of such mean gradients, they provide a natural tool with which to examine the scattering process and its dependence on atmospheric conditions. It should be emphasized that such parameterizations may not be fully accurate under all conditions. In what follows, we exploit such a model (using the parameterization of Brost and Wyngaard, 1978) to examine the sensitivity of scattering to Ri and the effect of changing atmospheric flow conditions on sodar echo patterns. With these caveats, we show in Fig. 21 the dependence of the backscatter on the potential temperature gradient for two values of Ri. Under weakly stable conditions, the backscatter depends strongly on both temperature gradient and shear. Further inspection of Fig. 21 shows that with strong stability, backscatter depends most strongly on the shear. (For example, an increase in shear of 15 % decreases Ri from 0.20 to 0.15 and produces more than an order-of-magnitude increase in the backscatter for a wide range of stability.)

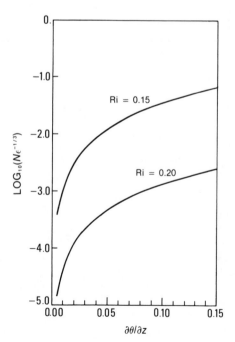

Fig. 21 Sensitivity of the backscatter cross section to Ri and potential temperature gradient derived from a second-order moment turbulence closure hypothesis. (From Neff, 1980.)

5.1.2.2 Mechanisms Changing Ri in the Stable Atmosphere

Changes in Richardson's number can arise from several sources. Roach (1970), for example, proposed the following equation describing the evolution of Ri in large-scale deformation fields:

$$\text{Ri}^{-1}(D\text{Ri}/Dt) = (2\text{Ri} - 1)[(\partial \mathbf{V}_H/\partial p) \cdot \boldsymbol{\nabla}_\theta p] \\ + [(\partial V/\partial y)_p - (\partial U/\partial x)_p] \quad (47)$$

where $\mathbf{V}_H$ is the horizontal wind in pressure coordinates, $D/Dt = (\partial/\partial t)_p + (\mathbf{V}_H \cdot \boldsymbol{\nabla})_p$, partial derivatives are evaluated with respect to constant-pressure surfaces, and the horizontal pressure gradient is taken along isentropic surfaces. In this equation, the first term on the right side represents the effect of tilting isentropic surfaces relative to pressure surfaces, and the second term arises from changes in shear because of stretching of vortex lines. On a smaller scale the Ri tendency equation has the form (Neff, 1980)

$$\frac{\partial \text{Ri}}{\partial t} = \left[-\frac{g}{TS^2}\frac{\partial^2 \overline{\theta w}}{\partial z^2} + \frac{2\text{Ri}}{S}\frac{\partial^2 \overline{uw}}{\partial z^2} \right] - \frac{g}{TS^2}\frac{\partial}{\partial z}\left(W\frac{\partial \Theta}{\partial z} \right) \\ + \frac{2\text{Ri}}{S}\frac{\partial}{\partial z}\left(W\frac{\partial U}{\partial z} \right) \quad (48)$$

for rectilinear flow with a mean horizontal component U in the absence of Coriolis forces, where $S = |\partial \mathbf{V}/\partial z|$, W is the vertical velocity, and only terms involving W and its perturbations w have been retained. We see that the curvature of the turbulent flux profiles as well as subsidence in the presence of a temperature inversion and shear can lead to changes in Ri.

Presumably, large-scale processes described by Eq. (47) can have an effect in the low-level inversion layer as well. This could occur because of a spin-down process: the mass flow in the Ekman layer of a stratified cyclonic vortex will produce a mass convergence in its center, tilting isentropic surfaces until the low-level and synoptic pressure gradients are balanced, cutting off the inward mass flux. The stratification and strength of this secondary circulation determine at what height shear increases and reduces Ri. Neff (1980) has described the possibility of such events in Antarctic observations. In the absence of adequate observing networks such speculations must remain somewhat tentative. Therefore, in subsequent discussion we restrict ourselves to the implications of Eq. (48).

5.1.2.3 Turbulence in Elevated Scattering Layers

If we consider an elevated shear layer characterized by a thickness ΔH, a speed change of ΔU, and a potential temperature jump of $\Delta \Theta$, the bulk Richardson number Rb for the layer can be written as

$$\text{Rb} = (g/T)[(\Delta \Theta \cdot \Delta H)/(\Delta U)^2] \quad (49)$$

If the effect of turbulence is to redistribute heat and momentum through a thicker layer, ΔH must increase, and therefore so does Rb. Since turbulence in general diminishes as Rb increases, one would expect the turbulence to fade. To conclude such a heuristic argument, one must therefore expect the maintenance of echo layers to require a decrease in $\Delta \Theta$ or an increase in ΔU. Since the above finite-difference argument may mislead us about the details of internal structure in the layer, we examine, following Neff (1980), the characteristics of such a scattering layer using a second-order moment closure model. In one such model run, shown in Fig. 22, we created an elevated shear layer with Ri below its critical value and then observed the evolution of C_T^2 and ϵ, as seen in Fig. 23. We concluded that the model predicts a rapid rate of decrease for ϵ and a somewhat slower rate for C_T^2. In both cases, however, the values predicted are much less than those that sodars sense. If one now imposes a slight acceleration of the flow that spans the isolated shear layer, considerably different values result. In the case shown in Fig. 24, the acceleration was increased linearly with height from 0 m/s^2 at 10 m to 0.0028 m/s^2 at 100 m. This increase produced a slight increase in shear through the elevated layer. As can be seen, C_T^2 and ϵ are much larger, comparable to the values actually observed, as listed in Table 1. Peak values of C_T^2 increase by an order of magnitude and ϵ increases by two orders of magnitude when differential acceleration is added to the flow. We conclude that even a small differential acceleration of the flow is sufficient to produce and maintain elevated scattering layers. In the absence of such differential acceleration, echo layers fade rapidly.

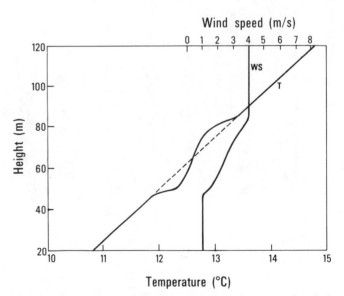

Fig. 22 Model-computed wind and temperature profiles associated with the creation of an elevated shear layer. The stepped temperature structure is associated with turbulent mixing within the shear layer. The dashed line represents the initial temperature profile.

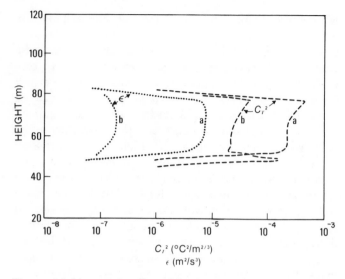

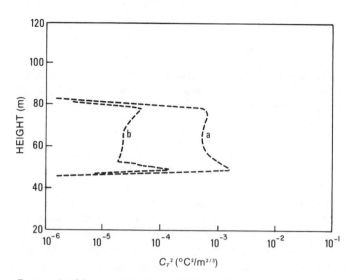

Fig. 23   Model-computed profiles of C_T^2 and ϵ 10 min (a) and 60 min (b) after creation of the shear layer in the absence of external accelerations.

Fig. 24   Model-computed profiles of C_T^2 60 min after creation of the elevated shear layer with (a) and without (b) a differential acceleration.

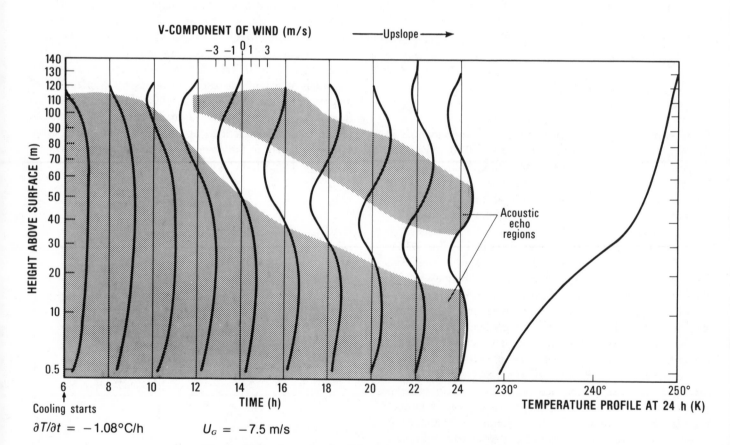

Fig. 25   Results of a numerical model simulation showing the relationship between elevated scattering layers and inertial oscillations (shown in the crossisobaric component) of the wind. In this case, cooling over a sloped surface produces a highly baroclinic boundary layer in which the pressure gradient produced by the sloping inversion layer counters the synoptic-scale pressure gradient. (From Neff, 1980.)

Therefore, we conclude that in the interpretation of elevated echo layers, one must first ask what meteorological processes may exist to accelerate the flow. In the following sections, we will encounter several examples that illustrate this point.

5.2 Temperature Structure, Inversion Heights, and Boundary-Layer Depths

One of the most practical applications of monostatic acoustic sounders is to enable researchers to infer temperature structure in the first kilometer of the atmosphere. As we showed above, the dependence of the scattering on the distribution of turbulence complicates such inferences. This was illustrated in Fig. 3, where a digitized sodar profile was compared with fixed-level tower data from the BAO. As can be seen, scattering layers only occasionally coincide with variations in the temperature inversion. The reason for this is determined by the mechanisms changing Ri within the flow as well as the time history of the development of the inversion. Figure 25 illustrates this point using a model calculation from Neff (1980). In this case of an inversion developing from an initial adiabatic state over a terrain slope of 0.001, the increasing stability suppresses turbulence at the upper part of the boundary layer. This limits the downward transfer of heat, and the continuing effect of surface cooling is confined to a thinner boundary layer near the surface. The actual boundary layer then occupies only a small portion of the inversion layer that has built up over time. In addition, the cessation of turbulence aloft leads to inertial oscillations above the boundary layer. The phase of this oscillation varies with height, leading to an inertially oscillating vector shear that can either add to or subtract from the shear of the ambient flow. This can reduce Richardson's number locally and produce an isolated elevated turbulent layer. Such an effect on scattering layers can be seen clearly in Fig. 26, obtained at the South Pole for three subsequent inertial periods (12 h). Even with such complications, reasonable comparisons between sodar-determined depths and temperature profiles can be found, as from Wyckoff et al. (1973), Goroch (1976), and von Gogh and Zib (1978).

Acoustic data have also been used to test a variety of stable-boundary-layer scaling laws. For example, Figs. 27 and 28 show results from Neff (1980). In the first case a relatively uncomplicated sodar-determined boundary-layer depth, taken as the top of the first continuous echo layer on the record, is compared with a rather highly uncertain prediction that uses the surface stress and heat flux. In the second case (Fig. 28), boundary-layer depth determined by a different scaling law, using bulk stability and surface stress, is compared with sodar data, here displayed in a scatter plot. In this case the comparison is much better. The reason for this difference appears to be the greater sensitivity of surface heat flux, relative to surface stress, to variations in surface cooling

rate. Again, this is a case in which sodar measurements provide a much more sensitive measure of turbulent-boundary-layer depth than either balloon-borne temperature measurements or inferences from in situ surface turbulence instrumentation.

5.3 Examples of Special Flows

5.3.1 Gravity-Driven Flows

Gravity-driven flows are common in the lower troposphere. These range from dramatic events, such as those resulting from thunderstorm cold air outflows, to less dramatic drainage flows and pooling in complex terrain, to the subtle effects of baroclinity over what would normally be considered uniform surfaces.

Flows in which there are considerable temperature gradients are usually well defined by acoustic techniques. In each of the examples below, horizontal temperature gradients provide a means of accelerating the flow, and vertical temperature gradients combined with shear-induced turbulent mixing provide tracers for acoustic backscatter.

5.3.1.1 Thunderstorm Outflows

Density currents flowing out from thunderstorms are often well defined on sodar records when the associated wind does not produce excessive background noise, i.e., generally less than 10 m/s at the surface. Such events are described in some detail by Hall et al. (1976).

5.3.1.2 Drainage Flows

Drainage flows originating from the cooling of sloped surfaces produce a variety of characteristic patterns on sodar records, depending on the topography, the size of the drainage area, the surface composition, and the magnitude and direction of the ambient wind. Depending on the size of the drainage area, the onset and evolution of the drainage flow at downstream sites often follow a characteristic course. Figure 29 shows the transition from afternoon convection to nocturnal drainage at three different sites. In each case the upstream collection area is greater than 10 km². The Big Sulphur Creek site is at the bottom of a deep, V-shaped valley; the site at Diamond-D is in the outflow region from a collecting valley; and the Upper Brush Creek site is at the bottom of a long, narrow, U-shaped valley. In the first case, the drainage flow develops in opposition to the prevailing westerly wind. In the second, obtained under slack synoptic conditions, an abrupt front appears to mark the onset of the drainage flow. This is similar to the front seen on drainage flows that originate from the Rocky Mountains and spread over the plains to the east (Hootman and Blumen, 1983). In the last case, the transition from heating to cooling is abrupt because of the steep side walls and orientation of the canyon with respect to the sun.

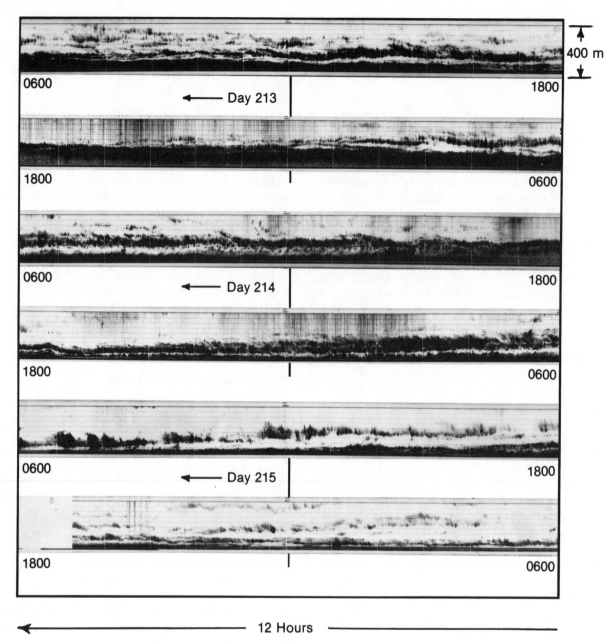

Fig. 26   Example from the South Pole showing the occurrence of elevated scattering layers reappearing during three successive inertial cycles (of 12 h each). (From Neff, 1980.)

5.3.1.3   Cold Air Accumulation in Valleys

Sodar records from basins in which cold air can accumulate often show a multilayered echo structure, as has been described by Brown and Hall (1978). Such turbulent layers may result from the local reduction of Richardson's number by internal waves or depend in some complex fashion on the time history of the development of the flow and local accelerations. In very stable conditions, it appears that Ri maintains a value close to its critical value. Model calculations under these conditions (Neff, 1980) suggest that turbulence quantities are quite small. Given a variety of source regions for drainage flow into a basin, one can postulate a relatively small frictional retardation and mixing of each flow as it enters the basin circulation. Figure 30 shows an example of such a flow within a basin where there are a number of adjacent drainage areas. Figure 31 shows a more schematic representation of Doppler-sodar-derived winds at the edge of this same basin. Monostatic sodar records, as well as a smoke tracer release, confirmed a highly layered structure associated with the wind profile in this case. Such layering processes are relevant, for example, to pollutant dispersal in high mountain valleys. However, detailed data sets from which to analyze the mixing of such drainage flows are lacking.

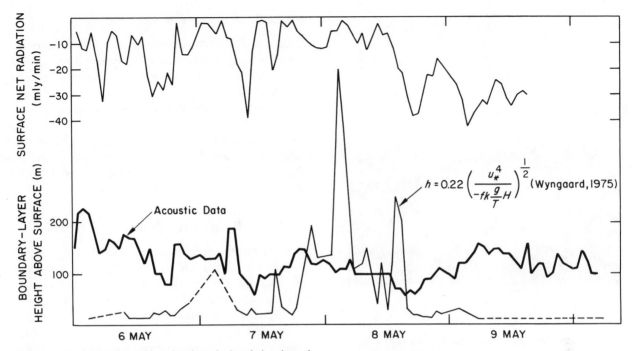

Fig. 27 *Comparison of sodar-observed boundary-layer depth with that obtained using dimensional arguments together with surface temperature flux and surface friction velocity (u_*) measurements. Surface net radiation (millilangleys per minute) is shown for comparison. (From Neff, 1980.)*

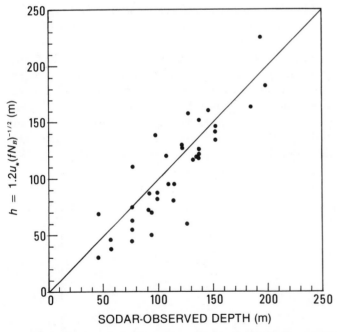

Fig. 28 *Comparison of sodar-observed boundary layer depth with that predicted using a bulk depth-scaling expression (f = Coriolis parameter). (From Neff, 1980.)*

Fig. 29 *Examples of sodar displays of drainage flow initiation in complex terrain. In each case, the progression from daytime convection to nocturnal drainage is noted.*

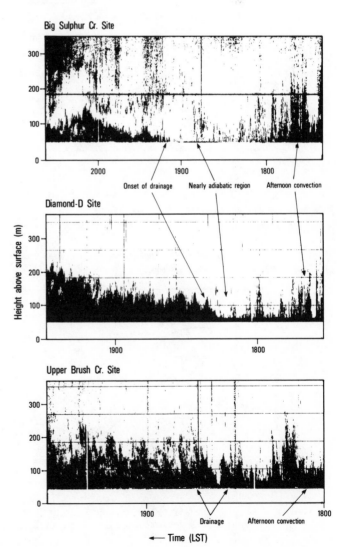

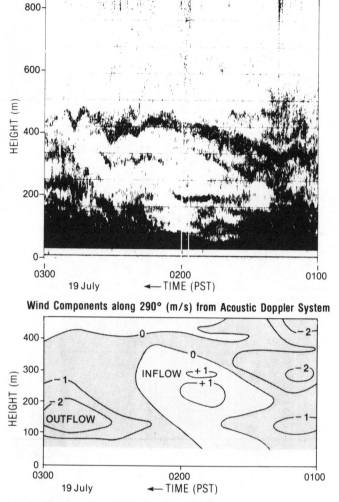

Wind Components along 290° (m/s) from Acoustic Doppler System

Fig. 30 Comparison of Doppler-derived wind component contours with a monostatic sodar facsimile recording in a mountain valley.

Fig. 31 Example of use of Doppler sodar wind profiles to examine flows in complex terrain under nighttime stable conditions.

5.3.1.4 Frontal Structure and Layering

Lu et al. (1983) describe the wave and turbulence structure associated with a small-scale front observed at the BAO. On a longer time scale, frontal passages are often followed by the formation of multilayered echo layers. Neff (1980) shows several cases of such multiple layers developing in response to increasing baroclinity in the boundary layer. Again, associated model calculations suggest that the slopes of the isentropic surfaces in these cases produce sufficient differential, horizontal accelerations to produce and maintain such elevated layers.

5.3.2 Sea Breeze/Lake Breeze Circulations

Sodar returns before, during, and after the passage of a lake breeze "front" at Argonne National Laboratory are shown in Fig. 32. The lake breeze front, a miniature example of the passage of a cold front, is characterized by the contrast of cold, relatively stable air from (in this case) Lake Michigan and heated, unstable air over land. The turbulent temperature fluctuations resulting from the sharp temperature gradients and wind shear across the front are evident in the narrow band of strong returns that passed over the site at 1710 CST; passage at higher levels was slightly delayed, depending on the rate of movement of the marine air inland and the slope of the frontal interface.

Along the coast of California, a strong sea breeze often develops in the afternoon in response to the heating of the interior valleys contrasted with the relatively cool Pacific Ocean. This results in a strong, westerly flow over the coastal mountains. The use of a Doppler sodar through the summer and fall in one of these mountain areas shows this sea breeze effect quite clearly, as in Fig. 33, where histograms of wind directions and speeds are shown between heights of 100 and 600 m. Such sea breezes also interact with the local drainage flows described earlier. This effect is shown in Fig. 34, where

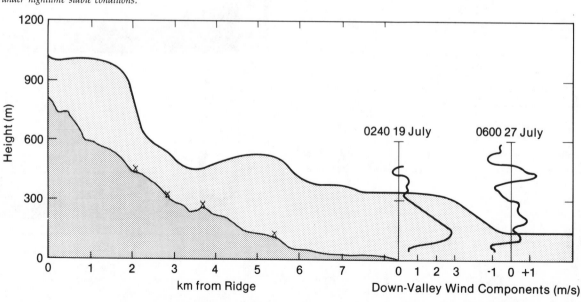

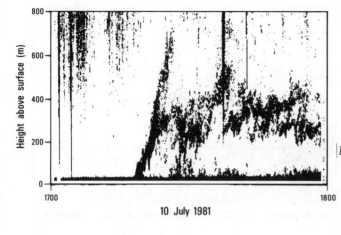

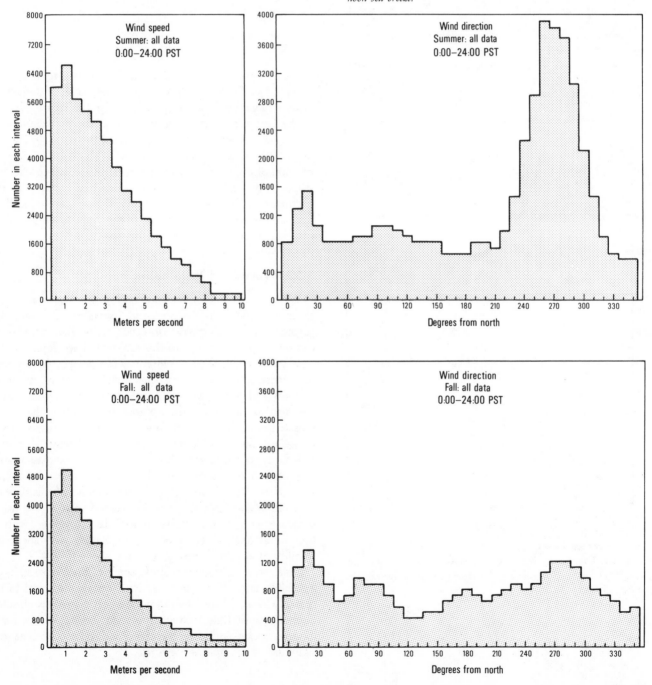

Fig. 32 Example of a sea breeze front on 10 July 1981.

Fig. 33 Examples of wind statistics that can be developed from unattended Doppler sodar operation taken over a five-month period in the coastal mountains of California. Strong summer peak in the direction arises from a persistent afternoon sea breeze.

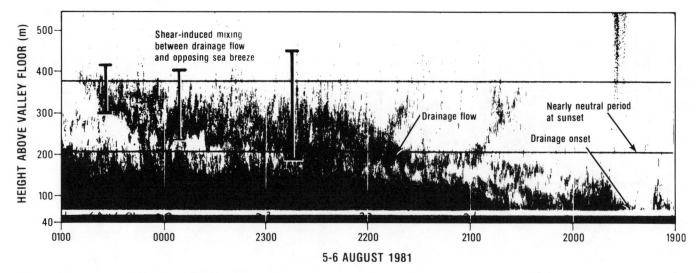

Fig. 34 Sodar facsimile recording showing echo patterns that develop with two opposing flows. In this case, a sea breeze opposes the nocturnal drainage down a narrow valley. Characteristic echo regions are noted on the figure. Most notable is the region of shear between the two flows.

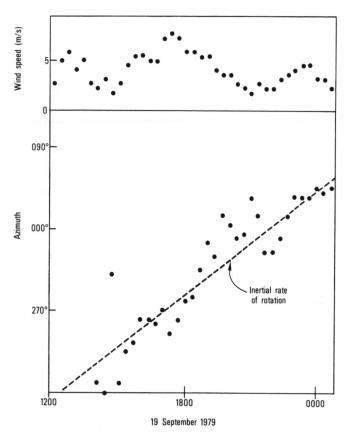

Fig. 35 Doppler-derived wind measurements at 500 m above the surface in complex terrain. These can be used to detect winds that rotate inertially (in this case, a sea breeze) by measuring the rate of rotation of the wind from the continuous sodar data.

the sea breeze opposes the drainage flow until after midnight. The increased shear results in a fairly deep layer of dynamical instability above the drainage, as noted in this figure. The use of the Doppler sodar in this area also allows an examination of the rotation of the sea breeze during the night. In Fig. 35, Doppler wind measurements at 500 m show a clear inertial rotation of the westerly wind after the sea breeze forcing has stopped. By early morning, the winds have rotated to the northeast. In the west-facing valleys, the ambient wind then has a component in the direction of the drainage, while in the east-facing ones, it opposes the drainage.

5.3.3 Waves and Instabilities

With the advent of remote sensing, internal waves and shear instabilities in the nocturnal inversion have received considerable attention. We observed earlier that echo layers can develop in response to the differential acceleration of a flow with height. Such accelerations can develop in response to ageostrophic winds associated with moving weather systems and other mesoscale processes, but internal waves also provide a mechanism for locally increasing the shear. Gossard et al. (1973), motivated by FM-CW radar observations, postulated that internal waves could account for multiple layers of turbulent echo layers and that wave-associated increases in the shear could lead to dynamical instability when Ri < Ri$_{cr}$. Sodars often show examples of this type of shear instability. Merrill (1977) analyzed such a case using sodar, tower, and microbarograph data together with a theoretical analysis to associate the echoes with shear instability of an otherwise dynamically stable flow.

Sodar records in general show some distinct separation of instability and wave processes. First, instabilities tend to be spatially localized and of short period, less than a few minutes in their transit time through the sodar beam. Waves have longer periods, and often the associated sodar echoes have a signature of a simple wave form.

A number of references describe wave and instability phenomena in great detail (e.g., Gossard and Hooke, 1975; Turner, 1973). Some of the features of waves relevant to sodar observations are:

• Waves only exist (in a nonevanescent model) when their frequency is less than that of the buoyancy frequency N_B. Larger values of N_B thus allow a broader range of wave motions (Turner, 1973).

• Depending on the horizontal and vertical wave numbers of the internal wave, regions of convergence and divergence occur within the flow. When these occur near the lower boundary, the depth of the surface inversion layer as seen by the sodar may oscillate in response to the wave.

• Internal waves can propagate arbitrarily with respect to the mean flow except when encountering a critical level, a region where the wave horizontal phase velocity equals the velocity of the local mean flow.

Some of the properties of shear instability are:

• Detailed calculations can be carried out using the Taylor-Goldstein equation (Merrill, 1977). However, some simple results are available. For example, Turner (1973) reports that the most unstable wavelength within a layer of thickness H is given by $\lambda = 7.5H$. Sodar observations show typical thicknesses of 50 to 200 m, corresponding to horizontal wavelengths of 375 to 1,500 m.

• Initial disturbances tend to be two-dimensional and align themselves normal to the vector shear but move with the mean wind over the layer in which they are embedded (Turner, 1973).

Sodar records, as noted above, often reveal the presence of shear instability in the statically stable atmosphere. Identification of these events, while of intrinsic scientific interest, can also give an indication of the character of the wind profile from a simple monostatic record, because the tilted echo pattern associated with an instability moves with the mean wind. An expanded sodar record is shown in Fig. 36 along with a schematic illustration of a series of Kelvin-Helmholtz instabilities. In this case, the strong echo region on the facsimile record has a slope that corresponds to winds increasing with height. If the wind decreased with height, the slope would be opposite, as in Fig. 37: in this figure, the presence of a jet produces sloping structures of opposite sense on either side of the wind-speed maximum, as suggested by the record displayed on the right of the figure. The minimum in

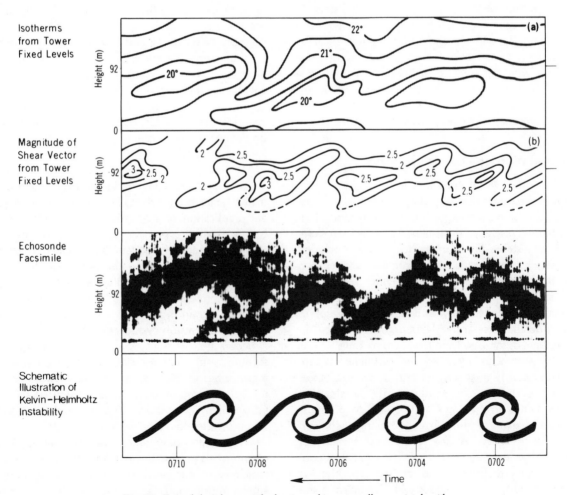

Fig. 36 Expanded sodar record showing echoes normally associated with Kelvin-Helmholtz instability. (Adapted from Neff, 1975.)

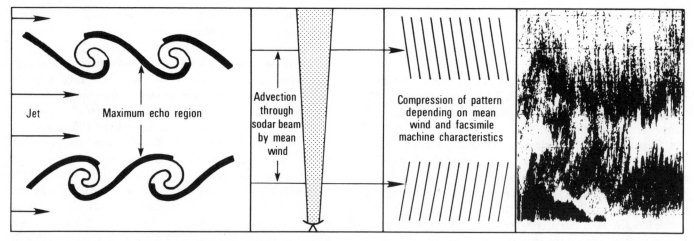

Fig. 37 Example showing how shear-induced dynamical instability leads to characteristic echo patterns in sodar facsimile recordings. By examining the slope of the echo structures, one can determine a qualitative indication of the wind profile. In this example, a low-level jet is shown.

scattering between the two echo patterns can arise from the lowered shear and larger Ri in this region. Given this identification procedure, Fig. 38 shows a descending jet, as can be verified by inspection of the associated tower wind data. In this case, the clear echo regions correspond closely with the location of the wind maximum and with minimum shear.

Monostatic records are thus of value in interpreting the evolution of the wind field when only intermittent wind profiles are available, as from a tethered balloon system. Increases in wind shear or the development of jets can thus be deduced in the absence of frequent wind profiles.

5.3.4 Interpretation of Complex Records

Figures 29 through 38 illustrate a series of physical processes as seen in sodar facsimile recordings. These include sea breezes, inertial rotations, drainage flows, and dynamical instability. Earlier we discussed the role of differential accelerations of statically stable flows; in particular, we concluded that persistent echo layers required a continuing acceleration of the flow in some region. Figure 39 provides an example reflecting these varied processes. In the figure we have identified different regions of the flow as deduced from Doppler sodar data. In detail, these are: (A) Drainage flows occur in these regions. Speeds increase from near-zero to 2 m/s. (B) As the flow decelerates, echo layers fade and turbulence is suppressed. (C) Elevated drainage layers intrude from a larger valley system located to the southeast. (D) Accelerating northwesterly winds (from 3 m/s at 0000 LT to 5 m/s at 0200 LT) create dynamical instability and mixing at the top of the nocturnal inversion. As the winds drop, the echoes fade (between 0200 and 0300 LT). (E) As the wind aloft continues to rotate towards the northeast and again accelerates (from 3 m/s at 0300 to 7 m/s by 0500 LT), dynamical instability and turbulence again increase, and the external flow erodes into the nocturnal inversion.

6 CONCLUSIONS

This article has addressed the uses, possibilities, advantages, and problems of acoustic remote sensing of the lower atmosphere. Due to its usefulness and relatively low cost, sodar is commercially available from several firms. Mean wind profiles and a facsimile-type display of signal intensity are the main outputs. We have attempted to describe some methods of obtaining and interpreting these outputs, which include estimating turbulence, shear, dynamic instabilities, and areal-averaged surface parameterizations. We have sought to emphasize the basic principles involved in sodar, explaining their relationship to the measurement of atmospheric variables and indicating how these measurements can be utilized to investigate physical processes important in the lower atmosphere.

Many advances have been made in sodar technology and applications using conventional instruments (those operating in the range from 1 to 3 kHz); however, new development and application of high-frequency microsodars might well follow from the pioneering work of Moulsley and Cole (1979). In addition, sodars have operated in diverse environments ranging from the polar regions to the tropical oceans, and new opportunities are still appearing. Application of conventional sodars and microsodars to studies in complex terrain in recent years is only one example of the potential expansion of sodar technology.

Sodars are commonly found at universities and research institutions and in commercial institutions using environmental monitoring. In spite of, or perhaps because of, this, only a limited amount of research is committed to develop and utilize them more fully. The integration of sodar-derived wind profiles directly into operational models and the development of new models that are tailored to use the type of information directly available from sodars are areas in which very little work has been done at present. For example, automatic, reliable values of the mixing height on a continuous basis can potentially be achieved through analysis of the signal intensity, wind profile, and surface temperature.

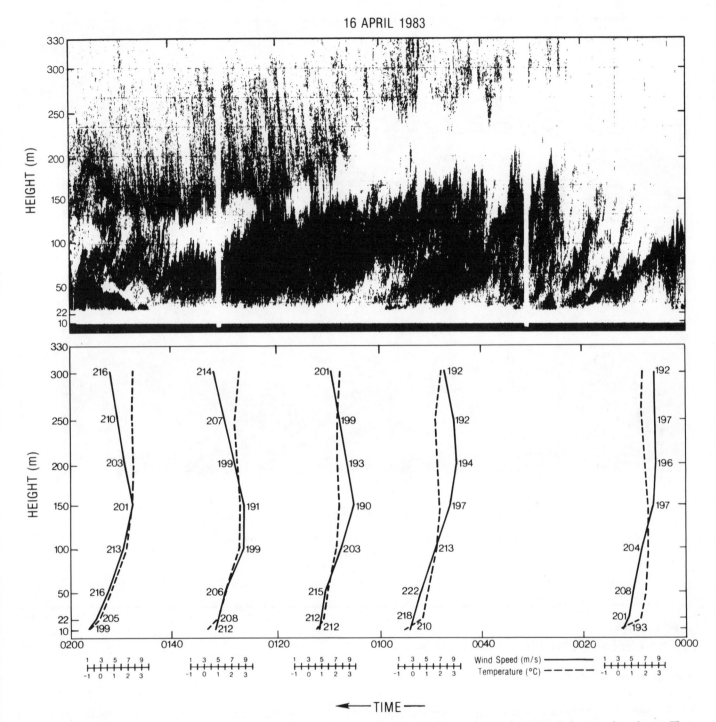

Fig. 38 A sodar record obtained at the BAO showing a descending jet. The interpretation from Fig. 37 is used. Tower data confirm that the clear echo region generally corresponds to the maximum in the wind speed.

One reason for this deficiency is that numerical modelers are not familiar enough with the peculiarities, pitfalls, and promise of remote sensors to use them to advantage. On the other hand, the number of people adequately trained in remote-probing techniques is at present too small to generate many new approaches to solve contemporary problems. Continued education and interaction among instrumentalists, physical and numerical modelers, and the public-oriented user community can only lead to significant interdisciplinary advances.

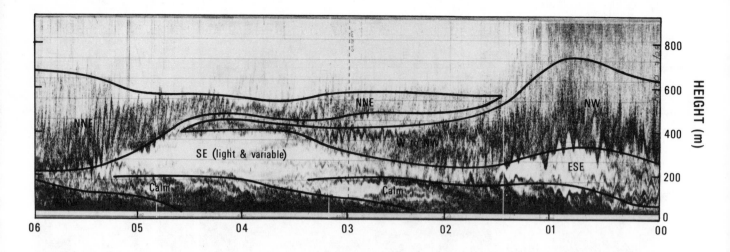

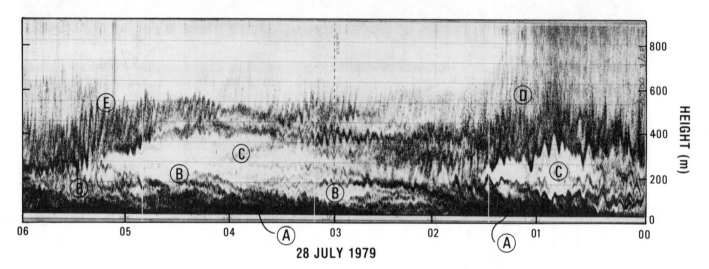

Fig. 39   Complex echo pattern observed in a mountain basin together with patterns of wind directions measured by a Doppler sodar. Key to identification is in text.

APPENDIX:
THE SYSTEM FUNCTIONS S_V AND S_T

Leif Kristensen, NCAR[1]

The derivation of the sodar equation is a rather lengthy but trivial geometrical problem. This appendix is a somewhat shortened account of the discussion by Kristensen et al. (1978) and Kristensen (1978). A slightly different but equivalent approach was taken by Moulsley and Cole (1980).

A bistatic sodar consists of a transmitter antenna T and a receiver antenna R placed a distance $2d$ apart. The two antennas must be pointed so that their axes intersect within the domain in which we wish to measure. The plane defined by the base line between the antennas and any of the two axes is used as a reference plane, which in most applications is approximately vertical.

A monostatic sodar has only one antenna that both transmits and receives. Since the geometry of a monostatic sounder is a special case of the much more complicated bistatic geometry, the monostatic sodar equation will be derived after the bistatic sodar equation has been established.

The acoustic wave from the transmitter is assumed to be a spherical wave (i.e., the equiphase surfaces are spheres[2]). The energy density flux at a given distance from the transmitter is here assumed to be a function of the angle, ψ, from the antenna axis only, but the approach can easily be generalized to so-called fan-beams with an azimuthal dependence as well.

[1]On leave from Risoe National Laboratory, Roskilde, Denmark.

[2]Far-field approximation.

This function $F_t(\psi)$ is the so-called directivity of the antenna. Apart from absorption, the power density in a given direction is inversely proportional to the square of the distance from the transmitter.

The receiver also has a directivity $F_r(\psi')$ where ψ' is the angle from the antenna axis.

The power of the transmitter is defined in terms of the peak power P_0, which is related to the energy density flux $W(r,0)$ in the direction of the axis at distance r by

$$W(r,0) = P_0/r^2 \qquad (A1)$$

The energy density flux at distance r and angle ψ from the axis is

$$W(r,\psi) = (P_0/r^2)F_t(\psi) \qquad (A2)$$

We see that our definitions imply that

$$F_t(0) = F_r(0) = 1 \qquad (A3)$$

The total power P_{tot} is

$$
\begin{aligned}
P_{tot} &= r^2 \int_0^\pi \sin\psi \, d\psi \int_0^{2\pi} d\alpha \, W(r,\psi) \\
&= 2\pi P_0 \int_0^\pi F_t(\psi)\sin\psi \, d\psi
\end{aligned}
\qquad (A4)
$$

Let us imagine that the transmitter starts emitting at time t_1 and stops at time t_2. At time t the scattered power $\delta P(t)$ arrives at the receiver only from the scattering volume between the two axially symmetric (oblong) ellipsoids with the half major axes

$$a_1 = (c/2)(t - t_1) \qquad (A5)$$

and

$$a_2 = (c/2)(t - t_2) \qquad (A6)$$

and the common distance $2d$ between the foci (see Fig. A1). We see that

$$t_1 < t_2 \qquad (A7)$$

implies that

$$a_1 > a_2 \qquad (A8)$$

In the following we will assume that

$$t_2 - t_1 \blacktriangleleft t - [(t_1 + t_2)/2] \qquad (A9)$$

or, equivalently, that

$$a_1 - a_2 \blacktriangleleft [(a_1 + a_2)/2] \qquad (A10)$$

We will then find the power scattered into the receiver from an infinitesimally thin ellipsoid shell with thickness

$$\delta a = a_1 - a_2 \qquad (A11)$$

and a major half axis

$$a = (a_1 + a_2)/2 \qquad (A12)$$

A point on the ellipsoid is characterized by the distance r from the transmitter, the angle ϕ from the base line between the transmitter and the receiver, and the angle θ between the reference plane and the plane defined by the base line and the two lines from the foci to the point in question. Since the points we are considering are confined to the ellipsoid, the three spherical coordinates are not independent. There is a relation between ϕ and r, namely, the equation for the ellipsoid

$$r = (a^2 - d^2)/(a - d\cos\phi) \qquad (A13)$$

Sound scattered from this point into the receiver antenna has been scattered by the angle

$$\gamma = \phi + \phi' \qquad (A14)$$

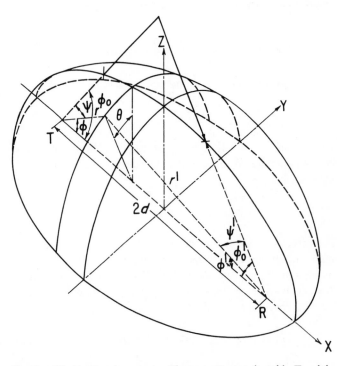

Fig. A1 The bistatic sodar geometry. The transmitter is indicated by T and the receiver by R. The angles defined in the text are indicated. A coordinate system XYZ has been chosen so that the coordinates of T and R are $(-d,0,0)$ and $(d,0,0)$, respectively. The reference plane is the XZ plane.

where ϕ' is the angle from the base line to the line connecting the point and the receiver antenna.

The relation between ϕ and ϕ' can be written

$$\cos \phi' = [2ad - (a^2 + d^2)\cos \phi]/(a^2 - 2ad \cos \phi + d^2) \quad (A15)$$

In order to calculate the power scattered at the point (r,ϕ,θ) and received by the receiver antenna, we must know the distance r' from the point to the receiver antenna and also the two angles ψ and ψ'. Application of Eq. (A13) and a well-known property of the ellipse yields

$$r' = 2a - r = (a^2 - 2ad \cos \phi + d^2)/(a - d \cos \phi) \quad (A16)$$

and the cosine relation from spherical geometry leads to

$$\cos \psi = \cos \phi_0 \cos \phi + \sin \phi_0 \sin \phi \cos \theta \quad (A17)$$

and

$$\cos \psi' = \cos \phi_0' \cos \phi' + \sin \phi_0' \sin \phi' \cos \theta \quad (A18)$$

where ϕ_0 and ϕ_0' are the angles between the base line and the transmitter axis and the receiver axis, respectively. As we wish to integrate the specific cross section η over the ellipsoid shell, we finally need the volume element δV corresponding to the increments δa, $\delta \phi$, and $\delta \theta$. Simple but somewhat lengthy calculations lead to the following expression for δV:

$$\delta V = J(a,\phi,\theta) \, \delta a \, \delta \phi \, \delta \theta = [(a^2 - d^2)^2/(a - d \cos \phi)^4] \times (a^2 - 2ad \cos \phi + d^2) \sin \phi \, \delta a \, \delta \phi \, \delta \theta \quad (A19)$$

The energy density flux at the point (r,ϕ,θ) is

$$P_0 \times F_t(\psi)/r^2 \quad (A20)$$

The energy density flux per solid angle, scattered from the volume element around (r,ϕ,θ) in the direction of the receiver antenna, is

$$P_0 \times (F_t(\psi)/r^2)\eta(\gamma) \, \delta V \quad (A21)$$

The effective area of the receiver antenna is A_e, so the solid angle subtended by the receiver antenna multiplied by the antenna directivity becomes $(A_e/r'^2)F_r(\psi')$. The energy density flux received from the scattering in the volume δV around (r,ϕ,θ) is therefore

$$P_0 \times [F_t(\psi)/r^2]\eta(\gamma) \, \delta V \times (A_e/r'^2)F_r(\psi') \quad (A22)$$

This expression must now be integrated over the ellipsoid to give the total received power $\delta P(t)$ at the time t. As all the scattered energy has traveled the same distance $2a$, we can account for extinction if this can be described by a constant extinction coefficient α, by multiplying the integral with $e^{-2a\alpha}$,

$$\delta P(t) = e^{-2a\alpha}P_0\delta a \int_0^\pi d\phi \int_{-\pi/2}^{\pi/2} d\theta \, J(a,\phi,\theta)\eta(\gamma) \frac{F_t(\psi)}{r^2} \frac{F_r(\psi)}{r'^2} A_e \quad (A23)$$

From Eqs. (A13), (A16), and (A19) we have

$$[J(a,\phi,\theta)]/(r^2r'^2) = \sin \phi/(a^2 - 2ad \cos \phi + d^2) \quad (A24)$$

and Eq. (A20) can be written

$$\frac{1}{P_0}\frac{dP}{da} = A_e e^{-2a\alpha} \int_0^\pi \frac{I(\phi)\eta(\gamma)\sin \phi}{a^2 - 2ad \cos \phi + d^2} \, d\phi \quad (A25)$$

where $\eta(\gamma)$, γ, ψ, and ψ' are given by Eqs. (11), (A14), (A15), (A17), (A18), and

$$I(\phi) = \int_{-\pi/2}^{\pi/2} F_t(\psi)F_r(\psi') \, d\theta \quad (A26)$$

Note that only ψ and ψ' are dependent on θ.

According to Eq. (11), $\eta(\gamma)$ consists of two terms, proportional to (C_T^2/T_0) and (C_V^2/c^2), respectively. We define

$$\eta_T(\gamma) = 0.0041\cos^2\gamma \, [\sin(\gamma/2)]^{-11/3} \quad (A27)$$

and

$$\eta_V(\gamma) = 0.030\cos^2\gamma \, [\sin(\gamma/2)]^{-11/3} \cos^2(\gamma/2) \quad (A28)$$

and rewrite Eq. (A25) as

$$\frac{1}{P_0}\frac{dP}{da} = A_e e^{-2a\alpha}k^{1/3} \int_0^\pi \frac{I(\phi)\sin \phi}{a^2 - 2ad \cos \phi + d^2}$$

$$\times \left[\frac{C_T^2}{T_0^2}\eta_T(\gamma) + \frac{C_V^2}{c^2}\eta_V(\gamma)\right]d\phi \quad (A29)$$

$$= \frac{A_e}{d^2}e^{-2a\alpha}k^{1/3}\left[\frac{C_T^2}{T_0^2}S_T + \frac{C_V^2}{c^2}S_V\right]$$

We have introduced the two dimensionless system functions

$$S_T = \int_0^\pi \frac{I(\phi)\sin \phi}{a^2/d^2 - 2(a/d)\cos \phi + 1}\eta_T(\gamma) \, d\phi \quad (A30)$$

and

$$S_V = \int_0^\pi \frac{I(\phi)\sin \phi}{a^2/d^2 - 2(a/d)\cos \phi + 1}\eta_V(\gamma) \, d\phi \quad (A31)$$

The geometry for the monostatic sodar is a special case of the geometry for the bistatic sodar. In other words, Eq. (A25) applies if we set $d = 0$, $F_r(\psi') = F_t(\psi) = F(\psi)$, and $\eta(\gamma) = \eta(\pi)$.

Since

$$\eta_V(\pi) = 0 \qquad (A32)$$

only temperature fluctuations contribute to the scattering.

The monostatic sodar is assumed to possess axial symmetry, and it is easier to integrate over polar coordinates with the antenna axis as reference axis than to use a "degenerate" form of Eq. (A25).

$$\frac{1}{P_0}\frac{dP}{da} = \frac{A_e}{a^2}\, e^{-2a\alpha}\eta(\pi) \int_0^{\pi/2} \sin\theta\; d\theta \int_0^{2\pi} d\phi\; F^2(\theta)$$

$$= \frac{A_e}{a^2}e^{-2a\alpha}k^{1/3}\frac{C_T{}^2}{T_0{}^2}M \qquad (A33)$$

where M is a constant given by

$$M = 0.0256 \int_0^{\pi/2} F^2(\theta)\sin\theta\; d\theta \qquad (A34)$$

Since

$$a = (c/2)[t - (t_1 + t_2)/2] \qquad (A35)$$

the system functions are functions of the time elapsed after the pulse leaves the transmitter antenna.

Glossary

Many of the terms used to describe the technology discussed in this book are assigned to different quantities in different chapters, reflecting the prevailing usage in the particular area being discussed. We have standardized the symbols and notation for these quantities throughout each chapter. However, the duplication of the same symbols in different chapters means that a general glossary covering all the terms would be unwieldy.

Therefore, instead of attempting to present a complete glossary, we list here those quantities that are used throughout the book. We have omitted terms that are used only within a single article, those that are used only a few times, and those whose definitions are obvious from their use.

$\mathscr{P} = P + p$	any variable ($\mathscr{P}$) can be expressed as the sum of a mean P and a fluctuation about the mean p
$\mathscr{V} = \mathbf{V} + \mathbf{v}$	vector quantities (in this case, velocity) are denoted by boldface type
$C_q{}^2$	humidity structure parameter
$C_T{}^2$	temperature structure parameter
$C_{Tq}{}^2$	temperature-humidity crossed structure parameter
$C_V{}^2$	wind velocity structure parameter
f	frequency
u_*	friction velocity scale
w_*	convective velocity scale
x, y, z, t	spatial coordinates and time
z_i	height of the convective boundary layer
δ_{ij}	Kronecker delta
ϵ	viscous dissipation of turbulent kinetic energy
η_c	Corrsin length scale
η_k	Kolmogoroff length scale
$\varkappa$	scalar wave number
$\boldsymbol{\varkappa}$	vector wave number
ν	kinematic molecular viscosity and frequency of electromagnetic radiation
ω	angular frequency

References

Abed-Navandi, M., A. Berner, and O. Preining, 1976: The cylindrical aerosol centrifuge. In *Fine Particles: Aerosol Generation, Measurement, Sampling and Analysis* (B.Y.H. Liu, Ed.), Academic Press, New York, N.Y., 447–464.

Adams, G.W., and M.J. Hadfield, 1984: Breaking the precision barrier for aircraft inertial systems. *Navigation 30,* 301–308.

Airco Industrial Gases, 1976: *Spectra Seal Aluminum Cylinders.* Airco, Murray Hill, N.J.

Aitken, J., 1923: *Collected Scientific Papers* (C.G. Knott, Ed.). Cambridge University Press, Cambridge, England.

Albrecht, B.A., S.K. Cox, and W.H. Schubert, 1979: Radiometric measurements of in-cloud temperature fluctuations. *J. Appl. Meteorol. 18,* 1066–1071.

Allee, P.A., T.B. Harris Jr., and R.H. Proulx, 1970: Atmospheric constituents near Lake Erie. In *Proc. 13th Conference on Great Lakes Research,,* International Association for Great Lakes Research, Ann Arbor, Mich.

American Conference of Governmental Industrial Hygienists, 1977: *Air Sampling Manual,* Fifth ed. American Conference of Governmental Industrial Hygienists, Cincinnati, Ohio.

Andre, J.C., G. De Moor, P. Lacarrere, G. Therry, and R. du Vachat, 1978: Modeling the 24-hour evolution of the mean wind and turbulent structures of the planetary boundary layer. *J. Atmos. Sci. 35,* 1861–1883.

Andreas, E.L., 1981: The effects of volume averaging on spectra measured with a Lyman-alpha humidiometer. *J. Appl. Meteorol. 20,* 467–475.

Angell, J.K., and A.B. Bernstein, 1976: Evidence for a reduction in wind speed on the upwind side of a tower. *J. Appl. Meteorol. 15,* 186–188.

Antonia, R.A., 1975: Influence of velocity sensitivity on the statistics of small scale temperature fluctuations. *Phys. Fluids 18,* 1584–1585.

Appleton, E.V., and M.A.F. Barnett, 1925: On some direct evidence for downward atmospheric reflection of electric rays. *Proc. R. Soc. London A109,* 621–640.

Armstrong, J.T., and P.R. Buseck, 1975: Quantitative chemical analysis of individual microparticles using the electron microprobe. *Theoretical Anal. Chem. 47,* 2178.

Arnold, A., 1976: Observation of the development of individual clear air convective cells. In Preprint Vol., 17th Conference on Radar Meteorology, Seattle, Wash., 26–29 October 1976. AMS, Boston, Mass., 338–341.

Asimakopoulos, D.N., R.S. Cole, S.J. Caughey, and B.A. Crease, 1976: A quantitative comparison between acoustic sounder returns and the direct measurement of atmospheric temperature fluctuations. *Boundary-Layer Meteorol. 10,* 137–148.

————, ————, B.A. Crease, and S.J. Caughey, 1978: A comparison of acoustic Doppler vertical velocity power system spectra with direct measurements. *Atmos. Environ. 12,* 1951–1956.

Atlas, D., K.R. Hardy, and T.G. Konrad, 1966: Radar detection of the tropopause and clear air turbulence. In *Proc. 12th Weather Radar Conference,* Norman, Okla., 17–20 October 1966. AMS, Boston, Mass., 279–284.

———— and C.L. Korb, 1981: Weather and climate needs for lidar observations from space and concepts for their realization. *Bull. Am. Meteorol. Soc. 9,* 1270–1285.

Axford, D.N., 1968: On the accuracy of wind measurements using an inertial platform in an aircraft, and an example of a measurement of the vertical mesostructure of the atmosphere. *J. Appl. Meteorol. 7,* 645–666.

Baker, P.W., 1983: Atmospheric water vapor differential absorption measurements on vertical paths with a CO_2 lidar. *Appl. Opt. 22,* 2257–2264.

Balsley, B.B., W.L. Ecklund, D.A. Carter, and P.E. Johnston, 1980: The MST radar at Poker Flat, Alaska. *Radio Sci. 15,* 213–223.

Barrett, E.C., and L.F. Curtis, 1982: *Introduction to Environmental Remote Sensing,* second ed. Chapman and Hall, London, England, 352 pp.

———— and D.W. Martin, 1981: *The Use of Satellite Data in Rainfall Monitoring.* Academic Press, New York, N.Y., 340 pp.

————, F.P. Parungo, and R.F. Pueschel, 1979: Cloud modification by urban pollution: A physical demonstration. *Meteorol. Resch. 32,* 136.

Bartlett, K.G., and C.Y. She, 1977: Single-particle correlated time-of-flight velocimeter for remote wind-speed measurement. *Opt. Lett. 1,* 175–177.

Barton, D.K., 1964: *Radar Systems Analysis*. Prentice-Hall, New York, N.Y., 608 pp.

Barton, I.J., and J.F. Le Marshall, 1979: Differential-absorption lidar measurements in the oxygen A band using a ruby lidar and stimulated Raman scattering. *Opt. Lett. 4*, 78–80.

Bassett, M.R., C.-Y. Shaw, and R.G. Evans, 1981: An appraisal of the sulfur hexafluoride decay technique for measuring air infiltration rates in buildings. *ASHRAE Trans. 87*, 361–373.

Batchelor, G.K., 1957: Wave scattering due to turbulence. In *Symposium on Naval Hydrodynamics* (F.S. Sherman, Ed.), NAS-NRC Publication No. 515, National Research Council, Washington, D.C., 409–430.

Battan, L.J., 1973: *Radar Observation of the Atmosphere*. University of Chicago Press, Ill., 324 pp.

Baumgardner, D., 1983: An analysis and comparison of five water droplet measuring instruments. *J. Clim. Appl. Meteorol. 22*, 891–910.

Baxter, R.A., D. Pankratz, and I. Tombach, 1983: *An Advanced Continuous SF₆ Analyzer for Real-Time Tracer Gas Dispersion Measurements from Moving Platforms*. Report No. AV-TP-83/504, Aerovironment, Inc., Pasadena, Calif.

Bean, B.R., and E.J. Dutton, 1966: *Radio Meteorology*. Monograph 92, National Bureau of Standards, Washington, D.C., 435 pp.

——— and C.B. Emmanuel, 1980: Aircraft. In *Air-Sea Interaction: Instruments and Methods* (F. Dobson, L. Hasse, and R. Davis, Eds.), Plenum Press, New York, N.Y., 571–588.

———, R.E. McGavin, R.B. Chadwick, and B.D. Warner, 1971: Preliminary results of utilizing the high resolution FM radar as a boundary layer probe. *Boundary-Layer Meteorol. 1*, 466–473.

Bearman, P.W., 1971: Corrections for the effect of ambient temperature drift on hot-wire measurements in incompressible flow. *DISA Information 11*, 25–30.

Belinskii, V.A., and V.A. Pobiyakho, 1962: *Aerology* (Israel Program for Scientific Translations, translator). U.S. Department of Commerce, Clearinghouse for Federal Scientific and Technical Information, Springfield, Va., 384 pp.

Bello, P.A., 1965: Some techniques for the instantaneous real-time measurement of multipath and Doppler spread. *IEEE Trans. Commmun. Technol. 13*, 285–292.

Belyaev, S.P., and L.M. Levin, 1974: Techniques for collection of representative aerosol samples. *J. Aerosol Sci. 5*, 325.

Benedetti-Michelangeli, G., F. Congeduti, and G. Fiocco, 1974: Determination of vertical eddy diffusion parameters by Doppler optical radar. *Atmos. Environ. 8*, 793–799.

——— and G. Fiocco, 1974: Active and passive optical Doppler techniques for the determination of atmospheric temperature. Part 2: A highly coherent laser radar. In *Structural Dynamics of the Upper Atmosphere* (F. Verniani, Ed.), Elsevier, Amsterdam, The Netherlands, 211–219.

Beuttel, R.G., and A.W. Brewer, 1949: Instruments for the measurement of the visual range. *J. Sci. Instrum. 26*, 357.

Bevington, P.R., 1969: *Data Reduction and Error Analysis for the Physical Sciences*. McGraw-Hill, New York, N.Y., 56–60.

Bigg, E.K., 1957: A new technique for counting ice-forming nuclei in aerosols. *Tellus 9*, 394.

———, S.C. Mossop, R.T. Meade, and N.S.C. Thorndyke, 1963: The measurement of ice nuclei concentration by means of Millipore filters. *J. Appl. Meteorol. 2*, 266.

———, A. Ono, and J.A. Williams, 1974: Chemical tests for individual submicron aerosol particles. *Atmos. Environ. 8*, 1.

Bilbro, J., G. Fichtl, D. Fitzjarrald, M. Krause, and R. Lee, 1984: Airborne Doppler lidar wind field measurements. *Bull. Am. Meteorol. Soc. 65*, 348–359.

Bingham, G.E., R.O. Gilmer, S.M. Maish, and K. Hansen, 1983: Area CO₂ flux studies: Aircraft measurements over open ocean and sea ice. In *Annual Report to the Department of Energy, CO₂ and Climate Office*, Lawrence Livermore Laboratory, Livermore, Calif.

Blackwelder, R.F., 1981: Hot-wire and hot-film anemometers. In *Fluid Dynamics: Methods of Experimental Physics*, Vol. 18, Part A (R.J. Emrich, Ed.), Academic Press, New York, N.Y., 259–314.

Blanc, T.V., 1983: A practical approach to flux measurements of long duration in the marine atmospheric surface layer. *J. Clim. Appl. Meteorol. 22*, 1093–1110.

Blanchard, R.L., 1971: A new algorithm for computing inertial altitude and vertical velocity. *IEEE Trans. Aerosp. Electron. Syst. AES-7*, 1143–1146.

Bodhaine, B.A., 1983: Aerosol measurements at four background sites. *J. Geophys. Res. 88*, 10753.

——— and B.G. Mendonca, 1974: Preliminary four wavelength nephelometer measurements at Mauna Loa Observatory. *Geophys. Res. Lett. 1*, 119.

Booker, H.G., and W.E. Gordon, 1950: A theory of radio scattering in the troposphere. *Proc. IEEE 38*, 401–412.

Born, M., and E. Wolf, 1965: *Principles of Optics*. Pergamon, New York, N.Y., Chap. 10.

Boumans, P.W.J.M., and F.J. DeBoer, 1975: Studies of an inductively coupled high-frequency argon plasma for optical emission spectrometry. Part II: Compromise conditions for simultaneous multielement analysis. *Spectrochim. Acta 30B*, 309.

Bradley, J.T., M. Lefkowitz, and R. Lewis, 1978: Automating prevailing visibility. In *Proc. Conference on Weather Forecasting and Analysis and Aviation Meteorology,* Silver Spring, Md., 16–19 October 1978. AMS, Boston, Mass., 332–338.

Braham, R.R., 1974: Cloud physics and urban weather modification. *Bull. Am. Meteorol. Soc. 55,* 100.

Breit, G., and M.A. Tuve, 1925: A test of the existence of the conducting layer. *Phys. Rev. 28,* 554–575.

Brenneis, H.J., 1931: Uberqualitative Mikroelektrolysen mittels kleiner Elecktroden. *Mikrochemie 9,* 385.

Brock, F.V., and G.H. Saum, 1983: Portable automated mesonet II. In Preprint Vol., Fifth Symposium on Meteorological Observations and Instrumentation, Toronto, Ontario, Canada, 11–15 April 1983. AMS, Boston, Mass., 314–320.

Brookner, E., 1977: How to look like a genius in detection without really trying. In *Radar Technology* (E. Brookner, Ed.), Artech House, Dedham, Mass., 1977.

Brost, R.A., and J.C. Wyngaard, 1978: A model study of the stably stratified planetary boundary layer. *J. Atmos. Sci. 35,* 1427–1440.

Browell, E.V., A.F. Carter, S.T. Shipley, R.J. Allen, C.F. Butler, M.N. Mayo, J.H. Siviter Jr., and W.M. Hall, 1983: NASA multipurpose airborne DIAL system and measurements of ozone and aerosol profiles. *Appl. Opt. 22,* 522–534.

———, ———, and T.D. Wilkerson, 1980: An airborne water vapor lidar system. In *Atmospheric Water Vapor* (A. Deepak, T.D. Wilkerson, and L.H. Ruhnke, Eds.), Academic Press, New York, N.Y., 461–476.

Brown, A., E.L. Thomas, R. Foord, and J.M. Vaughan, 1978: Measurements on a distant smoke plume with a CO_2 laser velocimeter. *J. Phys. D11,* 137–145.

Brown, E.H., and S.F. Clifford, 1976: On the attenuation of sound by turbulence. *J. Acoust. Soc. Am. 60,* 788–794.

——— and F.F. Hall, 1978: Advances in atmospheric acoustics. *Rev. Geophys. Space Phys. 16,* 47–110.

Brown, E.N., 1982: Ice detector evaluation for aircraft hazard warning and undercooled water content measurements. *J. Aircraft 19,* 980–983.

———, C.A. Friehe, and D.H. Lenschow, 1983: The use of pressure fluctuations on the nose of an aircraft for measuring air motion. *J. Clim. Appl. Meteorol. 22,* 171–180.

Brown, H.A., 1980: *Automation of Visual Weather Observations.* AFGL-TR-80-0097, Air Force Geophysics Laboratory, Hanscom Air Force Base, Mass., 34 pp.

Browning, K.A., 1971: Structure of the atmosphere in the vicinity of large-amplitude Kelvin-Helmholtz billows. *Q.J.R. Meteorol. Soc. 97,* 283–299.

——— and R. Wexler, 1968: The determination of kinematic properties of a wind field using Doppler radar. *J. Appl. Meteorol. 7,* 105–113.

Broxmeyer, C., 1964: *Inertial Navigation Systems.* McGraw-Hill, New York, N.Y., 254 pp.

Bruner, F., G. Bertoni, and G. Crescentini, 1978: Critical evaluation of sampling and gas chromatographic analysis of halocarbons and other organic air pollutants. *J. Chromatogr. 167,* 399–407.

———, G. Crescentini, F. Mangani, E. Brancaleoni, A. Cappiello, and P. Ciccioli, 1981: Determination of halocarbons in air by gas chromatography–high resolution mass spectrometry. *Anal. Chem. 53,* 798–801.

Brunner, F.K., 1982: Determination of line averages of sensible heat flux using an optical method. *Boundary-Layer Meteorol. 22,* 193–207.

Buck, A.L., 1975: *Error Sensitivity of Fixed- and Variable-Path Lyman-alpha Hygrometers.* Technical Note NCAR-TN/EDD-103, NCAR, Boulder, Colo., 45 pp.

———, 1976: The variable-path Lyman-alpha hygrometer and its operating characteristics. *Bull. Am. Meteorol. Soc. 57,* 1113–1118.

———, 1983: The Lyman-alpha absorption technique for fast humidity measurement. In Preprint Vol., Fifth Symposium on Meteorological Observations and Instrumentation, Toronto, Ontario, Canada, 11–15 April 1983. AMS, Boston, Mass., 16–20.

Bunker, A.F., 1955: Turbulence and shearing stresses measured over the North Atlantic by an airplane acceleration technique. *J. Meteorol. 12,* 445–455.

Burk, S.D., 1980: Refractive index structure parameters: Time-dependent calculations using a numerical boundary layer model. *J. Appl. Meteorol. 19,* 562–576.

———, 1981: Comparison of structure parameter scaling expressions with turbulence closure model predictions. *J. Atmos. Sci. 38,* 751–761.

Burnham, D.C., 1983: *Evaluation of Visibility Sensors at the Eglin Air Force Base Climatic Chamber.* DOT-TSC-FAA-82, Department of Transportation, Transportation Systems Center, Cambridge, Mass., 126 pp.

——— and D.F. Collins, 1982: *AWOS Sensor Evaluation.* DOT-TSC-FAA-82-6, Department of Transportation, Transportation Systems Center, Cambridge, Mass., 169 pp.

Busch, N.E., 1973: On the mechanics of atmospheric turbulence. In *Workshop on Micrometeorology* (D.A. Haugen, Ed.), AMS, Boston, Mass., 1–66.

————, O. Christensen, L. Kristensen, L. Lading, and S.E. Larsen, 1980: Cups, vanes, propellers and laser anemometer. In *Air-Sea Interaction: Instruments and Methods* (F. Dobson, L. Hasse, and R. Davis, Eds.), Plenum Press, New York, N.Y., 11–46.

———— and L. Kristensen, 1976: Cup anemometer overspeeding. *J. Appl. Meteorol. 15*, 1328–1332.

Butcher, S.S., and R.J. Charlson, 1972: *An Introduction to Air Chemistry.* Academic Press, New York, N.Y., 241 pp.

Byers, H.R., 1965: *Elements of Cloud Physics.* University of Chicago Press, Ill., 191 pp.

Cahen, C., G. Msegie, and P. Flamant, 1982: Lidar monitoring of the water vapor cycle in the troposphere. *J. Appl. Meteorol. 21*, 1506–1515.

Camp, D.W., and J.W. Kaufman, 1970: Comparison of tower influence on wind velocity for NASA's 150-m meteorological tower and a wind tunnel model of the tower. *J. Geophys. Res. 75*, 1117–1121.

Carswell, A.I., 1983: Lidar measurements of the atmosphere. *Can. J. Phys. 61*, 378–395.

Cassedy, E.S., and J. Fainberg, 1960: Back scattering cross-sections of cylindrical wires of finite conductivity. *IRE Trans. Antennas Propag. 8*, 1–7.

Caughey, S.J., 1982: Observed characteristics of the atmospheric boundary layer. In *Atmospheric Turbulence and Air Pollution Modelling* (F.T.M. Nieuwstadt and H. Van Dop, Eds.), D. Reidel, Dordrecht, The Netherlands, 107–158.

———— and S.G. Palmer, 1979: Some of the aspects of turbulence structure through the depth of convective boundary layer. *Q.J.R. Meteorol. Soc. 105*, 811–827.

————, J.C. Wyngaard, and J.C. Kaimal, 1979: Turbulence in the evolving stable boundary layer. *J. Atmos. Sci. 36*, 1041–1052.

Chadwick, R.B., J. Jordan, and T.R. Detman, 1983: Radar detection of wingtip vortices. In Preprint Vol., Ninth Conference on Aerospace and Aeronautical Meteorology, Omaha, Neb., 6–9 June 1983. AMS, Boston, Mass., 235–240.

————, K.P. Moran, and W.C. Campbell, 1979: Design of a wind shear radar for airports. *IEEE Trans. Geosci. Electron. GE-14(4)*, 137–142.

————, ————, R.G. Strauch, G.E. Morrison, and W.C. Campbell, 1976a: A new radar for measuring winds. *Bull. Am. Meteorol. Soc. 57(9)*, 1120–1125.

————, ————, ————, ————, and ————, 1976b: Microwave radar wind measurements in the clear air. *Radio Sci. 11*, 795–802.

Champagne, F.H., 1979: The temperature sensitivity of hot wires. In *Proc. Dynamic Flow Conference*, Baltimore, Md., 18–21 September 1978. Published in Skovlunde, Denmark, 101–114.

————, C.A. Friehe, J.C. LaRue, and J.C. Wyngaard, 1977: Flux measurements, flux estimation techniques, and fine-scale turbulence measurements in the unstable surface layer over land. *J. Atmos. Sci. 34*, 515–530.

Chanin, M.-L., 1982: Lidar measurements of temperature and global winds in the atmosphere. In *Collection of Extended Abstracts Presented at the Symposium on the Application of Lidar to Atmospheric Radiation and Climate Studies* (V. Derr, Ed.), IAMAP Third Scientific Assembly, Hamburg, FRG, 17–28 August 1981. UCAR, Boulder, Colo., 109–116.

Charlson, R.J., 1980: The integrating nephelometer. *Atmos. Technol. 12*, NCAR, Boulder, Colo., 10–14.

————, D.S. Covert, T.V. Larson, and A.P. Waggoner, 1978: Chemical properties of tropospheric sulfur aerosols. *Atmos. Environ. 12*, 39.

Chaussee, D.S., P.G. Buning, and D.B. Kirk, 1983: *Convair 990 Transonic Flow-Field Simulation about the Forward Fuselage.* Paper No. 1785, AIAA Sixth Computational Fluid Dynamics Conference, July 1983, Denver, Colo.

Chen, C.Y., 1955: Filtration of aerosols by fibrous media. *Chem. Rev. 55*, 595.

Cheng, R.J., V.A. Mohnen, T.T. Shen, M. Current, and J.B. Hudson, 1976: Characterization of particulates from power plants. *J. Air Pollut. Control Assoc. 26*, 787.

Chisholm, D.A., and L.P. Jacobs, 1975: *An Evaluation of Scattering Type Visibility Instruments.* AFCRL-TR-75-0411, Air Force Geophysics Laboratory, Hanscom Air Force Base, Mass., 31 pp.

————, R.H. Lynch, J.C. Weyman, and F.B. Geisler, 1980: *A Demonstration Test of the Modular Automated Weather System (MAWS).* AFGL-TR-80-0087, Air Force Geophysics Laboratory, Hanscom Air Force Base, Mass.

Clark, G.H., E. Charash, and E.O.K. Bendun, 1977: Pattern recognition studies in acoustic sounding. *J. Appl. Meteorol. 16*, 1365–1368.

Clements, W.E., Ed., 1979: *Experimental Design and Data of the April 1977 Multitracer Atmospheric Experiment at the Idaho National Engineering Laboratory.* Report No. LA-7795-MS, Los Alamos National Laboratory, Los Alamos, N. Mex.

Clemons, C.A., and A.P. Altshuller, 1966: Responses of electron-capture detector to halogenated substances. *Anal. Chem. 38(1)*, 133–136.

————, A.I. Coleman, and B.E. Saltzman, 1968: Concentration and ultrasensitive chromatographic determination of sulfur hexafluoride for application to meteorological tracing. *Environ. Sci. Technol. 2*, 551–556.

Clifford, S.F., and E.H. Brown, 1979: Excess attenuation in echosonde signals. *J. Acoust. Soc. Am. 67*, 1967–1973.

———— and L. Lading, 1983: Monostatic diffraction-limited lidars: The impact of optical refractive turbulence. *Appl. Opt. 22*, 1696–1701.

———— and S. Wandzura, 1981: Monostatic heterodyne lidar performance: The effect of the turbulent atmosphere. *Appl. Opt. 20*, 514–516.

Coantic, M., and C.A. Friehe, 1980: Slow-response humidity sensors. In *Air-Sea Interaction: Instruments and Methods* (F. Dobson, L. Hasse, and R. Davis, Eds.), Plenum Press, New York, N.Y., 399–411.

Cobb, W.E., 1982: The electrical conductivity of the environment in rural Boulder County, Colorado, for the years 1967–1980: An indication of deteriorating air quality. *Meteorol. Resch. 35*, 59.

———— and H.J. Wells, 1970: The electrical conductivity of oceanic air and its correlation to global atmospheric pollution. *J. Atmos. Sci. 27*, 814.

Cochran, W.G., 1965: *Sampling Techniques.* Wiley, New York, N.Y.

Cole, H.L., 1978: Air temperature and differential measurement using IC temperature sensors. In Preprint Vol., Fourth Symposium on Meteorological Observations and Instrumentation, Denver, Colo., 10–14 April. AMS, Boston, Mass., 25–30.

Collis, D.C., and M.J. Williams, 1959: Two-dimensional convection from heated wires at low Reynolds numbers. *J. Fluid Mech. 6*, 357–389.

Collis, R.T.H., 1969: Lidar. In *Advances in Geophysics,* Vol. 13 (H.E. Landsberg and J. van Mieghem, Eds.), Academic Press, New York, N.Y., 113–139.

———— and P.B. Russell, 1976: Lidar measurement of particles and gases by elastic backscattering and differential absorption. In *Laser Monitoring of the Atmosphere: Topics in Applied Physics,* Vol. 14 (E.D. Hinkley, Ed.), Springer-Verlag, Berlin, FRG, 71–151.

Comte-Bellot, G., 1976: Hot-wire anemometry. In *Ann. Rev. Fluid Mech. 8,* Annual Reviews, Inc., Palo Alto, Calif., 208–229.

Condon, P.E., D.T. Grimsrud, M.H. Sherman, and R.C. Kamerud, 1980: An automated controlled-flow air infiltration measurement system. In *Building Air Change Rate and Infiltration Measurements* (C.M. Hunt, J.C. King, and H.R. Treschel, Eds.), Report No. ASTM STP 719, American Society for Testing and Materials, Philadelphia, Pa., 60–72.

Congeduti, F., G. Fiocco, A. Adriani, and C. Guarrella, 1981: Vertical wind velocity measurements by a Doppler lidar and comparisons with a Doppler sodar. *Appl. Opt. 20*, 2048–2054.

Cooney, J.A., 1970: Remote measurement of atmospheric water vapor profiles using the Raman component of laser backscatter. *J. Appl. Meteorol. 9*, 182–184.

————, 1971: Comparison of water vapor profiles obtained by radiosonde and laser backscatter. *J. Appl. Meteorol. 10*, 301–308.

————, 1972: Measurement of atmospheric temperature profiles by Raman backscatter. *J. Appl. Meteorol. 11*, 108–112.

————, K. Petri, and A. Salik, 1980: Acquisition of atmospheric water vapor profiles by a solar blind Raman lidar. In *Atmospheric Water Vapor* (A. Deepak, T.D. Wilkerson, and L.H. Ruhnke, Eds.), Academic Press, New York, N.Y., 419–431.

Corrsin, S., 1951: On the spectrum of isotropic temperature fluctuations in an isotropic turbulence. *J. Appl. Phys. 22*, 469–473.

————, 1963: Turbulence: Experimental methods. In *Handbuch der Physik,* Vol. 8, Part 2, Springer-Verlag, Berlin, FRG, 524–590.

Coulter, R.L., 1979: A comparison of three methods for measuring mixing layer height. *J. Appl. Meteorol. 18*, 1495–1499.

———— and E. Roth, 1979: Sodar system function and error analysis. In *Argonne National Laboratory, Radiological and Environmental Research Division Annual Report,* ANL-79-65, Argonne National Laboratory, Ill., 30–34.

———— and K.H. Underwood, 1980: Some turbulence and diffusion parameter estimates within cooling tower plumes derived from sodar data. *J. Appl. Meteorol. 19*, 1395–1404.

———— and M.L. Wesely, 1980: Estimates of surface heat flux from sodar and laser scintillation measurements in the unstable boundary layer. *J. Appl. Meteorol. 19*, 1209–1222.

Courant, R., and D. Hilbert, 1953: *Methods of Mathematical Physics,* Vol. 1. Wiley Interscience, New York, N.Y., 561 pp.

Cowan, G.A., D.G. Ott, A. Turkevich, L. Masta, G.J. Ferber, and N.R. Daly, 1976: Heavy methanes as atmospheric tracers. *Science 191*, 1048–1050.

Crane, R.K., 1980: A review of radar observations of turbulence in the lower stratosphere. *Radio Sci. 15*, 177–194.

Crosby, P., and B.W. Koeber, 1963: Scattering of light in the lower atmosphere. *J. Opt. Soc. Am. 53*, 358.

Curcio, J.A., L.F. Drummeter, and T.H. Cosden, 1955: *The Absorption Spectrum of the Atmosphere from 4400 Å to 5500 Å.* NRL Report 4669, Naval Research Laboratory, Washington, D.C.

————, ————, and G.L. Knestrick, 1964: An atlas of the absorption spectrum of the lower atmosphere from 5400 Å to 8520 Å. *Appl. Opt. 3*, 1401–1409.

———— and G.L. Knestrick, 1958: Correlation of atmospheric transmission with backscattering. *J. Opt. Soc. Am. 48*, 686–689.

Cwilong, B.M., 1947: Sublimation in a Wilson chamber. *Proc. R. Soc. London, Ser. A 190*, 137.

Dabberdt, W.F., 1981: *Analyses, Experimental Studies, and Evaluations of Control Measures for Air Flow and Air Quality on and near Highways*, Vol. II, *User Guidelines and Application Notes for Estimating Air Quality for Alternative Roadway Configurations*. SRI Project 2761-1, FHWA Contract DOT-FH-11-8125, SRI International, Menlo Park, Calif.

————, 1983: *Ozone Transport in the North Central Coast Air Basin*. SRI Projects 1898 and 4637, CARB Contract A9-143-31, SRI International, Menlo Park, Calif.

————, R. Brodzinsky, B.C. Cantrell, R.E. Ruff, R.N. Dietz, and S. SethuRaman, 1982: *Atmospheric Dispersion over Water and in the Shoreline Transition Zone*. SRI Project 3450, SRI International, Menlo Park, Calif., for the American Petroleum Institute.

————, W.B. Johnson, R. Brodzinsky, and R.E. Ruff, 1984: *Central California Coastal Air Quality Model Validation Study: Data Analysis and Model Evaluation*. SRI Project 3868, Final Report, Minerals Management Service Contract 14-12-0001-29114, SRI International, Menlo Park, Calif.

————, E. Shelar, D. Marimont, and G. Skinner, 1981: *Analysis, Experimental Studies, and Evaluations of Control Measures for Air Flow and Air Quality on and near Highways*, Vol. I, *Experimental Studies, Analyses, and Model Development*. SRI Project 2761, FHWA Contract DOT-FH-11-8125, SRI International, Menlo Park, Calif.

Dams, R., K.A. Rahn, and J.W. Winchester, 1972: Evaluation of filter materials and impaction surfaces for nondestructive neutron activation analysis of aerosols. *Environ. Sci. Technol. 6*, 441.

Davey, R.F., 1976: A coherent acoustic Doppler radar for real-time wind measurement. In Preprint Vol., 17th Conference on Radar Meteorology, Seattle, Wash., 26–29 October 1976. AMS, Boston, Mass., 270–275.

Davies, C.N., 1973: *Air Filtration*. Academic Press, London, England, 171 pp.

Davis, D.D., 1980: Project GAMETAG: An overview. *J. Geophys. Res. 85*, 7285–7292.

Deacon, E.L., 1980: Slow-response temperature sensors. In *Air-Sea Interaction: Instruments and Methods* (F. Dobson, L. Hasse, and R. Davis, Eds.), Plenum Press, New York, N.Y., 255–267.

Deardorff, J.W., and G. Willis, 1982: Investigation of the frozen-turbulence hypothesis for temperature spectra in a convectively mixed layer. *Phys. Fluids 25*, 21–28.

Debye, P., 1909: *Ann. Physik 30*, 57.

Deirmendjian, D., 1969: *Electromagnetic Scattering on Spherical Polydispersions*. Elsevier, New York, N.Y., 290 pp.

Deley, G.W., 1970: Waveform design. In *Radar Handbook* (M.I. Skolnik, Ed.), McGraw-Hill, New York, N.Y.

Dennenberg, J.N., 1971: *The Estimation of Spectral Moments*. Technical Report No. 23, Laboratory for Atmospheric Probing, University of Chicago and Illinois Institute of Technology.

Derr, V.E., N.L. Abshire, R.E. Cupp, and G.T. McNice, 1976: Depolarization of lidar returns from virga and source cloud. *J. Appl. Meteorol. 15*, 1200–1203.

Desjardins, R.L., E.J. Brach, P. Alvo, and P.H. Schwepp, 1982: Aircraft monitoring of surface carbon dioxide exchange. *Science 216*, 733–735.

Dickson, L.D., 1970: Characteristics of a propagating Gaussian beam. *Appl. Opt. 9*, 1854–1861.

Dietz, R.N., 1970: Stack plume tracing with sulfur hexafluoride. In *The Atmospheric Diagnostics Program at Brookhaven National Laboratory: Third Status Report*. Report No. BNL 50280, Brookhaven National Laboratory, Upton, N.Y., 15–18.

———— and E.A. Cote, 1971: GC determination of sulfur hexafluoride for tracing air pollutants. *Div. Air Water Waste Chem. 11*, 208–215.

———— and ————, 1972: *Tracing Atmospheric Pollutants by Gas Chromatographic Determination of Sulfur Hexafluoride*. Report No. BNL 16642, Brookhaven National Laboratory, Upton, N.Y.

———— and ————, 1973: Tracing atmospheric pollutants by gas chromatographic determination of sulfur hexafluoride. *Environ. Sci. Technol. 7(4)*, 338–342.

———— and ————, 1982: Air infiltration measurements in a home using a convenient perfluorocarbon tracer technique. *Environ. Int. 8*, 419–433.

————, ————, and R.M. Brown, 1972: A portable electron capture GC for tracing air pollutants. *Div. Air Water Waste Chem. 12*, 29–35.

————, ————, and G.J. Ferber, 1973: Evaluation of an airborne gas chromatograph for long distance meteorological tracing. *Div. Air Water Waste Chem. 13*, 5–11.

————, ————, and R.W. Goodrich, 1976a: Air mass movements by real-time frontal chromatography of sulfur hexafluoride. In *Measurement, Detection and Control of Environmental Pollutants*, International Atomic Energy Agency, Vienna, Austria, 277–299.

————, ————, and ————, 1976b: *Development and Application of Sulfur Hexafluoride Measurement Capabilities at Brookhaven.* Report No. BNL 21087, Brookhaven National Laboratory, Upton, N.Y., 17 pp.

————, ————, G.I. Senum, and R.F. Wieser, 1981: *An Inexpensive Perfluorocarbon Tracer Technique for Wide-Scale Infiltration Measurements in Homes.* Report No. BNL 30032, Brookhaven National Laboratory, Upton, N.Y., 7 pp.

———— and R.W. Goodrich, 1980: *The Continuously Operating Perfluorocarbon Sniffer (COPS) for the Detection of Clandestine Tagged Explosives.* Report No. BNL 28114, Brookhaven National Laboratory, Upton, N.Y., 21 pp.

————, ————, and E.A. Cote, 1978: Detection of perfluorinated taggants in electric blasting caps by electron capture monitors. In *New Concept Symposium and Workshop on Detection and Identification of Explosives,* Reston, Va., October 1978. U.S. Treasury Department, Bureau of Alcohol, Tobacco, and Firearms, Washington, D.C., 281–288.

————, ————, J.D. Smith, and W. Vogel, 1976c: *Summary Report of the Brookhaven Explosive Tagging Program.* Report No. BNL 21041, Brookhaven National Laboratory, Upton, N.Y., 125 pp.

———— and J.D. Smith, 1976: Calibration of permeation and diffusion devices by an absolute pressure method. In *Calibration in Air Monitoring,* American Society for Testing and Materials, Philadelphia, Pa., 164–179.

DiMarzio, C., C. Harris, J.W. Bilbro, E.A. Weaver, D.C. Burnham, and J.N. Hallock, 1979: Pulsed laser Doppler measurements of wind shear. *Bull. Am. Meteorol. Soc. 60,* 1061–1066.

Dobson, F.W., 1980: Air pressure measurement techniques. In *Air-Sea Interaction: Instruments and Methods* (F. Dobson, L. Hasse, and R. Davis, Eds.), Plenum Press, New York, N.Y., 231–254.

————, L. Hasse, and R. Davis, Eds., 1980: *Air-Sea Interaction: Instruments and Methods.* Plenum Press, New York, N.Y., 801 pp.

Dorman, R.G., 1960: The role of diffusion, interception and inertia in the filtration of airborne particles. In *Aerodynamic Capture of Particles* (E.G. Richardson, Ed.), Pergamon Press, New York, N.Y., 112–122.

————, 1966: Filtration. In *Aerosol Science* (C.N. Davies, Ed.), Pergamon Press, London, England, 195–222.

Douglas, C.A., and R.L. Booker, 1977: *Visual Range: Concepts, Instrumental Determination, and Aviation Applications.* National Bureau of Standards Monograph 159, U.S. Government Printing Office, Washington, D.C., 362 pp.

Doviak, R.J., and M. Berger, 1980: Turbulence and waves in the optically clear planetary boundary layer resolved by dual-Doppler radars. *Radio Sci. 15,* 297–317.

———— and C.T. Jobson, 1979: Dual-Doppler radar observations of clear air wind perturbations in the planetary boundary layer. *J. Geophys. Res. 84,* 697–702.

Draxler, R.R., 1979: Estimating vertical diffusion from routine meteorological tower measurements. *Atmos. Environ. 13,* 1559–1564.

Drivas, P.J., and F.H. Shair, 1975: *Transport and Dispersion of Plumes Associated with Complex Coastal Meteorology.* Report No. ARB3-915, California Institute of Technology, Pasadena, Calif.

————, P.G. Simmonds, and F.H. Shair, 1972: Experimental characterization of ventilation systems in buildings. *Environ. Sci. Technol. 6,* 609–614.

Dyer, A.F., 1981: Flow distortion by supporting structures. *Boundary-Layer Meteorol. 20,* 243–251.

Eberhard, W.L., 1983: Eye-safe tracking of oil fog plumes by UV lidar. *Appl. Opt. 22,* 2282–2285.

———— and R.M. Schotland, 1980: Dual-frequency Doppler-lidar method of wind measurement. *Appl. Opt. 19,* 2967–2976.

Ecklund, W.L., K.S. Gage, B.B. Balsley, R.G. Strauch, and J.L. Green, 1982: Vertical wind variability observed by VHF radar in the lee of the Colorado Rockies. *Mon. Weather Rev. 110,* 1451–1457.

Elias, L., 1977: In-situ quantification of background halofluorocarbon levels. *Nat. Bur. Stand. Spec. Publ. 464,* 435–438.

Elterman, L., 1951: The measurement of stratospheric density distribution with the searchlight technique. *J. Geophys. Res. 56,* 509–520.

Endemann, M., and R.L. Byer, 1981: Simultaneous remote measurements of atmospheric temperature and humidity using a continuously tunable IR lidar. *Appl. Opt. 20,* 3211–3217.

Ensor, D., and A.P. Waggoner, 1970: Angular truncation error in the integrating nephelometer. *Atmos. Environ. 4,* 481.

Evans, W.E., and W. Viezee, 1982: *EPRI Automated Telephotometer: Field Test, Color-Measuring Capability, and Data Analysis.* EPRI EA-2386, Contract RP1630-10, Final Report, Electric Power Research Institute, Palo Alto, Calif., 128 pp.

Fassel, V.A., and R.N. Kniseley, 1974: Inductively coupled plasma: Optical emission spectroscopy and Inductively coupled plasma. *Anal. Chem. 46,* 1110A–1120A and 1155A–1164A.

Ferber, G., K. Telegadas, C.R. Dickson, P.W. Krey, R. Lagomarsino, and R.N. Dietz, 1983: ASCOT 1980 perfluorocarbon tracer experiments. In *ASCOT Data from the 1980 Field Measurement Program in the Anderson Creek Valley, California* (P.H. Gudiksen, Ed.), UCID-18874-80, Vol. III, 1202–1316.

———, ———, J.L. Heffter, C.R. Dickson, R.N. Dietz, and P.W. Krey, 1981: *Demonstration of a Long-Range Atmospheric Tracer System Using Perfluorocarbons.* NOAA Technical Memorandum ERL ARL-101, Contract DE-A101-80EV10081, Air Resources Laboratories, Silver Spring, Md.

Finkelstein, P.L., 1981: Measuring the dynamic performance of wind vanes. *J. Appl. Meteorol. 20,* 588–594.

Fiocco, G., G. Benedetti-Michelangeli, K. Maischberger, and E. Madonna, 1971: Measurement of temperature and aerosol to molecule ratio in the troposphere by optical radar. *Nature London Phys. Sci. 229,* 78–79.

——— and L.D. Smullin, 1963: Detection of scattering layers in the upper atmosphere. *Nature 199,* 1275–1276.

Fletcher, N.N., 1962: *The Physics of Rainclouds.* Cambridge University Press, Cambridge, England, 386 pp.

Flocchini, R.G., D.J. Shadoan, T.A. Cahill, R.A. Eldred, P.J. Feeney, and G. Wolfe, 1975: Energy, aerosols and ion-excited X-ray emission. In *Advances in X-ray Analysis,* Vol. 18 (W.L. Pickles, C.S. Barrett, J.B Newkirk, and C.O. Ruud, Eds.), Plenum Press, New York, N.Y., 579–587.

Fowler, M.M., 1979: *The Use of Heavy Methane as Long Range Atmospheric Tracers.* Report No. LA UR-80-1342, Los Alamos National Laboratory, Los Alamos, N. Mex.

——— and S. Barr, 1983: A long range atmospheric tracer field test. *Atmos. Environ. 17(9),* 1677–1685.

Frank, R.L., 1983: Current developments in Loran-C. *Proc. IEEE 71,* 1127–1139.

Frankel, M.S., N.J.F. Chang, and M.J. Sanders Jr., 1977: A high-frequency radio acoustic sounder for remote measurement of atmospheric winds and temperature. *Bull. Am. Meteorol. Soc. 58,* 928–934.

Fredriksson, K., B. Galle, K. Nystrrom, and S. Svanberg, 1979: Lidar system applied in atmospheric pollution monitoring. *Appl. Opt. 18,* 2998–3003.

———, ———, ———, and ———, 1981: Mobile lidar system for environmental probing. *Appl. Opt. 20,* 4181–4189.

Freeman, B.E., 1977: Tensor diffusivity of a trace constituent in a stratified boundary layer. *J. Atmos. Sci. 34,* 124–136.

Freymuth, P., 1983: History of thermal anemometry. In *Handbook of Fluids in Motion* (N.P. Cheremismoff and R. Gupta, Eds.), Ann Arbor Science, Ann Arbor, Mich., 79–91.

Friedlander, S.K., 1978: A review of the dynamics of sulfate-containing aerosols. *Atmos. Environ. 12,* 187–195.

Friehe, C.A., 1982: *Path-Length Sensitivity of the Lyman-Alpha Humidiometer.* Technical Note NCAR/TN-190+EDD, NCAR, Boulder, Colo., 7 pp.

——— and C.D. Winant, 1982: Observations of wind and sea surface temperature structure off the northern California coast. In Preprint Vol., First International Conference on Meteorology and Air/Sea Interaction of the Coastal Zone (H. Tennekes and C.N.K. Mooers, Eds.), The Hague, Netherlands, 10–14 May 1982. AMS, Boston, Mass., 209–214.

Friis, H.T., A.B. Crawford, and D.C. Hogg, 1957: A reflection theory for propagation beyond the horizon. *Bell System Tech. J. 36,* 627–644.

Frisch, A.S., R.B. Chadwick, W.R. Moninger, and J.M. Young, 1976: Observations of boundary layer convection cells measured by dual-Doppler radar and echosonde and by microbarograph array. *Boundary-Layer Meteorol. 10,* 55–68.

——— and S.F. Clifford, 1974: A study of convection capped by a stable layer using Doppler radar and acoustic echo sounders. *J. Atmos. Sci. 31,* 1622–1628.

Fuchs, N.A., 1964: *The Mechanics of Aerosols* (R.E. Daisley and M. Fuchs, translators). Pergamon Press, Oxford, England, 408 pp.

———, 1975: Sampling of aerosols. *Atmos. Environ. 9,* 697.

Fujita, T.T., and F. Caracena, 1977: An analysis of three weather-related aircraft accidents. *Bull. Am. Meteorol. Soc. 58,* 1164–1181.

Gage, K.S., and J.L. Green, 1979: Tropopause detection by partial specular reflection using VHF radar. *Science 203,* 1238–1240.

Gal-Chen, T., 1978: A method for the initialization of the anelastic equations: Implications for matching models with observations. *Mon. Weather Rev. 106,* 587–606.

——— and R.A. Kropfli, 1983: Deduction of thermodynamic properties from dual-Doppler radar observations in the PBL. In *Proc. 21st Radar Meteorology Conference,* Edmonton, Alberta, Canada, 19–23 September 1983. AMS, Boston, Mass., 33–38.

——— and J. Wyngaard, 1982: Effects of volume averaging on the line spectra of vertical velocity from multiple-Doppler radar observations. *J. Appl. Meteorol. 21,* 1881–1890.

Garratt, J.R., 1975: Limitations of the eddy-correlation technique for the determination of turbulent fluxes near the surfaces. *Boundary-Layer Meteorol. 8*, 255–259.

——— and R.A. Brost, 1981: Radiative cooling effects within and above the nocturnal boundary layer. *J. Atmos. Sci. 38*, 2730–2746.

Gaynor, J.E., 1977: Acoustic Doppler measurements of atmospheric boundary layer velocity structure functions and energy dissipation rates. *J. Appl. Meteorol. 16*, 148–155.

George, D.H., and M. Lefkowitz, 1972: A new concept: Sensor equivalent visibility. In *International Conference on Aerospace and Aeronautical Meteorology*, Washington, D.C., 22–26 May 1972. AMS, Boston, Mass., 243–250.

Gerber, H.E., 1971: On the performance of the Goetz aerosol spectrometer. *Atmos. Environ. 5*, 1009.

Giever, P.M., 1968: Analysis of number and size of particulate pollutants. In *Air Pollution*, second ed., Vol. 2 (A.C. Stern, Ed.), Academic Press, New York, N.Y., 249–280.

Gill, G.C., L.E. Olsson, J. Sela, and M. Suda, 1967: Accuracy of wind measurements on towers or stacks. *Bull. Am. Meteorol. Soc. 48*, 665–674.

Gill, R., K. Geller, J. Farina, J. Cooney, and A. Cohen, 1979: Measurement of atmospheric temperature profiles using Raman lidar. *J. Appl. Meteorol. 18*, 225–227.

Gilman, G.W., H.B. Coxhead, and F.H. Willis, 1946: Reflection of sound signals in the troposphere. *J. Acoust. Soc. Am. 18*, 274–283.

Glover, K.M., and K.R. Hardy, 1966: Dot angels: Insects and birds. In *Proc. 12th Weather Radar Conference*, Norman, Okla., 17–20 October 1966. AMS, Boston, Mass., 264–268.

Goetz, A., H.J.R. Stevenson, and O. Preining, 1960: The design and performance of the aerosol spectrometer. *J. Air Pollut. Control Assoc. 10*, 378.

Goldstein, H., 1980: *Classical Mechanics*. Addison-Wesley, Reading, Mass., 672 pp.

Goroch, A.K., 1976: Comparison of radiosonde and acoustic echo sounder measurements of atmospheric thermal structure. *J. Appl. Meteorol. 15*, 520–521.

Gossard, E.E., 1960: Power spectra of temperature, humidity, and refractive index from aircraft and tethered balloon measurements. *IEEE Trans. Antennas Propag. AP-3*, 186–201.

———, R.B. Chadwick, W.D. Neff, and K.P. Moran, 1982: The use of ground-based Doppler radars to measure gradients, fluxes and structure parameters in elevated layers. *J. Appl. Meteorol. 21*, 211–226.

——— and W.H. Hooke, 1975: *Waves in the Atmosphere*. Elsevier, Amsterdam, The Netherlands, 456 pp.

———, J.H. Richter, and D. Atlas, 1970: Internal waves in the atmosphere from high-resolution radar measurements. *J. Geophys. Res. 75*, 3523–3536.

———, ———, and D.R. Jensen, 1973: Effect of wind shear on atmospheric wave instabilities revealed by FM/CW radar observations. *Boundary-Layer Meteorol. 4*, 113–131.

——— and R.G. Strauch, 1983: *Radar Applications in Cloud and Clear Air*. Developments in Atmospheric Science, No. 14, Elsevier, Amsterdam, The Netherlands, 280 pp.

——— and K.C. Yeh, Eds., 1980: Radar investigations of the clear air. Special issue of *Radio Sci. 15(2)*, 147–242.

Grams, G.W., 1978: Laser atmospheric studies: An overview of recent work and potential contributions to the atmospheric sciences. *Bull. Am. Meteorol. Soc. 59*, 1160–1164.

———, 1981: In-situ measurements of scattering phase functions of stratospheric aerosol particles in Alaska during July 1979. *Geophys Res. Lett. 8*, 13.

———, I.H. Blifford Jr., D.A. Gillette, and P.B. Russell, 1974: Complex index of refraction of airborne soil particles. *J. Appl. Meteorol. 13*, 459.

———, A.J. Dascher, and C.M. Wyman, 1975: *Opt. Eng. 14*, 85.

Grant, D.R., 1965: Some aspects of convection as measured from aircraft. *Q.J.R. Meteorol. Soc. 91*, 268–281.

Grant, L.O., Ed., 1971: *The Second International Workshop on Condensation and Ice Nuclei*. Colorado State University, Fort Collins.

Grant, W.B., and R.D. Hake Jr., 1975: Calibrated remote measurements of SO_2 and O_3 using atmospheric backscatter. *J. Appl. Phys. 46*, 3019–3023.

——— and R.T. Menzies, 1983: A survey of laser and selected optical systems for remote measurement of pollutant gas concentration. *J. Air Pollut. Control Assoc. 33*, 187–194.

Green, J.L., K.S. Gage, and T.E. VanZandt, 1979: Atmospheric measurements by VHF pulsed Doppler radar. *IEEE Trans. Geosci. Electron. GE-17*, 262–280.

Haberl, J.B., 1975: A stratospheric Aitken nuclei counter. *Rev. Sci. Instrum. 46*, 443.

Hall, F.F. Jr., J.G. Edinger, and W.D. Neff, 1975: Convective plumes in the planetary boundary layer, investigated with an acoustic echo sounder. *J. Appl. Meteorol. 14*, 513–523.

———, W.D. Neff, and T.V. Frazier, 1976: Wind shear observations in thunderstorm density currents. *Nature 264*, 408–411.

——— and J.W. Wescott, 1974: Acoustic antennas for atmospheric echo sounding. *J. Acoust. Soc. Am. 40*, 1376–1382.

Hanafusa, T., Y. Kobori, and Y. Mitsuta, 1980: Single-head sonic anemometer-thermometer. In *Instruments and Observing Methods Report No. 3*, World Meteorological Organization, Geneva, Switzerland, 7–13.

Hanna, S.R., 1982: Natural variability of observed hourly SO₂ and CO concentrations in St. Louis. *Atmos. Environ. 16*, 1435–1440.

Hardesty, R.H., R.J. Keeler, M.J. Post, and R.A. Richter, 1981: Characteristics of coherent lidar returns from calibration targets and aerosols. *Appl. Opt. 20*, 3763–3769.

Hardy, K.R., D. Atlas, and K.M. Glover, 1966: Multi-wavelength backscatter from the clear atmosphere. *J. Geophys. Res. 71*, 1537–1552.

———— and I. Katz, 1969: Probing the clear atmosphere with high power, high resolution radars. *Proc. IEEE 27*, 468–480.

———— and H. Ottersten, 1969: Radar investigations of convective patterns in the clear atmosphere. *J. Atmos. Sci. 26*, 666–672.

Harris, C.M., 1966: Absorption of sound in air versus humidity and temperature. *J. Acoust. Soc. Am. 40*, 148–159.

Harris, F.J., 1978: On the use of windows for harmonic analysis with the discrete Fourier transform. *Proc. IEEE 66*, 51–83.

Harrje, D.T., D.S. Dutt, and J. Beyea, 1979: Locating and eliminating obscure but major energy losses in residential housing. *ASHRAE Trans. 85*, 521–534.

————, C.M. Hunt, S.J. Treado, and N.J. Malik, 1975: *Automated Instrumentation for Air Infiltration Measurements in Buildings.* Report No. 13, Princeton University, Center for Environmental Studies, Princeton, N.J.

Hasse, L., and M. Dunckel, 1980: Hot wire and hot film anemometers. In *Air-Sea Interaction: Instruments and Methods* (F. Dobson, L. Hasse, and R. Davis, Eds.), Plenum Press, New York, N.Y., 47–63.

Haugen, D.A., 1978: Effects of sampling rates and averaging periods on meteorological measurements. In Preprint Vol., Fourth Symposium on Meteorological Observations and Instrumentation, Denver, Colo., 10–14 April. AMS, Boston, Mass., 15–18.

———— and J.C. Kaimal, 1978: Measuring temperature structure parameter profiles with an acoustic sounder. *J. Appl. Meteorol. 17*, 895–899.

————, ————, and E.F. Bradley, 1971: An experimental study of Reynolds stress and heat flux in the atmospheric surface layer. *Q.J.R. Meteorol. Soc. 97*, 168–180.

————, ————, J.C. Readings, and R. Rayment, 1975: A comparison of balloon-borne and tower-mounted instrumentation for probing the atmospheric boundary layer. *J. Appl. Meteorol. 14*, 540–545.

Hay, D.R., 1980: Fast-response humidity sensors. In *Air-Sea Interaction: Instruments and Methods* (F. Dobson, L. Hasse, and R. Davis, Eds.), Plenum Press, New York, N.Y., 413–432.

Heintzenberg, J., and H. Quenzel, 1973: Calculations on the determination of the scattering coefficient of turbid air with integrating nephelometers. *Atmos. Environ. 7*, 509.

Heiss, W.H., 1976: *Highway Fog: Visibility Measures and Guidance Systems.* National Cooperative Highway Research Program Report 171, Transportation Research Board, National Research Council, Washington, D.C.

Henry, R.C., J.F. Collins, and D. Hadley, 1981: Potential for quantitative analysis of uncontrolled routine photographic slides. *Atmos. Environ. 15(10/11)*, 1859–1864.

Henry, W.M., and E.R. Blosser, 1970: *A study of the nature of the chemical characteristics of particulates collected from ambient air.* Contract CPA 22-69-153, EPA Report PB 220 40114, Durham, N.C.

Hicks, B.B., and R.T. McMillen, 1984: A simulation of the eddy accumulation method for measuring pollutant fluxes. *J. Clim. Appl. Meteorol. 23(4)*, 637–643.

Hildebrand, P.H., 1977: A radar study of turbulent diffusion in the lower atmosphere. *J. Appl. Meteorol. 16*, 493–510.

Hilst, G.R., and N.E. Bowne, 1971: Diffusion of aerosols released upwind of an urban complex. *Environ. Sci. Technol. 5*, 327–333.

Hinkley, E.D., Ed., 1976: *Laser Monitoring of the Atmosphere: Topics in Applied Physics*, Vol. 14. Springer-Verlag, Berlin, FRG, 380 pp.

Hinze, J.O., 1975: *Turbulence.* McGraw-Hill, New York, N.Y., 790 pp.

Hirschfeld, T., E.R. Schildkraut, H. Tannenbaum, and D. Tannenbaum, 1973: Remote spectroscopic analysis of ppm-level air pollutants by Raman spectroscopy. *Appl. Phys. Lett. 22*, 38–40.

Hochrainer, D., 1971: A new centrifuge to measure the aerodynamic diameter of aerosol particles in the submicron range. *J. Colloid Interface Sci. 36*, 191.

———— and P.M. Brown, 1969: Sizing of aerosol particles by centrifugation. *Environ. Sci. Technol. 3*, 830.

Hochreiter, F., 1973: *Videograph Calibration.* Laboratory Report No. 4-73, National Weather Service, Test and Evaluation Division, Sterling, Va., 51 pp.

Hoff, R.M., and F.A. Froude, 1979: Lidar observation of plume dispersion in northern Alberta. *Atmos. Environ. 13*, 35–43.

Hogg, D.C., M.T. Decker, F.O. Guiraud, K.B. Earnshaw, D.A. Merritt, K.P. Moran, W.B. Sweezy, R.G. Strauch, E.R. Westwater, and C.G. Little, 1983a: An automatic profiler of the temperature, wind and humidity in the troposphere. *J. Clim. Appl. Meteorol. 22*, 807–831.

————, F.O. Guiraud, J.B. Snider, M.T. Decker, and E.R. Westwater, 1983b: A steerable dual-channel microwave radiometer for measurement of water vapor and liquid in the troposphere. *J. Clim. Appl. Meteorol. 22*, 789–806.

Högström, U., 1982: A critical evaluation of the aerodynamical error of a turbulence instrument. *J. Appl. Meteorol. 21*, 1838–1844.

Hooke, W.H., Ed., 1979: *Project PHOENIX: The 1978 Field Operations.* NOAA/NCAR Boulder Atmospheric Observatory Report No. 1, available from NOAA/ERL, Boulder, Colo., 281 pp.

Hootman, B.W., and W. Blumen, 1983: Analysis of nighttime drainage winds in Boulder, Colorado, during 1980. *Mon. Weather Rev. 111*, 1052–1061.

Horst, T.W., 1973: Corrections for response errors in a three-component propeller anemometer. *J. Appl. Meteorol. 12*, 716–725 (cited in Kaimal article).

————, 1973: Spectral transfer functions for a three-component sonic anemometer. *J. Appl. Meteorol. 12*, 1072–1075 (cited in Wyngaard article).

Horvath, J., and K.E. Noll, 1969: The relationship between atmospheric light scattering coefficient and visibility. *Atmos. Environ. 3*, 543.

Hrohn, D.H., 1969: Depolarization of a laser beam at 6328 Å due to atmospheric transmission. *Appl. Opt. 8*, 367–369.

Hudson, J.G., C.F. Rodgers, and G. Keysers, 1981: Simultaneous operation of three CCN counters and an isothermal haze chamber at the 1980 International CCN Workshop. *J. Rech. Atmos. 15*, 271.

———— and P. Squires, 1976: An improved continuous flow diffusion cloud chamber. *J. Appl. Meteorol. 15*, 776.

Hunt, C.M., 1980: Air infiltration: A review of some existing measurements techniques and data. In *Building Air Change Rate and Infiltration Measurements* (C.M. Hunt, J.C. King, and H.R. Treschel, Eds.), ASTM STP 719, American Society for Testing and Materials, Philadelphia, Pa., 3–23.

Hunt, J.C.R., 1973: A theory of turbulent flow round two-dimensional bluff bodies. *J. Fluid Mech. 61*, 625–706.

Huschke, R.E., Ed., 1959: *Glossary of Meteorology.* AMS, Boston, Mass., 638 pp.

Hyson, P., and B.B. Hicks, 1975: Single beam infrared hygrometer for evaporation measurement. *J. Appl. Meteorol. 14*, 301–307.

Iberall, A.S., 1950: Attenuation of oscillatory pressures in instrument lines. *J. Res. Nat. Bur. Stand. 45*, 85–108.

Inaba, H., 1976: Detection of atoms and molecules by Raman scattering and resonance fluorescence. In *Laser Monitoring of the Atmosphere: Topics in Applied Physics*, Vol. 14 (E.D. Hinkley, Ed.), Springer-Verlag, Berlin, FRG, 153–237.

Izumi, Y., and M.L. Barad, 1970: Wind speeds measured by cup anemometers and influenced by tower structure. *J. Appl. Meteorol. 9*, 851–856.

Jaffer, M., V.A. Dutkerwicz, and L. Husain, 1981: Trichlorofluoromethane as a tracer of urban air masses. *Atmos. Environ. 15*, 775–779.

Jaklevic, J.M., and R.L. Walter, 1976: Comparison of minimum detectable limits among X-ray spectrometers. In *X-ray Fluorescence Methods for Analysis of Environmental Samples*, Ann Arbor Science, Ann Arbor, Mich.

Johnson, H.D., D.H. Lenschow, and K. Danninger, 1978: A new fixed vane for air motion sensing. In Preprint Vol., Fourth Symposium on Meteorological Observations and Instrumentation, Denver, Colo., 10–14 April 1978. AMS, Boston, Mass., 467–470.

Johnson, P.N., and J.L. Fink, 1982: Multiple aircraft tracking system for coordinated research missions. *Bull. Am. Meteorol. Soc. 63*, 487–491.

Johnson, W.B., 1983: Meteorological tracer techniques for parameterizing atmospheric dispersion. *J. Clim. Appl. Meteorol. 22*, 931–946.

Junge, C., 1955: The size distribution and aging of natural aerosols as determined from electrical and optical data on the atmosphere. *J. Meteorol. 12*, 13.

————, 1963: *Air Chemistry and Radioactivity.* Academic Press, New York, N.Y., 382 pp.

Kaganov, E.I., and A.M. Yaglom, 1976: Errors in wind speed measurement by rotating anemometers. *Boundary-Layer Meteorol. 10*, 15–34.

Kaimal, J.C., 1969: Measurement of momentum and heat flux variations in the surface boundary layer. *Radio Sci. 4*, 1147–1153.

————, 1973: Turbulent spectra, length scales and structure parameters in the stable surface layer. *Boundary-Layer Meteorol. 4*, 289–309.

————, 1975: Sensors and techniques for the direct measurement of turbulent fluxes and profiles in the atmospheric surface layer. *Atmos. Technol. 7*, NCAR, Boulder, Colo., 7–14.

————, 1980: Sonic anemometers. In *Air-Sea Interaction: Instruments and Methods* (F. Dobson, L. Hasse, and R. Davis, Eds.), Plenum Press, New York, N.Y., 81–96.

————, N.L. Abshire, R.B. Chadwick, M.T. Decker, W.H. Hooke, R.A. Kropfli, W.D. Neff, F. Pasqualucci, and P.H. Hildebrand, 1982: Estimating the depth of the daytime convective boundary layer. *J. Appl. Meteorol. 21*, 1123–1129.

———— and J.E. Gaynor, 1983: The Boulder Atmospheric Observatory. *J. Clim. Appl. Meteorol. 22*, 863–880.

———— and D.A. Haugen, 1969: Some errors in the measurement of Reynolds stress. *J. Appl. Meteorol. 8*, 460–462.

———— and ————, 1977: An acoustic Doppler sounder for measuring wind profiles in the lower boundary layer. *J. Appl. Meteorol. 16*, 1298–1305.

————, J.C. Wyngaard, and D.A. Haugen, 1968: Deriving power spectra from a three-component sonic anemometer. *J. Appl. Meteorol. 7*, 827–837.

————, ————, ————, O.R. Coté, Y. Izumi, S.J. Caughey, and C.J. Readings, 1976: Turbulence structure in the convective boundary layer. *J. Atmos. Sci. 33*, 2152–2169.

Kallistratova, M.A., 1959: Procedure for investigating sound scattering in the atmosphere. *Sov. Phys. Acoust. 5*, 512–514 (English translation).

————, 1961: Experimental investigation of sound wave scattering in the atmosphere. *Tr. Inst. Fiz. Atmos., Atmos. Turbulentmost 4*, 203–256 (English translation, U.S. Air Force FTD TT-63-441).

Kalshoven, J.E. Jr., C.L. Korb, G.K. Schwemmer, and M. Dombrowski, 1981: Laser remote sensing of atmospheric temperature by observing resonant absorption of oxygen. *Appl. Opt. 20*, 1967–1971.

Katz, I., 1972: The detection and study of gravity waves with microwave radar. In *AGARD Conference Proceedings No. 115*, Wiesbaden, FRG, 21-1–21-9.

Kawall, J.G., M. Shokr, and J.F. Keffer, 1983: A digital technique for the simultaneous measurement of streamwise and lateral velocities in turbulent flows. *J. Fluid Mech. 133*, 83–112.

Kayton, M., and W.R. Fried, 1969: *Avionics Navigation Systems.* Wiley, New York, N.Y., 666 pp.

Keeler, R.J., 1976: *A Frequency Discriminator vs. FFT Doppler Extraction.* Technical Memorandum ERL WPL-17, NOAA, Boulder, Colo., 23 pp.

————, 1977: Acoustic Doppler extraction by adaptive linear-prediction filtering. *J. Acoust. Soc. Am. 61*, 1218–1227.

Kelly, T.J., and D.H. Lenschow, 1978: Thunderstorm updraft measurements from aircraft. In Preprint Vol., Fourth Symposium on Meteorological Observations and Instrumentation, Denver, Colo., 10–14 April 1978. AMS, Boston, Mass., 474–478.

Kelton, G., and P. Bricout, 1964: Wind velocity measurements using sonic techniques. *Bull. Am. Meteorol. Soc. 45*, 571–580.

Kerker, M., 1969: *The Scattering of Light.* Academic Press, New York, N.Y., 666 pp.

Kessinger, C., M. Hjelmfelt, and J. Wilson, 1983: Low level microburst wind structure using Doppler radar and PAM data. In *Proc. 21st Conference on Radar Meteorology*, Edmonton, Alberta, Canada, 19–23 September 1983. AMS, Boston, Mass., 609–615.

Kethley, T.W., M.T. Gordon, and C. Orr Jr., 1952: A thermal precipitator for aerobacteriology. *Science 116*, 368.

King, W.D., D.A. Parkin, and R.J. Handworth, 1978: A hot-wire liquid water device having fully calculable response characteristics. *J. Appl. Meteorol. 17*, 1809–1813.

Knighton, W.B., and E.P. Grimsrud, 1983: Linearization of electron capture detector response to strongly responding compounds. *Anal. Chem. 55(4)*, 713–718.

Knollenberg, R.G., 1970: The optical array: An alternative to extinction and scattering for particle size measurements. *J. Appl. Meteorol. 9*, 86.

————, 1972: Measurements of the growth of the ice budget in a persisting contrail. *J. Atmos. Sci. 29*, 1367.

————, 1975: An active scattering aerosol spectrometer. *Atmos. Technol. 2*, NCAR, Boulder, Colo., 80–81.

————, 1976: Three new instruments for cloud physics measurement: The 2-D spectrometer, the forward scattering spectrometer probe and the active scattering aerosol spectrometer. In Preprint Vol., International Conference on Cloud Physics, Boulder, Colo., 26–30 July 1976. AMS, Boston, Mass., 554–561.

————, 1981: Techniques for probing cloud microstructure. In *Clouds, Their Formation, Optical Properties, and Effects* (P.V. Hobbs and A. Deepak, Eds.), Academic Press, New York, N.Y., 495 pp.

———— and R.E. Luehr, 1976: Open cavity laser "active" scattering particle spectrometry from 0.05 to 5 microns. In *Fine Particles: Aerosol Generation, Measurement, Sampling, and Analysis* (B.Y.H. Liu, Ed.), Academic Press, New York, N.Y., 669–696.

Kolmogoroff, A.N., 1941: Ueber das logarithmisch normale Verteilungsgesetz der Dimensionen der Teilchen bei der Zerstueckelung. *Dokl. Akad. Nauk SSSR 31*, 99.

Konrad, T.G., 1968: The alignment of clear-air convective cells. In *Proc. International Conference on Cloud Physics*, Toronto, Ontario, Canada, 26–30 August 1968. 539–543.

———, 1970: The dynamics of the convective process in clear air as seen by radar. *J. Atmos. Sci. 27*, 1138–1147.

Kopp, F., R.L. Schwiesow, and C. Werner, 1984: Remote measurements of boundary-layer wind profiles using a CW Doppler lidar. *J. Clim. Appl. Meteorol. 23*, 148–154.

Korb, C.L., and C.Y. Weng, 1982: A theoretical study of a two-wavelength lidar technique for the measurement of temperature profiles. *J. Appl. Meteorol. 21*, 1346–1355.

Koschmieder, H., 1924: Theorie der horizontalen Sichtweite. *Beitr. Phys. Atmos. 12*, 33.

Kristensen, L., 1978: *On Sodar Techniques*. Report No. 381, Risoe National Laboratory, Denmark, 48 pp.

———, R.L. Coulter, and K.H. Underwood, 1978: Sodar geometry. In *Proc. Fourth Symposium on Meteorological Observations and Instrumentation*, Denver, Colo., 10–14 April 1978. AMS, Boston, Mass., 391–395.

——— and D.R. Fitzjarrald, 1984: The effect of line averaging on scalar flux measurements with a sonic anemometer near the surface. *J. Atmos. Oceanic Technol. 1*, 138–146.

Kronvall, J., 1981: Tracer gas techniques for ventilation measurements: A 1981 state of the art review. In *Studies in Building Physics* (A.-S. Anderson, Ed.), Report TVBH-3007, Lund Institute of Technology, Lund, Sweden.

Kropfli, R.A., 1977: *A Dual-Doppler Radar Study of the Urban Boundary Layer: A Summary of METROMEX Results*. Technical Memorandum ERL WPL-26, NOAA, Boulder, Colo., 1–22.

——— and P.H. Hildebrand, 1980a: Doppler measurements in the planetary boundary layer during PHOENIX. In Preprint Vol., 19th Conference on Radar Meteorology, Miami Beach, Fla., 15–18 April 1980. AMS, Boston, Mass., 637–644.

——— and ———, 1980b: Three-dimensional wind measurements in the optically clear planetary boundary layer with dual-Doppler radar. *Radio Sci. 15*, 283–296.

———, I. Katz, T.G. Konrad, and E.B. Dobson, 1968: Simultaneous radar reflectivity measurements and refractive index spectra in the clear atmosphere. *Radio Sci. 3* (New Series), 991–994.

——— and N.M. Kohn, 1978: Persistent horizontal rolls in the urban mixed layer as revealed by dual-Doppler radar. *J. Appl. Meteorol. 17*, 669–676.

Kuettner, D.P., 1971: Cloud bands in the Earth's atmosphere. *Tellus 23*, 404–425.

Kunkel, E.E., E.W. Eloranto, and S.T. Shipley, 1977: Lidar observations of the convective boundary layer. *J. Appl. Meteorol. 16*, 1306–1311.

Kuritsky, M.M., and M.S. Goldstein, 1983: Inertial navigation. *Proc. IEEE 71*, 1156–1176.

Kyle, T.G., S. Barr, and W.E. Clements, 1982: Fluorescent particle lidar. *Appl. Opt. 21*, 14–15.

———, W.R. Sand, and D.J. Musil, 1976: Fitting measurements of thunderstorm updraft profiles to model profiles. *Mon. Weather Rev. 104*, 611–617.

Labitt, M., 1979: *Some Basic Relations Concerning the Radar Measurement of Turbulence*. ATC Working Paper No. 46WP-5001, MIT Lincoln Laboratory, Cambridge, Mass.

Lading, L., 1980: Remote measurement of wind velocity: The time-of-flight laser anemometer. In *Proc. Symposium on Long and Short Range Optical Velocity Measurements* (H.J. Pfeifer, Ed.), Saint-Louis, France. Report No. R 117/80, German-French Research Institute (ISL), IX-1–IX-11.

———, A. Skov Jensen, C. Fog, and H. Anderson, 1978: Time-of-flight laser anemometer for velocity measurements in the atmosphere. *Appl. Opt. 17*, 1486–1488.

———, ———, and R.L. Schwiesow, 1980: Remote measurement of wind at moderate ranges: Comparison of different systems. In *Technical Digest Topical Meeting on Coherent Laser Radar*, Aspen, Colo., Optical Society of America, Washington, D.C., TuB4-1–TuB4-4.

Lamb, B.K., A. Lorenzen, and F.H. Shair, 1978a: Atmospheric dispersion and transport within coastal regions. Part I: Tracer study of power plant emissions from the Oxnard Plain. *Atmos. Environ. 12*, 2089–2100.

———, F.H. Shair, and T.B. Smith, 1978b: Atmospheric dispersion and transport within coastal regions. Part II: Tracer study of industrial emissions in the California delta region. *Atmos. Environ. 12*, 2101–2118.

Landsberg, H., 1938: Atmospheric condensation nuclei. *Ergeb. Kosm. Phys. 3*, 155.

Landstrom, D.K., and D. Kohler, 1969: *Electron microprobe analysis of atmospheric aerosols*. EPA Report PB 180-2821 BE, EPA, Durham, N.C.

Lane, J.A., and R.W. Meadows, 1963: Simultaneous radar and refractometer soundings of the troposphere. *Nature 197*, 35–36.

Langer, G., 1965: An acoustic particle counter: Preliminary results. *J. Colloid Sci. 20*, 602.

———, J. Rosinski, and C.P. Edwards, 1967: A continuous ice nucleus counter and its application to tracking in the atmosphere. *J. Appl. Meteorol. 6*, 114.

Lappe, U.O., B. Davidson, and C.B. Notess, 1959: *Analysis of Atmospheric Turbulence Spectra Obtained from Concurrent Airplane and Tower Measurements.* Report No. 59-44, Institute of Aerospace Sciences, New York, N.Y.

Large, W.G., and S. Pond, 1981: Open ocean momentum flux measurements in moderate to strong winds. *J. Phys. Oceanogr. 11,* 324–336.

———— and ————, 1982: Sensible and latent heat flux measurements over the ocean. *J. Phys. Oceanogr. 12,* 464–482.

Larsen, S.E., and N.E. Busch, 1974: Hot-wire measurements in the atmosphere. Part I: Calibration and response characteristics. *DISA Information 16,* 15–34.

———— and ————, 1976: Hot-wire measurements in the atmosphere. Part II: A field experiment in the surface layer. *DISA Information 20,* 5–21.

———— and J. Højstrup, 1982: Spatial and temporal resolution of a thin wire resistance thermometer. *J. Phys. E15,* 471–477.

————, ————, and C.H. Gibson, 1980: Fast-response temperature sensors. In *Air-Sea Interaction: Instruments and Methods* (F. Dobson, L. Hasse, and R. Davis, Eds.), Plenum Press, New York, N.Y., 269–292.

LaRue, J.C., T. Deaton, and C.H. Gibson, 1975: Measurement of high-frequency turbulent temperature. *Rev. Sci. Instrum. 46,* 757–764. (Note: Fig. 5 in this paper is incorrect.)

Latimer, D.A., R.W. Bergstrom, C.D. Johnson, and J.P. Killus, 1980: Modelling visibility. In *Proc. Second Joint Conference on Applications of Air Pollution Meteorology,* New Orleans, La., 24–27 March 1980. AMS, Boston, Mass., 346–360.

————, H. Hogo, and T.C. Daniel, 1981: The effects of atmospheric optical conditions on perceived scenic beauty. *Atmos. Environ. 15(10/11),* 1865–1874.

Lawrence, R.S., G.R. Ochs, and S.F. Clifford, 1972: Use of scintillations to measure average wind across a light beam. *Appl. Opt. 11,* 239–243.

———— and J.W. Strohbehn, 1970: A survey of clear-air propagation effects relevant to optical communications. *Proc. IEEE 58,* 1523–1545.

Leitz, E. 1976: *Leitz-Textur Analysensystem, GMBH.* Leitz, D-6330 Wetzlar, FRG.

LeMone, M.A., and W.T. Pennell, 1980: A comparison of turbulence measurements from aircraft. *J. Appl. Meteorol. 19,* 1420–1437.

Lenschow, D.H., 1965: Airborne measurements of atmospheric boundary layer structure. In *Studies of the Effects of Variations in Boundary Conditions on the Atmospheric Boundary Layer,* Sec. 4, Department of Meteorology, University of Wisconsin, Final Report, Contract DA-36-039-AMC-00878(E), U.S. Army Electronic Research and Development Agency, Ft. Huachuca, Ariz.

————, 1970: Airplane measurements of planetary boundary layer structure. *J. Appl. Meteorol. 9,* 874–884.

————, 1971: Vanes for sensing incidence angles of the air from an aircraft. *J. Appl. Meteorol. 10,* 1339–1343.

————, 1972: *The Measurement of Air Velocity and Temperature Using the NCAR Buffalo Aircraft Measuring System.* Technical Note NCAR/EDD-74, NCAR, Boulder, Colo., 39 pp.

————, 1976: Estimating updraft velocity from an aircraft response. *Mon. Weather Rev. 104,* 618–627.

————, 1982: Reactive trace species in the boundary layer from a micrometeorological perspective. *J. Meteorol. Soc. Jpn. 60,* 471–480.

————, C.A. Cullian, R.B. Friesen, and E.N. Brown, 1978a: Status of air motion measurements on NCAR aircraft. In Preprint Vol., Fourth Symposium on Meteorological Observations and Instrumentation, Denver, Colo., 10–14 April 1978. AMS, Boston, Mass., 433–438.

————, C.A. Friehe, and J.C. LaRue, 1978b: Development of an airborne hot-wire anemometer system. In Preprint Vol., Fourth Symposium on Meteorological Observations and Instrumentation, Denver, Colo., 10–14 April 1978. AMS, Boston, Mass., 463–466.

————, R. Pearson Jr., and B.B. Stankov, 1981: Estimating the ozone budget in the boundary layer by use of aircraft measurements of ozone eddy flux and mean concentration. *J. Geophys. Res. 86,* 7291–7297.

———— and W.T. Pennell, 1974: On the measurement of in-cloud and wet-bulb temperatures from an aircraft. *Mon. Weather Rev. 102,* 447–454.

————, J.C. Wyngaard, and W.T. Pennell, 1980: Mean-field and second-moment budgets in a baroclinic, convective boundary layer. *J. Atmos. Sci. 37,* 1313–1326.

Levy, J.D., 1976: An image analysis system. *Int. Lab. 81.*

Lhermitte, R.M., and L.J. Miller, 1970: Doppler radar methodology for observation of convective storms. In Preprint Vol., 14th Conference on Radar Meteorology, Tucson, Ariz., 17–20 November 1970. AMS, Boston, Mass., 133–138.

Ligthart, L.P., 1980: System considerations of the FM-CW Delft atmospheric research radar (DARR). In *Proc. IEEE International Conference,* Washington, D.C. IEEE, New York, N.Y., 38–43.

Lillian, D., H.B. Singh, A. Appleby, and L.A. Lobban, 1976: Gas chromatographic methods for ambient halocarbon measurements. *J. Environ. Sci. Health A11*, 687–710.

List, R.L., 1971: *Smithsonian Meteorological Tables*. Smithsonian Institution, Washington, D.C., 527 pp.

Little, C.G., 1969: Acoustic methods for the remote probing of the lower atmosphere. *Proc. IEEE 57*, 571–578.

———, 1972: On the detectability of fog, cloud, rain, and snow by acoustic echo sounding methods. *J. Atmos. Sci. 28*, 748–755 (cited in Neff and Coulter article).

———, 1972: Status of remote sensing of the troposphere. In *Remote Sensing of the Troposphere* (V.E. Derr, Ed.), U.S. Government Printing Office, Washington, D.C., 30-1-30-16 (cited in Schwiesow overview article).

Liu, B.Y.H., and L.W. Lee, 1976: Efficiency of membrane and Nuclepore filters for submicrometer aerosols. *Environ. Sci. Technol. 10*, 345.

——— and D.Y.H. Pui, 1975: On the performance of the electrical aerosol analyzer. *J. Aerosol Sci. 6*, 249.

———, ———, and A. Kapadia, 1979: Electrical aerosol analyzer: History, principle and data reduction. In *Aerosol Measurement* (D.A. Lundgren, Ed.), University Presses of Florida, Gainesville, 341.

Lockhart, T.J., J.E. Gaynor, and J.T. Newman, 1983: Field comparison of in-situ meteorological measurements. In Preprint Vol., Fifth Symposium on Meteorological Observations and Instrumentation, Toronto, Ontario, Canada, 11–15 April 1983. AMS, Boston, Mass., 270–273.

Lopez, A., J. Fontan, and J. Servant, 1974: Mesoscale determination of Aitken nuclei flux near the ground. In *Atmospheric-Surface Exchange of Particulates and Gases*, ERDA Symposium-Conference Series No. 740921, 171.

Lovelock, J.E., 1974: The electron capture detector: Theory and practice. *J. Chromatogr. 99*, 3–12.

——— and G.J. Ferber, 1982: Exotic tracers for atmospheric studies. *Atmos. Environ. 16*, 1467–1471.

——— and N.L. Gregory, 1962: *Gas Chromatography*. Academic Press, New York, N.Y., 219 pp.

Lu, N.-P., W.D. Neff, and J.C. Kaimal, 1983: Wave and turbulence structure in a disturbed nocturnal inversion. *Boundary-Layer Meteorol. 26*, 141–155.

Ludwig, F.L., E.M. Liston, and L.J. Salas, 1983: Tracer techniques for estimating emissions from inaccessible ground level sources. *Atmos. Environ. 17(11)*, 2167–2172.

Lumley, J.L., 1965: Interpretation of time spectra measured in high-intensity shear flows. *Phys. Fluids 8*, 1056–1062.

——— and H.A. Panofsky, 1964: *The Structure of Atmospheric Turbulence*. Wiley Interscience, New York, N.Y., 239 pp.

Mach, W.H., and A.B. Fraser, 1979: Inversion of optical data to obtain a micrometeorological temperature profile. *Appl. Opt. 18*, 1715–1723.

Maggiore, C.J., and I.B. Rubin, 1973: Optimization of an SEM X-ray spectrometer system for the identification and characterization of ultramicroscopic particles. In *Scanning Electron Microscopy*, Part 1, Illinois Institute of Technology, Chicago.

Malissa, H., 1951: Ueber die Empfindlichkeit mikroanalytischer Reaktionen. *Mikrochim. Acta 38*, 33.

——— and A.A. Benedett-Pichler, 1958: *Anorganische Qualitative Mikroanalyse*. Springer, Vienna, Austria.

———, J. Kaltenbrunner, and M. Grasserburner, 1974: Ein Beitrag zur Gefugeanalyse mit der Mikrosonde. *Mikrochim. Acta, Suppl. 5*, 453.

Malm, W., K. Kelley, J. Molenar, and T. Daniel, 1980: Human perception of visual air quality (uniform haze). *Atmos. Environ. 15(10/11)*, 1875–1890.

———, M. Pitchford, and A. Pitchford, 1982: Site specific factors influencing the visual range calculated from teleradiometer measurements. *Atmos. Environ. 16(10)*, 2323–2333.

Mamane, Y., and R.G. DePena, 1978: A quantitative method for the detection of individual submicrometer size sulfate particles. *Atmos. Environ. 12*, 69.

——— and R.F. Pueschel, 1980a: A method for the detection of individual nitrate particles. *Atmos. Environ. 14*, 629.

——— and ———, 1980b: Formation of sulfate particles in the plume of the Four Corners power plant. *J. Appl. Meteorol. 19*, 779.

Manton, M.J., 1978: The impaction of aerosols on a Nuclepore filter. *Atmos. Environ. 12*, 1669–1675.

———, 1979: Brownian diffusion of aerosols to the face of a Nuclepore filter. *Atmos. Environ. 13*, 525–531.

Marshall, J.S., R.C. Langille, and W.M. Palmer, 1947: Measurement of rainfall by radar. *J. Meteorol. 4*, 186–192.

Mason, B.J., 1945: *The Physics of Clouds*. Clarendon Press, Oxford, England.

———, 1975: Lidar measurement of temperature: A new approach. *Appl. Opt. 14*, 76–78.

Mastrantonio, G., and G. Fiocco, 1982: Accuracy of wind velocity determinations with Doppler sodars. *J. Appl. Meteorol. 21*, 823–830.

May, K.R., 1945: The cascade impactor. *J. Sci. Instrum. 22*, 187.

McAllister, L.G., 1968: Acoustic sounding of the lower troposphere. *J. Atmos. Terr. Phys. 30*, 1439–1440.

————, J.R. Pollard, A.R. Mahoney, and P.J.R. Shaw, 1969: Acoustic sounding: A new approach to the study of atmospheric structure. *Proc. IEEE 57*, 579–587.

McBean, G.A., 1982: Microscale temperature fluctuations in the atmospheric surface layer. *Boundary-Layer Meteorol. 23*, 185–196.

McCarthy, J., R. Roberts, and W. Schreiber, 1983: JAWS data collection, analysis highlights, and microburst statistics. In *Proc. 21st Conference on Radar Meteorology*, Edmonton, Alberta, Canada, 19–23 September 1983. AMS, Boston, Mass., 596–601.

McCartney, E.J., 1976: *Optics of the Atmosphere: Scattering by Molecules and Particles*. Wiley, New York, N.Y., 408 pp.

McClatchey, R.A., R.W. Fenn, J.E.A. Selby, F.E. Voltz, and J.S. Garing, 1971: *Optical Properties of the Atmosphere*. Report AFCRL-71-0279, Air Force Cambridge Research Laboratory, Bedford, Mass.

———— and A.P. O'Agati, 1978: *Atmospheric Transmission of Laser Radiation: Computer Code LASER*. Report AFGL-TR-78-0029, Air Force Geophysics Laboratory, Bedford, Mass.

McCrone, W.C., and J.G. Delly, 1973: *The Particle Atlas*, second ed., Vols. 1-3. Ann Arbor Science, Ann Arbor, Mich., 794 pp.

McElroy, J.L., and F. Pooler Jr., 1968: *St. Louis Dispersion Study*, Vol. II, *Analysis*. AP-53, U.S. Department of Health, Education, and Welfare, Arlington, Va., 51 pp.

McGillem, C.D., and G.R. Cooper, 1974: *Continuous and Discrete Signal and System Analysis*. Holt, Rinehart, and Winston, New York, N.Y., 395 pp.

McKay, D.J., 1978: A sad look at commercial humidity sensors for meteorological applications. In Preprint Vol., Fourth Symposium on Meteorological Observations and Instrumentation, Denver, Colo., 10–14 April 1978. AMS, Boston, Mass., 7–14.

Measures, R.M., 1984: *Laser Remote Sensing: Fundamentals and Applications*. Wiley, New York, N.Y., 510 pp.

Melfi, S.H., J.D. Lawrence Jr., and M.P. McCormick, 1969: Observations of Raman scattering by water vapor in the atmosphere. *Appl. Phys. Lett. 15*, 295–297.

Melling, H., and R. List, 1978a: Acoustic Doppler sounding of falling snow. *J. Appl. Meteorol. 17*, 1267–1273.

———— and ————, 1978b: Doppler velocity extraction from atmospheric echoes using a zero-crossing technique. *J. Appl. Meteorol. 17*, 1274–1285.

———— and ————, 1980: Characteristics of vertical velocity fluctuations in a convective urban boundary layer. *J. Appl. Meteorol. 19*, 1184–1195.

Merceret, F.J., 1976: Measuring atmospheric turbulence with airborne hot-film anemometers. *J. Appl. Meteorol. 15*, 482–490.

Merrill, J.T., 1977: Observational and theoretical study of shear instability in the airflow near the ground. *J. Atmos. Sci. 34*, 911–921.

Middleton, P., R.L. Dennis, and T.R. Stewart, 1981: Urban visual air quality: Modelled and perceived. In Preprint Vol., 12th International Technical Meeting on Air Pollution Modelling and Its Application, NATO Committee on Challenges of Modern Society, SRI International, Menlo Park, Calif., 144–158.

Middleton, W.E.K., 1958: *Vision through the Atmosphere*. University of Toronto Press, Toronto, Ontario, Canada, 250 pp.

————, 1964: The early history of the visibility problem. *Appl. Opt. 3*, 599.

———— and A.F. Spilhaus, 1953: *Meteorological Instruments*. University of Toronto Press, Toronto, Ontario, Canada.

Mie, G., 1908: Beitrage zur Optik truber Medien, speziell kolloidaler Metallosungen. *Ann. Physik 25*, 377.

Miller, J.M., Ed., 1975: *Geophysical Monitoring for Climatic Change, No. 3*. Summary Report 1974, NOAA/ERL, Boulder, Colo., 197 pp.

Miller, K.S., and M.M. Rochwarger, 1970: On estimates of spectral moments in the presence of colored noise. *IEEE Trans. Inf. Theory IT-16*, 303–308.

Miller, L.J., and R.G. Strauch, 1975: A dual-Doppler radar method for the determination of wind velocities within precipitating weather systems. *Remote Sens. Environ. 3*, 219–235.

Mitsuta, Y., 1974: Sonic anemometer-thermometer for atmospheric turbulence measurements. In *Flow: Its Measurement and Control in Science and Industry*, Vol. 1, Instrumentation Society of America, Pittsburgh, Pa., 341–348.

Mollo-Christensen, E., 1979: Upwind distortion due to probe support in boundary layer measurements. *J. Appl. Meteorol. 18*, 367–370.

Monin, A.S., 1962: Characteristics of the scattering of sound in a turbulent atmosphere. *Sov. Phys. Acoust. 7*, 370–373 (English translation).

Moninger, W.R., W.L. Eberhard, G.A. Briggs, R.A. Kropfli, and J.C. Kaimal, 1983: Simultaneous radar and lidar observations of plumes from continuous point sources. In *Proc. 21st Conference on Radar Meteorology*, Edmonton, Alberta, Canada, 19–23 September 1983. AMS, Boston, Mass., 246–250.

———— and R.A. Kropfli, 1982: Radar observations of a plume from an elevated continuous point source. *J. Appl. Meteorol. 21*, 1685–1697.

Monna, W.A.A., 1978: *Comparative Investigation of Dynamic Properties of Some Propeller Vanes.* Scientific Report WR78-11, Royal Netherlands Meteorological Institute, DeBilt, 30 pp.

Moroz, E.Y., 1977: *Investigations of Sensors and Techniques to Automate Weather Observations.* AFGL-TR-77-0041, AD A040747, Air Force Geophysics Laboratory, Hanscom Air Force Base, Mass.

Moses, H., 1968: Meteorological instruments for use in the atomic energy industry. In *Meteorology and Atomic Energy* (D.H. Slade, Ed.), U.S. Atomic Energy Commission, Division of Technical Information, Oakridge, Tenn., 257–300.

———— and H.G. Daubek, 1961: Errors in wind measurements associated with tower-mounted anemometers. *Bull. Am. Meteorol. Soc. 42*, 190–194.

Moulsley, T.J., D.N. Asimakopoulos, R.S. Cole, B.A. Crease, and S.J. Caughey, 1981: Measurement of boundary layer structure parameter profiles by acoustic sounding and comparison with direct measurements. *Q.J.R. Meteorol. Soc. 107*, 203–230.

———— and R.S. Cole, 1979: High frequency atmospheric acoustic sounders. *Atmos. Environ. 13*, 347–350.

———— and ————, 1980: A general radar equation for the bistatic acoustic sounder. *Boundary-Layer Meteorol. 19*, 359–372.

Mroz, E.J., 1983: *Tracking Antarctic Winds.* Report No. LA-UR-83-1419, Los Alamos National Laboratory, Los Alamos, N. Mex., 6 pp.

Mueller, C., and P.H. Hildebrand, 1983: The structure of a microburst: As observed by ground-based and airborne Doppler radar. In *Proc. 21st Conference on Radar Meteorology,* Edmonton, Alberta, Canada, 19–23 September 1983. AMS, Boston, Mass., 602–608.

Murray, E.R., D.D. Powell, and J.E. van der Laan, 1980: Measurement of average atmospheric temperature using a CO_2 laser radar. *Appl. Opt. 19*, 1794–1797.

Myrup, L.O., 1969: Turbulence spectra in stable and convective layers in the free atmosphere. *Tellus 21*, 341–354.

Name, Y., and R.F. Pueschel, 1979: Oxidation of SO_2 on the surface of flyash particles under low relative humidity conditions. *Geophys. Res. Lett. 6*, 109.

Nathanson, F.E., 1969: *Radar Design Principles.* McGraw-Hill, New York, N.Y., 626 pp.

Neff, W.D., 1975: *Quantitative Evaluation of Acoustic Echoes from the Planetary Boundary Layer.* Technical Report ERL 322-WPL, NOAA, Boulder, Colo., 38–34.

————, 1978: Beamwidth effects on acoustic backscatter in the planetary boundary layer. *J. Appl. Meteorol. 17*, 1514–1520.

————, 1980: *An Observational and Numerical Study of the Atmospheric Boundary Layer Overlying the East Antarctic Ice Sheet.* Ph.D. dissertation, University of Colorado, Boulder, 272 pp.

————, H.E. Ramm, and C. Wendt, 1980: The WPL Doppler sounder in the Boulder low-level intercomparison experiment. Preprint of World Meteorological Organization Report, Boulder Atmospheric Observatory Report No. 2, NOAA/NCAR, available from NOAA/ERL, Boulder, Colo.

Nelson, L.D., 1983: Field tests of a precision non-contact optical thermometer. In Preprint Vol., Fifth Symposium on Meteorological Observations and Instrumentation, Toronto, Ontario, Canada, 11–15 April 1983. AMS, Boston, Mass., 68–74.

Nicholls, S., 1983: *An Observational Study of the Mid-latitude, Marine Atmospheric Boundary Layer.* Ph.D. dissertation, University of Southampton, England, 307 pp.

————, W. Shaw, and T. Hauf, 1983: An intercomparison of aircraft turbulence measurements made during JASIN. *J. Clim. Appl. Meteorol. 22*, 1637–1648.

Nieuwstadt, F.T.M., and A.G.M. Driedonks, 1979: The nocturnal boundary layer: A case study compared with model calculations. *J. Appl. Meteorol. 18*, 1397–1405.

Nikiyama, S., and Y. Tanasawa, 1939: *Trans. Soc. Mech. Engrs. Jpn. 5*, 63.

O'Bannon, T., 1978: A study of dual-Doppler synthesized clear air wind fields. In Preprint Vol., 18th Conference on Radar Meteorology, Atlanta, Ga., 28–31 March 1978. AMS, Boston, Mass., 65–69.

Ochs, G.R., 1967: *Resistance Thermometer for Measurement of Rapid Air Temperature Fluctuations.* ESSA Technical Report IER 47-ITSA 46, Environmental Science Services Administration, Boulder, Colo., 17 pp.

Ottersten, H., 1969: Atmospheric structure and radar backscattering in clear air. *Radio Sci. 4*, 1179–1193 (cited in Neff article).

————, 1969: Radar backscatter from the turbulent clear atmosphere. *Radio Sci. 4*, 1251–1256 (cited in Chadwick and Gossard article).

Ozkaynak, H., R.G. Isaacs, and B.L. Murphy, 1979: Sensitivity analysis for models of local and regional visibility degradation. In *Proc. Fourth Symposium on Turbulence, Diffusion, and Air Pollution,* Reno, Nev., 15–18 January 1979. AMS, Boston, Mass., 257–268.

Pal, S.R., and A.I. Carswell, 1976: Multiple scattering in atmospheric clouds. *Appl. Opt. 15*, 1990–1995.

Panofsky, H., and G. Brier, 1958: *Some Applications of Statistics to Meteorology.* Pennsylvania State University, University Park, 224 pp.

———, H. Tennekes, D.H. Lenschow, and J.C. Wyngaard, 1977: The characteristics of turbulent velocity components in the surface layer under convective conditions. *Boundary-Layer Meteorol. 11*, 355–361.

Paranthoen, P., J.C. Lecordier, and C. Petit, 1983: Dynamic sensitivity of the constant-temperature hot-wire anemometer to temperature fluctuations. *TSI Quart. 9*, 3–8.

———, P., C. Petit, and J.C. Lecordier, 1982: The effect of the thermal prong-wire interaction on the response of a cold wire in gaseous flows (air, argon and helium). *J. Fluid Mech. 124*, 457–473.

Parkinson, B.W., and S.W. Gilbert, 1983: NAVSTAR: Global positioning system; ten years later. *Proc. IEEE 71*, 1177–1186.

Parungo, F., E. Ackerman, H. Proulx, and R.F. Pueschel, 1978: Nucleation properties of flyash in a coal-fired power plant plume. *Atmos. Environ. 12*, 929.

Pasqualucci, F., B.W. Bartram, R.A. Kropfli, and W.R. Moninger, 1983: A millimeter-wavelength dual-polarization Doppler radar for cloud and precipitation studies. *J. Clim. Appl. Meteorol. 22*, 758–765.

Patterson, E.M., B.A. Bodhaine, A. Coletti, and G.W. Grams, 1982: Volume scattering ratios determined by the polar and the integrating nephelometer: A comparison. *Appl. Opt. 21*, 394.

Pearson, R. Jr., and D.H. Stedman, 1980: Instrumentation for fast response ozone measurements from aircraft. *Atmos. Technol. 12*, NCAR, Boulder, Colo., 51–55.

Perry, A.E., 1982: *Hot-Wire Anemometry.* Oxford University Press, Oxford, England, 184 pp.

Persha, G., and W.C. Malm, 1980: Automated vertical scanning multi-wavelength telephotometer. In Preprint Vol., Symposium on Plumes and Visibility: Measurements and Model Components, Grand Canyon, Ariz., 10–14 November 1980. EPA, Durham, N.C.

Petit, C., P. Paranthoen, and J.C. Lecordier, 1981: Influence of Wollaston wire and prongs on the response of cold wires at low frequencies. *Lett. Heat Mass Transfer 8*, 281–291.

Pich, J., 1966: Theory of aerosol filtration by fibrous and membrane filters. In *Aerosol Science* (C.N. Davies, Ed)., Academic Press, London, England, 223–285.

Pietzner, H., and A. Schiffers, 1972: *Mineralogical and Chemical Investigations of Firing Residues and Depositions from the Furnace of Pulverized Coal-fired Power Plants.* Sondeft der VGB (Vereinigung Grosstechnischer Betriebe), Essen, FRG.

Plank, V., 1956: *A Meteorological Study of Radar Angles.* Paper 52, Geophysical Research Directorate, Air Force Cambridge Research Laboratory, Bedford, Mass., 117 pp.

Pollak, L.W., 1957: Methods of measuring condensation nuclei. *Geofis. Pura Appl. 36*, 44.

Post, M.J., 1978: Experimental measurements of atmospheric aerosol inhomogeneities. *Opt. Lett. 2*, 166–168.

Pourny, J.C., D. Renaut, and A. Orszag, 1979: Raman-lidar humidity sounding of the atmospheric boundary layer. *Appl. Opt. 18*, 1141–1148.

Pratte, J.F., and R.J. Clarke, 1983: PROFS mesonet: Description and performance. In Preprint Vol., Fifth Symposium on Meteorological Observations and Instrumentation, Toronto, Ontario, Canada, 11–15 April 1983. AMS, Boston, Mass., 303–307.

Priestley, J.T., and W.D. Cartwright, 1982: *Frequency Response Measurements on Lyman-Alpha Humidiometers.* Technical Memorandum ERL WPL-92, NOAA, Boulder, Colo., 9 pp.

Probert-Jones, J.R., 1962: The radar equation in meteorology. *Q.J.R. Meteorol. Soc. 88*, 485–495.

Pueschel, R.F., 1976: Aerosol formation during coal combustion: Condensation of sulfates and chlorides on flyash. *Geophys. Res. Lett. 3*, 651.

——— and E.W. Barrett, 1982: Sulfate in the atmospheric boundary layer: Concentration and mechanisms of formation. In *Heterogeneous Chemistry*, AGU Geophysical Monograph Series 26, American Geophysical Union, Washington, D.C., 241.

———, ———, D.L. Wellman, and J.A. McGuire, 1981: Cloud modification by man-made pollutants: Effects of a coal-fired powerplant on cloud drop spectra. *Geophys. Res. Lett. 8*, 221.

——— and Y. Mamane, 1979: Mechanisms and rates of formation of sulfur aerosols in power plant plumes. In *Proc. Symposium on Long Range Transport of Pollutants*, WMO Publication No. 538, World Meteorological Organization, Geneva, Switzerland.

——— and C.C. Van Valin, 1978: Cloud nucleus formation in a power plant plume. *Atmos. Environ. 12*, 307.

Rabin, R.M., 1983: Radar reflectivity in the clear boundary layer and its relation to surface fluxes. In *Proc. 21st Conference on Radar Meteorology*, Edmonton, Alberta, Canada, 19–23 September 1983. AMS, Boston, Mass., 646–649.

Rabinoff, R., and B. Herman, 1973: Effect of aerosol size distributions on the accuracy of the integrating nephelometer. *J. Appl. Meteorol. 12*, 184.

Radke, L.F., and P.V. Hobbs, 1969: An automatic cloud condensation nuclei counter. *J. Appl. Meteorol.*, 105.

Ragaini, R.C., R.E. Heft, and D. Garvis, 1976: *Neutron Activation Analysis at the Livermore Pool-type Reactor for the Environmental Research Program.* Report UCRL-52095, Lawrence Livermore Laboratory, Livermore, Calif.

Rasmussen, R.A., M.A.K. Khalil, A.J. Crawford, and P.J. Fraser, 1982: Natural and anthropogenic trace gases in the Southern Hemisphere. *Geophys. Res. Lett. 9(6),* 704–707.

Raymond, D.J., and M.H. Wilkening, 1982: Flow and mixing in New Mexico mountain cumuli. *J. Atmos. Sci. 39,* 2211–2228.

Raynor, G.S., 1978: A radio-controlled air sampling system for diffusion experiments. *J. Appl. Meteorol. 17,* 1619–1624.

Readings, C.J., E. Golton, and K.A. Browning, 1973: Fine scale structure and mixing within an inversion. *Boundary-Layer Meteorol. 4,* 275–287.

Reidiger, G., 1972: Teilchenzahlung und Teilchengrossenanalyse mit dem quantitativen Ferseh-Mikroskop Quantiment 720. *Staub 32,* 3.

Reiter, E.R., 1978: *Atmospheric Transport Processes.* Part IV: *Radioactive Tracers.* Technical Information Center, U.S. Department of Energy, 605 pp.

Remiarz, R.J., and E.M. Johnson, 1984: A new diluter for high concentration measurements with the aerodynamic particle sizer. *TSI Quart. 10,* 7–12.

Renaut, D., J.C. Pourny, and R. Capitini, 1980: Daytime Raman-lidar measurements of water vapor. *Opt. Lett. 5,* 233–235.

Richardson, L.F., 1922: *Weather Prediction by Numerical Process.* University Microfilms, Ann Arbor, Mich., 236 pp.

Richter, J.H., 1969: High-resolution tropospheric radar sounding. *Radio Sci. 4,* 1261–1268.

————, D.R. Jensen, V.R. Noonkester, J.B. Kreasky, M.W. Stimman, and W. Wolf, 1973: Remote radar sensing: Atmospheric structure and insects. *Science 180,* 1176–1178.

Rihaczek, A.W., 1969: *High Resolution Radar.* McGraw-Hill, New York, N.Y.

Roach, W.T., 1970: On the influence of synoptic development on the production of high level turbulence. *Q.J.R. Meteorol. Soc. 96,* 413–429.

Robinson, E., R.A. Rasmussen, and J. Krasnec, 1977: Halocarbon measurements in the Alaskan troposphere and lower stratosphere. *Atmos. Environ. 11,* 215–223.

Rose, H.E., and A.J. Wood, 1956: *An Introduction to Electrostatic Precipitation in Theory and Practice.* Constable, London, England.

Rose, W.G., 1962: Some corrections to the linearized response of a constant-temperature hot-wire anemometer operated in a low speed flow. *Trans. ASME 29E,* 554–558.

Rosin, P., and E. Rammler, 1933: The laws governing the fineness of powdered coal. *J. Inst. Fuel 7,* 29.

Rottger, J., 1980: Reflection and scattering of VHF radar signals from atmospheric refractivity structures. *Radio Sci. 15,* 259–276.

———— and G. Schmidt, 1979: High resolution VHF radar sounding of the troposphere and stratosphere. *IEEE Trans. Geosci. Electron. 4,* 182–189.

Rowland, J.R., 1973: Intensive probing of clear air convective field by radar and instrumented drone aircraft. *J. Appl. Meteorol. 12,* 149–155.

Rummler, W.D., 1968: *Two Pulse Spectral Measurements.* Technical Memorandum MM-68-4121-15, Bell Laboratories, Whippany, N.J.

Ruping, G., 1968: Die Bedeutung der geschwindigkentsgleichen Absaugung bei der Staubstraulmessung mittels Entnahmesonden. *Staub Reinhalt. Luft 28,* 137.

Russell, J.W., and L.A. Shadoff, 1977: The sampling and determination of halocarbons in ambient air using concentration on porous polymers. *J. Chromatogr. 134,* 375–384.

Ryan, J.S., S.R. Pal, and A.I. Carswell, 1979: Laser backscattering from dense water-droplet clouds. *J. Opt. Soc. Am. 69,* 60–67.

Sackinger, P.A., D.D. Reible, and F.H. Shair, 1982: Uncertainties associated with the estimation of mass balances and Gaussian parameters from atmospheric tracer studies. *J. Air Pollut. Control Assoc. 32(7),* 720–724.

Saltzman, B.E., A.I. Coleman, and C.A. Clemons, 1966: Halogenated compounds as gaseous meteorological tracers. *Anal. Chem. 38(6),* 753–758.

Sandborn, V.A., 1972: *Resistance Temperature Transducers.* Metrology Press, Fort Collins, Colo., 545 pp.

Sanders, P.A., 1979: *Handbook of Aerosol Technology,* second ed. Krieger, Melbourne, Fla., 540 pp.

Sasano, Y., H. Hirohara, T. Yamasaki, H. Shimizu, N. Takeuchi, and T. Kawamura, 1982: Horizontal wind vector determination from the displacement of aerosol distribution patterns observed by a scanning lidar. *J. Appl. Meteorol. 21,* 1516–1523.

Sassen, K., 1980: An initial application of polarization for orographic cloud seeding operations. *J. Appl. Meteorol. 19,* 298–304.

Saxton, J.A., J.A. Lane, R.W. Meadows, and P.A. Mathews, 1964: Layer structure of the troposphere: Simultaneous radar and microwave refractometer investigations. *Proc. Inst. Elec. Engr. (London) 3,* 275–283.

Schacher, G., and C.W. Fairall, 1976: Use of resistance wires for atmospheric turbulence measurements in the marine environment. *Rev. Sci. Instrum. 47*, 703–707.

Schaefer, G.W., 1976: Radar observations of insect flight. In *Proc. Symposium of the Royal Entomological Society of London*, 157–197.

Schaefer, V.J., 1948: *The Detection of Ice Crystals in the Free Atmosphere.* Proj. Cirrus Occasional Report No. 9, General Electric Company, Schenectady, N.Y.

———, 1976: *The Air Quality Patterns of Aerosols on the Global Scale*, Parts I and II. Atmospheric Sciences Research Center Publication No. 406, SUNY, Albany, N.Y.

Schehl, R., R.S. Ergun, and A. Headrich, 1973: Size spectrometry of aerosols using light scattering from the cavity of a gas laser. *Rev. Sci. Instrum. 44*, 1193.

Schlesinger, R.J., 1961: *Principles of Electronic Warfare.* Prentice-Hall, New York, N.Y., 213 pp.

Schmitt, K.F., C.A. Friehe, and C.H. Gibson, 1978: Humidity sensitivity of atmospheric temperature sensors by salt contamination. *J. Phys. Oceanogr. 8*, 151–161.

Schnell, R.C., 1979: A new technique for measuring atmospheric ice nuclei active at temperatures from −20 °C to approaching O °C with results. In *Proc. Seventh Conf. on Inadvertent and Planned Weather Modification*, Banff, Alberta, Canada, 9–12 October 1979. AMS, Boston, Mass., 110.

Schuster, B.G., and R.G. Knollenberg, 1972: Detection and sizing of small particles in an open cavity gas laser. *Appl. Opt. 11*, 1515.

——— and T.G. Kyle, 1980: Pollution plume transport and diffusion studies using fluorescence lidar. *Appl. Opt. 19*, 2524–2528.

Schutz, L., 1978: Analysis of atmospheric aerosol particles using a scanning electron microscope. In *Berichte zur Elektronemikroskopischen Dikektabbildung von Oberflachen (BEDO)*, Vol. 9, Remy, Münster, FRG.

Schwiesow, R.L., 1981: Horizontal velocity structure in waterspouts. *J. Appl. Meteorol. 20*, 349–360.

———, 1983: Potential for a lidar-based, portable, 1 km meteorological tower. *J. Appl. Meteorol. 22*, 881–890.

——— and R.F. Calfee, 1979: Atmospheric refractive effects on coherent lidar performance at 10.6 μm. *Appl. Opt. 18*, 3911–3917.

——— and R.E. Cupp, 1981: Offset local oscillator for CW laser Doppler anemometry. *Appl. Opt. 20*, 579–582.

———, ———, V.E. Derr, E.W. Barrett, and R.F. Pueschel, 1981a: Aerosol backscatter coefficient profiles measured at 10.6 μm. *J. Appl. Meteorol. 20*, 184–194.

———, ———, P.C. Sinclair, and R.F. Abbey Jr., 1981b: Waterspout velocity measurements by airborne Doppler lidar. *J. Appl. Meteorol. 20*, 1972–1979.

——— and L. Lading, 1981: Temperature profiles by Rayleigh-scattering lidar. *Appl. Opt. 20*, 1972–1979.

——— and R.S. Lawrence, 1982: Effects of a change of terrain height and roughness on a wind profile. *Boundary-Layer Meteorol. 22*, 109–122.

Seeley, B.K., 1952: Detection of micron and submicron chloride particles. *Anal. Chem. 24*, 576.

Sehmel, G.A., 1982: *An Atmospheric Tracer Investigation of Fugitive Emissions Transport in the Colorado Oil Shale Region.* Report No. PNL-SA-9834, Pacific Northwest Laboratory, Richland, Wash.

Senum, G.I., 1981: *Prediction of the Relative Response of the Electron Capture Detector for Various Compounds Including Ambient Atmospheric Species, Tracers, and Potential Tracers.* Internal report, Brookhaven National Laboratory, Upton, N.Y.

———, R.P. Gergley, E.M. Ferreri, M.W. Greene, and R.M. Dietz, 1980: *Final Report of the Evaluation of Vapor Taggants and Substrates for the Tagging of Blasting Caps.* Report No. BNL 51232, Brookhaven National Laboratory, Upton, N.Y., 6-1–6-2.

Shaw, W.J., and J.E. Tillman, 1980: The effect of and correction for different wet-bulb and dry-bulb responses in thermocouple psychrometry. *J. Appl. Meteorol. 19*, 90–97. (See also Corrigenda, *J. Appl. Meteorol. 19*, 1339.)

Sheppard, B.E., 1978: Comparison of scattering coefficient type visibility sensors for automated weather station applications. In Preprint Vol., Fourth Symposium on Meteorological Observations and Instrumentation, Denver, Colo., 10–14 April 1978. AMS, Boston, Mass., 201–206.

Shettle, E.P., and R.W. Fenn, 1976: Models of the atmospheric aerosols and their optical properties. In *Optical Propagation in the Atmosphere*, NATO, Advisory Group for Aerospace Research and Development Conference Proceedings, AGARD-CP-183 (NTIS ADA 028-615), 2-1–2-16.

Sievering, H., 1982: Profile measurements of particle dry deposition velocity at an air-land interface. *Atmos. Environ. 16*, 301.

Simmonds, P.G., A.J. Lovelock, and J.E. Lovelock, 1976: Continuous and ultrasensitive apparatus for the measurement of airborne tracer substances. *J. Chromatogr. 126*, 3–9.

———, G.R. Shoemaker, J.E. Lovelock, and H.C. Lord, 1972: Improvements in the determination of sulfur hexafluoride for use as a meteorological tracer. *Anal. Chem. 44*, 860–863.

Singh, H.B., L.J. Silas, and R.E. Stiles, 1983: Selected man-made halogenated compounds in the air and oceanic environment. *J. Geophys. Res. 88(C6)*, 3675–3683.

Sirmans, D., and B. Bumgarner, 1975: Numerical comparison of five mean frequency estimators. *J. Appl. Meteorol. 14*, 991–1003.

Sivertsen, B., 1983: Estimation of diffuse hydrocarbon leakages from petrochemical factories. *J. Air Pollut. Control Assoc. 33(4)*, 323–327.

Sliney, D.H., and B.C. Freasier, 1973: Evaluation of optical radiation hazards. *Appl. Opt. 12*, 1–24.

Smith, S.D., 1980: Dynamic anemometers. In *Air-Sea Interaction: Instruments and Methods* (F. Dobson, L. Hasse, and R. Davis, Eds.), Plenum Press, New York, N.Y., 65–80.

Smith, T., 1981: *A Study of the Origin and Fate of Air Pollution in California's Sacramento Valley.* Report No. A-1-FR-1842, Meteorology Research, Inc., Altadena, Calif., for California Air Resources Board.

Spizzichino, A., 1974: Discussion of the operating conditions of a Doppler sodar. *J. Geophys. Res. 79*, 5585–5591.

Spurny, K., and J. Pich, 1965: Analytical methods for determination of aerosols by means of membrane ultrafilters. *Collect. Czech. Chem. Commun. 30*, 2276.

Spyers-Duran, P., and D. Baumgardner, 1983: In-flight estimation of the time response of airborne temperature sensors. In Preprint Vol., Fifth Symposium on Meteorological Observations and Instrumentation, Toronto, Ontario, Canada, 11–15 April 1983. AMS, Boston, Mass., 352–357.

Squires, P., 1966: An estimate of the anthropogenic production of cloud nuclei. *J. Rech. Atmos. 2*, 297.

———, J.P. Lodge Jr., E.R. Frank, and D.C. Sheesley, 1969: Aerosol filtration by means of Nuclepore filters. *Environ. Sci. Techol. 3*, 453.

Sroga, J.T., E.W. Eloranta, and T. Barber, 1980: Lidar measurement of wind velocity profiles in the boundary layer. *J. Appl. Meteorol. 19*, 598–605.

Stafford, R.G., and H.J. Ettinger, 1972: Filter efficiency vs. particle size and velocity. *Atmos. Environ. 6*, 353.

Start, G.E., N.F. Hukari, J.F. Sagendorf, J.H. Cate, and C.R. Dickson, 1980: *EOCR Building Wake Effects on Atmospheric Diffusion.* NOAA Technical Memorandum ERL ARL-91, Air Resources Laboratory, Idaho Falls, Idaho, 220 pp.

Stevens, J.J., 1967: Filtering responses of selected distance-dependent weight functions. *Mon. Weather Rev. 95*, 45–46.

Stevenson, C.M., 1968: An improved Millipore filter technique for measuring the concentrations of freezing nuclei in the atmosphere. *Q.J.R. Meteorol. Soc. 94*, 35.

Stober, W., and H. Flachsbart, 1969: Size separating precipitation of aerosols in a spinning spiral duct. *Environ. Sci. Technol. 3*, 1280.

Stoss, P.W., and D. Atlas, 1968: Wind shear and reflectivity gradient effects on Doppler radar spectra. In Preprint Vol., 13th Conference on Radar Meteorology, Montreal, Quebec, Canada, 20–23 August 1968. AMS, Boston, Mass., 297–302.

Strapp, J.W., and R.S. Schemenauer, 1982: Calibrations of Johnson-Williams liquid water content meters in a high-speed icing tunnel. *J. Appl. Meteorol. 21*, 98–108.

Strauch, R.G., W.C. Campbell, R.B. Chadwick, and K.P. Moran, 1976: Microwave FM-CW Doppler radar for boundary-layer probing. *Geophys. Res. Lett. 3*, 193–196.

———, V.E. Derr, and R.E. Cupp, 1971: Atmospheric temperature measurement using Raman backscatter. *Appl. Opt. 10*, 2665–2669.

———, ———, and ———, 1972: Atmospheric water vapor measurement by Raman lidar. *Remote Sensing Environ. 2*, 101–108.

Swerling, P., 1970: Radar measurement accuracy. In *Radar Handbook* (M.I. Skolnik, Ed.), McGraw-Hill, New York, N.Y.

Taconet, O., and A. Weill, 1981: *Vertical Velocity Field and Convective Plumes in the Atmospheric Boundary Layer as Observed with an Acoustic Doppler Sodar.* Technical Note CRPE/98, Centre de Recherches en Physiques de l'Environment Terrestre et Planetaire, Issy-les-Moulineaux, France.

Tatarskii, V.I., 1961: *Wave Propagation in a Turbulent Medium* (R.A. Silverman, translator). McGraw-Hill, New York, N.Y., 285 pp.

———, 1967: *The Effects of the Turbulent Atmosphere on Wave Propagation.* Israel Program for Scientific Translations, Jerusalem, Israel; National Technical Information Service No. TT 68-50464, U.S. Department of Commerce, Springfield, Va., 472.

Taylor, G.I., 1938: The spectrum of turbulence. *Proc. R. Soc. London A164*, 476–490.

Telford, J.W., and P.B. Wagner, 1974: The measurement of horizontal air motion near clouds from aircraft. *J. Atmos. Sci. 31*, 2066–2080.

———, ———, and A. Vaziri, 1977: The measurement of air motion from aircraft. *J. Appl. Meteorol. 16*, 156–166.

——— and J. Warner, 1962: On the measurement from an aircraft of buoyancy and vertical air velocity in clouds. *J. Atmos. Sci. 19*, 415–423.

Tennekes, H., and J.L. Lumley, 1972: *A First Course in Turbulence.* MIT Press, Cambridge, Mass., 300 pp.

Thompson, D.W., and R.L. Coulter, 1974: Analysis and limitation of phase coherent ACDAR sounding measurements. *J. Geophys. Res. 79*, 5541–5549.

———, ———, and Z. Warhaft, 1978: Simultaneous measurements of turbulence in the lower atmosphere using sodar and aircraft. *J. Appl. Meteorol. 17*, 723–734.

Tillman, J.E., 1965: Water vapor density measurements utilizing the absorption of vacuum ultraviolet and infrared radiation. In *Humidity and Moisture*, Vol. 1 (A. Wexler, Ed.), Reinhold, New York, N.Y., 428–433.

Tolgyessy, J., and S. Varga, 1974: *Nuclear Analytical Chemistry*, Vol. 3. University Park Press, Baltimore, Md., 57.

Tsvang, L.R., 1969: Microstructure of temperature fields in the free atmosphere. *Radio Sci. 4*, 1175–1177.

Turner, D.B., 1970: *Workbook of Atmospheric Dispersion Estimates*. Publication No. AP-26, EPA, Office of Air Programs, Research Triangle Park, N.C., 81 pp.

Turner, J.S., 1973: *Buoyancy Effects in Fluids*. Cambridge University Press, Cambridge, England, 367 pp.

Tutu, N.K., and R. Chevray, 1975: Cross-wire anemometry in high intensity turbulence. *J. Fluid Mech. 71*, 785–800.

Twomey, S., 1954: The composition of hygroscopic particles in the atmosphere. *J. Meteorol. 11*, 334–338.

Tyndall, J., 1875: *Selected Works of John Tyndall: Sound*. D. Appleton, New York, N.Y., 306–320.

Uberoi, M.S., and L.S.G. Kovasznay, 1953: *Q. Appl. Math. 10*, 379–393.

Underwood, K.H., 1981: *Sodar Signal Processing Methods and the Riso '78 Experiment*. Ph.D. dissertation, Pennsylvania State University, University Park, 175 pp.

——— and R.L. Coulter, 1983: Vertical velocity spectra from a Doppler sodar. Presented at the Second International Symposium on Acoustic Remote Probing of the Atmosphere and Oceans, Rome, Italy, 27 August–1 September 1983.

Uthe, E.E., B.M. Morely, and N.B. Nielsen, 1982: Airborne lidar measurements of smoke plume distribution, vertical transmission, and particle size. *Appl. Opt. 21*, 460–463.

———, N.B. Nielsen, and W.L. Jimison, 1980: Airborne lidar plume and haze analyzer (ALPHA-1). *Bull. Am. Meteorol. Soc. 61*, 1035–1043.

Vali, G., Ed., 1976: *The Third International Workshop on Ice Nucleus Measurements*. University of Wyoming, Laramie.

Van de Hulst, H.C., 1957: *Light Scattering by Small Particles*. Wiley, New York, N.Y., 470 pp.

Van Duuren, H., G.D. Krijt, and A.J. Elshout, 1974: *Sulfur Hexafluoride as a Tracer for the Dispersion of Air Polluting Components from a Point Source*. No. IV 8903-74, N.V. tot Kenring van Electrotechnische Materialen, Arnhem, The Netherlands.

Vaughan, J.M., 1979: Remote wind measurement in the atmosphere using laser Doppler methods. In *Proc. LASER 79*, Munich, FRG. International Publishing Corporation Scientific and Technical Press, London, England.

Vidal-Madjar, C., F. Parey, J.-L. Excoffier, and S. Bekassy, 1981: Quantitative analysis of chlorofluorocarbons: Absolute calibration of the electron-capture detector. *J. Chromatogr. 203*, 247–261.

Viezee, W., and W.E. Evans, 1979: Development of a special visibility monitor for pristine air quality areas. In *Proc. Fourth Symposium on Turbulence, Diffusion, and Air Pollution*, Reno, Nev., 15–18 January 1979. AMS, Boston, Mass., 279–282.

——— and ———, 1980: *Development and Evaluation of a Prototype Automated Telephotometer System*. EPRI EA-1434, RP862-14, Final Report, Electric Power Research Institute, Palo Alto, Calif.

——— and ———, 1983: Automated measurements of atmospheric visibility. In *Proc. 21st American Institute of Aeronautics and Astronautics Aerospace Sciences Meeting*, Reno, Nev., 10–13 January 1983. AIAA, New York, N.Y.

Von Gogh, R.G., and P. Zib, 1978: Comparison of simultaneous tethered balloon and monostatic acoustic sounder records of the statically stable lower atmosphere. *J. Appl. Meteorol. 17*, 34–39.

Vonnegut, B., 1950: Vortex thermometer for measuring true air temperature and true airspeeds in flight. *Rev. Sci. Instrum. 21*, 136–141.

Waggoner, A.P., A.H. Vanderpol, R.J. Charlson, S. Larsen, L. Granat, and C. Tragardh, 1976: Sulfate-light scattering ratio as an index of the role of sulfur in tropospheric optics. *Nature 261*, 120.

——— and R. Weiss, 1980: Comparison of fine particle mass concentration and light scattering extinction in ambient aerosol. *Atmos. Environ. 14*, 623.

Walkenhorst, W., 1962: Ein neuer Thermalprazipitator mit heizbad und seine Leistung. *Staub 22*, 103.

Wang, T.-I., G.R. Ochs, and S.F. Clifford, 1978: A saturation-resistant optical scintillometer to measure C_n^2. *J. Opt. Soc. Am. 68*, 334–338.

Warner, J., and J.W. Telford, 1965: A check of aircraft measurements of vertical heat flow. *J. Atmos. Sci. 22*, 463–465.

Watson, H.H., 1936: The thermal precipitator. *Trans. Inst. Min. Met. 46*, 176.

Wechter, S.G., 1976: Preparation of the stable pollution gas standard using treated aluminum cylinders. In *Calibration in Air Monitoring*, STP598, American Society for Testing and Materials, Philadelphia, Pa., 40–54.

Weickmann, H., 1957: Recent measurements of the vertical distribution of Aitken nuclei. In *Artificial Stimulation of Rain*, Pergamon Press, New York, N.Y., 81–88.

Weil, J.C., and R.P. Brower, 1983: *Estimating Convective Boundary-Layer Parameters for Diffusion Applications*. Report PPSP-MP-48, Martin Marietta Corp., Baltimore, Md.

Weill, A., F. Baudin, J.P. Goutorbe, P. van Grundebeeck, and P. LeBerre, 1978: Turbulence structure in temperature inversion and in convection fields as observed by Doppler sodar. *Boundary-Layer Meteorol. 15*, 375–390.

————, C. Klapisz, B. Strauss, F. Baudin, C. Jaupart, P. van Grundebeeck, and J.P. Goutorbe, 1980: Measuring heat flux and structure function of temperature fluctuations with an acoustic Doppler sodar. *J. Appl. Meteorol. 19*, 199–205.

Weinman, J.A., 1976: Effects of multiple scattering on light pulses reflected by turbid atmospheres. *J. Atmos. Sci. 33*, 1763–1771.

Werner, C., 1981: Slant range visibility determination from lidar signatures by the two-point method. *Opt. Laser Technol. 13*, 27–36.

———— and H. Herrmann, 1981: Lidar measurements of the vertical absolute humidity distribution in the boundary layer. *J. Appl. Meteorol. 20*, 476–481.

Wesely, M.L., 1976: The combined effect of temperature and humidity on refractive index. *J. Appl. Meteorol. 15*, 43–49.

Westwater, E.R., M.T. Decker, A. Zachs, and K.S. Gage, 1983: Ground-based remote sensing of temperature profiles by a combination of microwave radiometry and radar. *J. Clim. Appl. Meteorol. 22*, 126–133.

Wexler, A., 1965: *Humidity and Moisture*, Vol. 1. Reinhold, New York, N.Y., 687 pp.

————, 1970: Measurement of humidity in the free atmosphere near the surface of the earth. In *Meteorological Observation and Instrumentation* (S. Teweles and J. Giraytys, Eds.), Meteorological Monographs, Vol. 11, No. 30, AMS, Boston, Mass., 262–282.

Wexler, R., 1947: Radar detection of a frontal storm 18 June 1946. *J. Meteorol. 4*, 38–44.

Whitby, K.T., 1978: The physical characteristics of sulfur aerosols. *Atmos. Environ. 12*, 135.

———— and W.E. Clark, 1966: Electric aerosol particle counting and size distribution measuring system for the 0.015 to 1 μm size range. *Tellus 18*, 573.

White, H.J., 1963: *Industrial Electrostatic Precipitation*. Addison-Wesley, Palo Alto, Calif., 376 pp.

Wieringa, J., 1980: A reevaluation of the Kansas mast influence on measurements of stress and cup anemometer overspeeding. *Boundary-Layer Meteorol. 18*, 411–430.

Wigand, A., 1919: Die vertikale Verteilung der Kondensationskerne in der freien Atmosphare. *Ann. Phys. 59*, 689.

Williams, R.M., and C.A. Paulson, 1977: Microscale temperature and velocity spectra in the atmospheric boundary layer. *J. Fluid Mech. 83(3)*, 547–567.

Willis, G.E., and J.W. Deardorff, 1976a: A laboratory model of diffusion into the convective planetary boundary layer. *Q.J.R. Meteorol. Soc. 102*, 427–445.

———— and ————, 1976b: Visual observations of horizontal platforms of penetrative convection. In *Proc. Third Conference on Turbulence, Diffusion, and Air Quality*, Raleigh, N.C., 19–22 October 1976. AMS, Boston, Mass., 9–12.

———— and ————, 1981: A laboratory study of dispersion from a source in the middle of a convectively mixed layer. *Atmos. Environ. 15*, 109–117.

Wills, J.A.B., 1980: *Hot-Wire and Hot-Film Anemometry*. NMI R82, National Maritime Institute, Middlesex, England.

Wilson, D.A., 1970: Doppler radar studies of boundary layer wind profiles and turbulence in snow conditions. In *Proc. 14th Conference on Radar Meteorology*, Tucson, Ariz., 17–20 November 1970. AMS, Boston, Mass., 191–196.

Wilson, J., and R. Roberts, 1983: Evaluation of Doppler radar for airport wind shear detection. In *Proc. 21st Conference on Radar Meteorology*, Edmonton, Alberta, Canada, 19–23 September 1983. AMS, Boston, Mass., 616–623.

Wittry, D.B., 1959: Resolution of electron probe microanalyzers. *J. Appl. Phys. 30*, 953.

Woodcock, A.H., 1940: Convection and soaring over the open sea. *J. Mar. Res. 3*, 248–253.

Woodman, R.F., 1980a: High-altitude-resolution stratospheric measurements with the Arecibo 430-MHz radar. *Radio Sci. 15*, 417–422.

————, 1980b: High-altitude-resolution stratospheric measurements with the Arecibo 2380-MHz radar. *Radio Sci. 15*, 423–430.

———— and A. Guillen, 1974: Radar observations of winds and turbulence in the stratosphere and mesosphere. *J. Atmos. Sci. 31*, 493–505.

Wucknitz, J., 1980: Flow distortion by supporting structures. In *Air-Sea Interaction: Instruments and Methods* (F. Dobson, L. Hasse, and R. Davis, Eds.), Plenum Press, New York, N.Y., 605–626.

Wyckoff, R.J., D.W. Beran, and F.F. Hall Jr., 1973: A comparison of the low-level radiosonde and the acoustic echo sounder for monitoring atmospheric stability. *J. Appl. Meteorol. 12*, 1196–1204.

Wyngaard, J.C., 1968: Measurement of small-scale turbulence structure with hot wires. *J. Sci. Instrum. Ser. 2*, 1105–1108.

————, 1969: Spatial resolution of the vorticity meter and other hot-wire arrays. *J. Sci. Instrum. Ser. 2*, 983–987.

————, 1971a: The effect of velocity sensitivity on temperature derivative statistics in isotropic turbulence. *J. Fluid Mech. 48*, 763–769 (cited in Friehe and Wyngaard articles).

————, 1971b: Spatial resolution of a resistance wire temperature sensor. *Phys. Fluids 14*, 2052–2054.

————, 1973: On surface layer turbulence. In *Workshop on Micrometeorology* (D.A. Haugen, Ed.), AMS, Boston, Mass., 101–149.

————, 1975: Modeling the planetary boundary layer: Extension to the stable case. *Boundary-Layer Meteorol. 9*, 441–460.

————, 1981a: Cup, propeller-vane and sonic anemometers in turbulence research. *Ann. Rev. Fluid Mech. 13*, 399–423.

————, 1981b: The effects of probe-induced flow distortion on atmospheric turbulence measurements. *J. Appl. Meteorol. 20*, 784–794 (cited in Kaimal and Wyngaard articles).

————, 1983: Lectures on the planetary boundary layer. In *Mesoscale Meteorology: Theories, Observations, and Models* (D.K. Lilly and T. Gal-Chen, Eds.), D. Reidel, Dordrecht, The Netherlands.

————, J.T. Bauman, and R. Lynch, 1974: Cup anemometer dynamics. In *Flow: Its Measurement and Control in Science and Industry*, Instrumentation Society of America, Pittsburgh, Pa., 701–708.

———— and S.F. Clifford, 1978: Estimating momentum, heat, and moisture fluxes from structure parameters. *J. Atmos. Sci. 35*, 1204–1211.

———— and O.R. Cote, 1971: The budgets of turbulent kinetic energy and temperature variance in the atmospheric boundary layer. *J. Atmos. Sci. 28*, 190–201.

————, J. Izumi, and S.A. Collins Jr., 1971: Behavior of the refractive index structure parameter near the ground. *J. Opt. Soc. Am. 61*, 1646–1650.

———— and M.A. LeMone, 1980: Behavior of the refractive index structure parameter in the entraining convective boundary layer. *J. Atmos. Sci. 37*, 1573–1585.

————, W.T. Pennell, D.H. Lenschow, and M.A. LeMone, 1978: The temperature-humidity covariance budget in the convective boundary layer. *J. Atmos. Sci. 35*, 47–58.

Yates, F., 1960: *Sampling Methods for Censuses and Surveys.* Griffin, London, England.

Young, A.T., 1981: Rayleigh scattering. *Appl. Opt. 20*, 533–535.

Yura, H., 1979: Signal-to-noise ratio of a heterodyne lidar system in the presence of atmospheric turbulence. *Optica Acta 26*, 627–644.

Zebel, G., 1977: Some problems in sampling of coarse aerosols. In *Recent Developments in Aerosol Science* (D. Shaw, Ed.), Wiley, New York, N.Y.

Zeiss, C., 1976: *Zeiss Mikro-Videomat 2, Informationsblatt Mikro No. 12.* Carl Zeiss, Oberkochen, FRG.

Zenker, P., 1971: Unterschungen zur Frage der nichtgeschwindigkeitsgleichen Teilstromentnahme bei der Staubgehaltsbestimmung in stromenden Gasen. *Staub Reinhalt. Luft 31*, 252.

Zrnic, D.S., 1977: Spectral moment estimates from correlated pulse pairs. *IEEE Trans. Aerosp. Electron. Syst. AES-13(4)*, 344–354.

Index